SEA SURFACE STUDIES

Climatologists and earth scientists agree that a major rise in global sea level of between 0.2–1.4 m is to be expected during the next century should the predicted rise in global temperature of 1.5°–4.5°C take place. This view reinforces earlier reports of the American National Academy of Sciences and the World Meteorological Organisation. Such statements may engender panic or disbelief. The question is, how much do we know about the causes, patterns and problems of sea-level change that underpin such statements? This book attempts an interdisciplinary exploration and major review of our state of knowledge in this field through the views of researchers studying the nature and applications of sea-level change and its consequences for the coastline. Apart from its linkage with climate the study of sea-level–coastal change is relevant to a wide range of issues, from geological problems to oceanographic, environmental and biological concerns. This book is therefore likely to be of particular interest to workers and students in geology, geography, marine sciences/engineering and technology, biology and related environmental sciences.

SEA SURFACE STUDIES

A GLOBAL VIEW

Edited by R.J.N. DEVOY

CROOM HELM
London • New York • Sydney

Croom Helm Ltd, Provident House, Burrell Row,
Beckenham, Kent BR3 1AT
Croom Helm Australia, 44-50 Waterloo Road,
North Ryde, 2113, New South Wales

Published in the USA by
Croom Helm
in association with Methuen, Inc.
29 West 35th Street
New York, NY 10001

British Library Cataloguing in Publication Data

Sea surface studies: a global view
   1. Sea level
   I. Devoy, R.J.N.
   551.46   GC89
   ISBN 0-7099-0871-7

Library of Congress Cataloging-in-Publication Data

Sea surface studies.

   Includes index.
   1. Sea level. 2. Coast changes. 3. Geodynamics.
I. Devoy, R.J.N.
GC89.S44 1987 551.46 87-5401
ISBN 0-7099-0871-7

Typeset in Times Roman by Leaper & Gard Ltd, Bristol, England

Printed and bound in Great Britain by
Biddles Ltd, Guildford and King's Lynn

Contents

Contributors

Andrews, John T., Professor, Department of Geological Sciences, University of Colorado, Campus Box 250, Boulder, Colorado 80309, USA.

Berryman, Kelvin, Dr, New Zealand Geological Survey, Department of Scientific and Industrial Research, P.O. Box 30 368, Lower Hutt, New Zealand.

Carter, R.W.G., Dr, Department of Environmental Studies, University of Ulster, Cromore Road, Coleraine, Northern Ireland, BT52 1SA.

Chappell, John, Professor, Research School of Pacific Studies (Biogeography and Geomorphology Section), The Australian National University, P.O. Box 4, Canberra, A.C.T. 2601, Australia.

Devoy, R.J.N., Dr, Department of Geography, University College Cork, College Road, Cork, Ireland.

Hopley, David, Professor, Sir George Fisher Centre for Tropical Marine Studies, James Cook University of North Queensland, Townsville, Post Office Q4811, Australia.

Lewis, A.W., Dr, Department of Civil Engineering, University College Cork, College Road, Cork, Ireland.

Mörner, Nils-Axel, Department of Geology, University of Stockholm, S-106 91, Stockholm, Sweden.

Orford, J., Dr, Department of Geography, The Queen's University of Belfast, Belfast, Northern Ireland BT7 1NN.

Ota, Yoko, Professor, Department of Geography, Yokohama National University, 156 Tokiwadai, Hodogaya-ku, Yokohama 240, Japan.

Peltier, W.R., Professor, Department of Physics, University of Toronto, 60 St George Street, Toronto, Ontario, M5S 1A7, Canada.

Pillans, B., Dr, Department of Geology, Research School of Earth Sciences, Victoria University of Wellington, Private Bag, Wellington, New Zealand.

Shennan, Ian, Dr, Department of Geography, University of Durham, Science Laboratories, South Road, Durham, DH1 3LE, England.

Sutherland, Donald G., Dr, Placer Analysis Ltd, 2 London Street, Edinburgh, EH3 6NA, Scotland.

Titus, J.G., (P.M.220) Strategic Studies Staff, US Environmental Protection Agency, Washington, DC 20460, USA.

Preface

The oceans are vast with two-thirds of our planet being covered by a thick layer of water, the depth of which can be likened to flying above the earth's surface at an altitude of 30,000 feet (9,800 m). Good to play in, essential for life but deadly to breathe, water is important to all organisms on the planet, and the oceans form its major reservoir containing approximately 97 per cent of all freely available surface water. In spite of this obvious importance mankind has still much to learn about this ocean environment. Study of the oceans has grown enormously since the eighteenth- and nineteenth-century voyages of scientific discovery, expanding greatly in the period post 1945. One of the subjects that has blossomed in this period has been the study of the ocean's surface, and in particular the study of sea-level and related sea-surface changes. Indeed this topic may even be termed 'popular', as reflected in the growing number of general geomorphology, physical geology and oceanography texts which now give space to the subject.

Stimuli to this growth have been varied. Concern since the 1960s over mankind's influence and interaction with the earth's environment has been important. Today a heightened awareness of the possible continued warming of the earth's atmosphere resulting from human induced variations in carbon dioxide (CO_2) and other 'greenhouse' gases, with its consequent linkage to ice melt and global sea-level rise, has gained especial prominence. The spectre of a 'massive' 5 to 7 m rise in global sea level has been raised, although the probability of a smaller 0.2 to 1.4 m rise during the next century is more likely. What we can do about this and how we manage our coastlines in response are matters which are only just beginning to be examined seriously. In this, as in other fields of earth science, the study of sea-level data has been recognised as being of great practical value: for geology and geophysics in understanding the structure and functioning of the earth, for geomorphology in coastal management and planning, for biogeography within the biological sciences in examining organism development and in commercial fields in hydrocarbon and mineral exploration.

As discussed in the book, these stimuli have prompted the setting-up of many national and international research projects to study sea-surface changes. This volume is in part a response to the activities of one of these programmes, the International Geological Correlation Programme's Project 200, 'Sea-level Correlations and Applications', to which the book forms a contribution. In other respects the book has grown from the editor's own long-term interest in this field, and from the need now to pull together the extensive knowledge about the subject that has been acquired over the last forty years. This is not to say that excellent texts dealing with various aspects of sea-level studies have not been produced in this time; they have, and reference to many of these has been made here in this volume. Rarely, however, have they considered the subject in a global context, as here, integrating the subject in the same text with treatment of relevant aspects of coastal process studies and the applications and value of such work.

Preparation for this book began in 1983, with the different chapters being contributed at various stages since that time, as follows: Chapters 7, 12 in 1984; 2, 3, 4, 6, 8, 9, 10, 11, 13, 14, 16, 17 in 1985 and 1, 5, 15, 18, 19 in 1986. Where necessary these have undergone update and revision of detail prior to publication. As such, the volume forms a multi-authored text representing an interdisciplinary approach, often by acknowledged authorities within the different fields covered, to areas of complex study within the subject. The book is not intended, however, to take the reader step by step through the detail of sea levels, but to provide a comprehensive coverage of the important concepts and principles, global spatial — temporal patterns of sea-level change and discussion of their current applications. Whilst many may feel a multi-authored text has its drawbacks treatment of this rapidly evolving subject at this scale demands, almost by definition, that such an approach be taken. Further, the book is far from being intended for the expert alone, although it is hoped that those involved in sea-level research will find it of value. Through its structure and organisation, extensive referencing within chapters and presentation of material themes, the book is designed as a 'way in' for non-specialists in all disciplines to learn about this complex, often confusing, but most important aspect of our changing environment.

R.J.N. Devoy
Cork

Acknowledgements

Many people have been involved in the production of this volume in many different countries, and to all these I am most grateful. I wish to thank particularly the contributing authors for their interest and enthusiasm in the project, and the referees for their valuable advice; Dr J.R. Hails for a chapter contribution to the preliminary draft of this volume which unfortunately, due to reasons of space, could not finally be included; Prof. L. Königsson and Prof. W. Newman for the preparation of preliminary chapter draft material; Dr Douglas Grant for his thoughtfulness and help in the provision of air photographs; Mr Connell Foley who produced many of the figures and to whose perseverance and patience I am indebted; Ms Ann Phelan, Ms Anne Foley, Ms Anne Reddy and typists in the Secretarial Services Centre, UCC for their preparation of a number of the final manuscripts; my colleagues in the Department of Geography, UCC for their support during the work; my wife, Patsy, for her encouragement and advice throughout. Finally I wish to thank Mr Tim Hardwick and Croom Helm for their guidance, flexibility and patience.

The publishers and I wish to thank the following for permission to reproduce copyright material.

Ch. 1
J.T. Andrews, *A Geomorphological Study of Postglacial Uplift*, © 1970 Institute British Geographers for Fig. 1.2.
O. van de Plassche, and the editor *Meded. Rijks Geol. Dienst*, vol. *36-1*, for Figs. 1.3, 1.5 and Table 1.1, reproduced by permission of the Geological Survey of the Netherlands.
Figure 1.4 reproduced from the *Proc. Roy. Soc. Victoria*, vol. *92*, by permission of the Royal Society of Victoria, Inc. Australia, and Emeritus Prof. S.W. Carey, University of Tasmania.

Ch. 2
Fig. 2.4 reprinted with permission from *Marine Geology*, P.H. Kuenen, © 1950 John Wiley & Sons, Inc.
Open University Press for Table 2.1 from S334 — Oceanography Unit 7, 1978, J. Stewart and D. Sharp (eds) and for Table 2.2 from S334 — Oceanography Unit 3, 1977, J. Stewart (ed.)

Ch. 3
Longman for Fig. 3.6 from *Reconstructing Quaternary Environments*, J.J. Lowe and M.J.C. Walker.

ACKNOWLEDGEMENTS

Ch. 5
B. Isacks, J. Oliver & L.R. Sykes, *J. Geophys. Res.*, vol. *73*, © 1968 American Geophysical Union for Fig. 5.1
D.P. McKenzie, *Scientific American*, vol. *235* for Fig. 4.2.
S. Kaizuka et al., *Geographical Reports of Tokyo Metropolitan University*, vol. *8*, for Fig. 5.6, reproduced by permission of Tokyo Metropolitan University.
D. Denham, *J. Geophys. Res.*, vol. *74*, © 1969 American Geophysical Union, and J. Chappell, *Bull. Geol. Soc. Am.*, vol. *85*, for Fig. 5.7.
J. Chappell, *Bull. Geol. Soc. Am.*, vol. *85*, for Fig. 5.8.
Fig. 5.11 reproduced from *Late Quaternary Geology of the Cape Kidnappers Region, Hawke's Bay, N.Z.* by permission of A. Hull.
G. Plafker for Fig. 5.13 reproduced from *U.S.G.S. Open File Report 78-943*.
G. Plafker, *U.S.G.S. Open File Report 78-943* and T. Matsuda et al., *Bull. Geol. Soc. Am.*, vol. *89*, for Fig. 5.14.
M. McNutt & H.W. Menard, *J. Geophys. Res.*, vol. *83*, © 1978 American Geophysical Union for Fig. 5.16.
G.B. Dalrymple et al., *Bull. Geol. Soc. Am.*, vol. *86*, R.D. Jarrard & D.L. Turner, *J. Geophys. Res.*, vol. *84*, © 1979 American Geophysical Union and N. Yonekura et al., *Int. Symp. on Coastal Evolution in the Holocene* for Fig. 5.17.
Y. Ota and T. Yoshikawa, *J. Phys. Earth*, vol. *26* (Suppl.) for Table 5.1.

Ch. 6
Table 6.2 and Fig. 6.4 reprinted with permission from *Quaternary Geology*, D.Q. Bowen, © 1978 Pergamon Books Ltd.
Fig. 6.5 reprinted by permission from *Nature*, vol. *281* (*5732*), p. 542, © Macmillan Journals Ltd.

Ch. 7
Edward Arnold Ltd. for Fig. 7.2 from *An Introduction to Coastal Geomorphology* 1984, J. Pethick.
Fig. 7.3 reproduced with permission from *Shorelines and Isostasy*, D. Smith & A. Dawson, © 1983 Institute British Geographers.
S. Jelgersma for Fig. 7.4 from *Acta Univ. Ups., Symp. Univ. Ups. Ann. Quing. Cel.*, vol. *2*.
O. van de Plassche, and the editor *Meded. Rijks Geol. Dienst.*, vol. *36-1*, for Fig. 7.5, reproduced by permission of the Geologican Survey of the Netherlands.
Fig. 7.7a-c reprinted by permission from *Nature*, vol. *302* (*5907*), p. 404-6, © 1983 Macmillan Journals Ltd.
Fig. 7.9 reproduced with permission from *Quaternary Coastlines and*

Marine Archaeology, P.M. Masters & N.C. Fleming, © 1983 Academic Press.

Ch. 9
J. Chappell for Fig. 9.8.

Ch. 10
Longman for Fig. 10.1 from *Geographical Variation in Coastal Development* 2nd ed. 1980, J.L. Davies.
Fig. 10.2 & 10.16 (inset) reproduced from *Trans Delaware Acad. Sci.'s,* vol. 7 by permission of J.C. Kraft, Board of Directors, The Iron Hill Museum, © 1979 Delaware Academy of Science.
Fig. 10.3 reproduced by permission of P. Ziegler and The Royal Society from *Phil. Trans. R. Soc. Lond. A,* vol. *305.*
Fig. 10.4 reproduced with permission from *Quaternary Research,* vol. *11,* © 1979 Academic Press.
Fig. 10.6a reproduced with permission of the Minister of Supply and Services Canada from *Current Research, Geol. Surv. Canada,* 1986 *Part A.*
Fig. 10.6 reproduced with permission from *Pleistocene Geology and Biology,* R.G. West, © 1968 Longman.
Figs. 10.7, 10.10 & 10.12 reproduced with permission from *Shorelines and Isostasy,* D. Smith and A. Dawson, © 1983 Institute British Geographers.
M. Eronen for Figs. 10.8 & 10.9.
Figs. 10.13 and 10.19 reprinted with permission from *Earth Rheology, Isostasy and Eustasy,* © 1980 John Wiley & Sons Ltd.
K.O. Emery, *Scientific American,* vol. *221* for Fig. 10.16.
Fig. 10.15 reprinted with permission from Quaternary Science Review, vol. *1* © 1982 Pergamon Journals Ltd.
Fig. 10.17 reproduced from *J. Sed. Petr.,* vol. *47(2)* by permission of J.C. Kraft.

Ch. 13
Fig. 13.1 reprinted with permission from *Quaternary Science Reviews,* vol. *1,* © 1983 Pergamon Journals Ltd.
George Allen & Unwin Ltd. for Fig. 13.2 from *Practical Foundations of Physical Geography,* 1981, B.J. Knapp.
Fig. 13.4 reproduced from *Boreas,* vol. *12* by permission of Norwegian University Press.
Fig. 13.5 reprinted with permission from the *Journal of Geology,* vol. *87,* p. 444, © 1979 University of Chicago Press.
Fig. 13.7 reproduced from *Marine Geology,* vol. *5* by permission of Elsevier Science Publ. bv.
Longman for Fig. 13.8 from *Geographical Variation in Coastal Development* 1st ed. 1972, J.L. Davies.

Figs. 13.11 and 13.12 reprinted with permission from *Handbook of Coastal Processes and Erosion,* © 1983 CRC Press Inc.

Fig. 13.15 reproduced from *Sed. Geol.,* vol. *14* by permission of Elsevier Science Publ. bv.

Fig. 13.16 reproduced from *Bull. Geol. Soc. Am.,* vol. *82,* p. 2154 by permission of J.C. Kraft.

Fig. 13.17 reproduced from *Marine Geology,* vol. *42* by permission of Elsevier Science Publ. bv.

Fig. 13.18 reprinted with permission from the *Journal of Geology,* vol. *78,* p. 105, © 1970 University of Chicago Press.

Society of Economic Palaeotologists and Mineralogists for Fig. 13.19 from Special Publ. 34 *Siliclastic Shelf Sediments,* R.W. Tillman & C.T. Siemens.

Fig. 13.20 reproduced by permission of K. Pye from *Coastal Research: UK Perspectives,* M.W. Clark, Geo Books.

Ch. 14

Fig. 14.3 reproduced by permission of A.J. Bowen and The Royal Society from *Phil. Trans R. Soc. Lond. A,* vol. *272.*

Ch. 15

Fig. 15.5 reprinted by permission from *Nature,* vol. *271* (5643), p. 323, © 1978 Macmillan Journals Ltd.

Fig. 15.3 reproduced by permission from *Science,* vol. *213* (4511), pp. 957-66, J. Hanson et al., © 1981 AAAS.

Fig. 15.4 reproduced by permission from *Science,* vol. *215* (4540), pp. 1611-14, V. Garnitz et al., © 1982 AAAS.

Ch. 16

Fig. 16.1 reproduced by permission of E.C. Goldman and The Royal Society from *Phil. Trans R. Soc. Lond. A,* vol. *209.*

Figs. 16.2 & 16.3 reprinted by permission from *Nature,* vol. *277,* p. 464-5 © 1979 Macmillan Journals Ltd.

Figs. 16.4 & 16.13 (part) reproduced by permission from *Facies Interpretation and the Stratigraphic Record,* A. Hallam, © 1981 W.H. Freeman & Co. Fig. 16.4 first adapted by WHF & Co. from Heckel, 1972 and 16.13 (part) by WHF & Co. from Hallam, 1978.

Figs. 16.5, 16.6, 16.7, 16.10b & 16.14 reproduced by permission from *Seismic Stratigraphy: Applications to Hydrocarbon Exploration. Memoir 26,* C.E. Payton, © 1977 The American Association of Petroleum Geologists.

Institute of Petroleum for Fig. 16.9 (part) from *Petroleum Geology of the Continental Shelf of Northwest Europe,* L.V. Illing & G.D. Hobson.

Fig. 16.10a reprinted by permission from *Nature,* vol. *246,* p. 21, © 1973 Macmillan Journals Ltd.

ACKNOWLEDGEMENTS

Figs. 16.11 & 16.12 reproduced from *Bull. Geol. Soc. Am.*, vol. *89* pp. 1396 & 1401 by permission of W.C. Pitman III.
L.F. Brown Jr. for Fig. 16.15 reproduced from *Seismic Stratigraphic Interpretation and Petroleum Exploration*, L.F. Brown Jr. & W.L. Fisher.

Ch. 18
Edward Arnold Ltd. for Fig. 18.1 & 18.2b from *An Introduction to Coastal Geomorphology*, 1984 J. Pethick.
Fig. 18.8 reproduced by permission from *Hydroelectric Engineering for Civil Engineers*, S. Leliavsky, © 1982 Chapman & Hall.
Fig. 18.9 reproduced from *L'Usine Marémotrice de la Rance*, p. 4, Electricité de France by permission of M. Mourier.
Escher Wyss for Fig. 18.11
Figs. 18.14a, 18.15 & 18.17 reproduced by permission from *Wave Energy Utilization*, C.O.J. Grove-Palmer, © 1982 Tapir Publishers.
T. Whittaker for Fig. 18.14b.
Fig. 18.16 reproduced from *"SEA" CLAM, Wave Energy Converter* by permission of Sea Energy Associates Ltd.
The National Engineering Laboratory for Fig. 18.18.

Ch. 19
Fig. 19.19 reproduced from *Impact of Sea-level Rise on Society Conference* Delft 1986, by permission of H.G. Wind, Delft Hydraulics Laboratory.
Fig. 19.12 reproduced from *Geophys. J. Roy. Astr. Soc.*, vol. *87 (1)* by permission of T.E. Pyle and Blackwell Scientific Publications Ltd.

1

Introduction: First Principles and the Scope of Sea-surface Studies

R.J.N. Devoy

And God said, 'Let the waters under the heavens be gathered together into one place, and let the dry land appear.' And it was so. God called the dry land Earth, and the waters that were gathered together he called Seas. (Genesis 1: 9-10)

HISTORY AND DEVELOPMENT

The concept of land/sea-level change is not a new one. Strabo (68 BC to AD 24) concluded that volcanism in the Mediterranean basin was associated with the rise and fall of land levels adjacent to the sea and that this process appeared to be the cause of phases of marine inundation and retreat. In the fifteenth century Leonardo da Vinci asserted that the presence of marine and other fossil organisms at high elevations is proof of datum changes in both land and sea levels through time (Chorley *et al.*, 1964: p. 7). From the Baltic region an early published reference to sea-level changes comes from Ericus Erici (1625), who noted (in translation), 'the water has been lowered in many places so that skerries and rocks that had been covered before, and no man had known the existence of, now are visible and appear high above the water'.

In Fennoscandia changes in the level of land and sea have been a decisive force in the development of land surface form. It is thus not surprising that the region has formed a rich source of sea-level obser-vations, such as those of Erici, and of consequent hypotheses on shoreline formation. It is worth examining briefly, therefore, further ideas from the region on sea-level change as an example of the evolution of European thought on the subject held prior to the nineteenth century. The first major scientific approach here to such problems came with Urban Hjärne. In 1702 and 1706 he published the results of an earlier survey of 1694 on the physical conditions of the region. Here he described the noted seaward movement of shorelines and compared these with published shoreline data from other sources. He concluded that the observed lowering of relative sea level was restricted to the Baltic Sea basin. Further, since similar features had not been reported from Denmark, he proposed that the Baltic had once been isolated from the oceans. Hjärne's observations were generally ignored. However, Swedenborg discussed the observed features again in 1719 and concluded that these former higher Baltic sea-level stands evidenced the occurrence of the Noaharcian 'Deluge'; a theory held widely by his contemporary European society as the explanation of what were later to be interpreted as Quaternary glacial deposits and features (John, 1964; Boylan, 1981).

In 1733-4 Celsius, in reference to thoughts by Newton on a postulated desiccation of the planet, treated the problem anew and suggested that the amount of global water was diminishing due to 'evaporation and the amount of water used by plants to form humus'. Celsius, who believed the phenomenon was worldwide, prepared a timetable for desiccation asserting that the Baltic could be expected to dry out within 3,000 to 4,000 years. A contemporary, Linnaeus, also considered the problem, adjusting his results to fit the Creation narrative, although he also favoured the theory of a continuously diminishing global water supply as responsible for the observed changes in sea level. The latter part of the eighteenth century subsequently witnessed a vigorous discussion for and against these theories of falling sea level and desiccation. Despite these arguments an important new contribution came, however, from Ephraim Runeberg, director of the cadastral survey of Finland. Study of the problems of land emergence around the Gulf of Bothnia, resulting from the need to allot newly emerged land to owners, led Runeberg to suggest in 1765-9 that small movements of the earth's crust were responsible for much of the observed land-level changes (Wegmann, 1969). His thinking concerning problems of crustal deformation and its effects upon sea-level change appears now as remarkably modern. Runeberg's view was later supported by others in Europe, for example by Playfair, von Buch and Lyell, who suggested that the cause of shoreline deformation may lie with a secular cooling of the earth and an accompanying shrinkage of the crust.

Such early observations of land/sea-level changes are not confined, however, to Europe, and care must be taken not to associate the concept of sea-level change solely with countries bordering the North Atlantic. Records from Japan and China, for example, are common. The Chinese were actively investigating and recording the phenomenon of raised shorelines at least 2,200 years ago (Lin, 1983 and 1986). Similarly, non-literate societies must also have noted changing land levels and associated shoreline positions. Early societies, such as the hunter–gatherer communities of Starr Carr (Clark, 1954) and the Cumberland lowland (Huddart and Tooley, 1972) in England or the South Carolina coastal plain in the USA (Brooks and Canouts, 1981; Lepionka, 1981), dependent as they were for

2

survival upon a 'harmony' with nature, must have been very much aware of shoreline changes and how these affected the location of food resources. Andrews (1970) gives the more contemporary example of place name evidence from the Swampy Cree Indians of South Hudson Bay, Canada, dating back perhaps some 200 years, as recording the impact of land emergence here on local shoreline changes.

In western scientific literature the fundamental concepts of sea-level studies covering glacio-eustasy and glacio-isostasy were developed in the nineteenth century. Maclaren is regarded widely as the father of the first of these ideas, whereby water mass exchanges between land and ocean, formed by the build-up and decay of ice sheets, is seen as the dominant controlling factor in sea-level fall and rise respectively during the Quaternary. (See Appendix I, Geological timescale.) The term eustasy itself was coined much later, by Suess (1906), to describe *globally* and *simultaneously* recorded vertical displacements of the sea surface, or sea-level changes (cf. Daly, 1934: p. 41). Maclaren (1842), noting the publication of Agassiz's new and controversial book on the 'glacial theory' (1840), wrote, 'if we suppose the region from the 35th parallel to the North Pole to be invested with a coat of ice thick enough to reach the summits of the Jura, ... it is evident that the abstraction of such a quantity of water from the ocean would immediately affect its depth.' The allied concept of tectono-eustasy, the influence of structural changes in the earth's crust on sea-level, has been ascribed by Fairbridge (1961) to the work of Darwin (1842) on the mechanism of coral reef development, and to the later research of Chambers.

A few years later Jamieson (1865, 1882), working in Scotland, put forward the idea that land uplift associated with glaciated regions may result from the release of pressure on the earth's crust exerted by the former ice load. He wrote (1865: p. 178), 'It has occurred to me that the enormous weight of ice thrown upon the land may have something to do with this (land level) depression.' On this basis Jamieson may be regarded as the founder of the second of the major sea-level concepts, the theory of glacio-isostasy, although Nansen (1922) noted that Shaler (1874) in America had independently arrived at the same conclusion. Terminology and the geophysical concepts of isostasy (Gr. *isostasios*, 'in equipoise') itself are first associated with Pratt 1855, Airy 1855 and Dutton 1889 (cf. Holmes, 1965).

In Europe, though to a lesser extent in America, dissemination and acceptance of these ideas were slow. An indication of this comes from Suess, who published the first part of his influential work *Anliz der Erde* in 1865, the authoritarian tone of which considerably affected contemporary geological thought. Here, Suess argued that changes of sea level ought to be related to the cooling and subsequent shrinkage of the globe. Geological evidence showing a sea-level rise would thus result directly from shrinkage

3

of the ocean basins. Further, Suess did not believe in the dipping shorelines, indicative of land uplift, described by Bravais (1840) from the Altenfjord area of northern Norway, which show a dip from the inner fjord to the coast. This widespread 'dipping shoreline problem' was re-studied in Norway by De Geer (1888 and 1890), and by other researchers in Scandinavia and Britain, confirming Bravais's observations. From this and from his investigations of the marine limit (the 'MG') in Sweden De Geer was able to demonstrate that an irregular land uplift had occurred in Fennoscandia. He constructed the first isobase maps for the region on the basis of the observations, and pointed out a central area along an axis from Oslo to Haparanda where the greatest uplift had occurred. De Geer concluded that the upheaval constituted the restoration of the earth's crust following its former depression by the Fennoscandian ice cap and that restoration was still incomplete, a view supporting the earlier proposals of Jamieson and Shaler.

The linking of observation–theory to prediction (e.g. shoreline-isobase mapping) is seen by Andrews (1970) as indicative of the advanced progression in terms of the stages of scientific thought of sea-level studies in Europe, and particularly in Scandinavia, by this time (see also Mörner, 1980). During the 1920s and 1930s concept and methodology had proceeded apace with workers such as Nansen, Hyppä, Tanner and Sauramo developing important shoreline analytical techniques, such as the Shoreline Relation diagram (Tanner, 1933) and time/elevation graphs showing the form of postglacial land emergence. In North America, by comparison, work concentrated upon data collection and little synthesis–analysis of shoreline information had been attempted prior to Wright's publication (1914) of the *Quaternary Ice Age*. De Geer (1892), however, had written a pioneering paper on land-/sea-level changes in eastern North America, whilst observational studies of raised shorelines from this region were common during the late nineteenth and early twentieth centuries as part of larger geological projects, for example, by Tyrell, Upham, Daly, Bell and Low (cf. Andrews, 1970, 1974). Further, study was not confined alone to areas of land uplift. The work of Mudge (1858), Dawson (1868), Shaler (1885) and later Davis (1910) and Johnson (1925) on submerging Atlantic coastal marshes/sediment sequences provided fundamental insights into the value of sedimentary data in studying sea-level changes.

During this period of growth remarkably little attention was paid to the repercussions of glacial mass–water volume transfers as an explanation of sea-level movements, despite the earlier stimulus of Maclaren and later Penck (1882). As outlined, studies concentrated upon the effects of ice loading and crustal delevelling. Not until the twentieth century, and Daly's work (1910, 1925) on the 'glacial control' on coral reef development, did the glacio-eustatic concept gain wider acknowledgement. Wright's *Isokinetic theory* (1914) forms an important milestone here in the linking of eustatic and isostatic effects in the explanation of postglacial land-/sea-level change. At a similar time Holmsen (1918) and Nansen (1922) in Norway also succeeded in showing more precisely the relationship between land uplift — shoreline displacement and eustasy. In a wider, international context Daly's book, *The Changing World of the Ice Age* (1934) brings many of these ideas into focus. Daly (1934) clearly stated that surficial redistribution of both ice and water loads involved immediate crustal elastic responses as well as delayed deep-seated plastic deformation and mass transfer. As such, the work provides an important formalising of many of the concepts of earth rheology, ice marginal crustal forebulge, geoidal changes and ice-water surface gravitational attraction, that have become influential in sea-level/shoreline thinking since 1970 (see Chapters 3, 4 and 10).

As glacio-isostasy may be said to have dominated nineteenth-century thought so eustasy has preoccupied many researchers in the mid-twentieth century. Kuenen, in his authoritative text *Marine Geology* (1950), synthesised views on the subject in his treatment of ocean water volume change, listing the following reasons for eustatic changes of sea level:

(1) Alteration in the shape of ocean basins (sedimentation, isostatic and orogenic movements of the ocean floor).
(2) Alteration of the quantity of water on the continents (growth and decay of glaciers).
(3) Alteration of the total mass of water on the surface of the earth (the production of water by volcanoes and from other internal sources, balanced in part by the absorption of water by minerals during weathering and hydration).

Despite the re-definition of the concept these principles haven't changed. Kuenen further calculated values for former glacial and interglacial sea levels, concluding that −60 m to −80 m was a likely eustatic level for the sea during the last ice age, with calculations for an ice-free earth showing global sea level some 20 to 50 m above its present position. Subsequent work on ice and sea-level data have led to several conflicting estimates, although most fall within Kuenen's figures (cf. Flint, 1971; Bloom, 1971, 1983; Clark *et al.*, 1978).

The reason for the interest in eustasy lies in its possible value in 'transferring sequences and chronologies from one part of the world to another' (Godwin, 1943: p. 209), a prospect described recently by Carey (1981), in response to the growing commercial–geological value of sea-level studies (see Part Five, this vol.), as opening 'a new era for eustatism'. (Godwin's own contribution to sea-level research through methodological innovation and data gathering has been summarised by Tooley, 1986.) The introduction of palynology, in which Godwin was influential, facilitated such correlations. It provided a new means of dating sedimentary changes in time and the correlation of events. Its usage, building on the pioneering work from Fennoscandia on the Litorina Sea (cf. Eronen, 1974; Devoy, this vol., Ch. 10), brought about a surge of biostratigraphic research on sea-level studies within Europe that has subsequently widened geographically and continued to the present day (see Tooley, 1978a, b). From about 1950, the application of Libby's radiocarbon method of age determination provided an even greater revolution of the study of Late Quaternary sea levels (post ~ 20,000 BP), allowing the generation of time/depth–elevation plots of relative sea-level change. This is well exemplified in the expansion in shoreline data from many parts of North America after this time (cf. Andrews, 1970; Bloom, 1983, 1984).

For eustasy, use of these techniques led to an intensified search for the 'Holy Grail', the identification of a single universally valid sea-level curve, an eustatic curve. In particular attention focused on establishing the form and pattern of Holocene (postglacial) eustatic sea-level recovery, these quests setting the tone for the interpretation of the worldwide growth in data during the 1950s-1960s and for the ensuing sea-levels debate. The controversial opinions that grew up on these subjects were summed up by Jelgersma (1961, 1966) as being divided into 'three groups which could nearly be called three parties, each supported by dedicated followers'. The first group supported an oscillatory pattern of global sea-level recovery, showing sea level as rising rapidly to a high point above present levels during the expected postglacial thermal maximum, the Hypsithermal, reaching ~ + 3 m psl by ~ 5,000 BP (Fairbridge, 1961). After this time, sea level was seen as 'oscillating' through an amplitude of 6 m around its present position. The concept of a 'higher than present' sea level grew from Daly's work (1920, 1934) on coral terraces/platforms in the Indian and Pacific oceans. Here, apparently 'fresh' coral terraces elevated to ~ + 6 m psl were interpreted as being Mid-Holocene in age, formed during the postulated Hypsithermal increased ice-melt — expansion in ocean volume, the succeeding fall in sea level seen as occurring with subsequent global cooling. Later, protagonists reaffirmed the concept with complementary evidence of elevated shoreline data from other coral islands in the Pacific, from New Zealand (Schofield, 1960, 1964, 1973), Australia (Gill, 1961, 1965; Ward, 1965) and the Mediterranean (Fairbridge, 1961). Disagreement with this view centred on the validity of the dating of these features and in particular on the likelihood of tectonic crustal changes influencing their observed position.

The second group favoured the concept of a 'standing' sea level after ~ 3,600 BP, with global sea level shown as rising to its present position between 5,000 to 3,600 BP (Fisk, 1951; Le Blanc and Bernard, 1954; Gould and McFarlan, 1959; McFarlan, 1961; Coleman and Smith, 1964). Data in support of this view came predominantly from the Gulf coast of the

USA, critics pointing again to problems in dating and in this case to the long-term subsidence of the study region. The curve of Godwin, Suggate and Willis (1958) also showing a 'stationary' sea level after ~ 5,000 BP received the criticism subsequently levelled at Fairbridge (1961): that of a cavalier use of data from around the world, irrespective of data type, provenance and reliability. The third school of thought found no evidence for sea level rising above present levels, with recovery represented by a 'smooth' exponential decay curve, sea level reaching its present position only in the last few thousand years (Shepard, 1963, 1964; Scholl and Stuiver, 1967; Jelgersma, 1961, 1966). Initial data supporting this concept came predominantly from low-lying, long-term depositional coastlines. Criticism that tectonic subsidence in such regions would distort significantly the pattern of observed sea-level change from a true eustatic record was discounted (Jelgersma, 1966). Studies from these zones, based largely upon biostratigraphic data, were proposed as being the 'best test' of eustasy, in distinct opposition to the arguments for the use of zones of quantifiable uplift as being the most reliable for determining an eustatic sea-level curve (Mörner, 1969, 1971, 1976a, 1980b).

These arguments about the 'high' 'stillstand' or 'steadily rising' schools of thought were to a large degree subsumed, with the generation of greater data volume and reliability, by the 'smooth' or 'oscillating' pattern of sea-level recovery controversy which, although less pronounced, still continues (McLean, 1984). The point of this argument lies in the eventual application of sea-level studies. Accurate characterisation of the form of Holocene sea-level change is essential to an understanding of the future changes in the world's coastlines; for their planning and management. Conceptual, methodological and technical stagnation in attempts to tackle this problem has often led to harangues between protagonists, one group firmly in the belief that they are right the other equally fervid in opposition (Fig. 1.1).

This period of growth and debate in studying sea-surface movements culminated in the International Geological Correlation Programme (IGCP) Project 61 (1974-82), focusing sea-level work from around the world (see Shennan, this vol., Ch. 7). Contributions from researchers in the thirty-one participant countries ranged from detailed empirical examination of local data to theoretical global studies. The aim of the project, 'to establish a graph of the trend of mean sea level during the past 15,000 years' (or deglacial hemicycle — see Farrand, 1965), led to the compilation of an *Atlas of Sea-level Curves* and later Supplement (Bloom, 1977, 1982; see also Ota *et al.*, 1981). Publication of this, however, only served to underline the growing awareness of many researchers during the project (Mörner, 1976b; Faure *et al.*, 1980) that sea-level variations are modified by many local, regional and global factors; further, that spatially uniform changes of sea level, characterised by a single, global curve, represent an unrealistic response of the earth's crust to water mass transfers, and that no

7

Figure 1.1: 'As you can see there are no "oscillations" in sea-level recovery.'

point on the earth's surface can be regarded as having provided a stable datum for recording an eustatic sea level.

The *Atlas of Sea-level Curves* showed a wide spread of sea-level age and altitude data from around the world. Some patterns for Holocene sea-level recovery did emerge, however. For many locations in the Northern Hemisphere a rapid rise after ~ 10,000 BP from levels below −40 m to −50 m msl gave way between 4,000 and 5,000 BP to a slower, decelerating rate of rise, with levels close to the present sea-surface position being reached in the last 1,000-2,000 years (Greensmith and Tooley, 1982; Bloom, 1977). The record from Australia and locations in the Southern Hemisphere indicates that sea-level rose rapidly from a position ~ 25 m below present at ~ 10,000 BP to reach its present level by between 6,000 and 7,000 BP, with wide local variations in subsequent sea-level behaviour (Hopley and Thom, 1983; McLean, 1984; see also Hopley, this vol., Ch. 12). Tooley (1985) views this lack of concurrence in data as stemming partly from the 'failure to employ a unified methodology and an homogenous data base', making 'correlation of marine events at best elusive and at worst erroneous and misleading'. However, this lack of agreement doesn't lie solely with the methodological approach. As Tooley (1985) indicates, fundamental problems remain with understanding the nature of sea-level change itself; problems which, in the context of the global eustatic concept, had already been established by papers in the 1960s and 1970s. Fairbridge (1961) had outlined all those factors which would diminish the possibility of identifying an eustatic (postglacial) sea-level curve. Continuing the lead of Daly (1934), he was perhaps the first to delimit:

(1) The effects of glacier growth and decay on hydro-isostasy (a factor of change later defined more precisely by Bloom, 1967, 1971).
(2) The influences of variations in the geoid on the sea surface (including asthenosphere channel flow, the gravitational anomalies resulting from both surficial and deep-seated mass transfer), ideas subsequently developed by Mörner (1976b, 1980b).
(3) The isostatic effects on level changes of sediment loading.

Apart from the work of IGCP Project 61 many international groups have been active in studying sea-level and related coastal problems since 1950. Because of the significance of climate and consequent shoreline changes in the Quaternary, the International Union for Quaternary Research (INQUA) established in 1953 its *Commission on Quaternary Shorelines.* Organised on a regional basis, the Commission serves to promote studies aimed at understanding palaeosea-levels in a global context. As such, it acts as a major co-ordinator and initiator of many aspects of sea-level research, from numerical modelling, geophysical–geodetic studies through to its major area of concern, the collection of basic geological data. Further, it also serves as a platform for interaction with many of the international organisations now working on sea-surface studies (see Chapter 19 for further discussion) and with the activities of related groups, such as INQUA's Commissions on Neotectonics and the Holocene, and with UNESCO's International Geological Correlation Programme. Stemming from later work at the end of the 1960s and a working group on the 'Dynamics of Shoreline Erosion' (1972) there grew up the present International Geographical Union's *Commission on the Coastal Environment,* begun in 1976. The purpose of this multidisciplinary group is to develop coastal research and its application to seaside management (Paskoff, 1984). Its work parallels aspects of the follow-up project to IGCP-61, namely that of Project 200, *Sea-Level Correlations and Applications.* This project, mounted in sixty-three countries (1984), is aimed at identifying and quantifying

the processes of sea-level change by producing detailed local histories that can be analysed and correlated for tectonic, climatic, tidal and oceanic fluctuations. The ultimate purpose is to provide a basis for predicting near-future changes for application to a variety of coastal problems, with particular reference to densely populated low-lying coastal areas. (Shennan and Pirazzoli, 1984)

These aims perhaps best represent the purpose of many of the groups now working on sea-level research. Whether its goals are attainable remains to be seen!

DEFINITION OF TERMS AND CONCEPTS

Today's concepts of sea-level, the position at which the sea surface intersects the land, and its change over time are complex, the nature and magnitude of sea-level change being conditioned by a number of components. These include eustatic changes related to the accumulation and wasting of land-based ice (glacio-eustasy), the distribution of ocean water under the influence of gravitational forces (geoidal eustasy), earth movements (tectono-eustasy), and isostatic changes following shifts in surface loads such as those of ice (glacio-isostasy), water (hydro-isostasy) and sediment (Mörner, 1976b; Kidson, 1982). Short-term disturbances of the sea surface through changes in physical oceanographic, meteorological and land based hydrological factors (Lisitzin, 1974; Rossiter, 1962, 1967; Woodworth, 1985) are also an integral part of changing sea level (see Mörner, this vol., Ch. 8, Fig. 8.3, and also Fig. 1.3). Variations in air pressure, ocean temperature or salinity, for example, can affect the dynamic height of the ocean surface by $\leqslant 2$ m, or even by > 3 m in confined epicontinental seas (Mörner, 1981). Thus sea-level change at a particular location is the sum of one or more of these components, the proportional contribution of each varying locally or regionally.

These short-term factors contributing to sea-level variation are now readily observable and quantifiable through the study of land-fixed tide gauges (PSMSL, 1976-9; Emery, 1980; Woodworth, 1986), satellite based geodesy (Gaposchkin, 1973; Marsh and Vincent, 1974; Geodynamics Program Office, 1984) and physical atmospheric and oceanographic studies (Klige, 1985; Hoffman, 1984). Regional and global analyses of tide gauge–satellite records (Gutenberg, 1941; Fairbridge and Krebs, 1962; Emery, 1980; Gornitz et al., 1982; Emery and Aubrey, 1985) together confirm a picture of a continuing worldwide rise in relative sea level. Gornitz et al. (1982) indicate an averaged rise of ~ 1.2 mm a^{-1} based on a network of 193 stations, and Emery (1980), working from 247 stations, an even larger rise of 3.0 mm a^{-1}. Apart from doubts as to the size and statistical reliability of these data bases, opinion is divided upon the causes and the rates of these trends (Bird and Koike, 1985; Pirazzoli, 1986). The probability is that given the wide range of regional trends (± 13 mm) observed (Mörner, 1976b; Emery and Aubrey, 1985) that climate–human influences, coupled with isostatic, tectonic and oceanographic controls, today outweigh eustatic ones.

The nature of the components contributing to this relative rise, such as eustasy and earth movements, has already been touched on and will be examined in more detail in following chapters. It is important, however, to have the connotations of the terms clear. Yet, as has been shown in the discussion of eustasy, a problem of definition in a geological study such as that of sea level is that it is constantly subject to revision as new fieldwork is undertaken. Of all the factors concerned perhaps isostasy has remained the least altered, and is one in which major advances in quantification have been made. The process–response model of Andrews (1970) for glacio-isostatically controlled crustal delevelling demonstrates clearly the factors involved (see Peltier, this vol., Ch. 3, 1980; Walcott, 1973, 1980; Cathles, 1975, 1980) (Fig. 1.2a, b). The initial phase in the model (Fig. 1.2b) shows an immediate elastic, and a slower non-elastic response of the crust to a loss in ice mass as melting begins at $t_0 = 0$, termed a period of *restrained rebound* (Ur). As melting continued during a lateglacial time the pace of non-elastic recovery accelerates, reaching maximum rates of rebound during the postglacial phase. The cutting of the first shorelines (marine limit) occurs at $t_1 = 0$ and from this time the course of postglacial uplift (Up) is recorded in the form of raised shorelines and associated sediments (see Figs 3.3 and 10.9). Rebound continues after $t_2 = 0$ at a declining rate, forming a final phase of *residual recovery* (Urr) which may still be operational today in many formerly glaciated areas (see Devoy, this vol., Ch. 10).

The rheological properties of the earth that control the form of isostatic crustal rebound are discussed by Peltier (this vol., Ch. 3). Variations in earth rheology, geological structure, tectonic and loading history may lead to important modifications of the uplift pattern shown in Figure 1.2b for postglacial time (Hillaire-Marcel, 1980; Flemming, 1982; Cathles, 1980; Walcott, 1980; Lambeck and Nakada, 1985). In the different models of mantle structure proposed (Cathles, 1980) the simplest shows a uniform mantle viscosity of $\sim 10^{22/23}$ P in which the stress to strain relationship is linear. Most simply, postglacial crustal rebound here would take place at approximately a constant rate in each locality, but with the constant varying between locations. This vertical displacement would conform to:

$$Z = f(T) + g(x,y) \times T \qquad \text{(Eq. 1.1)}$$

where Z is the relative vertical displacement, T is the age in thousands of years, x and y are geographical co-ordinates and f and g are unknown polynomial functions (Flemming, 1982). In models where earth rheology is linear–non-linear, applicable to areas of crustal bulge peripheral to loading as in North America and Europe, the approach to isostatic equilibrium is

11

Figure 1.2: (a) A process–response model for glacio-isostatic behaviour; (b) Pattern of crustal recovery following the start of ice melt

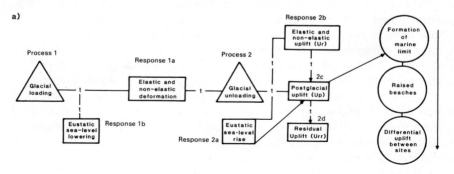

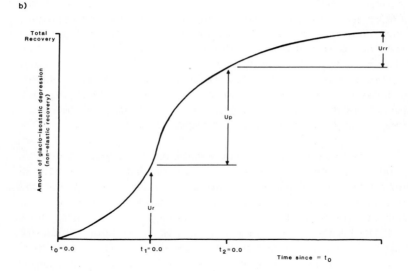

Source: After Andrews (1970).

best represented by a curve of damped exponential decay.

Based on a knowledge of the rates of land uplift over time, through ^{14}C and other techniques for shoreline dating (see Sutherland, this vol., Ch. 6), together with an estimate for vertical movement of the water column (eustatic change), the amount of isostatic uplift/depression (I) at a point relative to present sea level since the time of shoreline formation can be calculated from:

$$I = S + E \qquad \text{(Eq. 1.2)}$$

where S and E form the differences between the present sea level and the marine limit, or shoreline of a known age, and the contemporary eustatic sea level respectively (Andrews, 1970; Mörner, 1969, 1971, 1980c). Inclusion of additional geodynamic factors, such as geoidal (G) or tectonic (T) changes, expands the equation to:

$$I = S + E + G + T \qquad \text{(Eq. 1.3)}$$

Where the magnitude of land uplift is known from independently derived time–gradient curves, from zones of ice loading, modification of Eq. 1.2 to give $E = I - S$ has led to the plotting of postulated regional changes in eustatic sea level (Mörner, 1971, 1976a; Chappell, 1974).

Consideration of the other factors involved in sea-level change shows greater controversy in definition, or if not in definition then in quantification for water-level change. This is shown clearly in terms of the influences of earth–tectonic movements. These change the volume of the ocean basins and hence the vertical ocean level (Fig. 1.3). Quantification of their impact on sea level is less easy to agree, due to uncertainty as to the nature and interlinkage of the geophysical processes involved over Phanerozoic time, or to the scale of changes in sea level that would result (Carey, 1981; Donovon and Jones, 1979; see also Chappell, this vol., Ch. 2). Changes in ocean basic capacity below the geoid are summarised by Carey (1981) as the result of changes in area and/or mean depth below the geoid. For the latter, he points to the problem in computing depth changes over the long timescales involved in the earth's crust adjusting overall to hydrostatic equilibrium following continental spreading. Thus, 'the relaxation time for isostatic disequilibrium of continental dimensions is in the order of 1 ka, whereas the relaxation time of departures from hydrostatic equilibrium is of the order of 1,000 Ma'. Other causes of depth change result from sediment infill, seen by Mörner (1976b) as of minor significance, and the deepening of ocean basins with age by several kilometres following the initial formation of oceanic crustal rocks at ridge spreading centres. In terms of the area factor the linkage to sea-level change is either simple or equally complex. For earth models which assume a fixed earth size the total area remains approximately the same, and forms a minor variable for sea-level changes. A change in area becomes a major variable by comparison with models assuming earth expansion (Egyed, 1969), the growth of existing ocean floors since the Palaeozoic being taken in this case to imply an increase in total earth surface area.

As shown in Figure 1.3 changes in ocean basin and ocean water volume, the latter affected mainly by the climate–ice volume linkage, together control vertical movements in ocean level, or real eustasy (glacio-eustasy +

13

Figure 1.3: Factors and processes involved in the movement of relative sea level

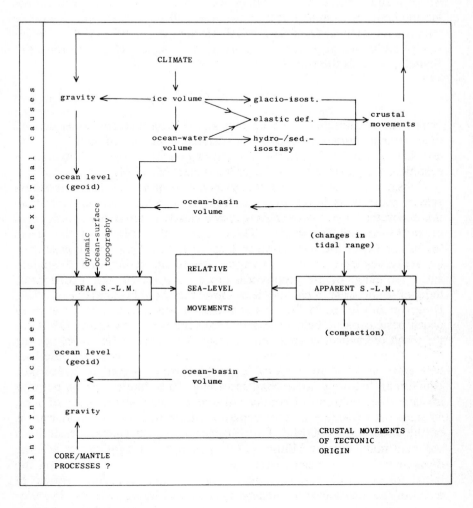

Source: After van de Plassche (1982).

tectono-eustasy). Use of the term eustasy is now very different from that of its first usage, namely, representing worldwide simultaneous changes in sea level resulting from the ocean water's own changes in level, as opposed to those of tectonic or isostatic movements. Definitions of the linkages shown in Figure 1.3 has led to the re-definition of eustasy by Mörner (1967b, 1980a) as simply 'ocean level changes', irrespective of their cause and

14

'implying vertical movements of the ocean surface'. Important in this re-definition is the role of gravity induced effects on ocean level or geoidal eustasy. Inclusion of this factor as a component of eustasy is based upon the concept of extra local–global affects of gravitational changes on vertical ocean levels which would be indistinguishable from glacio-/tectono-eustatic ones.

The geoid, the surface of the non-perturbed sea, is an equilibrium surface. The shape of this may be modelled best as an oblate spheroid with a marked north–south asymmetry (Carey, 1981), represented in gravity terms as:

$$Y = c_1 + c_2 \sin^2 \phi + c_3 \cos^2\phi + c_4\sin^2 2\phi$$
$$+ c_5\sin(\lambda - C_6) \sin \phi \qquad \text{(Eq. 1.4)}$$

Where ϕ and λ are the latitude and longitude variables, c is the mean gravity at the equator, c_2 the centrifugal acceleration at the equator, c_3 north–south gravity asymmetry, c_4 the mid-latitude correction, c_5 the gravity difference between major and minor axes of ellipticity and c_6 the longitude of a minor axis of ellipticity. Models of the geoid show this equi-potential surface, or geodetic sea level, to be irregular in form (Fig. 1.4), having a spatially varied pattern of swells and depressions and with a range in relief of up to 200 m (Marsh and Martin, 1982; Mörner, 1976b). This

Figure 1.4: The geoid

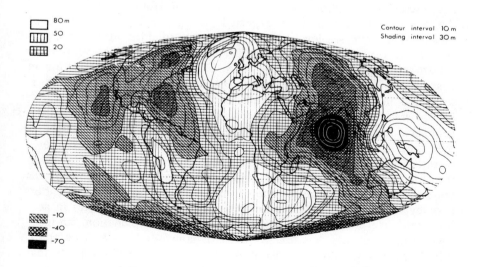

Source: Carey (1981).

surface is seen as a function of the earth's gravity, as determined by differences in earth structure, density, rheology and rotation, and of astronomical gravity (planetary forces) together with the influence of the universal gravitational constant (G). Tectono-/glacio-eustatic and isostatic factors amongst others will also affect the geoid surface, particularly through the influence of material flows within the earth in response to changes in surface-mass (ice, water, sediment) distribution. However, the main configuration of the geoid will remain essentially separate from the longer-term influences of mountain building and related earth movements.

As indicated by Mörner (1976b) measurement of the geoidal influence in eustatic sea-level change is difficult. Because of the interlinkages between the geoid and other components of sea-level change the plotting and evaluation of relative sea-level indicators (see Chapters 6 and 7) will not represent changes in geoid elevation alone, or act as a measure of the palaeogeoid *per se*. In this context Newman *et al.*'s (1980, 1981) use of the term palaeogeoid in association with such data is therefore misleading. The data used represent changes rather in the palaeosea surface, although geoidal variations may be an important conditioning factor here. Further confusion in the application of the geoid concept to sea-level studies is the assumption that the oceanic geoid must be regarded as unstable both in space and time. Movements are seen as being of two main forms, vertical variation in the magnitude of sea-surface relief and secular migration of the sea-surface topography, recorded at the coastline as long amplitude sea-level oscillations. Support for this idea stems largely from work by Mörner (1976b, 1980d, 1980e, 1981) on the analysis of sea-level records of differing timescales throughout the Phanerozoic. In the Holocene, for example, the convergence rather than expected divergence at ~ 7,000 BP of presumed eustatic sea-level curves, from areas of differing tectonic and isostatic history, is cited by Mörner (1976b) as a problem that can only be solved satisfactorily by geoid migration. The 'out-of-phase' oscillation in eustatic sea-level records between Brazil and Northwest Europe is explained in similar terms (Mörner, 1981, 1983; Suguio *et al.*, 1980; Martin *et al.*, 1985). All the evidence to date rests on interpretation of the geological records, with their wide variation in data quality and reliability over time. In the case of the Brazilian and Northwest European sea-level records, for example, the observed anomalies are only to be expected given the differences in methodological approach and materials used; conclusions about palaeogeoid changes on the existing data base would be premature (Tooley, 1985). Mörner has pointed to apparent inconsistencies in sea-level data for which geoid instability may provide an explanation, but measurable geoid change has not yet been demonstrated; data error apart, other explanations may also exist. It may be, as Carey (1981) suspects, that humanly long-term westward migration of the geoid configuration occurs, but it has yet to be tested. Equally, equating the cause of the assumed instability to earth rheology and magnetic/core variations is controversial. The viscosity of the earth's mantle used in many of the models of crustal behaviour would be incompatible with the operation of geoid migration over short timescales, of the order of Quaternary time, or even over possibly longer time periods (Lambeck, 1985).

THE METHODOLOGY OF SEA-LEVEL STUDIES

Within the context of discussing the history, development and definition of the terms of reference of sea-level studies, a study of sea-level methodology, namely the sources, sampling and evaluation of sea-level data, must form a subject of primary significance. The validity of the discussion of sea-level change problems depends upon data reliability. In this volume, this section together with Chapters 9 to 12 examine various aspects of this topic. For fuller treatment, readers are referred to van de Plassche (1977, 1986) and Hopley and Thom (1983).

The central question in considering methodology is how to recognise, and subsequently determine the value of, a change in sea level? The natural earth materials in which marine events are recorded provide the basic data source, or rather the *index points* (sea-level indicators), from which sea-level changes are identified. Every marine feature or organism which has a quantifiable vertical relationship to a water level (*indicative meaning*), the height of which is ultimately controlled by tidal amplitude at the coast (Jelgersma, 1961; van de Plassche, 1982), can be regarded as a sea-level indicator. The value of the indicator's indicative meaning or height range relative to the reference water level is based upon modern observation of this relationship, and generally the narrower the range the more valuable the indicator will be. However, the height, age and palaeoenvironmental data represented by these materials are subject to alterations in time which require careful evaluation and quantification. As Tooley (1985) states most clearly, 'palaeosea-level index points are affected by changes in the rate of sedimentation, by the type of sediment and its consolidation characteristics' and by factors of erosion. 'Changes in palaeotidal conditions, the position of the coast in relation to wave and tidal environments and the nature of the palaeocoast will also affect the interpretation of the index point.' Further, although a sea-level indicator by definition contains information on the height of a water level, not all will allow definition of the age, and even the most precise height indicator is greatly reduced in value without this information. On a global basis many different categories of sea-level indicator are used, varying through morphological features (i.e. raised beaches/shorelines) to data (chemical, fossil material) contained in sediments. Table 1.1 provides a summary of the range of potentially useful indicators.

17

Table 1.1: Broad classification of water-level indicators (wli's)

(1) *wli's consisting of datable material*

    (a) *Physico-chemical*

        1. Beachrock
        2. Ooids
        3. Calcareous mud (biochemical)

    (b) *Biotic* (found in growth position, or correct stratigraphic context)

        1. Vegetation (peat)
        2. Wood
        3. Remains of organic matter (other than 1 or 2)
        4. Shells — Cheniers
        5. Coral
        6. Coraline algae
        7. Stromatolites
        8. Bones (in relevant position)

    (c) *Archaeological*

        1. Human built structures (irp)
        2. Traces of human settlement (irp)

(2) *wli's not consisting of datable material*

    (a) *Physico-chemical*

        1. Morphological (destructional)

            1. cliff-face
            2. shore platform
            3. notch
            4. marine washing limits, etc.

        2. Sedimentological and morphological (constructional)

            1. sediments (sedimentary environments)

                1. tidal flats with morphological features
                2. estuary with morphological features
                3. lagoon with morphological features
                4. delta with morphological features
                5. beach with morphological features

            2. sedimentary structures

    (b) *Biotic*

        1. Stromatolites
        2. Structures and forms produced by animals (other than 1)

    (c) *Archaeological*

        1. Human built structures (irp)
        2. Traces of human settlement (irp)

Source: After van de Plassche (1977).

Given the global variations in dating methods and sea-level indicators used, differences imposed in part by indicator provenance, it is not surprising that wide variations in methodological approach to data collection and interpretation have evolved, criticism of which has been referred to earlier (Tooley, 1985). Despite these disparate approaches, growth in an understanding of the nature of sea-level change and of the ways in which solutions to problems raised can be solved, has perhaps resulted in its own uniformity of approach, aided by greater international research exchange and the setting-up of data banks through projects such as IGCP-61 and 200 (Shennan, 1986). An illustration of this is provided in Figure 1.5, showing the suggested stages involved in the collection and evaluation of age-height information. Although not necessarily universally accepted, the model does provide an approach relevant to most types of sea-level indicator found. Similarly, the following six main criteria, incorporating some of the points shown in Figure 1.5, are now commonly used by sea-level researchers (Tooley, 1978a, 1978b; Devoy, 1979, 1982; Shennan, 1982, 1983, 1986) in assessing the suitability of an indicator for plotting sea-level change:

(1) In order to reduce errors resulting from differences in tidal regime, coastal shape, earth movements and geoid variations, index points should be drawn from small, clearly definable areas. Nevertheless, even within small geographical units great spatial variability in tidal range and the recording of sea-level change can still occur, especially on meso-macrotidal, high-energy coastlines (Gerrard *et al.*, 1984; Carter, 1983).

(2) Clear understanding of the palaeogeography of the study area from which the sea-level indicator(s) came. The complexity of many coastal environments with beach-barrier lagoons, reefs or offshore blocking sedimentary structures, resulting in the restriction of marine access and tidal changes on the coastline, may lead to the incorrect interpretation of isolated index points (Devoy, 1982, 1985; Jennings and Smyth, 1982).

(3) Sea-level index points used in constructing the time-altitude trend of sea-level change should be of the same type, representing similar palaeoenvironments and, most importantly, the same indicative meaning (Tooley, 1978b; Laborel; 1979-80a, b; Hillaire-Marcel, 1980). Many sea-level studies have highlighted the wide variation in sea-level record that can occur in different, though spatially close, environmental situations (Devoy, 1979; Morzadec-Kerfourn, 1974; van de Plassche, 1980, 1982), for example from an estuary compared to the open coast. The practice of using isolated records, such as publications in *Radiocarbon,* or spatially separate and different indicator types, is now seen as unacceptable (Tooley, 1985).

19

Figure 1.5: Stages in the evaluation of sea-level age — altitude data

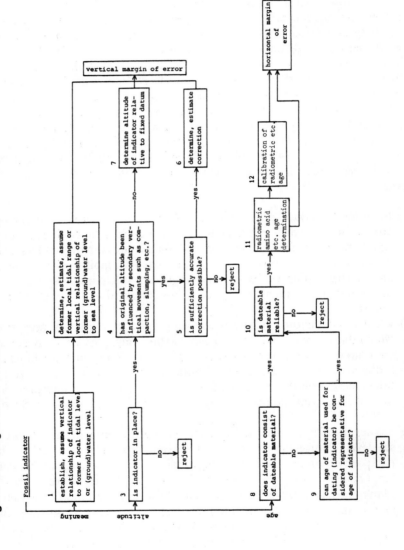

Source: After van de Plassche (1982).

(4) Radiocarbon/radiometric dating techniques used should be independently verifiable (Sutherland, 1983; see also this vol., Ch. 6) through use of a different dating method. Where this is not possible, then several samples from the same section/core should be dated to provide a check on consistency.

(5) Application of a repeatable methodology in sample retrieval using accurate sampling techniques that ensure minimal stratigraphic disturbance or vertical distortion of the samples taken. This is perhaps the least satisfactory of the criteria as sampling is dependent on many factors, such as site type and location, equipment and personnel availability. However, its importance cannot be dismissed as the validity of all subsequent analyses depend upon the reliability of this initial fieldwork.

(6) Determination of the sea-level indicator's height relationship to a definable datum, ideally a geodetic datum. Prior to the development of satellite geodesy this was not feasible technically in many parts of the world. However, the confusion that results from attempts to analyse information based upon different data, even at the local scale, has perhaps been as great an inhibitor to correlation as the use of different indicator types (Devoy, 1983).

The increased concern with sea-level methodology and terminology, evidenced in these criteria, has both resulted from and fuelled a reappraisal of the interpretation and interregional correlation of sea-level data (Tooley, 1982; Chappell et al., 1983; Shennan, 1983). Realisation of the inadequacy of using sea-level curves (Kidson, 1982), in the form of a single line drawn through index points, as a basis for the analysis and comparison of data has led to a search for alternatives (Shennan, 1983, see also this vol., Ch. 7). The use of time-altitude error boxes around variates and bands of change, although limiting the degree of misrepresentation, does not open any new avenues to the problems of data correlation. One approach that does take a different line, however, has been the introduction of the concept of 'tendencies of sea-level movement' based on the indicative meaning of an indicator (Shennan, 1982). This allows the identification of the trends in water level movement, inundation (positive) or marine removal (negative), at a site. Placed on an areal level, recognition of a general tendency pattern between sites is seen as providing a basis for correlation between zones of differing environmental and crustal history (Shennan et al., 1983). A more detailed discussion of the approach is given in Chapter 7.

21

Palaeotides

By way of conclusion to both Chapter 1 and the theme of methodology, it is worth drawing the reader's attention to the subject of palaeotides and their relation to sea-level change. Unfortunately, it has not been possible to present in this book detailed treatment of this important topic, although a brief reference is given to tidal changes and the sea surface in Chapter 18. The study of palaeotides highlights a number of problems in sea-level methodology, not least being an assumption in many studies that tidal amplitude has remained approximately constant in time; a possibly incorrect assumption in terms for example of the vertical scale of Holocene sea-level recovery.

The processes of tidal generation have long been established and readers are referred to Lisitzin (1974) and Wood (1985) for detail. The attraction of the sun and moon on the water of the oceans causes tides. The spatial variation in the height of the tide or standing wave produced results from the interaction of earth rotation, the shape and depth of the oceans−coastal waters and finally from coastal−offshore zone configuration. Consequently changes in the altitude of the sea surface, and thus in water depth, will have a fundamental effect on tidal amplitude. Studies by Davis on sites in western Britain showed that significant changes in the amplitude of semi-diurnal (M_2) tides occurred when water depth (sea-surface height) was reduced by 20 m (~ 9,500 BP), 10 m (~ 8,500 BP) and 5 m (~ 7,500 BP). In Liverpool Bay tidal range was shown to be reduced from ~ 2.8 m to 1.8 m with a 20 m drop in sea level. At the head of the Bristol Channel, the reduction was from 4.3 m to 1.5 m (Tooley, 1985). Further, within a confined coastal environment local tidal range may differ considerably from that at the open coast. In a funnel-shaped embayment (estuary) tidal amplitude often rises landward, due to confinement of the tidal wave as it progresses up-estuary (the 'estuary effect', Fairbridge, 1961). Conversely a reduction in tidal range takes place when the tidal wave, after passing through a confined coastal inlet, enters a flanking basin with a large storage capacity ('flood-basin effect', Van Veen, 1950). Dissipation of tidal energy due to water-bed interface friction also results in a decrease in tidal amplitude (Allen *et al.*, 1980).

Despite knowledge of the importance of changes in water depth on tidal range and the existence of numerical models to calculate tidal variation (Rossiter, 1963; Pinless, 1975; Greenberg, 1979), the application of tide analysis to the quantification of past sea-surface positions has been limited. An exception occurs in studies of the Bay of Fundy, Canada (Grant, 1970; Garrett, 1974; Scott and Greenberg, 1983). Here, palaeowater-depths based on sea-level data covering the period 0-7,000 BP were incorporated into models of contemporary tidal variations (Greenberg, 1979), allowing generation of the different tidal regimes through time for varying coastal−

basin configurations. Scott and Greenberg (1983) show that tidal amplitude here increased in time in a non-linear fashion, rising more rapidly in the period 7,000-4,000 BP than in 4,000-0 BP. Between 7,000 and 2,500 BP tidal range may have increased by as much as 5.5 m in some locations, with an averaged increase of 1.2 per cent per 1 m rise in sea level. They concluded also that changes in water depth over Georges Bank, south of the Gulf of Maine, conditioned by relative sea-level rise, was the dominant factor controlling tides in the Bay of Fundy and that a strongly macrotidal regime had been established here by ~ 4,000 BP.

By comparison, work on palaeotidal changes elsewhere in the world has been essentially non-numerical, based purely on geological inference. In the Netherlands, van de Plassche (1980) has tentatively identified the influence of a 'river gradient' effect upon water levels — sea-level data in the Rhine–Meuse delta. This shows the convergence in height toward the present of inner estuary indicators with those from more open coastal sites (see Fig. 7.5 and accompanying discussion). Convergence probably results from a decline in the amplification of water levels within the estuary (river gradient effect) as flood depression zones expand in size consequent upon sea-level rise. A similar pattern is shown for the Thames estuary in England (Devoy, 1979). Further, the assumption that tidal range along the Netherlands coast has been approximately constant since ~ 8,000BP (Jelgersma, 1961) is supported in part in work by Roep et al. (1975). This showed, through the study of the altitude of sedimentary and worm burrow structures and the position of shell concentrations and clay seams in ancient beach barrier deposits, that tidal ranges have changed little on this coastline since ~ 4,800 BP, supporting incidentally the broad trends from the Bay of Fundy.

Overall, the value of palaeotidal work lies in aiding the precise evaluation on meso-macrotidal coasts of the relationship between a sea-level indicator's height and a former sea-level position. On many such coasts the most reliable sea-level indicators occur in sediments from the upper quarter of the tidal range (Scott and Greenberg, 1983). Changes in tidal amplitude over time will have greatest effect here, variably distorting the 'true picture' and the reconstruction of sea-level movements derived from such zones. The application of numerical studies of tidal amplitude is essential, therefore, in aiding quantification of this distortion and obtaining a more accurate prediction of future changes in coastal position. To date, the disciplines of physics, mathematics, engineering and oceanography have been most involved in the examination of tides and related phenomena affecting the contemporary sea surface and its interaction with the land at the coast. It is perhaps the interfacing of these approaches with the more traditional geological ones that hold the most promise for future advances in sea-level research.

REFERENCES

Agassiz, L. (1840) *Etudes sur les glaciers*, privately published, Neuchâtel.

Allen, G.P., Salomon, J.C., Bassoullet, P., Penhoat, Y. du and Grandpré, C. (1980) 'Effects of tides on mixing and suspended sediment transport in macrotidal estuaries', *Sed. Geol., 26*, 69-90.

Andrews, J.T. (1970) *A Geomorphological Study of Postglacial Uplift, with Particular Reference to Arctic Canada*, Spec. Publ. No. 2, Inst. Br. Geogr., London.

—— (1974), *Glacial Isostasy*. Dowden, Hutchinson & Ross, Stroudsberg, Pa.

Boylan, P.J. (1981) 'The role of William Buckland (1784-1856) in the recognition of glaciation in Great Britain', in J. Neale and J. Flenley (eds), *The Quaternary in Britain*, Pergamon Press, Oxford, pp. 1-8.

Bird, E.C.F. and Koike, K. (1985) 'Man's impact on sea-level changes: a review' (unpublished paper). Abstract, *Proc. Fifth Int. Coral Reef Congr., Tahiti*, vol. 3, pp. 91-2.

Bloom, A.L. (1967) 'Pleistocene shorelines: a new test of isostasy', *Bull. Geol. Soc. Am., 78*, 1477-94.

—— (1971) 'Glacial-eustatic and isostatic controls of sea level since the last glaciation', in K.K. Turekian (ed.), *The Late Cenozoic Ice Ages*, Yale University Press, New Haven, Conn., pp. 355-79.

—— (1977) *Atlas of Sea-level Curves*, IGCP Project 61, Cornell University, Ithaca, NY.

—— (1982) *Atlas of Sea-level Curves. Zone A, Curves and Partial Bibliography*, IGCP Project 61, Cornell University, Ithaca, NY.

—— (1983) 'Sea-level and coastal morphology of the United States through the Late Wisconsin glacial maximum', in H.E. Wright (ed.), *Late Quaternary Environments of the United States*, Longman, London, vol. 1, pp. 215-29.

—— (1984), 'Sea-level and coastal changes', in H.E. Wright (ed.), *Late Quaternary Environments of the United States, vol. 2, The Holocene*, Longman, London, pp. 42-51.

Bravais, A. (1840) 'Sur les lignes d'ancien niveau de la mer dans le Finmark', in *Voyages en Scandinavie, en Lapponie, au Spitzberg et aux Feröe, pendant les années 1838, 1839 et 1840 sur la corvette 'La Recherche'*, vol. I, pt 1, pp. 57-137.

Brooks, M.J. and Canouts, V. (1981) 'Environmental and subsistence change during the Late Prehistoric period in the interior Lower Coastal Plain of South Carolina', in D.J. Colquhoun (ed.), *Variation in Sea-level on the South Carolina Coastal Plain*, Dept. of Geology, University of South Carolina, Columbia, SC, pp. 45-72.

Carey, W.S. (1981) 'Causes of sea-level oscillations', *Proc. Roy. Soc. Vict., 92*, 13-17.

Carter, R.W.G. (1983) 'Raised coastal landforms as products of modern process variations, and their relevance in eustatic sea-level studies: examples from Eastern Ireland', *Boreas, 12*, 167-82.

Cathles, L.M. (1975) *The Viscosity of the Earth's Mantle*, Princeton University Press, Princeton, NJ.

—— (1980) 'Interpretation of postglacial isostatic adjustment phenomena in terms of mantle rheology', in N.-A. Mörner (ed.), *Earth Rheology, Isostasy and Eustasy*, Wiley, Chichester and New York, pp. 11-43.

Chappell, J. (1974) 'Geology of coral terraces, Huon peninsula, New Guinea: a study of Quaternary tectonic movements and sea-level changes', *Bull. Geol. Soc. Am., 85*, 553-70.

—— Chivas, A., Wallensky, E., Polach, H. and Aharon, P. (1983) 'Holocene palaeoenvironmental changes, central to north Great Barrier Reef inner zone', *BMR J. Austr. Geol. Geophys.*, *8*, 223-35

Chorley, R.J., Dunn, A.J. and Beckinsale, R.P. (1964) *The History of the Study of Landforms*, Wiley, New York.

Clark, J.A., Farrell, W.E. and Peltier, W.R. (1978) 'Global changes in postglacial sea-level: a numerical calculation', *Quat. Res., 9*, 265-87.

Clark, J.G.D. (1954) *Excavations at Starr Carr; An Early Mesolithic Site at Seamer, Near Scarborough, Yorkshire*, Cambridge University Press, Cambridge.

Coleman, J.M. and Smith, W.G. (1964) 'Late recent rise of sea level', *Bull. Geol. Soc. Am., 75*, 833.

Daly, R.A. (1910) 'Pleistocene glaciation and the coral reef problem', *Am. J. Sci., 30*, 297-308

—— (1920) 'A recent worldwide sinking of ocean level', *Geol. Mag., 57*, 246-61.

—— (1925) 'Pleistocene changes of level', *Am. J. Sci., 10*, 281-93.

—— (1934) *The Changing World of the Ice Age*, Yale University Press, New Haven, Conn.

Darwin, C.R. (1842) *The Structure and Distribution of Coral Reefs*, Smith, Elder, London.

Davis, C.A. (1910) 'Saltmarsh formation near Boston and its geological significance', *Econ. Geol., 5*, 623-39.

Dawson, J.W. (1868) *Acadian Geology*, 2nd edn, Macmillan, London.

De Geer, G. (1888 and 1890) 'Om Skandinaviens nivå förändringar under Quartär-perioden', *Geol. Fören. Stockh. Förh., 10*, 366-79 and *12*, 61-110.

—— (1892) 'On Pleistocene changes of level in eastern North America', *Proc. Boston Soc. Nat. Hist., 25*, 454-77.

Devoy, R.J.N. (1979) 'Flandrian sea-level changes and vegetational history of the lower Thames estuary', *Phil. Trans. Roy. Soc. Lond., 13, 285*, 355-410.

—— (1982) 'Analysis of the geological evidence for Holocene sea-level movements in southeast England', *Proc. Geol. Ass. 93*, 65-90.

—— (1983) 'Late Quaternary shorelines in Ireland: an assessment of their implications for isostatic land movement and relative sea-level changes', in D.E. Smith, and A.G. Dawson (eds), *Shorelines and Isostasy*, Academic Press, London and New York, pp. 227-54.

—— (1985) 'Holocene sea-level changes and coastal processes on the south coast of Ireland: corals and the problems of sea-level methodology in temperate waters', in *Proc. Fifth Int. Coral Reef Congr., Tahiti*, vol. 3, pp. 173-8.

Donovan, D.T. and Jones, E.J.W. (1979) 'Causes of world-wide changes in sea level', *J. Geol. Soc. Lond., 136*, 187-92.

Egyed, L. (1969) 'The slow expansion hypothesis', in K. Runcorn (ed.), *Application of Modern Physics to Earth and Planetary Interiors*, Wiley Interscience, New York, pp. 65-75.

Emery, K.O. (1980) 'Relative sea levels from tide-gauge records', *Proc. Nat. Acad. of Sci., 77.*, 6968-72.

—— and Aubrey, D.G. (1985) 'Glacial rebound and relative sea levels in Europe from tide-gauge records', *Tectonophys., 120*, 239-55.

Eronen, M. (1974) 'The history of the Litorina Sea and associated Holocene events', *Comment. Phys. — Math., 44*, 80-195.

Fairbridge, R.W. (1961) 'Eustatic changes in sea level', in L.H. Ahrens, F. Press, K. Rankama, and S.K. Runcorn (eds), *Physics and Chemistry of the Earth* Pergamon Press, London, vol. 4, pp. 99-185.

—— and Krebs, O.A., Jr (1962), 'Sea level and the Southern Oscillation',

Geophys. J. Roy. Astron. Soc., 6, 532-45.

Farrand, W.R. (1965) 'The deglacial hemicycle', *Geol. Rundschau*, 54, 385.

Faure, H., Fontes, J.C., Hébrard, L., Monteillet, J. and Pirazolli, P.A. (1980) 'Geoidal change and shore level tilt along Holocene estuaries: Senegal river area, West Africa', *Science*, 210, 421-3.

Fisk, H.N. (1951) 'Loess and Quaternary geology of the Lower Mississippi valley', *J. Geol.*, 59, 333-56.

Flemming, N.C. (1982) 'Multiple regression analysis of earth movements and eustatic sea-level change in the United Kingdom in the past 9,000 years', *Proc. Geol. Assoc.*, 93, 113-25.

Flint, R.F. (1971) *Glacial and Quaternary Geology*, Wiley, Chichester and New York.

Gaposchkin, E.M. (1973) 'Satellite dynamics', in E.M. Gaposchkin (ed.), *Smithsonian Standard Earth (III)*, Spec. Rep. 353, Smithsonian Astron. Obs., pp. 85-192.

Garrett, C. (1974) 'Normal resonance in the Bay of Fundy and the Gulf of Maine', *Can. J. Earth Sci.*, 11, 549-56.

Geodynamics Program Office (1984) *NASA Geodynamics Program: Fifth Annual Report*, NASA Technical Memorandum 87359, Washington, D.C.

Gerrard, A.J., Adlam, B.H. and Morris, L. (1984) 'Holocene coastal changes: methodological problems', *Quaternary Newsletter*, 44, 7-14.

Gill, E.D. (1961) 'Change in the level of the sea relative to the land in Australia during the Quaternary era', *Zeit. Geomorph. Suppl.*, 3, 73-9.

—— (1965) 'Radiocarbon dating of past sea levels in S.E. Australia', in *INQUA VII Congress Abstracts*, p. 167.

Godwin, H. (1943) 'Coastal peat beds of the British Isles and North Sea', *J. Ecol.*, 31, 199-247.

—— Suggate, R.P. and Willis, E.H. (1958) 'Radiocarbon dating of the eustatic rise in ocean level', *Nature*, 181, 1518-19.

Gornitz, V., Lebedeff, S. and Hansen, J. (1982) 'Global sea-level trend in the past century', *Science*, 215, 1611-14.

Gould, H.R. and McFarlan, E. (1959) 'Geologic history of Chenier plain, south western Louisiana', *Bull. Am. Ass. Petrol. Geol.*, 43, 2520.

Grant, D.R. (1970) 'Recent coastal submergence of the Maritime Provinces, Canada', *Can. J. Earth Sci.*, 7, 679-89.

Greenberg, D.A. (1979) 'A numerical model investigation of tidal phenomena in the Bay of Fundy and Gulf of Maine', *Mar. Geodesy*, 2, 161-87.

Greensmith, J.T. and Tooley, M.J. (eds) (1982) 'IGCP Project 61: Sea-level movements during the last deglacial hemicycle (about 15,000 years): Final Report of the UK Working Group', *Proc. Geol. Ass.*, 93, 1-125.

Gutenberg, B. (1941) 'Changes in sea level, postglacial uplift, and mobility of the earth's interior', *Bull. Geol. Sci. Am.*, 52, 721-72.

Hillaire-Marcel, C. (1980) 'Multiple component postglacial emergence, Eastern Hudson Bay, Canada', in N.-A. Mörner (ed.), *Earth Rheology, Isostasy and Eustasy*, Wiley, Chichester and New York, pp. 215-30.

Hoffman, J.S. (1984) 'Estimates of future sea-level rise', in M.C. Barth, and J.G. Titus (eds), *Greenhouse Effect and Sea-Level Rise*, Van Nostrand Reinhold, New York, pp. 79-103.

Holmes, A. (1965) *Principles of Physical Geology*, Nelson, London.

Holmsen, G. (1918) 'Strandlinjernas vidnesbryd om landets isostase — bevaegelse', *Geol. Fören Stockh. Förh.*, 40, 521-8.

Hopley, D. and Thom, B.G. (1983) 'Australian sea levels in the last 15,000 years:

an introductory review', in D. Hopley (ed.), *Australian Sea 15,000 years: A Review*, Monogr. Ser., Occ. Paper No. 3, De James Cook University of North Queensland, pp. 3-26.

Huddart, D. and Tooley, M.J. (1972) *The Cumberland Lowland,* Quaternary Research Association, UK.

Jamieson, T.F. (1865) 'On the history of the last geological change *Quart. J. Geol. Soc. Lond., 21,* 161-203.

—— (1882) 'On the course of the depression and re-elevation of th ____ during the glacial period', *Geol. Mag., 9,* 400-7 and 457-66.

Jelgersma, S. (1961) 'Holocene sea-level changes in the Netherlands', *Meded. Geol. Sticht., Serie C. VI, 7,* 1-100.

—— (1966) 'Sea-level changes during the last 10,000 years', in J.S. Sawyer (ed.), *World climate 8,000 to 0 BC, Roy. Met. Soc., 229,* London, pp. 54-69.

Jennings, S. and Smyth, C. (1982) 'A preliminary interpretation of coastal deposits from East Sussex', *Quaternary Newsletter, 37,* 12-19.

John, B.S. (1964) 'A description and explanation of glacial till in 1603', *J. Glaciol., 5,* 369-70.

Johnson, D.W. (1925) *New England Acadian Shoreline,* Wiley, New York.

Kidson, C. (1982) 'Sea-level changes in the Holocene', *Quat. Sci. Rev., 1,* 121-51.

Klige, R.K. (1985) *Variations of Global Water Exchange,* Reidel, Dordrecht.

Kuenen, P.H. (1950) *Marine Geology,* Wiley, New York.

Laborel, J. (1979-80b) 'Utilisation des cnidaires hermatypiques comme indicateurs de niveau', *Oceanis, 5,* 241-9.

—— (1979-80a) 'Les Gasteropodes vermetidés: leur utilisation comme marqueurs biologiques de rivages fossiles', *Oceanis, 5,* 221-38.

Lambeck, K. (1985) Personal communication resulting from discussion at Fifth Int. Coral Reef Congr., Tahiti.

—— and Nakada, M. (1985) 'Holocene fluctuations in sea-level constraints on mantle viscosity and meltwater sources', in *Proc. Fifth Int. Coral Reef Congr., Tahiti,* vol. 3, pp. 79-84.

Le Blanc, R.J. and Bernard, H.A. (1954) 'Résumé of late recent geological history of the Gulf Coast', *Geol. Mijnb., N.S. 16,* 185-94.

Lepionka, L. (1981) 'The second refuge site, sea level, palaeogeography and cultural attributes', in D.J. Colquhoun (ed.), *Variation in Sea Level on the South Carolina Coastal Plain,* Dept. of Geology, University of South Carolina, Columbia, SC, pp. 73-84.

Lin, D. (1983) 'Progress of sea-level variations research in China during the Late Quaternary', in *Abstracts of Papers: Int. Symp. on Coastal Evolution in the Holocene, Tokyo,* pp. 68-77.

—— (1986) 'A Chinese method of geo-research; abstract', in R.W. Carter and R.J. Devoy (eds), *The Hydrodynamics and Sedimentary Consequences of Sea-Level Change,* A Conference, University College, Cork, Ireland.

Lisitzin, E. (1974) *Sea-level Changes, Oceanogr. Ser. 8,* Elsevier, Amsterdam.

McFarlan, E. (1961) 'Radiocarbon dating of Late Quaternary deposits, South Louisiana', *Bull. Geol. Soc. Am., 72,* 129-58.

MacLaren, C. (1842) 'The glacial theory of Professor Agassiz', *Am. J. Sci., 42,* 346-65.

McLean, R. (1984) 'Coastal landforms: sea-level history and coastal evolution', *Prog. Phys. Geogr., 8,* 431-42.

Marsh, J.G. and Martin, T.V. (1982) 'The SEASAT altimeter mean sea surface model', *J. Geophys. Res., 87,* C5 3269-80.

—— and Vincent, S. (1974) 'Global detailed geoid computation and model analysis', *Geophys. Surveys*, *1*, 481-511.

Martin, L., Flexor, J.-M., Blitzkow, D. and Suguio, K. (1985) 'Geoid change indications along the Brazilian coast during the last 7,000 years', in *Proc. Fifth Int. Coral Reef Congr., Tahiti*, Vol. 3, pp. 85-90.

Mörner, N.-A. (1969) 'Eustatic and climatic changes during the last 15,000 years', *Geol. Mijnb.*, *48*, 389-99.

—— (1971) 'Eustatic changes during the last 20,000 years and a method of separating the isostatic and eustatic factors in an uplifted area', *Palaeogeography, Palaeoclimatol., Palaeoecol.*, *9*, 153-81.

—— (1976a) 'Eustatic changes during the last 8,000 years in view of the radiocarbon calibration and new information from the Kattegatt region and other Northwestern European coastal areas', Palaeogeography, Palaeoclimatol., Palaeoecol., *19*, 63-85.

—— (1976b) 'Eustasy and geoid changes', *J. Geol.*, *84*, 123-51.

—— (1980a) 'The Northwestern European "sea-level laboratory" and regional Holocene eustasy', *Palaeogeography, Palaeoclimatol., Palaeoecol.*, *29*, 281-300.

—— (ed.) (1980b) *Earth Rheology, Isostasy and Eustasy*, Wiley, Chichester and New York.

—— (1980c) 'The Fennoscandian uplift: geological data and their geodynamic implication', in N.-A. Mörner, (ed.), *Earth Rheology, Isostasy and Eustasy*, Wiley, Chichester and New York, pp. 251-84.

—— (1980d) 'Eustasy and geoid changes as a function of core/mantle changes', in N.-A. Mörner (ed.), *Earth Rheology, Isostasy and Eustasy*, Wiley, Chichester and New York, pp. 535-53.

—— (1980e) 'Relative sea-level, tectono-eustasy, geoidal-eustasy and geodynamics during the Cretaceous', *Cretaceous Res.*, *1*.

—— (1981) 'Space geodesy, palaeogeodesy and palaeogeophysics', *Ann. de Géophys.*, *37*, 69-76.

—— (1983) 'Differential Holocene sea-level changes over the globe: evidence from glacial eustasy, geoidal eustasy and crustal movements', in *Int. Symp. on Coastal Evolution in the Holocene*, Tokyo, pp. 93-6.

Morzadec-Kerfourn, M.-T. (1974) 'Variations de la ligne de rivage Armoricaine au Quaternaire', *Mém. Soc. géol. mineral. Bretagne*, *17*, 1-208.

Mudge, B.F. (1858) 'The salt marshes of Lynn', *Proc. Essex Inst.*, *2*, 117-19.

Nansen, F. (1922) *The Strandflat and Isostasy*, Kristiania, Christiana Vidensk-Selsk. Skr. Mat. — Naturv. K1. 1921, 11, 1-315.

Newman, W.S., Marcus, L.F., Pardi, R.R., Paccione, J.A. and Tomecek, S.M. (1980) 'Eustasy and deformation of the geoid: 1000-6000 radiocarbon years BP', in N.-A. Mörner, (ed.), *Earth Rheology, Isostasy and Eustasy*, Wiley, Chichester and New York, pp. 555-67.

—— Marcus, L.F. and Pardi, R.R. (1981) 'Palaeogeodesy: Late Quaternary geoidal configurations as determined by ancient sea levels', in I. Allison, (ed.), *Sea Level, Ice and Climatic Change, Int. Ass. Hydrol. Sci.*, Publ. No. 131, Washington, DC, pp. 263-75.

Ota, Y., Matsushima, Y. and Moriwaki, H. (1981) *Atlas of Holocene Records in Japan*, IGCP Project 61, Yokohama.

Paskoff, R. (1984) 'International Geographical Union — Commission on the Coastal Environment', *Litoralia Newsletter*, *10*, 6.

Peltier, W.R. (1980) 'Ice sheets, oceans, and the earth's shape', in N.-A. Mörner (ed.), *Earth Rheology, Isostasy and Eustasy*, Wiley, Chichester and New York, pp. 45-63.

Penck, A. (1882) *Die Vergletscherung der Deutschen Alpen*, J.A. Barth, Leipzig.

Pinless, S.J. (1975) 'The reduction of artificial boundary reflexions in numerically modelled estuaries', *Proc. Inst. Civ. Eng.*, *59*, 255-64.

Pirazzoli, P.A. (1986) 'Secular trends of relative sea-level (RSL) changes indicated by tide-gauge records', *J. Coastal Res.*, *S1(1)*, 1-26.

Plassche, O. van de (1977) *A Manual for Sample Collection and Evaluation of Sea-Level Data*, Inst. for Earth Sciences, Free University, Amsterdam.

—— (1980) 'Holocene water-level changes in the Rhine–Meuse delta as a function of changes in relative sea level, local tidal range and river gradient', *Geol. Mijnb.*, *59*, 343-51.

—— (1982) 'Sea-level change and water-level movements in the Netherlands during the Holocene', *Meded. Rijks Geol. Dienst.*, *36*, 1-93.

—— (1986), *Sea-Level Research: a Manual for the Collection and Evaluation of Data*, Geo Books, Norwich.

PSMSL (1976-9), *The Permanent Service for Mean Sea Level*, vols. I-III, Inst. Oceanogr. Sci., Bidston, UK.

Roep, Th. B., Beets, D.J. and Ruegg, G.H.J. (1975) 'Wavebuilt structures in sub-recent beach barriers of the Netherlands', in *IXme Congrès Int. de Sed., Extraits des Publ. du Congrès*, pp. 141-5.

Rossiter, J.R. (1962) 'Long term variations in sea level', in M.N. Hill (ed.), *The Sea: Ideas and Observations in Progress in the Study of the Seas*, Interscience, New York, vol. 1, pp. 590-610.

—— (1963) 'Tides', in H. Barnes, (ed.) *Oceanography and Marine Biology: An Annual Review*, *1*, 11-25.

—— (1967) 'An analysis of annual sea-level variations in European waters', *Geophys. J. Roy. Astron. Soc.*, *12*, 259-99.

Schofield, J.C. (1960) 'Sea-level fluctuations during the past four thousand years', *Nature*, *185*, 836.

—— (1964) 'Postglacial sea levels and isostatic uplift', *NZ J. Geol. Geophys.*, *7*, 359-70.

—— (1973) 'Postglacial sea levels and isostatic uplift', *NZ J. Geol. Geophys.*, *16*, 359-66.

Scholl, D.W. and Stuiver, M. (1967) 'Recent submergence of southern Florida: a comparison with adjacent coasts and other eustatic data', *Bull. Geol. Soc. Am.*, *78*, 437-54.

Scott, D.B. and Greenberg, D.A. (1983) 'Relative sea-level rise and tidal development in the Fundy tidal system', *Can. J. Earth Sci.*, *20*, 1554-64.

Shaler, N.S. (1874) 'Preliminary report on the recent changes of level on the coast of Maine', *Mem. Boston Soc. Nat. Hist.*, *2*, 320-40.

—— (1885)'Preliminary report on sea coast swamps on the eastern United States', *US Geol. Surv. 6th Ann. Rept.*, 359-98.

Shennan, I. (1982) 'Interpretation of Flandrian sea-level data from the Fenland, England', *Proc. Geol. Ass.*, *93*, 53-63.

—— (1983) 'Flandrian and Late Devensian sea-level changes and crustal movements in England and Wales', in D.E. Smith and A.G. Dawson (eds), *Shorelines and Isostasy*, Academic Press, London and New York, pp. 255-83.

—— (ed.) (1986) *IGCP Project 200: Newsletter and Annual Report*, Dept. of Geography, University of Durham.

—— (1986) 'Flandrian sea-level changes in the Fenland', *J. Quat. Sci.*, 1, 119-79.

—— and Pirazzoli, P. (comps.) (1984) *Directory of Sea-level Research*, printed for IGCP-200 by N.-A. Mörner, Geol. Inst., Stockholm University.

—— Tooley, M.J., Davis, M.J. and Hagpart, B.A. (1983) 'Analysis and interpretation of Holocene sea-level data', *Nature*, *302*, 404-6.

Shepard, F.P. (1963) 'Thirty five thousand years of sea level', in T. Clements (ed.), *Essays in Marine Geology in Honour of K.O. Emery*, University of Southern California Press, Los Angeles, pp. 1-10.

—— (1964) 'Sea-level changes in the past 6,000 years; possible archaeological significance', *Science, 143*, 574.

Suess, E. (1906) *The Face of the Earth*, vol. 2, Clarendon Press, Oxford.

Suguio, K., Martin, L. and Flexor, J.-M. (1980) 'Sea-level fluctuations during the past 6,000 years along the coast of the State of São Paulo, Brazil', in N.-A. Mörner (ed.), *Earth Rheology, Isostasy and Eustasy*, Wiley, Chichester and New York, pp. 471-86.

Sutherland, D.G. (1983) 'The dating of former shorelines', in D.E. Smith, and A.G. Dawson (eds), *Shorelines and Isostasy*, Academic Press, London and New York, pp. 129-57.

Tanner, V. (1933) 'L'étude des terrasses littorales en Fennoscandie et l'homotaxie intercontinental', in *Comptes Rendus Cong. Int. Géog., Paris 1931*, vol. 2, pp. 61-76.

Tooley, M.J. (1978a) *Sea-level Changes: Northwest England During the Flandrian Stage*, Clarendon Press, Oxford.

—— (1978b) 'Interpretation of Holocene sea-level changes', *Geol. Fören Stockh. Förh., 100*, 203-12

—— (1982) 'Sea-level changes in northern England', *Proc. Geol. Ass., 93*, 43-51.

—— (1985) 'Sea Levels', *Prog. Phys. Geogr., 9*, 113-20.

—— (1986) 'Sea Levels', *Prog. Phys. Geogr., 10*, 120-9.

Veen, Van (1950) 'Eb-en vloedschaar systemen in de Nederlandse getij wateren', *Tijdschr. Kon. Ned. Aardr. Gen., Tweede reeks, 67*, 303-50.

Walcott, R.I. (1973) 'Structure of the earth from glacier isostatic rebound', *Ann. Rev. Earth Planet. Sci., 1*, 15-37.

—— (1980) 'Rheological models and observational data of glacio-isostatic rebound', in N.-A. Mörner, (ed.), *Earth Rheology, Isostasy and Eustasy*, Wiley, Chichester and New York, pp. 3-10.

Ward, W.T. (1965) 'Eustatic and climatic history of the Adelaide area, South Australia', *J. Geol., 73*, 592.

Wegmann, E. (1969) 'Changing ideas about moving shorelines', in C.J. Scheer (ed.), *Toward a History of Geology*, MIT Press, Cambridge, Mass., pp. 386-414.

Wood, F.J. (1985) *Tidal Dynamics*, Reidel, Dordrecht.

Woodworth, P.L. (1985) 'A worldwide search for the 11 year solar cycle in mean sea-level records', *Geophys. J. Roy. Astron. Soc., 30*, 743-55.

—— (1986) 'Trends in UK mean sea level', *Mar. Geodesy* (to be published).

Wright, W.B. (1914) *The Quaternary Ice Age*, Macmillan, London.

Part One

Sea-surface (Sea-level) Changes: What Are They?

2

Ocean Volume Change and the History of Sea Water

John Chappell

INTRODUCTION

As might be expected, the history of the volume and composition of the oceans becomes less easy to identify the further one goes back in time. The picture should generally be good for the later Phanerozoic, where sea-floor spreading data provide a basis for estimating ocean volume changes, stratigraphic evidence for sea-level changes is quite widespread and marine evaporites and other deposits are indicators of ocean composition. However, reaching back towards the far Precambrian the detail becomes less certain. About thirty years ago, Mason (1958) wrote, 'a review ... reveals little agreement concerning the history and evolution of the ocean', and pointed out that views about the quantity of sea water ranged from that of an accelerating increase through geological time to that of essential constancy since condensation of the primeval atmosphere. Composition of the oceans through time also was the subject of a similar range of different views.

Knowledge has increased enormously over the last thirty years and even textbooks concerned with palaeo-oceanography have appeared during this time. The voluminous nature of the literature is indicated by Kennett (1983), who cites nearly 400 references in a short review article. Increased knowledge notwithstanding, however, the evolution of ocean volume and composition is still a matter for detective work. One problem has been to work out what trends might have occurred in the oceans and their constituents while the ocean floors were in the process of being recycled, many times, through the upper mantle as well as through rocks of the landmasses. Another has been to identify fluctuations in water volume and composition which might have occurred around these trends. A few years ago, Holland (1983), when speaking of the ocean as 'a reservoir engaged in geologically rapid exchange with the atmosphere and the biosphere, and reacting at a more leisurely rate with the lithosphere', observed that if the progress of the past fifty years is any indication, the next fifty years may bring the

33

solution of many outstanding problems about the past, present, and perhaps the future of the oceans.

The title of this chapter indicates that ocean volume changes are its main subject of interest. Although the composition of sea water is to a large extent independent of volume, it is perhaps through examining its chemistry that the best overview of the second area of interest, ocean history, is to be gained. Accordingly, we begin by reviewing past ocean chemistry and then move on to examine the effects upon ocean volume and continental freeboard of factors such as changing heat flow, sea-floor spreading and others.

CHEMICAL CONTROLS ON PAST OCEAN COMPOSITION

The classical mass balance argument for development of the chemistry of the present oceans is now out of date but is a useful starting point; detailed discussion of ocean chemistry and the complex cycling of salts in the oceans can be found in Broecker (1974), MacIntyre (1970) and Holland (1984). Goldschmidt (1933) estimated the amount of igneous rock which, when weathered, yields the known concentrations of various non-volatile constituents of sea water and of sedimentary rocks. Goldschmidt's figures reduce to about 0.6 kg of primary rock to supply the solutes in 1 litre of sea water plus the constituents of about 0.6 kg of sedimentary rock. However, very much more than 0.6 kg has passed through each litre of sea water during the earth's history. Although estimates of the total dissolved load of all rivers differ somewhat, that of Livingstone (1963) will suffice. Livingstone put the mean annual dissolved load at 3.9×10^{12} kg a^{-1}, giving a time of about 180 Ma for the development of 0.6 kg l^{-1} of sea water (ocean volume is about 1.2×10^{21} litres or ~ 1370 M km^3 (Kuenen, 1950)), which suggests that the amount of weathering and sedimentation required by Goldschmidt could be recycled about 20 times within the age of the earth. All solutes are continually moving (cycled) through the oceans (see Fig. 2.2) and, as Barth (1961) and other authors have shown, different solutes have very different residence times in ocean water. For example, figures given by Mackenzie and Garrels (1966) indicate that ocean chloride could be supplied from runoff in about 100 Ma, potassium in about 7 Ma and silica in about 20 ka (see also Table 2.1). Suggestions, therefore, by Halley (1715) and later Joly (1899) that residence times of salts could be used to give an age to ocean water are thus incorrect (King, 1974).

Barth (1961) considered that dynamic equilibrium between supply and removal would be attained in about 3 times the residence time of a solute. Leaving aside the questions of how and whether equilibrium is attained, the ocean will only be in a steady state with respect to a given constituent if sources and sinks also are constant through time (see Stewart and Sharp,

Table 2.1: River fluxes and residence times of dissolved constituents in sea water. Amounts shown differ perhaps in some cases due to the difficulty in accurately analysing constituent concentrations in sea water

Constituent	River flux[a] ($\times 10^8$ t a^{-1})	Mass in ocean ($\times 10^{14}$ t)	Residence time[a] (Ma)
Na	2.05	144	70.2
K	0.75	5	6.7
Ca	4.88	6	1.23
Mg	1.33	19	14.3
Cl	2.54	261	103
HCO$_3$	18.95	1.9	0.1
SO$_4$	3.64	37	10.2
SiO$_2$	4.26	0.08	0.02
Fe	0.22	0.000014	0.00006
Mn	0.001	0.000014	0.014
Cu	0.0007	0.000014	0.02
Co	0.001	0.000014	0.014
Zn	0.0007	0.000014	0.02

Note: [a]These values are not corrected for cyclic salts.
Source: After Stewart and Sharp (1978).

1978; MacIntyre 1970 for definitions and discussion). Before examining the geological records, therefore, for appropriate evidence, the identity and operation of sources and sinks must be known. This is not always the case. Taking magnesium as an example, Broecker (1971) wrote that 'failure to identify the main sink for Mg represents one of the major problems in marine geochemistry', as at that time the maximum amount of Mg which could be found to be accumulating in sediments was one-fifth of what is supplied each year by rivers. This particular problem appears to have been resolved by the finding that Mg is removed by reaction with hot basaltic rocks of new oceanic crust formed at mid-ocean ridges and related sites (Mottl and Holland, 1978). This reaction apparently also injects calcium, potassium and several trace elements into the oceans, adding another input to calculations of sea water solutes and sediment recycling. These results indicate that the composition of the oceans can be affected by variations of both the input from rivers and the rate of sea-floor spreading, in so far as the latter affects ocean floor vulcanism. Other factors which affect supply and removal will be outlined shortly.

Development of equilibrium of a given solute is more complex than the simple dependence on residence time which Barth suggested and depends on reactions within the oceans as well as with sediments and other sinks. Kinetics as well as concepts of thermodynamic equilibria have to be reckoned with. Twenty years ago, most discussions concerning the composition and evolution of sea water focused on thermodynamic equilibria between the water and mineral phases in the underlying sediments (Sillen,

1963; Holland, 1965; Mackenzie and Garrels, 1966). Sinks for the major cations Na, K, Mg and Ca must be found because ratios between these in sea water are different from terrestrial waters. The molar ratio of Na/K in average river discharge is 4.8 according to Mackenzie and Garrels (1966). Although there are other estimates, all are low compared with the sea water value of 47. The problem of disposing of the excess K was linked to silica which has a short residence time and high delivery rate (around 7 T moles a^{-1}, compared with Na at 9 T moles a^{-1}; Mackenzie and Garrels, 1966). Magnesium was an even larger problem, as has been mentioned. One hypothesis was that these ions equilibrate with clays passing through the ocean, adopting the thermodynamic equilibrium ratios dictated by these minerals (Sillen, 1963). Against this, however, is evidence that clay sediments do not show signs of equilibration (Broecker, 1971) and that radiogenic argon levels show that the bulk of potassium in the clays is of detrital and not marine chemical origin (Hurley et al., 1963). A second model, by Mackenzie and Garrels (1966), involves precipitation of authigenic silicate minerals which are sinks for excess cations. These have been elusive in young sediments but may be the best explanation of some of the major cation ratios in sea water.

Problems of sources and sinks aside, thermodynamic arguments do appear to set limits within which the ocean has altered in the past. Figure 2.1 shows the range within which some parameters of sea water can have varied, according to Holland (1972) (see Table 2.2 for present sea water composition). Within a frame bounded by HCO_3^- and Ca^{2+} concentrations, the series of triangles represent the ranges of these variables which are permitted by reasonable observations, under different SO_4^{2-} concentrations. The 'reasonable boundaries' for each triangle are based on ancient marine evaporites, from which Holland concludes that (i) gypsum ($CaSO_4$) saturation always precedes halite (NaCl), fixing the lower bound; (ii) molar Ca^{2+} must exceed half that of HCO_3^- otherwise alkali bicarbonates would precipitate, fixing the southeastern bound; (iii) Ca^{2+} concentration should not exceed the sum of SO_4^{2-} and half (HCO_3^-) concentrations, otherwise late evaporites would be strongly enriched in Ca and depleted in sulphate, fixing the upper bound, and (iv) ocean pH should not have exceeded a value of 9 otherwise brucite ($Mg(OH)_2$) would form. The position of present sea water (S) in this frame is shown in the diagram. The range of variation allowed by this diagram is rather large, but it illustrates the sorts of limits which can be set by considering one form of evidence — the marine evaporites — relevant to the composition of the ocean in the past. It can be noted that Holland (1972) considered that sea water composition probably has varied within a smaller range, represented by an imaginary ellipsoid circumscribed by the various planes within Figure 2.1, amounting to changes by less than a factor of 2 from present Ca^{2+}, HCO_3^-, and SO_4^{2-} concentrations.

Figure 2.1: Limits on the composition of sea water in a frame bounded by concentrations of calcium and bicarbonate, for different sulphate concentrations, and NaCl = 0.55 mole kg⁻¹. S = modern sea water

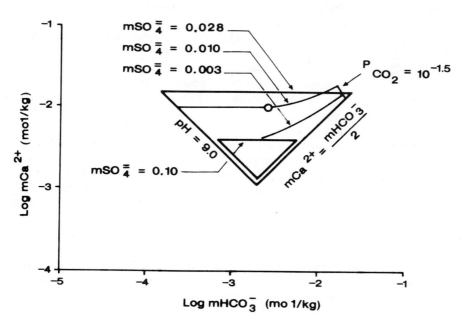

Source: After Holland (1972).

Table 2.2: Average concentration of the principal ions in sea water (parts per thousand by weight)

Ion	‰ by weight			
chloride, Cl^-	18.980			
sulphate, SO_4^{2-}	2.649			
bicarbonate, HCO_3^-	0.140			
bromide, Br^-	0.065	Negative ions (anions)	total	= 21.861‰
borate, $H_2BO_3^-$	0.026			
fluoride, F^-	0.001			
sodium, Na^+	10.556			
magnesium, Mg^{2+}	1.272			
calcium, Ca^{2+}	0.400	Positive ions (cations)	total	= 12.621‰
potassium, K^+	0.380			
strontium, Sr^{2+}	0.013			
			Overall total salinity	= 34.482‰

Source: After Stewart (1977).

The thermodynamic approach represented by Figure 2.1 is applicable only as far as equilibrium can be shown to apply between sea water and sea-floor deposits. Equilibrium does not always exist, particularly for constituents which are engaged in organic cycles in the oceans. Li *et al.* (1969) showed that the oceans are supersaturated with respect to $CaCO_3$ above the thermocline (see Fig. 2.2), and undersaturated towards their floors, particularly so in the case of the Pacific. Neither aragonite nor calcite $(CaCO_3)$ appear to crystallise spontaneously in the sea, and the $CaCO_3$ excess in the surface waters is removed by organisms. Undersaturation at depth occurs because $CaCO_3$ solubility increases with pressure and with decreasing temperature and as pCO_2 rises in the deep sea due to oxidation of organic debris settling from the surface. Other elements involved in organic cycles are similarly affected by the kinetics of this system, in which the fluxes of river inputs and of vertical water exchanges across the thermocline regulate concentrations as well as loss to sea-floor sediments. Figure 2.2 shows the model used by Broecker (1971, 1974) to examine the behaviour of elements such as N, P, C and Si which enter the marine life cycles.

Thermodynamic considerations give a guide to the mean oceanic concentration of a constituent but its value at any point can vary quite widely. The relationship between the fluxes in Figure 2.2 can be illustrated by comparing net biogenic production with sedimentation. In steady state, sediment output z = river input, that is, $z = v_r C_r$ where v_r is river volume input per year and C_r is concentration of a particular element. Net productivity measured as the flux of biogenic debris settling out of the surface water $= v_r C_r + v_d C_d - v_d C_s$, where v_d is the volume of water cycling from the surface to deep ocean each year and C_s, C_d are surface and deep-water concentrations of the element. The ratio of biogenic debris production to sea-floor sedimentation, for the element, is

$$b = 1 + \frac{v^d}{v_r} \frac{(C_d - C_s)}{C_r} \qquad \text{(Eq. 2.1)}$$

(Broecker, 1971, puts this in its reciprocal form using f instead of $1/b$.) High b values typify elements with low sedimentation rates relative to their rates of uptake and dissolution in the surface to deep water debris cycle. Data of Broecker (1971) suggest b values of about 100 for Si, P and N, and about 50 for C, indicating major participation of these elements in marine biological productivity (siliceous radiolaria in the case of Si, presumably), which demonstrates the importance of the biological factor in the status of such elements in the oceans.

Stability or instability of the system is well illustrated by the case of carbon. Carbon enters the inorganic cycle in the $CO_2-HCO_3^- - CO_3^{2-}$

Figure 2.2: Simple model for marine cycles of elements participating in biological production. Rivers form the main *source* in the model of constituents to sea water and precipitation of elements in sediments constitutes the major *sink* (see Stewart and Sharp, 1978 for further discussion)

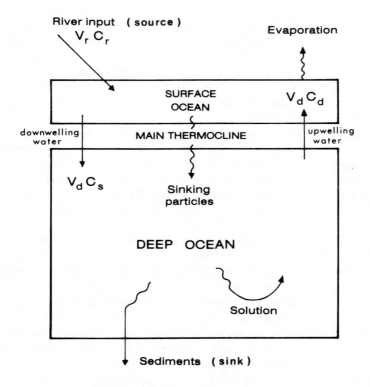

(V=water flux, C= concentration of given ion)

Source: After Broecker (1971).

system, vitally significant in ocean pH and the mineralogy of sediments (Fig. 2.1), as well as being central in the organic cycle. Oxidation of settling organic debris raises pCO_2 in the deep water, contributing to the deep water undersaturation of $CaCO_3$. Sedimentation of $CaCO_3$ depends on whether the sea floor locally lies above or below the calcite compensation depth (CCD), defined as the level at which calcite solution exceeds the input of settling calcitic debris. The ocean carbon cycles and the CCD interact in several ways. An increase of biogenic production in the surface waters, by an increased input of phosphorus from rivers, say, will produce more organic and $CaCO_3$ debris. The net carbon concentration in the ocean will fall and the CCD will tend to move up. Both C_s and C_d in (Eq. 2.1) will fall, tending to restore the value of b. If the increase of P from rivers is accompanied by a concomitant rise of carbon in greater proportion than required to match P in living tissue (about 106 C atoms for each P), then the CCD would tend to move down as $CaCO_3$ would be deposited more widely. Both C_r and C_d in Eq. 2.1 have then risen, again tending to restore b. Net rates of sedimentation, and the distribution of different sediments in different depths, will vary with inputs from the land, but, as Broecker argues, the homeostatic tendency of the system shown in Figure 2.2 and Eq. 2.1 ensure that changing inputs flow smoothly through the oceans despite disequilibrium effects.

One feature of marine sediments which reflects the nature of past ocean basins can now be discussed. This is the relationship between carbon-rich and carbonate-rich sediments. In some parts of the geological record, conspicuously the Lower Cretaceous of the north, central and south Atlantic as well as the palaeo-Indian Ocean (Weissert, 1981), black shale sequences overlie carbonate sequences. The transition evidently marks a change in position of the depositional sites relative to the CCD. The first point about these carbon-rich shales concerns ocean circulation when they were deposited. Organic material is preserved when oxygen supply to deep water is too low to support the quantity of benthic organisms needed to oxidise the settling organic debris. Less deep water oxygen implies either less oxygen in the atmosphere/ocean system or reduced vertical circulation (v_d in Eq. 2.1). A widespread increase in organic content through much of the Lower Cretaceous also implies an increase of the terrigenous carbon/ clay input ratio (increased $v_r C_r$ for carbon in Eq. 2.1), as ocean carbon reserves would be 'used up' in about 100 ka on the figures of Broecker (1971). It is notable that detailed organic geochemistry of these deposits shows a substantial terrigenous component in many, though not all, sites reviewed by Weissert (1981). If at the same time the surface water productivity does not increase, which would depend on inputs and sinks of phosporus, then the CCD should rise as CO_2 produced in the depths rises. Although this is a very simple sketch of a complex system it illustrates the importance of identifying sources as well as sinks of significant elements before inferring vertical circulation from sediments.

The second point arising from comparison of carbon-rich and carbonate-rich sediments concerns vertical movements of the sea floor. A change from one sediment type to the other indicates vertical movement of the CCD or the sea floor or both. Inspection of the Lower Cretaceous record shows how both can be involved. Figure 2.3 (after Tucholke, 1981) shows (a) locations of Deep Sea Drilling Project (DSDP) sites in the western North Atlantic (see DSDP, 1974; Davies et al., 1974 for details of

Figure 2.3: (a) Locations of west Atlantic DSDP cores passing through Cretaceous sediments. (b) Depth-time tracks of these cores with subsidence due to spreading. Dotted part of each core line = period of carbonate sedimentation, and solid part = period of black shale deposition. Stippled zone shows rise of CCD between about 120 and 105 Ma

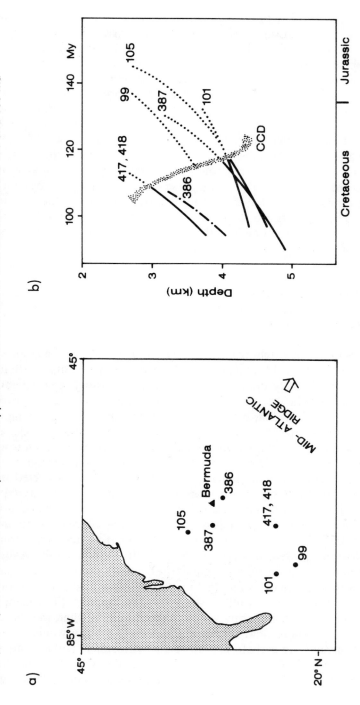

Source: After Tucholke (1981).

the project), and (b) age versus depth curves for these sites. The palaeo-depth increases as a site moves away from the ocean floor spreading centre (Sclater *et al.*, 1971), and the sloping curves in Figure 2.3b represent this progressive subsidence at each drill site taking sediment accumulation into account. Heavy lines trace the portions of each core in which organic shales predominate and dotted lines represent the core section dominated by carbonates. The depth of the carbonate/black shale transition evidently rises through the Lower Cretaceous in the western North Atlantic, repre-sented by the stippled curve in Figure 2.3b, showing a rise of CCD while the local sea floor is subsiding. It is not clear whether this was due to progressive increase of terrigenous carbon input (suggested above) or one or more other factors. The example serves to show that subsidence and chemical processes within the ocean independently affect the sedimentary record. These points affect our later discussion of ocean volume changes. Before turning to the detail of this, however, main aspects of the general history of the oceans are reviewed.

THE EVOLVING OCEAN

The origin of sea water

Theories upon the origin of sea water (Kitano, 1975) may be resolved into two schools of thought. One, that the hydrosphere and atmosphere formed within approximately the first 1000 Ma of earth history, consequent upon the release of the necessary volatile components through a process of degassing from the first, high energy earth mantle and crust (King, 1974; Stewart and Sharp, 1978; Holland, 1984). These volatile phases, such as water vapour, hydrogen, carbon dioxide, carbon monoxide and nitrogen, of otherwise abundant elements would have been excluded from the early, 'growing' earth. Such substances would, however, have occurred in non-volatile form, with water, for example, being included in clay-type and other hydrated silicate minerals; chlorine, carbon and nitrogen occurring in silicate and organic compounds respectively. None of these compounds, at least in the quantities found, can be released from the weathering and breakdown of present-day crustal rocks. Such compounds, however, have been found in certain types of meteorites called *carbonaceous chondrites* and it is thought that this material may have formed as a 'veneer' or outer layer to the earth in its initial stages of development (Turekian, 1972, 1976). If this volatile bearing veneer formed quickly, then the subsequent high release of energy and heating of the planet's interior would have led to rapid degassing and early formation of the hydrosphere and a proto-atmos-phere. Alternatively, the second school of thought suggests that the volatiles accumulated gradually through a process of continuous degassing of the earth's mantle material, leading to release of the volatile compounds from volcanic and associated mantle–crust fluxes, water being added to the oceans as juvenile water. The amount of juvenile (first time) water being added today by this process is probably small < 0.1 km^3 Ma^{-1} ($\sim 1.025 \times 10^5$ litres a^{-1}) (King, 1974). Extrapolation from this figure of the quantity of juvenile water added over the last 600 Ma gives a figure of 60 M km^3, compared to a total ocean volume of $\sim 1,370$ M km^3. Although such a calculation may be unrealistic, as it assumes uniform rates of volcanism, crustal spreading and organic activity, it does indicate that ocean water volume must have been close to its present figure for much of the Phanerozoic at least (Fig. 2.4). Other early viewpoints, showing more gradual increases in ocean water over time, based on assumed gradual degassing or other geological arguments, are also shown in Figure 2.4. The problem for both hypotheses is that there are no rocks for this first ~ 600 Ma in earth history, thus what happened remains a matter of conjecture.

Figure 2.4: Differing views on the changing volume of ocean water from permanency early in earth history to a more gradual (Twenhofel, 1932) or a late, rapid increase (Walther, 1926)

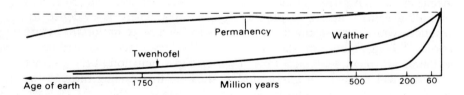

Source: After Kuenen (1950).

To summarise, the hydrosphere appears to have existed since the beginning of the geological record about 4 Ga ago, though the nature of the oceans in the first billion years is rather uncertain. From 4 to 3.5 Ga the crust was mainly granitic, thin and mobile, apparently without recognisable continental blocks (Shaw, 1976). It is not clear whether the quantity of water in the hydrosphere was still increasing at this early stage (Perry and Tan, 1972; Chase and Perry, 1972) but it has become conventional to regard the volume of water as having been constant at least for the last 2.5 Ga. Further, sediments become increasingly prominent from 3.5 to 3 Ga, including carbonates, with very early life indicated by stromatolites reaching back to 3.5 Ga (Groves *et al.*, 1981).

Chemical evidence

Both chemical evidence and physical reasoning are useful for interpreting the history of ocean basins in later Archean and earlier Proterozoic times. As ocean chemistry has been our focus so far, this evidence is outlined first.

The oxygen content of the atmosphere and the chemistry, particularly the biological component of the oceans, are opposite sides of the same coin. Possible oxygen-producing life forms are found back to at least 2.8 Ga ago and weathering is indicated by the mineralogy of very early sediments. There is, however, evidence that the partial pressure of oxygen was considerably lower than now in the Archean and early Proterozoic, and Cloud (1976) considered that present atmospheric levels were reached only in the last 1 Ga. Holland (1984) points out that the occurrence of detrital uranium in pebble conglomerate ores, and the development of huge banded iron formations that are not related to nearby vulcanism, is best explained if surface waters contained less oxygen than now and deep water was mildly anoxic. The degree to which lower oxygen levels in the early ocean/atmosphere system reflect lower rates of photosynthetic oxygen production is uncertain. Oxygen levels, as with other elements, reflect a balance between sources and sinks and in the view of Veizer (1984) the Archean may have been a time of more rapid recycling of ocean crust, with spreading centres acting as prolific oxygen sinks. There is evidence for reactions with freshly-formed sea floor having been more important before about 2.6 Ga, found in the records of $^{87}Sr/^{86}Sr$ ratios in carbonate sediments through geological times (Veizer and Compston, 1976). Other evidence which has been interpreted as showing that cycling through the sea floor and upper mantle became less important at this time (Veizer, 1984) includes the increased degree of sulphur isotope fractionation between sulphides and sulphates (Schidlowski et al., 1983).

Deposition of carbonates throughout the past 3.5 Ga, in the presence of sulphate and oxygen for at least most of this time, suggests that sea water composition throughout has been within the 'inner area' of Figure 2.1 indicated by Holland (1972). Despite the fact that the concentrations of some elements and their oceanic cycles (Fig. 2.2) have changed as life forms have multiplied, such as for Si with the appearance of siliceous radiolaria, which have probably reduced Si levels below that needed for widespread precipitation of cherts, the emerging picture is one of general constancy with only the degree of change now being debated. Holland (1984) argues for a greater degree of constancy than does Veizer (1984), who stresses the evidence for a substantially greater role for ocean floor interaction and cycling before 2.3 to 2.6 Ga. Since that time, perturbations of ocean chemistry such as occurred near the Permo–Triassic boundary when there was large-scale formation of marine evaporites (Claypool et al., 1980) and during the Lower Cretaceous 'black shale' period mentioned

44

above, may be as large as any which have occurred. Since the large cratonic blocks came into being in Proterozoic times, the now familiar elements of runoff from the land, restrictions on ocean circulations by palaeogeographic distributions of land and sea, and variation of sea-floor spreading rates, appear to be the factors which have modulated sedimentation and chemistry in an otherwise rather constant hydrosphere. There have been high frequency fluctuations, of course, particularly during periods of repeated glaciation such as occurred during the Quaternary, affecting both chemical (Broecker, 1982) and physical parameters of the oceans, but these lie beyond the first-order factors being discussed at present.

Physical evidence

We now turn to general physical characteristics of ocean basins since early times. Little can be said about this for earlier Archean times when, with a thin and mobile crust, basins as such might not have existed. Komatiite pillow lavas in Barberton Mountain Land point to substantial local sea-floor depths in the earliest Archean (Anhaeusser, 1978), but comparable data seem too sparse in this early period for conclusions to be drawn. It is with development of Lower Proterozoic fold belt sedimentation that cratonic blocks and ocean basins in their contemporary sense come into existence. One interesting line of thought is that the close juxtaposition of earth and moon calculated by Gerstenkorn may have occurred during this early period (Lambeck, 1980). Tides at this time would have been very large in any oceans with similar dimensions to those of today. Evidence for this has not been reported, as far as the writer is aware, which may have more bearing on the history of earth/moon distance than it has upon the dimensions of the early seas.

Freeboard and the influence of supercontinents on ocean basin volume

Earth heat flow is the first-order factor affecting the evolution of ocean basins since portions of the crust thickened into cratonic blocks. Heat flow from the mantle decreases as time passes, through gradual decline of radioactive heat sources. It is believed that this allows ocean basins to deepen as the density of the upper mantle beneath the ocean increases, while that of the sialic continents remains more constant. The 'freeboard', or extent to which the continents stand above the sea surface, depends on the ocean basin volume and therefore on the fraction of the earth's area under sea. Freeboard will change through time in response to both diminishing heat flow and changing area of continental crust. The dominance of runoff on ocean chemistry and sedimentation for the last 2.5 Ga shows that continents have been emergent throughout this time, placing a simple constraint on calculations linking continental area, heat flow and ocean depth.

45

Figure 2.5a shows the simple continent-ocean geometry used by Schubert and Reymer (1985) to analyse the freeboard problem. Assuming isostatic equilibrium, the difference between past freeboard h and that of the present h* is

$$h - h^* = \frac{(p_m - p_w)}{p_m} d_a^* + d_b^* \frac{q_s^*}{q_s} - \frac{V_o}{(Ae - V_c/d_c^*)} \qquad \text{(Eq. 2.2)}$$

where p_m and p_w are mantle and water densities, q_s is mean oceanic surface heat flow, A_e is area of the earth, V_o and V_c are ocean water volume and continental volume, and d_a, d_b and d_c are depths defined in Figure 2.5. Quantities marked * refer to present-day values. On the assumptions that the quantity of water (V_o) and average continental crustal thickness d_c have remained constant, results calculated by Schubert and Reymer for different degrees of continental growth are as shown in Figure 2.5b. The period before 2.5 Ga is not shown in Figure 2.5b as there are different models for surface heat flow (Lambert, 1980) as well as uncertainty about the thickness of proto-continents during this very early period. The heat flow curve used by Schubert and Reymer (1985) is shown in Figure 2.5c.

The main result of this calculation is that constant freeboard will be maintained if continental area increases by 25 per cent in the last 2.5 Ga (Fig. 2.5b), if V_o and d_c have been constant. Further calculations by Schubert and Reymer (1985) show that even a modest increase of mean continental thickness of about 5 per cent over this time requires that continental area doubles to maintain constant freeboard. Although geological control is not very good before the Phanerozoic, the reviews of Windley (1977a, b) indicate less than 50 per cent continental growth as well as relatively constant freeboard since the Archean. Consistency amongst these results supports the first-order heat flow model and the assumption of approximate constancy of continental thickness through the past 2.5 Ga. Simple area/volume relationships show that the ocean basins will have deepened by no more than 1000 m and more probably by about 500 m over the same interval.

At the next level of detail, freeboard is affected by the major process of continental drift. Rifting and dispersion of Pangaea-like supercontinents has occurred five times in the last 2 Ga (Windley, 1977a). Evidence for this comes from age grouping of mafic dyke swarms throughout the present landmasses. Age groupings of major orogenies occur somewhat before those of the dyke swarms (Condie, 1982) and have been suggested to represent the collision tectonics associated with supercontinent aggregation, at least in part. Ocean volume versus freeboard relationships are affected in two ways by alternating aggregation and dispersion of continents. Firstly, new ocean floor created by rifting is younger and has a

Figure 2.5: (a) Geometry of ocean basins, mid-ocean ridge (MOR) and continent, used by Schubert and Reymer (1985) to calculate freeboard as a function of continent growth and heat flow change. (b) Difference between present freeboard h* and past h for different values of continental growth over the last 2.5 billion years with constant continental thickness. (c) Ocean heat flow curve used by Schubert and Reymer in calculations

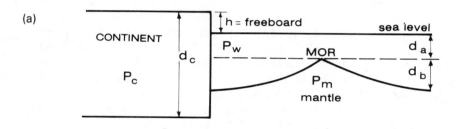

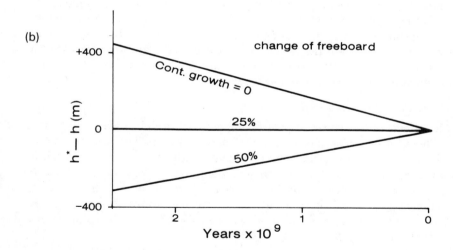

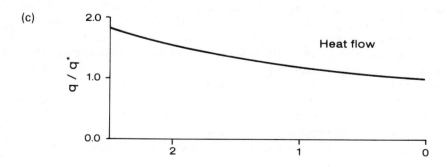

47

higher heat flow than the remainder of the oceans which surrounded a supercontinent. Secondly, it is an empirical fact that, at the present day, the larger the continent the greater is its average elevation, neglecting ice-covered Antarctica (Hay and Southam, 1977). If mean continental thickness has been roughly constant for the last 2.5 Ga, which seems likely as we have seen, then this implies that the larger the continent the lower is its crustal density. It has been suggested that this is due to a tendency for larger continents to be more stationary than smaller ones, thereby over-riding less subducted cool crust and trapping disproportionately more mantle heat flow (Worsley *et al.*, 1984). Whatever the cause, the data given by Hay and Southam (1977) closely fit the equation $H = 0.38 \ln(A) - 0.55$ (Eq. 2.3), where H is elevation in km and A is continent area in millions of km². On this basis, Upper Palaeozoic Pangaea with an area of about 180×10^6 km² could have had a mean elevation of ~ 1.4 km, about 700 m higher than the average of today's continents. Worsley *et al.* (1984) suggest a difference of about 400 m between the freeboard of Pangaea and today's mean, reasoning on the basis of a comparison between the most stationary post-Palaeozoic continent, Africa, and geological evidence for partial flooding of the other continents. It is noted that if the elevation/area relationship is due to decreasing crustal density with increasing continent size then Eq. 2.2 and the conclusions of Figure 2.5 are not affected. Any long-term trend of freeboard should be modulated by the effects of cycles of supercontinent formation and break-up.

Formation of new ocean floor effects

The effects of formation of post-rifting new ocean floor have to be added to those of supercontinent formation and break-up. This can be seen as follows. Before rifting, the supercontinent is surrounded by an ocean floor of mean age t which depends on spreading rate. As new ocean is created after rifting the mean age of the total ocean floor changes. If the mean age of the original ocean is constant (despite that it is now shrinking in size) and if the age of the new ocean floor t_n is proportional to its fractional area e_n, that is $t_n = ke_n$, then mean age $\bar{t}$ of the total $= t_o - t_o e_n + ke_n^2$ (Eq. 2.4). This quadratic relationship shows that the mean age at first diminishes following rifting and then rises until the point is reached where subduction occurs on the margins of the new sea floor and its age becomes constant. Fluctuations of spreading rate will alter the detail of this result but not its general sense. The mean depth changes with mean age through thermal shrinkage of oceanic crust as it moves away from spreading centres; Parsons and Sclater (1977) assess this as $h = 2500 + 350t^{0.5}$ (Eq. 2.5) on the basis of age-depth profiles of present-day sea floors, where h is mean depth and age t is in units of 1 Ma. If the mean age falls from 50 to 45 Ma say, then mean depth becomes 127 m shallower.

Worsley *et al.* (1984) have combined the continent size/elevation and post-rifting sea-floor factors to produce a quantitative model of freeboard changes which should occur during a cycle of rifting, dispersion, and re-aggregation of a supercontinent. This model is shown in Figure 2.6a, based on the following values used by Worsley et al.: (i) the difference of mean elevation between supercontinent and dispersed continents is 400 m, (ii) thermal relaxation of continents to their new lower positions takes 80 Ma after rifting, (iii) mean age of the total sea floor is 53 Ma before break-up and decreases to a minimum of 48 Ma after 80 Ma, at which stage the new 'Atlantic-type' ocean is 17 per cent of the world total and (iv) the 'Atlantic-type' ocean increases to become about 33 per cent of the total, with a mean age of 70 Ma, after about another 80 Ma. Re-aggregation is assumed to commence soon after this. The secondary effect of formation of 'Himalayan-type' welts during aggregation, which reduce net continental area and thereby increase freeboard by up to a few tens of metres (Berger and Winterer, 1974), is not included in Figure 2.6a. If this reasoning is correct, the pattern in Figure 2.6a should occur with every cycle of supercontinent formation and dispersal that has occurred during the earth's history. The pattern should also be reflected in the geochemical record of those elements which are affected by sinks and sources created by new sea floor. To examine this we turn to the Phanerozoic, being the period with the most data.

The Phanerozoic record

The model in Figure 2.6a has obvious appeal, as it resembles the pattern of sea-level changes over the last 200 Ma which has been suggested by various authors. Following the rifting of Pangaea in the Late Palaeozoic, sea level seems to rise (or the continents to subside) until Mid to Late Cretaceous times, after which the trend reverses. This pattern resembles the first part of the cycle in Figure 2.6a. Worsley et al. (1984) test the model across the Phanerozoic using sea level, percentage of platform flooding and the number of continents as test parameters. These data, compiled from various sources by Worsley et al., and shown here in Figure 2.6b, conform quite closely to the predicted pattern back to Cambrian times. The covariance between predicted freeboard and generalised sea level (taken from Vail et al., 1977) is 85 per cent, which is high, and it seems fair to conclude that cycles of supercontinent aggregation and dispersion have modulated freeboard and ocean basin volume. The essential factors in this model are the effects of sea-floor spreading in oceans with passive margins and of the geothermal 'trapping' by continents of different sizes and mobilities. The degree to which these have regulated sea level and other aspects of the oceans requires closer examination of the data, however, as much detail has been smoothed out in the averaged curves in Figure 2.6b. We will take sea level as a starting point.

Figure 2.6: (a) Effects of continental rifting, drifting, and reaggregation on sea level relative to continents, combining effects of break-up and reassembly (dashed line) and changing mean age of ocean floors (dotted line) to produce model curve (solid line). (b) Comparison of model of Worsley *et al.* (1984) in (a) with smoothed version of Vail *et al.* (1977) sea-level curve shown in (f). Curves (c), (d), (e) show various isotope records through the Phanerozoic, after Veizer (1984)

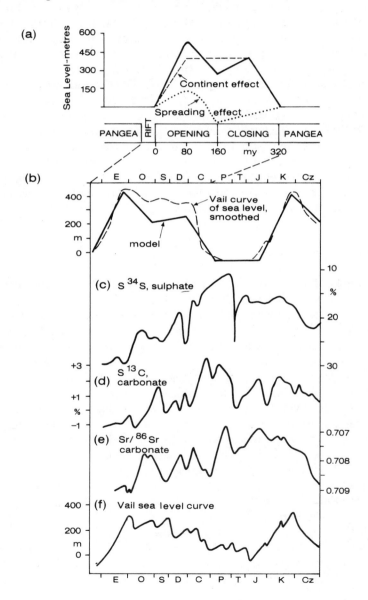

Problems in the 'global' sea-level model

Sea level in Figure 2.6b is based on the widely used curve of Vail *et al.* (1977) (see Devoy, this vol., Ch. 16), based primarily on evidence from north America for the pre-Jurassic part of the curve. Different continents have different histories, however. In Australia, for example, the extent of marine deposition in Early Permian times was as large as at any time through the Mesozoic, excepting for extensive inland marine flooding during the Late Neocomian to Early Albian (about 122 to 105 Ma) according to palaeogeographic maps of Veevers and Evans (1975). Data compiled by Jenkin (1984) show marine withdrawal from most Australian marginal basins in Late Cretaceous time, followed by several phases of Tertiary flooding. These facts obviously differ from the 'global' pattern shown in Figure 2.6b. While an explanation for Australia might be sought in terms of its late rifting from Antarctica about 90 Ma ago and accelerating drift about 50 Ma ago (Cande and Mutter, 1982), this example illustrates the problem of reducing the geological records of individual continents to a single global pattern. The work of Vail and his colleagues shows ample evidence that regional differences were recognised when they synthesised their global curves (cf. Vail *et al.*, 1977: p. 90). These differences become important at the next level of detail, where departures from the trend of Figure 2.6 are examined. Some of these departures are shown in chemical data.

Ocean chemistry is liable to change through a cycle of supercontinent aggregation and dispersal. In addition to the effects of reactions at spreading centres, the formation of marine evaporites during rifting may remove significant amounts of some elements and drowning of platforms may release others. Veizer (1984) and others argue that these processes have modulated ocean chemistry in the Phanerozoic and some of the data are examined here. Using the same time frame as Figure 2.6b, Figures 2.6c,d,e, show $\delta^{34}S$ in evaporite sulphates, ^{13}C in carbonates, and $^{87}Sr/^{86}Sr$ ratio in carbonates (from Veizer, 1984). The original sea-level curve of Vail *et al.* (1977) is shown in Figure 2.6f, as it has more detail than the smoothed version from Worsley *et al.* (1984) in Figure 2.6b. Sea-floor spreading and tectonism are considered by Veizer to be the fundamental cause of fluctuations in the strontium isotope ratio and, as far as there are correlations between the curves in Figure 2.6c to 2.6f, also to cause these other fluctuations. Some of these correlations are better than others. Covariances of the sulphur isotope values with carbon and with strontium are 76 per cent and 58 per cent respectively and 40 per cent for carbon-strontium. Covariances between the isotopic data and the sea-level curve are all insignificantly small (<10 per cent), contrary to the assessment of Veizer (1984) that 'these isotopic curves appear to correlate, at least in large scale features, with the sea-level stands'.

Several features of these results deserve comment, beginning with the strontium data. If the marine strontium is derived entirely from continental sources, its $^{87}Sr/^{86}Sr$ ratio is 0.711, whereas the value is 0.703 if the source is from the mantle via new ocean crust (Veizer, 1984). The present sea water value of 0.709 indicates that 75 per cent comes from continental sources. This may increase both with runoff and with drowning of subsiding platforms, but the ratio presumably will decrease with increased rifting and sea-floor spreading. The curve shown in Figure 2.6e is enigmatic as the lowest values (0.707, corresponding to 50 per cent continental contribution) occur in the Permian when Pangaea was at its pre-rifting maximum. Coupled with substantial fluctuations throughout the rest of the curve, this suggests that the effect of the mantle contribution may be masked by other factors. It is notable that substantial climatic changes occurred during the history of Pangaea, from wet in the Carboniferous through the widespread Gondwana glaciation in the Permian and extensive aridity in the Triassic (Frakes, 1979). Perhaps the Sr isotope record through this period is to be explained in these terms, through the agency of changing continental runoff. At the very least, this record indicates that the supercontinent formation and break-up cycle in Figure 2.6b is too general to predict this aspect of chemical behaviour of the oceans.

Similar conclusions are reached from sulphur and carbon isotope records. Low $\delta^{34}S$ (sulphate) values have been associated with an enhanced mantle contribution (Cameron, 1982), but the minimum in Figure 2.6c occurs in the Permian when Pangaea was fully together. This excursion has been associated with alteration of the ocean sulphur isotope pool by formation of widespread marine evaporites (Claypool et al., 1980), as has the $\delta^{13}C$ peak at the same time (Veizer et al., 1980). While a common cause for the Late Permian ^{34}S and $^{87}Sr/^{86}Sr$ minima as well as the $\delta^{13}C$ maximum may be sought in the combined effects of sudden but abortive 'Red Sea-like rifting' (as discussed by Windley, 1977a to account for this period of evaporite formation), this seems like dangerously special pleading, particularly as a different argument might be invoked for each peak and trough in Figure 2.6. Various mechanisms have been invoked to account for such fluctuations, such as variations of phosphorus input to the oceans affecting carbon isotope partitioning between organic and inorganic fractions (Broecker, 1982), which may carry through to the redox state of the sulphur cycle (Garrels and Lerman, 1981).

This commentary on Figure 2.6 shows that ocean history seems hard to resolve, at scales finer than perhaps 200 Ma, when proxy evidence such as chemical parameters or averaged data such as the Vail et al., sea-level curve are used. This seems to contrast with Quaternary records, where chemical, isotopic, sedimentary, and relative sea-level data seem to be interpreted with some facility and precision. The following conclusions explore this point and draw together the main points reviewed in this chapter.

CONCLUSIONS

There is a distinction to be made between chemical and physical measures of ocean history. Chemical indicators, except perhaps for measurements from trapped basins, tend to present averages of the condition of the ocean, with secondary regional differences such as in the carbonate/silicate ratio being interpretable in terms of circulation amongst the oceans (Broecker, 1971). Physical indicators of relative sea level do not present 'average', but hold at least as much information about vertical movements of individual landmasses as they do about global changes in landmass. This is reflected by both data (Newman, 1986) and theory (Clark *et al.*, 1978) at the Holocene timescale, as much as by the Tertiary records from different areas shown by Vail *et al.* (1977: p. 90). In this light it is not in the best interests of our science to use global sea-level averages unless we are sure that the 'noise' from local factors is much less than the global signal. This high signal/noise ratio seems assured in the case of Late Quaternary glacio-eustacy. Figure 2.6b suggests, however, that it is satisfactory of the timescale of continental aggregation and drift, which affects average continental freeboard. Between these two limits there seems little to be gained by averaging sea-level records from different continents, as shown earlier by reference to Australia. The influential Tertiary sea-level curve of Vail *et al.* (1977) is such an average and bears little comparison with the Australian Gippsland basin record (Vail *et al.*, 1977: p. 90) or with other Australian Tertiary sedimentary basins (Jenkin, 1984), except at the crudest level.

The tendency to average sea-level records from different continents is to be regretted. It appears as if we are trying to force patterns onto relative sea-level data, for the pre-Quaternary era, similar to those preserved by the ocean chemical records. The commentary here on Figure 2.6 shows that factors such as climate and rifting may modulate ocean chemistry on time scales of 10 to 100 Ma. This is the timescale of orogeny as well as of climatic change. While the global sea-level averaging of Vail *et al.* is justified in general terms by sea-floor spreading (Pitman, 1978) and marginal subsidence (Parkinson and Summerhayes, 1985; see Devoy, this vol., Ch. 16), good explanations at the level of individual continents are not provided. Even at the general level on a 10 Ma timescale, the relationships between 'global' sea level and ocean chemistry are very speculative (Berger *et al.*, 1981).

Two conclusions can be drawn. The last thirty years have been a period of extraordinary advance in palaeo-oceanography (Kennett, 1982, 1983). Rather convincing relationships have been shown between major continental drift cycles and average sea level models (or continental freeboards), but the influence of such global and long-term phenomena on the detailed condition of the oceans and on the freeboard of individual continents has

yet to be resolved. Setting aside the Quaternary, with its major glacial fluctuations, a theory of the oceans as influenced by continental drift and sea-
floor spreading on the 1-100 Ma scale has yet to be developed. It almost
seems as if data gathering has outrun development and theory. In the
context of the present concern with man-induced changes of our world, it is
important to explain changes of the past such as those shown in Figure 2.6.
'"Bad science" growled Alexandrov. "Correlations obtained after experiments is bloody bad. Only prediction in science"' (Fred Hoyle, *The Black
Cloud*, 1957). Predictive modelling is the task of the future.

REFERENCES

Anhaeusser, C.R. (1978) 'The geological evolution of the primitive earth —
    evidence from the Barberton Mountain Land', in D.H. Tarling (ed.), *Evolution
    of the Earth's Crust*, Academic Press, London and New York, pp. 71-106.
Barth, T.F.W. (1961) 'Abundance of the elements, areal averages and geochemical
    cycles', *Geochim. Cosmochim. Acta, 23*, 1-8.
Berger, W.H. and Winterer, E.L. (1974) 'Plate stratigraphy and the fluctuating
    carbonate line', in K. Hsu and H. Jenkyns (eds), *Pelagic Sediments on Land and
    under the Sea*, Spec. Publ., Int. Ass. Sedim., *1*, pp. 11-48.
—— Vincent, E. and Thierstein H.R. (1981) 'The deep sea record: major steps in
    Cenozoic ocean evolution', *SEPM Spec. Publ., 32*, 489-504.
Broecker, W.S. (1971) 'A kinetic model for the chemical composition of sea water',
    Quat. Res., 1, 188-207.
—— (1974) *Chemical Oceanography*, Harcourt Brace, New York.
—— (1982) 'Ocean chemistry during glacial times', *Geochim. Cosmochim. Acta,
    46*, 1689-705.
Cameron, E.H. (1982) 'Sulphate and sulphite reduction in early Precambrian
    oceans', *Nature, 296*, 145-8.
Cande, S.C., and Mutter, J.C. (1982) 'A revised identification of the oldest sea-floor
    spreading anomalies between Australia and Antarctica', *Earth Planet. Sci. Lett.,
    58*, 151-60.
Chase, C.G. and Perry, E.C., Jr (1972) 'The oceans: growth and oxygen isotope
    evolution', *Science, 177*, 992-4.
Clark, J.A., Farrell, W.E. and Peltier, W.R. (1978) 'Global changes in sea level: a
    numerical calculation', *Quat. Res., 9*: 265-87.
Claypool, G.E., Holser, W.T., Kaplan, I.R., Sakai, H. and Zak, I. (1980) 'The age
    curves of sulphur and oxygen isotopes in marine sulphate and their mutual inter-
    pretation', *Chem. Geol., 28*, 199-260.
Cloud, P.E. (1976) 'Beginnings of biospheric evolution and their biogeochemical
    consequences', *Palaeobiology, 2*, 351-87.
Condie, K.C. (1982) *Plate Tectonics and Coastal Evolution*, Pergamon Press,
    Oxford and New York.
Davies, T.A., Luyendyk, B.P. *et al.* (1974) *Initial Reports of the Deep Sea Drilling
    Project*, vol. 26, US Govt. Printing Office, Washington, DC.
DSDP (Deep Sea Drilling Project) (1974) *Ocean Sediment Coring Program*, US
    National Science Foundation.
Frakes, L.A. (1979) *Climate through Geologic Time*, Elsevier, Amsterdam.
Garrels, R.M. and Lerman, A. (1981) 'Phanerozoic cycles of sedimentary sulphur

and carbon', *Proc. Nat. Acad. Sci., 78*, 4652-6.

Goldschmidt, V.M. (1933) 'Grundlagen der quantitativen Geochemie', *Fortschrift der Mineral. Krist. Petrog., 17*, 112-56.

Groves, D.I., Dunlop, J.S.R. and Buick, R. (1981) 'An early habitat of life', *Sci. Am., 245*, 64-73.

Hay, W.W. and Southam, J.R. (1977) 'Modulation of marine sedimentation by the continental shelves', in N.R. Anderson and A. Malahoff (eds), *The Role of Fossil Fuel CO in the Oceans*, Plenum Press, New York, pp. 569-605.

Holland, H.D. (1965) 'The history of ocean water and its effects on the chemistry of the atmosphere', *Proc. Nat. Acad. Sci., 53*, 1173-82.

—— (1972) 'The geologic history of sea water: an attempt to solve the problem', *Geochim. Cosmochim. Acta, 36*, 637-51.

—— (1983) 'Large scale geochemistry', in P.G. Brewer (ed.), *Oceanography, the Present and the Future*, Springer Verlag, New York, pp. 219-30.

—— (1984) *The Chemical Evolution of the Atmosphere and Ocean*, Princeton University Press, Princeton, NJ.

Hoyle, Fred (1957) *The Black Cloud*. Heinemann, London.

Hurley, P.M., Heezen, C.B., Pinson, W.H. and Fairbairn, H.W. (1963) 'K-Ar values in pelagic sediments of the North Atlantic', *Geochim. Cosmochim. Acta, 27*, 393-9.

Jenkin, J.J. (1984) 'Evolution of the Australian coast and continental margin', in B.G. Thom (ed.), *Coastal Geomorphology in Australia*, Academic Press, Sydney, pp. 23-42.

Kennett, J.P. (1982) *Marine Geology*, Prentice-Hall, Englewood Cliffs, N.J.

—— (1983) 'Palaeo-oceanography: global ocean evolution', *Rev. Geophys. Space Phys., 21*, 1258-74.

King, C.A.M. (1974) *Introduction to Marine Geology and Geomorphology*, London, Edward Arnold.

Kitano, Y. (ed.) (1975) *Geochemistry of Water. Part 1, Origin and Evolution of Natural Water (Seawater)*, Benchmark Papers in Geology, vol. 16. Dowden, Hutchinson & Ross, Stroudsberg, Pa., pp. 8-138.

Kuenen, P.H. (1950) *Marine Geology*, Wiley, New York.

Lambeck, K. (1980) *The Earth's Variable Rotation: Geophysical Causes and Consequences*, Cambridge University Press, Cambridge.

Lambert, R.S.J. (1980) 'The thermal history of the earth in the Archean', *Precambrian Res., 11*, 199-213.

Li, T-H., Takahashi, T. and Broecker, W.S. (1969) 'The degree of saturation of $CaCO_3$ in the oceans', *J. Geophys. Res., 74*, 5507-25.

Livingstone, D.A. (1963) 'Chemical composition of rivers and lakes', in M. Fleischer (ed.), *Data of Geochemistry*, US Geol. Surv. Prof. Paper 440, G1-G64.

MacIntyre, F. (1970) 'Why the sea is salt', in *Oceanography*, Readings from Scientific American, W.H. Freeman, San Francisco, pp. 110-21.

Mackenzie, F.T. and Garrels, R.M. (1966) 'Chemical mass balance between rivers and oceans', *Am. J. Sc., 264*, 507-25.

Mason, B. (1958) *Principles of Geochemistry*, Wiley, New York.

Mottl, M.J. and Holland, H.D. (1978) 'Chemical exchange during hydrothermal alteration of basalt by sea water. I. Experimental results for major and minor components of sea water', *Geochim. Cosmochim. Acta, 42*, 1103-15.

Newman, W.S. (1986) 'Palaeogeodesy data bank', in I. Shennan (ed.), *IGCP Project 200 Newsletter and Report*, Durham University, Durham, p. 41.

Parkinson, N. and Summerhayes, C. (1985) 'Synchronous global sequence boundaries', *Am. Ass. Petrol. Geol., 69*, 685-7.

Perry, E.C. and Tan, F.C. (1972) 'Significance of oxygen and carbon isotope variations in Early Precambrian charts and carbonate rocks of southern Africa', *Bull. Geol. Soc. Am.*, *83*, 647-64.

Pitman, W.J., III (1978) 'Relationship between eustasy and stratigraphic sequences of passive margins, *Bull. Geol. Soc. Am.*, *89*, 1289-1403.

Schidlowski, M., Hayes, J.M. and Kaplin, I.R. (1983) 'Isotopic inferences of ancient biochemistries: carbon, sulphur, hydrogen and nitrogen', in J.W. Schopf (ed.), *Earth's Earliest Biosphere: Its Origin and Evolution*, Princeton University Press, Princeton, NJ.

Schopf, J.W. (1983) *The Earth's Earliest Biosphere: its Origin and Evolution*, Princeton University Press, Princeton, NJ.

Schubert. G. and Reymer, A.P.S. (1985) 'Continental volume and freeboard through geological time', *Nature*, *316*, 336-8.

Sclater, J.G., Anderson, R.N. and Bell, M.L. (1971) 'The elevation of ridges and the evolution of the central eastern Pacific', *J. Geophys. Res.*, *76*, 7883-915.

Shaw, D.M. (1976) 'Development of the early continental crust', in B.F. Windley (ed.), *The Early History of the Earth*, Wiley, New York, pp. 33-53.

Sillen, L.G. (1963) 'How has seawater got its present composition?', *Svensk Kemisk Tidschrift*, *75*, 161-77.

Stewart, J. (ed.) (1977) *Oceanography: Introduction to the Oceans*, Open University Press, Milton Keynes.

Stewart, J. and Sharp, D. (eds) (1978) *Oceanography: Chemical Processes*, Open University Press, Milton Keynes.

Tucholke, B.E. (1981) 'Geologic significance of seismic reflectors in the deep western North Atlantic basin', *SEPM Spec. Publ.*, *32*, 23-37.

Turekian, K.K. (1972) *Chemistry of the Earth*, Holt, Rinehart & Winston, New York.

—— (1976) *Oceans*, 2nd edn, Prentice-Hall, Englewood Cliffs, NJ.

Twenhofel, W.H. (1932) *Principles of Sedimentation*, McGraw-Hill, New York.

Vail, P.R., Mitchum, R.M. and Thompson, S., III (1977) 'Seismic stratigraphy and global changes of sea level', in C.E. Payton (ed.), *Seismic Stratigraphy: Applications to Hydrocarbon Exploration*, Am. Ass. Petrol. Geol., Tulsa, Okla., pp. 83-98.

Veevers, J.J. and Evans, P.R. (1975) 'Late Palaeozoic and Mesozoic history of Australia', in K.S.W. Campbell (ed.), *Gondwana Geology*, ANU Press, Canberra, pp. 579-607.

Veizer, J. (1984) 'The evolving earth: water tales', *Precambrian Res.*, *25*, 5-12.

—— and Compston, W. (1976) '$^{87}Sr/^{86}Sr$ in Precambrian carbonates as an index of crustal evolution', *Geochim. Cosmochim. Acta*, *40*, 905-15.

—— Holser, W.T. and Wilgus, C.K. (1980) 'Correlation of $^{13}C^{12}C$ and $^{34}S/^{32}S$ secular variations', *Geochim. Cosmochim. Acta*, *44*, 579-87.

Walther, J. (1926) *Die Methoden der Geologie als historischer und biologischer Wissenschaft*, Hamb. Biol. Arbeitsmeth. Abt. 10, Urban & Schwarzenburg, Berlin.

Weissert, H. (1981) 'The environment of deposition of black shales in the Early Cretaceous: an ongoing controversy', *SEPM Spec. Publ.*, *32*, 547-60.

Windley, B.F. (1977a) *The Evolving Continents*, Wiley, New York.

—— (1977b) 'Timing of continental growth and emergence', *Nature*, *270*, 426-8.

Worsley, T.R., Nance, D. and Moody, J.B. (1984) 'Global tectonics and eustasy for the past two billion years', *Mar. Geol.*, *58*, 373-400.

3

Mechanisms of Relative Sea-level Change and the Geophysical Responses to Ice-water Loading

W.R. Peltier

INTRODUCTION

The classification of mechanisms of sea-level change

The purpose of this chapter is firstly to provide a discussion of the variety of mechanisms which may cause changes in sea level, particularly those associated with the melting of continental ice sheets. Secondly, detailed consideration will also be given to some examples of what has been learnt about the earth and the operation of its climate system through the analysis of observations of such changes.

The physical mechanisms which are responsible for causing variations of sea level are clearly a strong function of the timescale on which the change occurs. Because most measurements of sea-level change, regardless of the timescale of the variability, are taken relative to the surface of the solid earth, it is useful in discussing mechanisms to distinguish the changes to which they give rise as belonging to one or the other of two classes. These classes are conventionally labelled *eustatic* and *isostatic*. The appearance of the root 'static' in both these words is meant to imply that the changes referred to are quasi-permanent and to be distinguished from the more ephemeral and quasi-periodic variations associated with oceanic tides and with the annual fluctuation of sea level which follows the seasonal climate cycle.

In the discussion which follows the word *eustatic* will be used to describe changes in sea level which are a consequence of a change in the volume of the water in the global ocean or in some local part of it. Such eustatic changes may be a consequence either of a change in ocean mass (caused, say, by the accretion or melting of continental ice sheets), or of a change in the volume which a fixed mass occupies through so-called *steric* effects associated for example with variations of temperature and the process of thermal expansion (caused, say, by a change of global climate). In the literatures of hydrodynamics and oceanography the words *barotropic* and *baroclinic* are also employed to denote respectively the non-steric and steric components of the eustatic sea-level change and these words will be used interchangeably in the following discussion.

The *isostatic* contribution to the local variation of sea level is completely distinct from this eustatic component with its barotropic and baroclinic constituents. Just as eustatic changes are associated with changes of ocean volume, so isostatic changes are caused by predominantly radial displacements of the surface of the solid earth. Since sea-level change is normally observed relative to this surface it is clear that every such observation is inherently ambiguous. If one observes an increasing relative sea level in some particular geographic location this could be due, in principle, either to an increase in the local height of the water column (a eustatic effect) or to a decrease in the local radius of the solid earth (an isostatic effect). Much of the challenge in the analysis of relative sea-level data lies in attempting to separate these two fundamentally different contributions to observations of relative sea-level change. From the eustatic component comes information on the nature of climate change, whilst from the isostatic contribution comes information concerning the nature of the solid earth. In what follows an attempt will be made to provide some examples of both types of information to illustrate the nature of sea-level data itself, as well as the specific questions which it may be employed to address.

THE MECHANISMS AND THEIR ASSOCIATED CHANGES OF SEA LEVEL

Short timescale mechanisms

Probably the most dramatic mechanism contributing to relative sea-level variations in a few special locations is that associated with earthquakes. Of course such events occur on an extremely short timescale, usually a small fraction of a minute, and significant effects on the sea-level record are found only when the earthquake is located offshore and has an appreciable 'dip-slip' component of its source mechanism. When these two conditions are satisfied, and the rupture actually extends to the earth's surface, then associated with the earthquake there is a relative vertical motion of the sea floor on either side of the fault break. This may extend onshore, where it will be registered as a change of relative sea level on a tide gauge attached to the surface of the solid earth. Besides this semi-permanent isostatic change in sea level, associated with the induced vertical displacement of the surface of the solid earth, such earthquakes also give rise to intense transient variations of relative sea level. These travel away from the epi-centre as water waves called *tsunamis* which can cause extreme damage when they propagate onshore at distant locations. Although the amplitude of the relative sea-level (RSL) change associated with the tsunami is only of the order of 1 cm in the open ocean, it may reach many tens of metres as it propagates into the shallow seas surrounding distant islands and coastal sites. Such events are relatively frequent in the Pacific Ocean since earthquakes of the required type are common occurrences in the vicinity of the Aleutian Islands and Japan. The semi-permanent isostatic variations of sea level associated with great earthquakes are large enough to be easily observable on tide gauge recordings (e.g. Wyss, 1976a, b; Lagios and Wyss, 1983). RSL changes resulting from such earthquakes may be of the order of 10 cm or more, but such effects are confined to the region in the immediate vicinity of the earthquake itself. The tide gauge data demonstrate that following the initial shock, relative sea level appears to relax back towards its original level over a period of years, as local stresses accumulate in reponse to the driving mechanism of plate tectonics (i.e. thermal convection in the mantle), until the accumulation is eventually released once more by a new earthquake. Much of the best information currently available concerning the dynamics of this earthquake cycle is contained in long time series of RSL change on tide gauge records from the islands of Japan. The effect itself is a good example of those isostatic changes of RSL which are often called *tectonic* in origin. Larger scale tectonic influences on relative sea level—ocean basin shape have been discussed elsewhere (see this vol., Chappell, Ch. 2 and Berryman, Ch. 5).

On somewhat longer timescales, of the order of hours to days, meteorological effects may induce rather substantial changes in relative sea level through the agency of the stress delivered by the wind to the sea surface. Depending upon the local direction of the wind relative to the coast, this stress field may induce either a rise or fall of relative sea level depending on whether the winds are onshore or offshore respectively. Clearly the RSL variations produced by such effects are essentially barotropic, since they are produced by an increase or decrease of the mass of the local water column. Such effects would also, however, be termed eustatic if they were associated with a sufficiently long timescale (climatological) change and aligned in the mean direction of the wind relative to the coast. There is currently a great deal of interest in the oceanographical community in using tide gauge observations of relative sea-level change to infer the strength of the currents through straights and channels as well as the nature of the circulation in the coastal ocean (Csanady, 1982; Thompson, 1981; Noble and Butman, 1979). Of course the most interesting recent application of sea-level data to the problem of inferring the nature of the circulation in the global ocean has been that involving the use of altimetry data from the artificial earth satellites SEASAT and GEOS 3 (Geodynamics Program Office, 1983, 1984). These observing platforms have provided maps of the geoid (the surface on which the gravitational potential is constant and which is coincident with the ocean where there is ocean) of unprecedented accuracy. These maps are contributing in an important way to advances in both geophysical and oceanographical science (e.g. Stanley, 1979; Rapp, 1979; Wunsch and Gaposhkin, 1980; Wunsch, 1981; Marsh and Martin, 1982; Schutz, Tapley and Shum, 1982; Carter *et al.*, 1986).

On even longer timescales, of the order of years to several years, important variations of relative sea level have been shown to be intimately related to the El Niño–Southern Oscillation (ENSO) phenomenon which is an extremely important source of climate system variability on these time-scales (Wyrtki and Nakahoro, 1984). In the ocean the El Niño represents an incursion of warm water from the equatorial zone southward into cooler higher latitude ocean water. In the popular literature ENSO is of course best known for its dramatic effect upon the Chilean anchovy harvest! The ENSO-related RSL fluctuations have ocean basin scale spatial coherence lengths of thousands of kilometres and amplitudes of tens of centimetres. According to Roemmich and Wunsch (1984) the El Niño-like fluctuations of sea level are largely baroclinic rather than barotropic in form and so are not associated with large-scale fluctuations of the local mass of the water column. Rather the relative sea-level variations observed in association with this phenomenon are driven by steric effects, due to the thermal expansion of sea water in response to a changing temperature. An example of the RSL variability associated with ENSO is provided in Figure 3.1. Although the characteristic timescale of ENSO of several years is reasonably long, it is still very significantly shorter than the timescales of hundreds to hundreds of thousands of years upon which most of the ensuing discussion will now focus.

Medium time scale mechanisms

The existence of tide gauge records, many extending as far back into the past as 100 years or more, also provide useful information as to the mechanisms of RSL change which are operative on these extended time-scales. On the timescale of 100 years, steric effects also seem to be an important contributor to the global scale secular increase of relative sea level, which appears to be characteristic of this period of earth history. Gornitz *et al.* (1982) have pointed to evidence of a warming of global surface temperature of about 0.4°C during the twentieth century which suggests a sterically induced rate of sea-level rise during this time of about 0.6 mm a^{-1} when interpreted in terms of a model of vertical thermal diffusivity of the ocean. This is sufficient to explain about 50 per cent of the global rate of present-day sea-level rise, inferred on the basis of analyses of the secular variations of relative sea level recorded on a 'global' array of tide gauges, the records of which extend many decades into the past (Barnett, 1983a, b). Examples of the tide gauge records which reveal this

Figure 3.1: Maps of sea-level anomaly for December 1975 and December 1977. Contours show sea-level anomalies in millimetres after removal of the seasonal cycle. These two cases were selected because of the large contrast which they reveal which is characteristic of the El Niño-Southern Oscillation effect

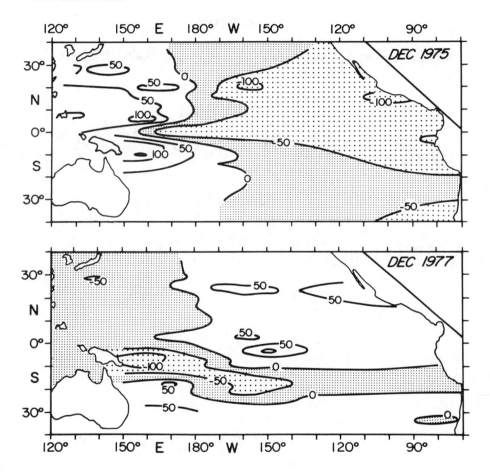

Source: Wyrtki and Nakahara (1984).

secular increase of RSL are provided in Figure 3.2. One of the most hotly debated issues in present-day sea-level research is whether or not the residual non-steric (barotropic) rate of relative sea-level rise of about 0.5 mm a^{-1} is statistically significant. Hansen *et al.* (1981) have suggested that this residual non-steric eustatic contribution could be caused by the melting of continental ice. They further suggest this to be an expected

Figure 3.2: Typical monthly sea-level series for the South Atlanic Bight (Charleston), Mid-Atlantic Bight (Sandy Hook) and Gulf of Maine (Boston), 1950-75

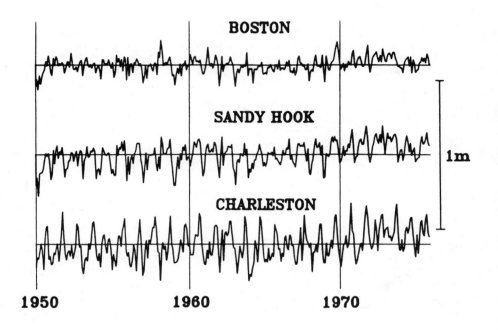

consquence of climate warming due to the increasing atmospheric load of CO_2 and other greenhouse gases, consequent upon the burning of fossil fuels and other side-effects of industrialisation (see Titus, this vol., Ch. 15 for further discussion). Meier (1984) has recently pointed out that although current glaciological evidence suggests that both the Greenland and Antarctic ice masses are stable, the melting back of the small ice sheets and glaciers of the world *is* suggested by the observational evidence. The volume of water released by this process is of an amount which could very nicely account for the increase of water mass in the global ocean, needed to balance the non-steric component of the RSL rise evident on the tide gauge records.

Central to the debate over interpretation of the tide gauge-inferred secular trends of sea level discussed above, is the question of whether or not even longer timescale isostatic processes may be contributing to the observed signals and thereby causing an isostatic signal to be misinterpreted as an eustatic one. This is quite conceivably an extremely serious problem since, as pointed out by Barnett (1983a, b), the global ocean is sampled by the present tide gauge network in a highly non-uniform fashion. Furthermore, many of the tide gauges from which extremely long records are available, are located in regions, such as along the eastern seaboard of the continental United States, which are known to be strongly influenced by isostatic effects. In the latter region the most important source of ultra long timescale variability of relative sea level, is that associated with the continuing isostatic adjustment of the surface of the solid earth, in reponse to the melting of the huge Laurentide ice sheet. This covered all of Canada and parts of the northern United States until ~ 18,000 BP, by which time it had reached its maximum extent and had begun to retreat, eventually disappearing completely by about 6,500 BP (see Andrews, this vol., Ch. 4). The reason why relative sea level continues to change in response to this cause, so many millennia after the ice sheet had completely disappeared, is the extremely high value of the effective viscosity of the earth's mantle, the viscosity governing the rate at which mantle material flows in the process of restoring the deformed shape of the earth to one of gravitational equilibrium. This process is called *glacial isostatic* adjustment. It is extremely fortunate for geophysical science in general that the earth has conspired to remember the history of relative sea-level variations caused by the melting of such large ice masses. For through careful analysis of these one is able, as will be shown later, to infer the value of the effective viscosity of the mantle. Knowing this number (Peltier, 1980, 1984a, 1985) one is then able to devise an objective test of the validity of the convection hypothesis of continental drift. That this is so makes clear the central place which relative sea-level data have come to play in modern geophysical research.

The way in which the planet remembers the history of deglaciation-induced relative sea-level change is illustrated in Figure 3.3, which shows a flight of raised beaches located in the Richmond Gulf in the southeast corner of Hudson Bay. Each of the horizons (shorelines) visible on the hillside represents a relict beach and therefore a past position of sea level. If one simply measures the height of each of these shorelines above present-day sea level and plots this height as a function of the age of the relict beach, determined by the application of ^{14}C dating to carbonaceous material from systems (e.g. shells) which were open at the time the shoreline was forming, then one obtains a relative sea-level curve (time-depth plot) such as that shown in Figure 3.4. Inspection of this figure shows that relative land emergence has been under way at this location for the last 7,000 years, at a rate which appears to be an exponentially decreasing function of time (see this vol., Andrews, Ch. 4, Sutherland, Ch. 6 and Devoy, Ch. 10 for further discussion of data and techniques). The relaxation time for this process is near 2,000 years and this value can be employed (e.g. Peltier, 1982) to fix the effective viscosity of the mantle beneath Hudson Bay to a value near 10^{21} Pa s. Outside the ice margin, relict beaches are drowned rather than raised, as illustrated by the series of

Figure 3.3: Raised marine shorelines in the Richmond Gulf on the southeastern shore of Hudson Bay

Figure 3.4: Relative sea-level curve obtained by ^{14}C dating of carbonaceous material from the relict shorelines at Richmond Gulf shown on Figure 3.3

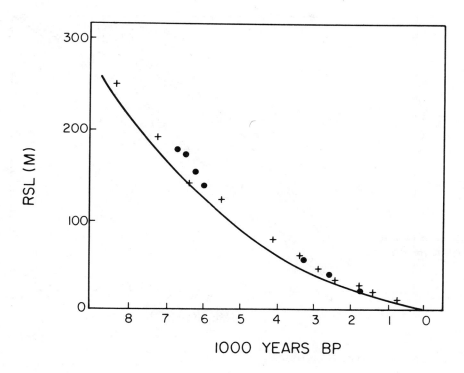

relative sea-level curves from sites along the US east coast shown in Figure 3.5. The reason for this characteristic variation in the RSL signature of glacial isostasy on either side of the ice margin is clear on the basis of a simple conservation of mass argument. At glacial maximum the surface of the earth is depressed under the weight of the ice load by the amount required for the 'Archimedes force' (buoyancy) to balance the weight of the load. This depression of the surface is accommodated by the flow of material out from under the ice sheet into the peripheral region where the surface is elevated above its equilibrium level, producing a forebulge zone (Fig. 3.6). When the ice sheet melts the surface of the earth sinks in the peripheral region as material flows back under the area which was ice covered, causing the surface there to rise out of the sea. Comparing the ^{14}C controlled relative sea-level curves shown on Figure 3.5 with the tide gauge-determined relative sea-level curves shown previously on Figure 3.2, which are for sites along the same coast, brings the problem of the interpretation of RSL data clearly into perspective. What part, if any, of the

Figure 3.5: Relative sea-level curves from four regions along the eastern seaboard of the continental United States

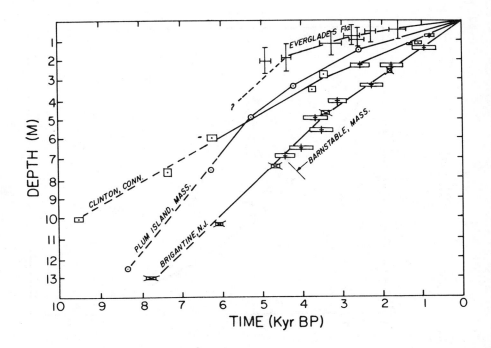

Source: Modified from Bloom (1967).

secular change seen on the tide gauges is not explicable in terms of the glacial isostatic adjustment effect? If such a eustatic component exists, is it entirely explicable as a steric effect, or must we invoke a change in ocean mass to understand it? In the next section of this chapter the theory of glacial isostasy will be used to illustrate, by a specific example, the way in which questions such as this may be objectively addressed.

Long timescale mechanisms

Before doing so, however, it will be useful to consider mechanisms of sea-level change which operate on timescales longer still than the roughly 10 ka timescale over which the glacial isostatic adjustment process has contributed dominantly to the record of RSL variability. On a timescale of 100 ka-1 Ma, the dominant contributor to the RSL record has been the

Figure 3.6: Model showing the effects of glacio-isostatic ice mass load on a land surface. Isostatic depression of the land surface A–B increases with ice load so that greater depression occurs towards the centre of an ice mass (h_1) than at the margin (h_2)

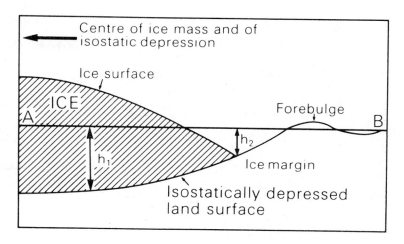

Source: Lowe and Walker (1984).

continuous process of accumulation and disintegration of large ice sheets, such as that which last covered Canada 18,000 years ago. In fact this ice sheet, along with its apparently inevitable Fennoscandian companion centred on the Gulf of Bothnia and a third sheet centred on West Antarctica, have been appearing and disappearing at regular 100 ka intervals for at least the past 700,000 years of earth history. Since the mass of ice bound in these complexes was extremely large when they were at their maximum extents, approximately 3.5×10^{19} kg in total, their growth caused a global eustatic fall of sea level (*glacio-eustasy*) of ~110 m, although opinions vary on this value (Bloom, 1983). Our knowledge that this has been the case derives from $\delta^{18}O$ vs. depth data obtained from sedimentary cores drilled from the floors of the major ocean basins, at sites well removed from the continental margins where turbidity current effects continually disturb the sedimentary record. The number $\delta^{18}O$ is simply a measure of the concentration of the heavy isotope of oxygen (^{18}O) relative to the more abundant light isotope (^{16}O). Examples of such data from four typical sedimentary cores are shown in Figure 3.7, using data from Shackleton and Opdyke (1973, 1976) and Imbrie *et al.* (1973). The importance of these data derives from the fact that $\delta^{18}O$ in the sediment is a direct measure of the amount of continental ice which exists in the climate system at the time of

Figure 3.7: Four typical $\delta^{18}O$ vs depth series obtained from long DSDP cores in three different oceans. M $+$ B = Brunhes–Matuyama boundary, at 730,000 $\pm$ 10,000 ka

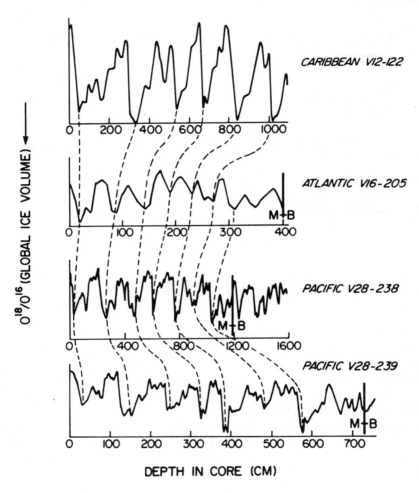

sediment deposition. This is because the process of evaporation of water at the sea surface, which provides the fuel through which later precipitation induces ice sheet growth, is a process which fractionates mass. Since the light isotope evaporates preferentially over the heavy, the precipitation which feeds ice sheets is always isotopically light, which is to say that it has lower $\delta^{18}O$ than normal sea water. Therefore, when ice accumulates on land this leads to an irreversible increase of the $\delta^{18}O$ of sea water, which is communicated to the organisms (e.g. foraminifera, coccoliths) whose shells (upon their death) contribute to the sediment which accumulates on the sea floor. When we later measure the $\delta^{18}0$ of the shells contained in the sediment at some depth in the core, corresponding to some time in the past, we are therefore able to infer how much ice must have been present on the continental surface at that time. Although we obviously cannot infer from these data where the ice was located, we can deduce, on the basis of the observation that indications of the presence of ice sheets (moraines, erratics, glacial striae) are noticeably absent in regions well removed from Laurentia and Fennoscandia, that the ice sheets must have continually reoccupied these same locations.

In order to make quantitative use of $\delta^{18}0$ data such as that shown in Figure 3.7, we must be able to calibrate the record so that we can quantitatively infer the volume of continental ice from the observed $\delta^{18}0$ measurement. This can be done relatively simply by using the known volumes of the Greenland and Antarctic ice sheets in conjunction with measurements of their oxygen isotopic signatures. We must also find a way to assign a specific age to a specific depth in each core. Clearly the same age will not correspond to the same depth in each core because the rate of sedimentation is a strong function of location on the ocean floor. For sedimentary cores which are sufficiently long, this problem may be rather elegantly solved by finding the depth in the core corresponding to the first horizon below which the remnant magnetisation of the sediments has the opposite direction from that above. This horizon corresponds to the time (down core) of the last reversal of the earth's magnetic field and is referred to as the Brunhes–Matuyama boundary. Since this reversal occurred 730,000 ± 10,000 BP (Cox and Dalrymple, 1967), if we can find the depth of the horizon corresponding to it in the core we can, assuming that the sedimentation rate has remained constant over the time spanned by the core, translate each specific depth into a specific time. Imbrie *et al.* (1984) have attempted to produce a 'best' $\delta^{18}0$ chronology from these data by stacking the time series from a number of such cores to discriminate against features which are not consistent from core to core. They have called the resulting record SPECMAP (Fig. 3.8). Obvious by inspection of the power spectrum (Fig. 3.8b) is the fact, first noted by Hays *et al.* (1976), that the dominant variability in the $\delta^{18}0$ data occurs at the periods of 100 ka (over 65 per cent of the variance), 41 ka, 23 ka and 19 ka. Since these are rather precisely the period of variation of the eccentricity of the earth's orbit around the Sun, the period of variation of the tilt of the spin axis with respect to the ecliptic and the dominant periods of the precession of the equinoxes, these data have been construed as verifying the validity of the Milankovitch (1941) theory of palaeoclimatic change. The largest (eustatic) changes in relative sea level of all, therefore, appear to be driven by the small changes of the effective intensity of (summer) solar insolation, caused by equally small changes in the values of the parameters which

Figure 3.8: (a) the SPECMAP time series of Imbrie *et al.* (1984) and (b) its power spectrum

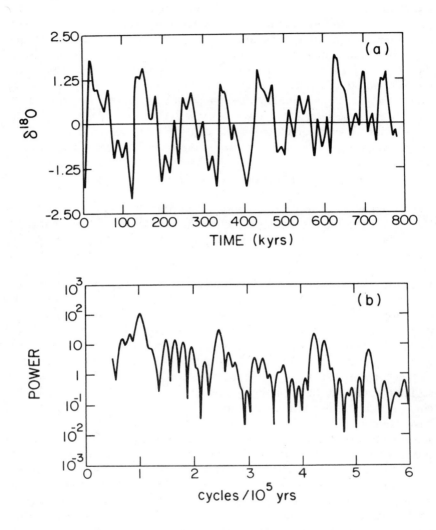

govern the geometry of the earth's orbit around the Sun. Much recent effort has been invested in explaining the detailed physical processes through which the astronomical input is translated into the ice volume fluctuations revealed by the $\delta^{18}O$ data (e.g. Peltier, 1982; Peltier and Hyde, 1984; Hyde and Peltier, 1985).

The extremely large amplitude glacio-eustatic variation of global sea level which accompanies the 'ice age cycle' suggests the operation of a further important mechanism of relative sea-level change. In any analysis of relative sea-level data during times of active ice sheet disintegration at sites near the glaciated regions, or at any time from sites well removed from the ice sheets, there will be an important contribution to the observed variation of relative sea level resulting from the direct effect of the water load applied to the ocean basins as the ice sheets melt. This water load will in turn induce its own deformation of the solid earth. This process is called *hydro-isostasy* (Bloom, 1967). It should be clear that as the earth deforms physically due to loads associated with ice sheet melting and ocean basin filling, this deformation will cause continuous changes in the gravitational potential field of the planet. This will require continuous redistribution of mass among and within the ocean basins in order to ensure that the surface of the sea (the ocean geoid) remains an equipotential surface. The actual history of relative sea-level change which is observed at any particular location on the earth's surface is therefore the consequence of a complex interaction, through the intermediary of the gravitational field, between the aquasphere, the cryosphere and the solid earth. This complex of interactions is illustrated by the schematic feedback loops shown in Figure 3.9. Over the past decade it has proved possible to develop a detailed mathematical model capable of accurate prediction of postglacial variations of relative sea level, including both glacial isostasy, hydro-isostasy and the full effects of the mutual gravitational attraction acting between the ice and sea water in a self consistent way. The next section of this chapter will now present a few of the detailed characteristics of this model, together with a discussion of the geophysical and oceanographical data which it has been shown to reconcile. Further, an application of the model is made in addressing the specific question raised earlier in this section, namely, the origin of the secular trend of rising relative sea level which seems to be characteristic of the last hundred years of earth history. It is hoped that by more sharply focusing the discussion to follow in this way the reader will be able to see the wealth of information that is contained in relative sea-level and associated data much more clearly than would be possible with further general commentary.

THE GLOBAL MODEL OF GLACIAL ISOSTASY AND POSTGLACIAL RELATIVE SEA-LEVEL CHANGE

The mathematical structure of the global model of glacial isostatic adjustment has been reviewed recently in Peltier (1982) to which the interested reader is referred for details. In this model the planetary interior is assumed to be radially stratified, with an elastic structure fixed by observations of the frequencies of the elastic gravitational modes of free oscillation, as described for example by Gilbert and Dziewonski (1975). The rheology of

Figure 3.9: Schematic diagram illustrating the interactions among ice loads, water loads, and the deformable earth: (A) The weight of the ice deforms the earth and (B) the ice mass attracts the water. (C) The transfer of matter within the earth distorts the geoid. Similarly, (D) the weight of the meltwater depresses the earth differentially and (E) more water flows into this depression, increasing the water load and (F) causing added deformation of the ocean floor. These processes are interrelated as indicated in (G), and all are included in the numerical model

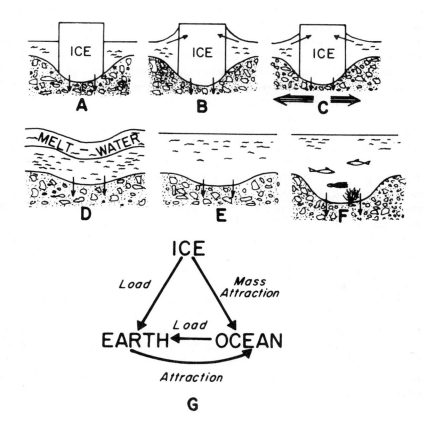

Source: Clark *et al.* (1978).

the interior is assumed to be linearly viscoelastic and either of the 'Maxwell' or 'Burger's' body type. In the former the initial response of the material to an applied shear stress is Hookean elastic, whilst the final response is Newtonian viscous and governed by a viscosity v_1 which is a function only of radial position r within the planet. The Burger's body differs from the Maxwell analogue in that it includes a transient as well as a

Newtonian steady state component in its viscosity spectrum (Peltier *et al.*, 1981; Peltier 1982, 1984a, 1985a, b). This requires specification of a second viscosity $v_2(r)$ to complete the description of the rheology. As demonstrated explicitly in Peltier (1985a), however, even if transient rheology is important in the glacial isostatic adjustment process, only models with large elastic defect are capable of simultaneously reconciling all the data. In this large defect limit the transient Burger's body model behaves as a Maxwell model with a new effective viscosity $v_{eff}(r) = v_1(r) v_2(r)/(v_1(r) + v_2(r))$. When we fit a Maxwell model to RSL data such as that shown in Figures 3.4 and 3.5 we are free to vary only the one parameter $v_{eff}(r)$ to achieve the fit.

In the actual prediction of such RSL variations what one does is to assume a melting history for all of the surface ice loads which exist at glacial maximum. A computation is then made, in a gravitationally self-consistent fashion, including the full effects of both glacial and hydro-isostasy and the ice-water attraction, of the manner in which the meltwater must be distributed over the global ocean. This is done in order to ensure that the instantaneous surface of the new ocean is maintained as an equi-potential surface at all times as the system evolves. This process requires inversion of an integral equation at every instant during, and subsequent to, the deglaciation event, and results in a direct prediction of the time dependent separation of the geoid and the surface of the solid earth at any point on the earth's surface where ocean and land meet. In the model, the geography of the oceans and continents is realistically described and the integral equation is inverted using a finite element discretisation of the surface. Further, a Green's function formalism is employed to describe the gravitational interactions between the aquasphere, cryosphere and solid earth components of the model, which are fundamental to the determination of sea-level change.

It is rather fortunate that it is possible to make reasonably accurate *a priori* estimates of the deglaciation histories of each of the three main ice-covered regions (Laurentia, Fennoscandia, and West Antarctica) subsequent to 18,000 BP. Without this information we would not be in any position to implement the relative sea-level model discussed above! Peltier and Andrews (1976) have described the way in which ^{14}C age controlled terminal moraine data can be combined with worldwide observations of RSL and ice physical–mechanical information to develop first order models of the glacial chronology. Their initial model, called ICE-1, has since been improved by Wu and Peltier (1983) who have called the new model ICE-2 (Fig. 3.10). This shows three time slices through the ice sheet topography for both the Laurentian and Fennoscandian regions. The initial problem for RSL prediction takes this deglaciation history as input to the model and, in conjunction with an assumed model of the planet's radial viscoelastic structure, produces as output a prediction of the RSL variation which should be

73

Figure 3.10: Three time slices through the ICE-2 melting chronologies for Laurentia (Canada), and Fennoscandia. The ice thickness contours are in metres

ICE 2

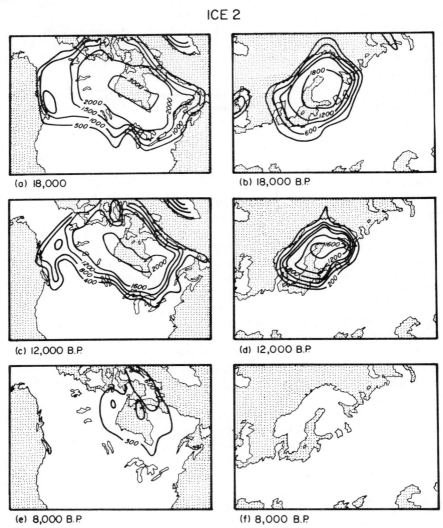

(a) 18,000

(b) 18,000 B.P.

(c) 12,000 B.P.

(d) 12,000 B.P.

(e) 8,000 B.P.

(f) 8,000 B.P.

observed at any site on the earth which might be of interest. In fact the primary output of the model consists of a sequence of maps, such as those shown on Figure 3.11, for four sample times within and subsequent to deglaciation. These maps show the increase or decrease of the local bathymetry of the ocean which has occurred by the time shown on each plate. This includes the combined effects of the glacial isostatic and hydro-isostatic

Figure 3.11: Four time slices through a typical solution of the sea-level equation. The relative sea-level rise is contoured in metres. Details of the model employed to make this prediction are provided in the text

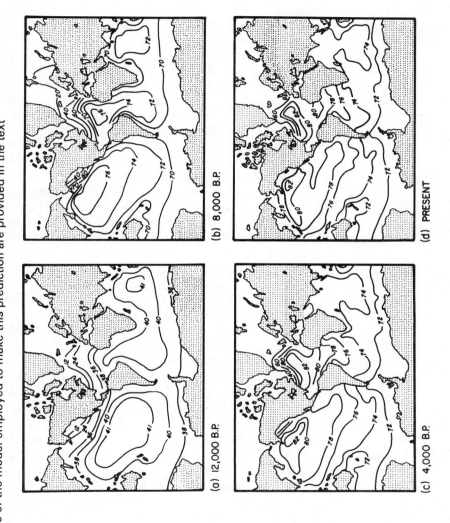

(a) 12,000 B.P.

(b) 8,000 B.P.

(c) 4,000 B.P.

(d) PRESENT

displacement of the surface of the solid earth, as well as the eustatic sea-level variation forced by the filling of the ocean basins by the meltwater produced in glacial decay. It also includes the full effects of the geoidal deformation associated with the redistribution of surface load and with the load-induced deformation of the solid earth. Inspection of the four plates of Figure 3.11 demonstrates that the increase of ocean bathymetry caused by ice sheet melting is a highly non-uniform function of position in the global ocean. There therefore does not exist a single 'eustatic curve' which could be used to characterise the effect of increasing water mass upon relative sea level at all locations remote from the main ice sheet centres.

Prior to the development of the model described here, which is based upon the theoretical description of glacial isostasy first advanced in Peltier (1974) and further refined in a number of subsequent papers (Peltier, 1976; Peltier and Andrews, 1976; Farrell and Clark, 1976; Peltier et al., 1978; Clark et al., 1978; Peltier, 1982; Wu and Peltier, 1983), it was generally believed that such a universal curve did exist and could be used to correct all relative sea-level data for the effect of glacial eustasy (e.g. Shepard, 1963). This was at least partly responsible for an important controversy which arose concerning the explanation of emerged Holocene beaches, which are found in a wide variety of locations at sites remote from the main glaciation centres. Several scientists had expressed opinions to the effect that these beaches were formed when postglacial eustatic sea levels were higher than they are at present and, therefore, inferred the occurrence of important climatic changes (e.g. Schofield, 1964; Gill, 1965; Fairbridge, 1976). Others disputed this contention and suggested either that the observed beaches were much older than Holocene in age, or were emerged as a consequence of tectonic activity, or simply the product of intense but infrequent storms (Shepard, 1963; Jelgersma, 1966; Bloom, 1970). The gravitationally self-consistent model of postglacial relative sea-level change has settled this controversy. It demonstrates that the emergence of shorelines at locations remote from ice sheet influence is an entirely expected consequence of the viscoelastic adjustment of the earth, in response to the changing ice and water load on its surface and the geoidal deformation which is thereby produced. This fact is demonstrated in Figure 3.12 which shows the surface of the earth divided into a number of zones, in each of which the variation of relative sea level is of a single characteristic form. The model which produced this result is described in detail in Wu and Peltier (1983, Figure 24). The earth is assumed to possess an elastic structure identical to that of model 1066B of Gilbert and Dziewonski (1975), the lithosphere is assumed to be 120.7 km thick and the mantle to have a constant viscosity of 10^{21} Pa s from the base of the lithosphere to the core-mantle boundary. The deglaciation history is essentially the ICE-2 model of Wu and Peltier (1983), although the melting from West Antarctica was neglected. In Zone I, which comprises the ice-covered regions, the

Figure 3.12: Zone boundaries separating the surface of the earth into six regions, in each of which the history of relative sea-level change since deglaciation has a characteristic form (see also Clark *et al.*, 1978)

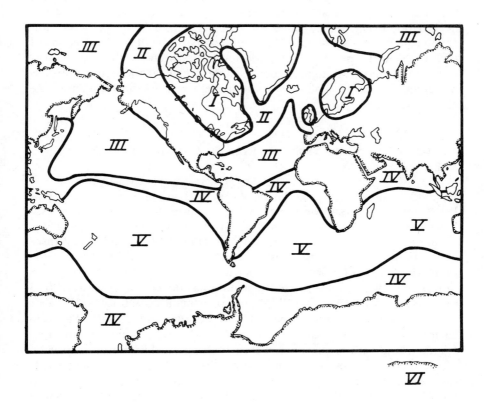

relative sea-level curves all display monotonic emergence, while in the peripheral bulge region of Zone II the predicted RSL signature is one of monotonic submergence. For this model Zone III is one in which raised beaches appear some finite time after the cessation of glacial decay, while in Zone IV the model predicts monotonic submergence. In Zone V raised beaches form immediately after the cessation of melting whilst in Zone VI, which comprises all continental shorelines removed from ice sheet influence, raised beaches are produced subsequent to glacial decay by the deformation associated with the offshore water load. Of course the existence of the zones and the locations of the boundaries separating them are a strong function of the earth and deglaciation models and can therefore be employed to discriminate between them.

The brief summary of the model presented here should be sufficient to demonstrate that when the full effects of glacial isostasy, hydro-isostasy and geoid deformation are included, the model is able fully to account for the occurrence of raised beaches distant from the effects of ice sheet development. There is no need to invoke further eustatic or tectonic contributions to the sea-level record subsequent to completion of the last deglaciation event, which ended ~6500 BP. The remaining sections of this chapter describe a sequence of detailed comparisons of the predictions of this model with specific relative sea-level curves and with several other geophysical and astronomical observations which are also explicable as representing various signatures of the planet's response to the Pleistocene glacial cycle.

POSTGLACIAL VARIATIONS OF RELATIVE SEA LEVEL AND RELATED EFFECTS

Figure 3.13 shows typical examples of observed and predicted relative sea-level variations at six locations on the North American continent (see Devoy, this vol., Ch. 10, for discussion of the sea-level record). Three occur at sites which were once ice covered (Fig. 3.13a, b and c) and three at locations along the east coast of the continental United States in the peripheral region of monotonic subsidence (Fig. 3.13d, e and f). The earth model employed to make these predictions has the elastic structure 1066B of Gilbert and Dziewonski (1975), an upper mantle viscosity of 10^{21} Pa s and a lower mantle viscosity beneath 670 km depth of 2×10^{21} Pa s. On each plate comparisons are shown for three different models which differ from one another only in terms of their lithospheric thicknesses. Inspection of these comparisons shows that the relative sea-level data at sites inside the ice margin are reasonably well fitted by the theoretical model and that the RSL variations at these sites are rather insensitive to changes of lithospheric thickness. This is entirely expected as the spatial scale of the Laurentian ice sheet (Fig. 3.10) is so large that the lithosphere is transparent to the response at locations which were once under the ice sheet centre. At sites in the peripheral region, on the other hand, the response is extremely sensitive to lithospheric thickness, as the deformation at such sites is significantly affected by relatively short horizontal wavelengths, which 'see the lithosphere clearly'. In Peltier (1984b) this sensitivity was first exploited to measure lithospheric thickness and a relatively high value in excess of 200 km was obtained. As reviewed in Peltier (1982), the totality of ^{14}C controlled RSL data also require an almost uniform profile of mantle viscosity, with little variation between the upper and lower mantles. Weertman (1978) has commented upon this result from the point of view of theoretical ideas concerning the microphysical basis of solid state creep in the earth. He has suggested that it might be taken to imply that the relaxation of the lower mantle, which occurs in postglacial rebound, is

Figure 3.13: Example comparisons of observed and predicted relative sea-level histories at six sites on the North American continent as a function of lithospheric thickness in a model with 1066B elastic structure, upper mantle viscosity of 10^{21} Pa s and lower mantle viscosity of 2×10^{21} Pa s. The first three sites (a, Churchill; b, Ottawa Islands; c, Southampton Islands) were ice covered whereas the last three (d, N.Y. City; e, Brigantine; f, Delaware) were beyond the ice margin on the US east coast

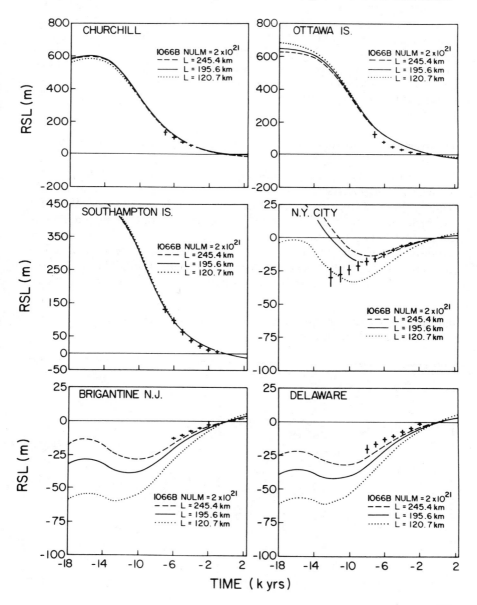

controlled by transient creep rather than the steady state creep which is assumed in the Maxwell analogue. This possibility may be tested using the simple Burger's body rheology derived in Peltier *et al.* (1981). It includes the transient component of the response via a single Debye peak governed by two additional physical constants. As previously mentioned, however, an analysis of the asymptotic properties of the Burger's body rheology demonstrates that when the elastic defect is large, and the short and long timescale viscosities sufficiently different, the Burger body rheology again behaves like a Maxwell solid, but with a viscosity equal to that which governs the short timescale transient response. Under these circumstances the lower mantle viscosity, inferred by analysis of rebound data based upon the Maxwell analogue, would be the transient viscosity as originally suggested by Weertman (1978).

The free air gravity anomaly over centres of postglacial rebound

Figure 3.14 shows maps of the free air gravity anomalies over the present-day centres of postglacial rebound in Laurentia (Fig. 3.14a) and Fennoscandia (Fig. 3.14b) respectively. Comparison of these maps with those for ice thickness at glacial maximum (Fig. 3.10) demonstrates a high degree of correlation between these two fields and provides strong support for the hypothesis that the observed free air anomalies are to be interpreted as measures of the currently existing degree of isostatic disequilibrium in these two regions. Figure 3.15 shows a comparison of observed and predicted

Figure 3.14: Observed free air gravity anomalies in milligals for (a) Laurentia and (b) Fennoscandia

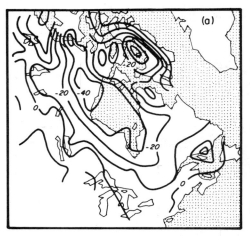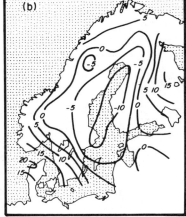

Figure 3.15: Observed (stippled) and predicted peak present-day free air gravity anomalies for Laurentia and Fennoscandia in models with a fixed upper mantle viscosity of 10^{21} Pa s as a function of the lower mantle viscosity. The predictions denoted □ are for a model with uniform mantle density (an adiabatic model), △ is for a model with a 6.2 per cent increase of density below 670 km depth, + is for a model with a 12.4 per cent increase at 670 km depth, while ▽ is for a model with a 6.2 per cent increase at 670 km and a 3.8 per cent increase at 420 km. The predictions denoted × and ○ are respectively for the seismically realistic models 1066B and PREM, in which the entire radial variation of density is treated as non-adiabatic

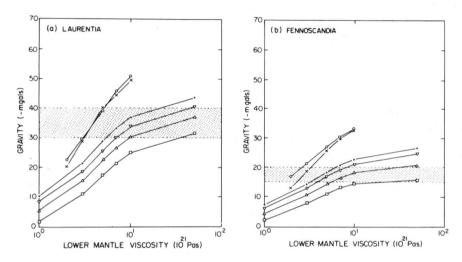

present-day peak free air anomalies for Laurentia and Fennoscandia, for a number of earth models having fixed lithospheric thickness of 120.7 km and an upper mantle viscosity of 10^{21} Pa s, as required by the sea-level data discussed previously. The earth models employed differ from one another only in terms of their elastic structures, and lower mantle viscosity is varied through the same sequence of values for all models. As described in the figure caption, four of the models are flat homogeneous layered approximations to the 1066B elastic structure. These have either 1 or 2 internal discontinuities of elastic parameters within the mantle at 420 km and/or 670 km depth, corresponding to the depths of the Olivine–Spinel and Spinel–Post Spinel phase boundary horizons. The remaining two curves are for the seismically realistic models 1066B of Gilbert and Dziewonski (1975) and PREM of Dziewonski and Anderson (1981). In these models all of the radial variation of density is treated as though it were non-adiabatic. Inspection of these results shows that to fit the observed free air gravity anomalies with a model with weak radial variation of viscosity, requires the presence in the model of significant internal buoyancy,

81

associated with a density structure which behaves non-adiabatically on the timescales of glacial rebound. This requires at least that the density variations across the phase boundaries at 420 and 670 km depth behave in this fashion, which is possible only to the extent that these transitions may be considered univariant (e.g. O'Connell, 1976; Mareschal and Gangi, 1977). Although such behaviour is not inconceivable, it may also prove possible to reconcile these data by appealing to other physical effects. This is clearly an extremely important issue in so far as the problem of mantle convection is concerned (Peltier, 1985b).

Pleistocene deglaciation and earth rotation

From about 1900 AD until 1982 the location of the earth's north pole of rotation was carefully monitored by the International Latitude Service (ILS), using a global network of photo-zenith tube equipped observatories. Since 1982 this observing system has been replaced by the much more accurate VLBI-based network of the IRIS earth orientation monitoring system, which routinely determines the pole position at 5-day intervals with a verified accuracy of 2 milliseconds of arc (Carter and Robertson, 1986). Although these new data will quickly replace the old as the industry standard, the duration of the time series is still sufficiently short that the ILS data remain the best source of information on the secular motion of the pole. These data are shown in Figure 3.16 as x and y components of the displacement relative to the axes shown on the inset polar projection. The origin of the co-ordinate system corresponds to the Conventional International Origin or CIO. Inspection of these data, which are based upon the reduction by Vincent and Yumi (1969, 1970), demonstrates that the dominant oscillatory signal, which consists of a 7-year periodic beat generated by the interference between the 14-month Chandler and 12-month Annual wobbles, is superimposed upon a secular drift at the rate of $0.95° \pm .15°$ Ma^{-1} towards Hudson Bay. The direction of this drift is shown by the arrow on the inset polar projection (Fig. 3.16).

In 1952 Munk and Revelle interpreted this observed secular drift of the rotation pole as requiring some present-day variation of surface mass load, and suggested that the cause of the apparently required variation might be found in melting of ice on Greenland and/or Antarctica. Their inference that such an effect was required to explain the data was, however, based upon a dynamical model. Here it was assumed that the earth could be treated as an homogeneous viscoelastic sphere, in so far as its rotational response to surface loading was concerned. To the extent that this approximation is valid the inference of Munk and Revelle is completely correct, since the theory then shows that the pole must be fixed at any instant of time in which the surface load is steady. As first demonstrated in Peltier

Figure 3.16: Polar motion data from the ILS records 1900-77, with the displacement shown as x and y components relative to a co-ordinate system with origin at the CIO (conventional international origin). The dominant oscillatory signal is due to the interference between the 12 month Annual and 14 month Chandler wobbles. This is superimposed upon a secular drift in the direction denoted by the arrow on the inset polar projection, where the locations of the major ice masses at 18 ka BP are shown stippled. This secular drift has been shown to be exactly that which is expected due to the disintegration of the Northern Hemisphere ice sheets

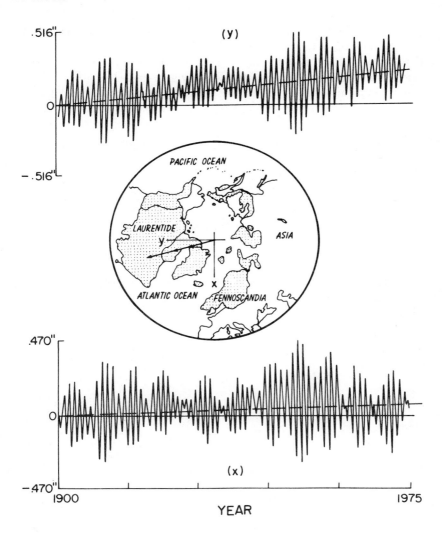

(1982) this holds true for homogeneous earth models, since in this limit the isostatic adjustment and rotational contributions to the rotational forcing counteract one another exactly. For radially stratified models, however, the dynamical symmetry which underlies this cancellation is broken and polar wander can occur even at a time when the surface load is steady. It therefore becomes plausible that the secular drift of the rotation pole shown on Figure 3.16, could simply be an effect due to the influence of planetary deglaciation, which began at ~ 18,000 BP and ended at about 7,000 BP. In Peltier (1982), Peltier and Wu (1983), Wu and Peltier (1984) and Peltier (1984) it was demonstrated that both the observed rate and direction of polar drift are just those to be expected if the earth has the viscoelastic stratification required by the previously discussed relative sea-level and free air gravity data, and if the only forcing to which the system has been subject is that due to a glaciation-deglaciation cycle which ended ~ 7,000 BP.

Figure 3.17 illustrates the nature of the fit to the observed polar wander

Figure 3.17: Predictions of polar wander speed for six different parameterisations of the radial viscoelastic structure. The observed speed is shown as the cross

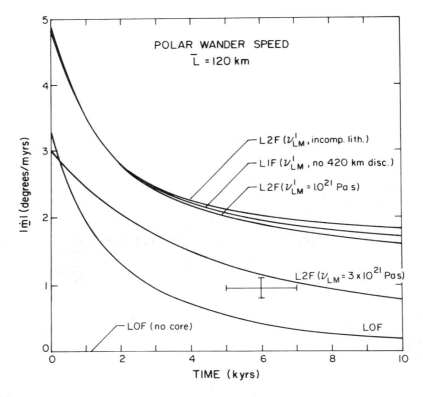

speed as a function of the viscoelastic model employed to make the prediction. Observations of oxygen isotope composition in deep-sea sedimentary cores (e.g. Fig. 3.8) are employed to constrain the cyclic variation of planetary ice cover which has occurred over the past 1 Ma. These data demonstrate that the major continental ice masses have appeared and disappeared in a highly periodic fashion, with a time interval of 100 ka separating successive interglacials (Fig. 3.8). Individual glaciation pulses in the sequence are each observed to have a saw-tooth form, with a slow glaciation period lasting ~ 90 ka followed by a fast collapse lasting 10 ka. The calculations illustrated in Figure 3.17 are based upon the assumption that seven such cycles have occurred and the observed polar wander speed of ~ 1° Ma^{-1} is shown at a time 6 ka following the last 10 ka disintegration event. Again this choice gives a best fit of the simple glaciation history model to the $\delta^{18}0$ data. Polar wander speed predictions are shown on this figure for six different viscoelastic models of the interior, each of which has the same lithospheric thickness $L = 120$ km. The prediction denoted by L0F (no core) is for the homogeneous model and verifies that the predicted speed following the end of the last deglaciation phase of the load cycle is essentially identical to zero, in accord with the theory of Munk and Revelle (1952). As radial structure is added to the model, however, the symmetry which enforces the null response in the homogeneous model is broken and the speeds predicted for times subsequent to the last glacial-deglacial pulse differ from zero. The effect of adding an inviscid high density core to the model is illustrated by the calculation denoted L0F, for which the elastic structure of the mantle is taken to be the average of model 1066B and the mantle viscosity is assumed to have the value 10^{21} Pa s. Model L1F includes in addition the influence of the density jump at 670 km depth in the earth, based upon the assumption that this discontinuity is capable of inducing a buoyant restoring force when it is displaced from equilibrium by the applied surface loads. The effect of this internal buoyancy in the mantle is further to increase the speed prediction in the model with 10^{21} Pa s uniform mantle viscosity. Adding a second density discontinuity at 420 km depth (model L2F with uniform viscosity) does not, however, produce a significant further increase in the speed prediction. The final calculation illustrated on Figure 3.17 (denoted L2F ($\nu_{LM} = 3 \times 10^{21}$ Pa s) demonstrates that the predicted speed can be reduced to the observed speed, simply by elevating the viscosity of the lower mantle to the same value required by the free air gravity and relative sea-level data discussed previously. As discussed in Wu and Peltier (1984) this model also correctly predicts the observed direction of polar wander. These results establish that the data shown in Figure 3.16 cannot be construed as requiring any currently ongoing variation in surface load, due to ice sheet disintegration on Greenland/Antarctica. Rather they are entirely explicable as a planetary memory of the last deglaciation event.

The final section of this chapter will focus, using the earth model whose properties have been fixed through analysis of the signatures of glacial isostasy discussed above, on the question of whether there is any evidence in modern tide gauge records of relative sea-level change for the operation of a current eustatic variation of sea level, due for example to ice sheet melting or thermal expansion.

SECULAR VARIATIONS OF RELATIVE SEA LEVEL WITH GLACIAL ISOSTASY REMOVED

It has been argued that the previously described global model of glacial isostasy is able to explain much of the observed variability in the record of relative sea-level change over the past 10 ka. Consequently it can be used to filter from the recent historical record of tide gauge observations of secular sea-level change that component which is due to this cause. One simply employs the data on the long timescale records of relative sea-level history (controlled by ^{14}C dating) to constrain the viscous component of mantle rheology. One then employs this earth structure to predict the present-day rate of sea-level rise/fall which should be observed at any location at which a tide gauge is installed, subtracts this prediction from the secular trend observed on the tide gauge and analyses the filtered data produced.

As a preliminary to this procedure it will be useful first to illustrate the continent scale variability of present-day relative sea-level variation predicted by the isostatic adjustment model. Figure 3.18 shows the present-day rate of sea-level rise/fall predicted for (a) North America and (b) Northwestern Europe, using an earth model with 1066B elastic structure. The viscous component of the model is one which has a lithospheric thickness of 200 km, an upper mantle viscosity of 10^{21} Pa s and a lower mantle viscosity of 2×10^{21} Pa s. This model provides a reasonably good fit to the ^{14}C record of relative sea-level rise along the US east coast when employed in conjunction with the ICE-2 deglaciation history of Wu and Peltier (1983) (Fig. 3 10). The meaning of the rates shown in the continental interior (where no ocean exists!) is that they represent the rates of separation between the surface of the solid earth and the geoid at such locations. Here the geoid is an imaginary surface continued inland from the oceans and on which the gravitational potential has the same value as obtains on the sea surface. Notable on these maps is the fact mentioned above, that the present-day maximum rates of relative sea-level fall in the regions which were once ice covered are near 1 cm a^{-1}. Surrounding each of the two main northern hemisphere centres of postglacial rebound, however, are ring-shaped regions in which relative sea-level is predicted to be rising at rates which may be as high as 2 mm a^{-1}; a maximum which obtains

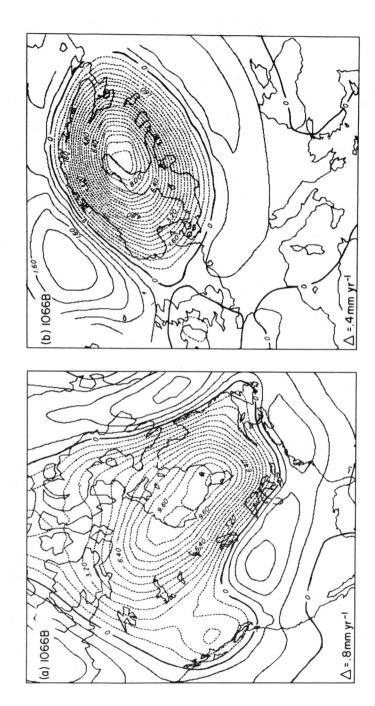

Figure 3.18: Predicted present-day rates of relative sea-level rise (solid contours) and fall (dashed contours) for both (a) North America and (b) Europe, based upon an earth model with 1066B elastic structure, a lithospheric thickness of 200 km, an upper mantle viscosity of 10^{21} Pa s and a lower mantle viscosity of 2×10^{21} Pa s. For each map the contour interval $\triangle$, in mm a^{-1}, is shown in the lower left-hand corner

along the passive continental margin of the US east coast. The variation with position along the coast is fairly extreme, however, with very low values obtaining both to the north of the maximum and to the south. Notable also on this map is the fact that the predicted rates of glacial isostatic submergence along the US west coast are rather different from those on the east coast. The former region is much further distant from the main Laurentian ice mass and is also quite strongly influenced by the separate Cordilleran ice sheet, which existed west of the Canadian Rocky Mountains. As a consequence of these effects, the predicted variations of the rate of present-day relative sea-level change along this active continental margin are more complex.

Examination of Figure 3.18b illustrates a similar degree of complexity of the pattern of present-day RSL change predicted by the model for Northwestern Europe. The maximum present-day rates of relative sea-level fall near the centre of uplift in the Gulf of Bothnia is again near 1 cm a^{-1}. Surrounding this central region of uplift is the region of peripheral submergence. Again the latter region is very strongly asymmetric with respect to the former, just as in North America, a consequence of the geometric complexity of the distribution of water and land. One important feature of these results is the fact that rates of present-day sea-level rise are predicted to be much lower along the coast of France, which extends to the southwest away from the centre of uplift in Fennoscandia, than those along the east coast of the US which is similarly located with respect to the larger Laurentian ice mass.

Perhaps the most interesting aspect of these model predictions of the rates of present-day relative sea-level variation, induced by the last deglaciation event, is that they may be compared to direct tide gauge observations of these rates at any point of interest on the earth's surface. By way of illustration Figure 3.19 summarises the comparisons for all tide gauge sites on the east and west coasts of the continental US. The tide gauge observations employed in this figure consist of the secular trends extracted from the individual time series of observations at each gauge over the time interval 1940-80, as published in the recent catalogue of the National Ocean Service (1983). On Figure 3.19a, US east coast data, each cross represents the secular trend at a specific gauge corrected for the secular trend expected due to the influence of glacial isostatic adjustment. These corrected data are plotted as a function of the distance (in radians) of each gauge from the station at Key West, Florida, which is the southernmost station along the coast. The northernmost data point is for the gauge at Eastport, Maine. Also shown on this figure is the secular drift which has been subtracted from the raw data to make the correction for isostatic disequilibrium. This correction attains a maximum near 2 mm a^{-1} about midway along the coast, with very small rates obtaining at sites in Florida and Maine to the south and north of the maximum respectively. To give

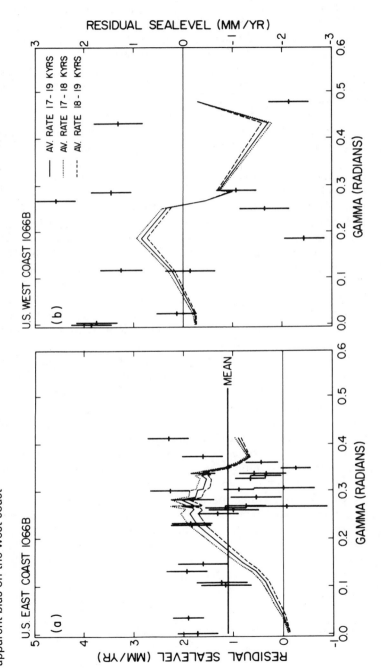

Figure 3.19: Tide guage observed rates of relative sea-level rise, corrected for the influence of glacial isostatic adjustment, are shown as the dark crosses as a function of distance in radians measured from the southernmost tide gauge along the US east (a) and west (b) coasts. The three lines show the correction which was applied to remove the isostatic effect at each gauge. Note that the residual is biased to positive values along the east coast whilst there is no apparent bias on the west coast

89

some indication of the error in the model predicted rates the predictions are shown not only for the present, 18 ka after glacial maximum, but also for 17.5 ka and 18.5 ka. Noticeable on these predicted curves are the several locations at which the predicted rates deviate from the smooth variation of rate with distance which otherwise obtains. These are sites at which errors due to the finite element discretisation occur (see Wu and Peltier, 1983, for a detailed discussion of the finite element discretisation employed to solve the sea-level equation).

The main point to note by inspection of the data (Fig. 3.19a) is that the correction, due to the influence of glacial isostasy, of the secular rate of relative sea-level rise along the US east coast accounts for as much as 100 per cent of the observed rate of rise at some sites. That is, at some sites along the US east coast, *all* of the observed secular rate of rise is explicable in terms of the influence of glacial isostatic adjustment. In general, however, there *is* a systematic misfit between the predicted rate of rise due to glacial isostasy and the rate of rise observed on an individual tide gauge, such that the observed rate of rise exceeds that predicted by the glacial isostatic adjustment model which fits the long timescale ^{14}C controlled relative sea-level histories. At no tide gauge along the US east coast is the observed rate of rise significantly slower than that predicted by the adjustment model. The average of the reduced rates of RSL rise at sites in this region is also shown on Figure 3.18a and is near 1.1 mm a^{-1}. The interpretation of this average in terms of a eustatic increase of sea level, either of steric or non-steric origin, is clearly made rather difficult by virtue of the fact that the residual trends vary erratically as a function of distance along the coast. Other physical processes must be contributing substantially to the secular variations of RSL which individual gauges are recording.

An identical treatment of the tide gauge data from the US west coast is shown on Figure 3.19b. Here the corrected tide gauge observed rates of RSL rise are shown as a function of position measured positive north from the southernmost gauge, which in this case is the one located at San Diego. The observations along this active continental margin are much more erratically scattered than those along the passive east coast continental margin, demonstrating the severe contamination of this record (for our present purposes) presumably as a result of local tectonic activity. Furthermore, the corrected data are scattered about zero and show no systematic bias toward positive rates as is evident for east coast sites. It would clearly be unreasonable to employ these data to make any inference whatsoever concerning eustatic sea-level variations.

CONCLUSIONS

In the previous sections of this chapter a number of the mechanisms responsible for producing changes in relative sea level have been discussed. The results of these changes have been classified as eustatic (deriving from the sea) or isostatic (deriving from the land). Although many changes in relative sea level are dominantly of either one type or the other, there do exist mechanisms which simultaneously induce both types of variation. The most important of these is that associated with the growth and decay of the vast continental ice sheets, whose appearance and disappearance has been a durable feature of the record of Pleistocene climatic change. Not only are there large eustatic variations of sea level associated with the transfer of mass from the oceans to the ice sheets and back again, but there are pronounced isostatic variations associated with the deformation of the earth induced by surface loading. These deformations in turn comprise both glacial isostatic (deriving from the ice) and hydro-isostatic (deriving from the water) contributions. Discussion in the chapter has tried to demonstrate the way in which a geophysical theory of the complex gravitational interactions, which ultimately determine postglacial relative sea-level change, can be exploited to infer important properties of the earth's interior on the basis of relative sea-level observations. These include not only aspects of the radial variation of viscosity, such as the thickness of the surface lithosphere and the value of the sub-lithospheric viscosity, but also the extent to which the density variation with depth in the mantle is adiabatic or non-adiabatic. Further, an attempt has been made to show how this model can be employed to filter deglaciation induced relative sea-level variations from modern tide gauge records of secular sea-level change, and so enable us to see more clearly any other climatological signal which may be present in these records. A great deal is left to be done before we will be in any position to claim a complete understanding of the full spectrum of sea-level variability. However, the record has already yielded an impressive collection of insights into the way in which the earth works.

REFERENCES

Barnett, T.P. (1983a) 'Possible changes in global sea level and their causes', *Climate Change, 5(1)*, 15-38.
—— (1983b) 'Long term changes in dynamic heights', *J. Geophys. Res., 88*, 9547-52.
Bloom, A.L. (1967) 'Pleistocene shorelines: A new test of isostasy', *Bull. Geol. Soc. Am., 78*, 1477-93.
—— (1970) 'Paludal stratigraphy of Truk, Ponape, and Kusaie, Eastern Caroline Islands', *Bull. Geol. Soc. Am., 81*, 1895-904.

—— (1983) 'Sea-level and coastal morphology through the Late Wisconsin glacial maximum', in S.C. Porter (ed.), *Late Quaternary Environments of the United States, Vol. 1 — The Late Pleistocene*, Longman, London, pp. 215-29.

Carter, W.E. and Robertson, D.S. (1986) 'Earth rotation from VLBI measurements', in A.J. Anderson and A. Cazanave (eds), *Space Geodesy and Geodynamics*, Academic Press, London and New York, pp. 85-96.

—— Robertson, D.S., Pyle, T.E. and Diamante, J. (1986) 'The application of geodetic radio interferometric surveying to the monitoring of sea level', *Geophys. J. Roy. Astron. Soc., 87*, 3-13.

Clark, J.A., Farrell, W.E. and Peltier, W.R. (1978) 'Global changes in postglacial sea level: A numerical calculation', *Quat. Res., 9*, 265-87

Cox, A. and Dalrymple, G.B. (1967) 'Statistical analysis of geomagnetic reversal data and the precision of potassium-argon dating', *J. Geophys. Res., 72*, 2603-14.

Csanady, G.T. (1982) *Circulation in the Coastal Ocean*, Reidel, Dordrecht.

Dziewonski, A.M. and Anderson D.L. (1981) 'Preliminary reference earth model', *Phys. Earth Planet. Int., 25*, 297-356.

Fairbridge, R.W. (1976) 'Shellfish-eating pre-ceramic Indians in coastal Brazil', *Science, 191*, 353-9.

Farrell, W.E. and Clark, J.A. (1976) 'On postglacial sea level', *Geophys. J. Roy. Astron. Soc., 46*, 647-67.

Geodynamics Program Office (1983), *The NASA Geodynamics Program, an Overview*, NASA Technical Paper No. 2147.

—— (1984), *NASA Geodynamics Program: Fifth Annual Report*, NASA Technical Memorandum, 87359

Gilbert, F. and Dziewonski, A.M. (1975) 'An application of normal mode theory to the retrieval of structural parameters and source mechanisms from seismic spectra', *Phil. Trans. Roy. Soc. Lond., A, 276*, 187-269.

Gill, E.D. (1965) 'Radiocarbon dating of past sea levels in SE Australia', *Abstracts, INQUA VII Congress*, Boulder, Col., p. 167.

Gornitz, V., Lebedeff, L. and Hansen, J. (1982) 'Global sea level trend in the past century', *Science, 215*, 1611-14.

Hansen, J., Johnson, D., Lacis, A., Lebedeff, S., Lee, P., Reid, D. and Russell, G. (1981) 'Climate impact of increasing atmospheric carbon dioxide', *Science, 213*, 957-66.

Hays, J.D., Imbrie, J. and Shackleton, N.J. (1976) 'Variations in the earth's orbit: Pacemaker of the ice ages', *Science, 194*, 1121-32.

Hyde, W.T. and Peltier, W.R. (1985). 'Sensitivity experiments with a model of the ice age cycle: the response to harmonic forcing', *J. Atmos. Sci.*, (September).

Imbrie, J., Van Donk, J. and Kipp, N.G. (1973) 'Paleoclimatic investigation of a Late Pleistocene Caribbean deep-sea core: Comparison of isotopic and faunal methods', *Quat. Res. (NY, 3*, 10-38.

Imbrie, J., Shackleton, N.J., Pisias, N.G., Morley, J.J., Prell, W.L., Martinson, D.G., Hays, J.D., McIntyre, A. and Mix, A.C. (1984) 'The orbital theory of Pleistocene climate: Support from a revised chronology of the marine $\delta^{18}O$ record', in A. Berger, J. Imbrie, J. Hays, G. Kukla and B. Saltzman (eds), *Milankovitch and Climate*, Reidel, Dordrecht, vol. I. pp. 269-305.

Jelgersma, S. (1966) 'Sea level changes during the last 10,000 years', in *Proceedings of the International Symposium on World Climate from 8000 to 0 BX*, Royal Meteorological Society, London, pp. 54-71.

Lagios, E. and Wyss, M. (1983) 'Estimates of vertical crustal movements along the coast of Greece, based upon mean sea level data', *PAGEOPH, 121*, 869-87.

Lowe, J.J. and Walker, M.J.C. (1984) *Reconstructing Quaternary Environments*, Longman, London.

Mareschall, J.-C and Gangi, A.F. (1977) 'Equilibrium position of phase boundary under horizontally varying surface loads', *Geophys. J. Roy. Astron. Soc., 49*, 757-72.

Marsh, J.G. and Martin, T.V. (1982) 'The SEASAT altimeter mean sea surface model', *J. Geophys. Res., 87*, 3269-80.

Meier, Mark F. (1984) 'Contribution of small glaciers to global sea level', *Science, 226*, 1418-21.

Milankovitch, M. (1941) *Canon of Insolation and the Ice-Age Problem*, K. Serb. Akad. Geogr., Spec. Publ. No. 132, translated by Israel Program for Scientific Translations, Jerusalem, 1976, US Department of Commerce.

Munk, W.H. and Revelle, R. (1952) 'On the geophysical interpretation of irregularities in the rotation of the Earth', *Mon. Not. Roy. Astron. Soc., Geophys. Suppl., 6*, 331-47.

National Ocean Service (1983) *Sea Level Variations for the United States 1855-1980*, US Department of Commerce, National Oceanic and Atmospheric Administration, Rockville, Md.

Noble, M. and Butman, B. (1979) 'Low frequency wind induced sea level oscillations along the east coast of North America', *J. Geophys. Res., 84*, 3227-36.

O'Connell, R.J. (1976) 'The effects of mantle phase changes on postglacial rebound', *J. Geophys. Res., 81*, 971-4.

Peltier, W.R. (1974) 'The impulse response of a Maxwell Earth', *Rev. Geophys. Space Phys., 12*, 649-69

—— (1976) 'Glacio-Isostatic adjustment–II. The inverse problem', *Geophys. J. Roy. Astron. Soc., 46*, 669-706.

—— (1980) 'Mantle convection and viscosity', in A.M. Dziewonski and E. Boschi (eds), *Physics of the Earth's Interior*, North Holland, Amsterdam, pp. 362-431.

—— (1981) 'Ice age geodynamics', *Ann. Rev. Earth Planet. Sci., 9*, 199-225.

—— (1982) 'Dynamics of the Ice Age Earth', *Adv. Geophys., 24*, 1-146.

—— (1983) 'Constraint on deep mantle viscosity from LAGEOS acceleration data', *Nature, 304*, 434-6.

—— (1984a) 'The rheology of the planetary interior', *Rheology, 28*, 665-97.

—— (1984b) 'The thickness of the continental lithosphere', *J. Geophys. Res., 89*, 11,303-16.

—— (1985a) 'The LAGEOS constraint on deep mantle viscosity: results from a new normal mode method for the inversion of viscoelastic relaxation spectra', *J. Geophys. Res., 90*, B11, 9411-21.

—— (1985b). 'Mantle convection and viscoelasticity', *Ann. Rev. Fluid. Mech., 17*, 561-608.

—— and Andrews, J.T. (1976) 'Glacial isostatic adjustment I: The forward problem', *Geophys. J. Roy. Astron. Soc., 46*, 605-46.

—— Farrell, W.E. and Clark, J.A. (1978) 'Glacial isostasy and relative sea level: a global finite element model', *Tectonophys., 50*, 81-110.

—— Wu, Patrick and Yuen, D.A. (1981) 'The Viscosities of the planetary mantle', in F.D. Stacey, A. Nicholas and M.S. Paterson (eds), *Anelasticity in the Earth*, American Geophysical Union, Washington, DC.

—— and Wu, Patrick (1982) 'Mantle phase transitions and the free air gravity anomalies over Fennoscandia and Laurentia', *Geophys. Res. Lett., 9*, 731-734.

—— and Wu, Patrick (1983). Continental lithospheric thickness and deglaciation induced true polar wander. *Geophys. Res. Lett., 10*, 181-4.

—— and Hyde, W.T. (1984) 'A model of the ice age cycle', in A. Berger, J.

93

Imbrie, J. Hays, G. Kukla and B. Saltzman (eds), *Milankovitch and Climate*, Reidel, Dordrecht, vol. II, pp. 565-80.

Rapp, R.H. (1979) 'Geos 3 data processing for the recovery of geoid undulations and gravity anomalies' *J. Geophys. Res.*, *84*, 3784-92.

Roemmich, D. and Wunsch, C. (1984) 'Apparent changes in the climatic state of the deep North Atlantic Ocean', *Nature*, *307*, 447-50.

Russell, R.J. (ed.) (1961) 'Pacific Island Terraces: Eustatic?' *Zeit. Geomorph. Suppl.*, *3*.

Sabadini, R. and Peltier, W.R. (1981) 'Pleistocene deglaciation and the earth's rotation: implications for mantle viscosity', *Geophys. J. Roy. Astron. Soc.*, *66*, 552-78.

Schofield, J.C. (1964) 'Post-glacial sea levels and isostatic uplift', *NZ J. Geol. Geophys.*, *7*, 359-70.

Schutz, E.B., Tapley, B.D. and Shum, C. (1982) 'Evaluation of the SEASAT altimeter time tag bias', *J. Geophys. Res.*, *87*, 3239-45.

Shackleton, N.J. and Opdyke, N.D. (1973) 'Oxygen isotope and paleomagnetic stratigraphy of equatorial Pacific core V28-238: Oxygen isotope temperatures and ice volumes on a 10 to 10^6 year timescale', *Quat. Res.*, *3*, 39-54.

—— and Opdyke, N.D. (1976) 'Oxygen isotope and paleomagnetic stratigraphy of Pacific core V28-239 late Pleistocene to latest Pleistocene', *Mem. Geol. Soc. Am.*, *145*, 449-64.

Shepard, F.P. (1963) 'Thirty-five thousand years of sea level', in T. Clements (ed.), *Essays in Marine Geology*, University of Southern California Press, Los Angeles, Calif., pp. 1-10.

Stanley, H.R. (1979) 'The Geos 3 project', *J. Geophys. Res.*, *84*, 3779-83.

Thompson, K.R. (1981) 'Monthly changes of sea level and the circulation of the North Atlantic', *Ocean Modelling*, *41*, 6-9.

Vincent, R.O. and Yumi, S. (1969, 1970) 'Co-ordinates of the pole (1899-1968), returned to the conventional international origin', *Publ. Int. Latitude Observ. Mizusawa*, *7*, 41-50.

Weertman, J. (1978) 'Creep laws for the mantle of the Earth', *Phil. Trans. Roy. Soc. Lond.*, *A288*, 9-26.

Wu, Patrick, and Peltier, W.R. (1983) 'Glacial isostatic adjustment and the free air gravity anomaly as a constraint on deep mantle viscosity', *Geophys. J. Roy. Astron. Soc.*, *74*, 377-449.

—— and W.R. Peltier (1984) 'Pleistocene deglaciation and the Earth's rotation: a new analysis', *Geophys. J. Roy. Astron. Soc.*, *76*, 753-92.

Wunsch, C. (1981) 'An interim relative sea surface for the North Atlantic ocean', *Mar. Geodesy*, *5*, 103-19.

—— and Gaposhkin, E.M. (1980) 'On using satellite altimetry to determine the general circulation of the oceans with applications to geoid improvement', *Rev. Geophys. Space Phys.*, *18*, 725-45.

Wyrtki, K. and Nakahara, S. (1984) *Monthly Maps of Sea Level Anomalies in the Pacific 1975-1981*, Hawaii Institute of Geophysics Report HIG-84-3.

Wyss, M. (1976a). 'Local sea level changes before and after the Hyuganada, Japan earthquakes of 1961 and 1968', *J. Geophys. Res.*, *81*, 5315-21.

—— (1976b) 'Local changes of sea level before large earthquakes in South America', *Bull. Seis. Soc. Am.*, *66*, 903-14.

4

Glaciation and Sea Level: A Case Study

John T. Andrews

INTRODUCTION

The topic 'Glaciation and Sea Level' potentially covers a wide range of problems. There is, for example, the vexed question of the actual magnitude of the volume of water retained at the maximum of the Late Quaternary ice sheets: was 'global' sea level lowered by 80 m, 100 m, or even 160 m? At the moment earth scientists simply have been unable to agree on a figure that fits the variety of methods that have been used to reconstruct past sea levels (see earlier discussion in Peltier, this vol., Ch. 3). This is unfortunate as it means we cannot yet use sea-level data to constrain glaciological/geological models of the ice sheets (e.g. of shape and size) and decide between maximum and minimum models of Northern Hemisphere glaciation (Miller and Dyke, 1974; Denton and Hughes, 1981; Andrews and Miller, 1984). Although this is a vital topic, if not *the* vital topic, in current Quaternary studies (e.g. Andrews, 1982a; Aharon, 1983) this chapter will concern itself more with the regional and local changes of sea level in the vicinity of the margins of former ice sheets.

The association between glaciation and changes of local sea level has been noted for over a century by field workers in Scandinavia, Britain, Greenland, Iceland and various areas of North America (see Devoy, this vol., Ch. 10). In a similar fashion, interpretations of glacially induced changes of local sea level have been used to infer rheological properties of the earth's interior for many decades. This platform of early observations and theory has been substantially added to in the last two to three decades by two major developments. The first was the application of radiocarbon dating to an accurate portrayal of local changes of sea level from sites within the maximum extent of the Late Quaternary ice sheets, as well as sites peripheral to the ice sheets. The second major development has been the joint advent of high-speed computers and suitable algorithms to take the results of local sea-level changes and incorporate these into models of earth rheology that operate at a global scale (Peltier, 1976; Peltier and

Andrews, 1976, 1983; Clark *et al.*, 1978; see Peltier, this vol., Ch. 3). Fortunately, in a personal sense, these two processes have not proceeded as individual enterprises but rather as a series of research questions that have shifted focus between the field observations and the content of the models.

This chapter will seek to reiterate some of the results that have been obtained on the interplay between glaciation and sea level for the Canadian Arctic. In addition it will examine the importance of changes of sea level on the inferred deglaciation history of the Laurentide Ice Sheet (Denton and Hughes, 1981; Ruddiman and McIntyre, 1981) and thus, in part, query the proposed link between global, eustatic sea level and glacial history that is suggested as a control on both Southern and Northern Hemisphere glaciation (Denton and Hughes, 1981, 1983). In this latter exercise an attempt will be made to assess the results of glacial isostatic modelling (discussed more specifically in Chapter 3) as well as proposing field tests for the proposed history of sea level. This chapter thus sets itself the task of reviewing and commenting on work that has largely been published. A final commentary in the chapter will evaluate the relationship between 'glacial divides' and glacial mass. This is a critical issue because the reconstruction of Holocene isobases on glacial isostatic recovery are frequently used to infer the position of *ice divides*. It will be argued that it is essential to keep mentally separate the two ways of reconstructing ice sheets: the one based on evidence of glacial flow (striations, erratics, etc.), the other based on glacial isostatic recovery measured by changes in local sea levels across the formerly glaciated region.

GLACIATION AND SEA LEVEL: ARCTIC CANADA

Since the advent of ^{14}C dating, the study of raised marine beach and sediment sequences has been an important element in Quaternary studies in Arctic Canada (e.g. Løken, 1962, 1965; Blake, 1970, 1975, 1976; England, 1976, 1983; Dyke, 1974, 1979, 1983, 1984; Miller, 1980; Miller, *et al.*, 1977; Mode *et al.*, 1983; Nelson, 1982; Vincent, 1982; Andrews, 1970, 1975, 1980). Studies of the raised beach sequences have resulted in a knowledge of changes in relative sea level throughout the interval since deglaciation, and the age of the marine limit has been used as a fundamental control on the rate of deglaciation from the Late Wisconsin glacial maximum (Prest, 1969; Bryson *et al.*, 1969; Dyke, 1974). Because of the aridity of the arctic climate, shells, wood, plant detritus and bone are well preserved and although in several key areas (such as the northern coast of Labrador) shell-bearing sediments are rare, in many areas materials suitable for ^{14}C dating are abundant and are found in clear stratigraphic context (e.g. Cape Storm, Ellesmere Island (Blake, 1975)).

Throughout much of Arctic Canada the vast majority of ^{14}C dates on raised beach/sediment sequences are based on marine shells. The same situation is true in Spitsbergen (Salvigsen, 1981; Miller, 1982; Boulton *et al.*, 1982) and in Norway (Mangerud *et al.*, 1979). Detailed research into the veracity of ^{14}C dates based on marine shells is well evident in this literature and has been investigated repeatedly by the Geological Survey of Canada (see Geol. Surv. Can. Radiocarbon Data Lists). Based on this literature and personal experience the author has found no reason to doubt that ^{14}C dates on marine shells in the < 14,000 BP range provide reliable estimates of the age of the enclosing sediment (see discussion in Sutherland this vol., Ch. 16). The so-called reservoir effect (Mangerud, 1972) and problems associated with how different laboratories report marine shell ages (Stuiver and Polach, 1977) may lead, however, to discrepancies of between 400 to 700 years. In the context of 'Glaciation and Sea level' such differences are relatively small, but may be significant for some problems given the rapid rate of isostatic recovery upon deglaciation. For example, estimated recovery rates of 1 to 10 m a^{-100} translate into 4 to 70 m elevational differences based on the above factors.

We can divide glaciated regions here into two basic units in terms of our interest. The *first* is the area that was adjacent to, and distal to the margin of the Late Wisconsin Laurentide/Innutian/Franklin ice complex (Prest, 1984). The *second* area lies within these boundaries. In terms of glacial isostatic modelling (e.g. Clark *et al.*, 1978) the first zone is termed the Zone I/II transition, whereas the area within the border of the former ice sheets is within their Zone I sea-level response. Figure 4.1 represents a picture of the North American ice sheets at the Late Wisconsin glacial maximum (Prest, 1984) with the boundary between Zones I and II drawn in. A great deal of the current and past debate on the extent of glaciation in the Canadian Arctic turns out to hinge on the interpretation of the raised sea-level history and *not* on glacial stratigraphy *per se*. Thus the arguments among and between such authors as England, Blake, Grosswald (1983), Denton and Hughes, Andrews, Miller, and Dyke are frequently based on the *inferred* relationship between glacial load and local sea-level response, as well as partly on the interpretation of the age of features associated with glacial erosion, transportation and deposition.

Zone I

In sea-level Zone I (Fig. 4.1) there exists the 'classic' relationship between ice retreat and the formation of the marine limit (Bird, 1954; Sim, 1960; Andrews, 1970). This relationship is illustrated in Flint (1971: p. 353). In Zone I glacial isostatic recovery commences prior to deglaciation of the specific site and is associated with regional unloading linked to deglaciation. Thus within this zone ice marginal features can be traced unequi-

97

Figure 4.1: Basic divisions of the Laurentide ice sheet based on ice flow and erratic indicators. The figure also shows the approximate boundary between the major sea-level zones of Clark *et al.* (1978), i.e. Zones I, I/II transition, and II (see also Devoy, this vol., Ch. 10).

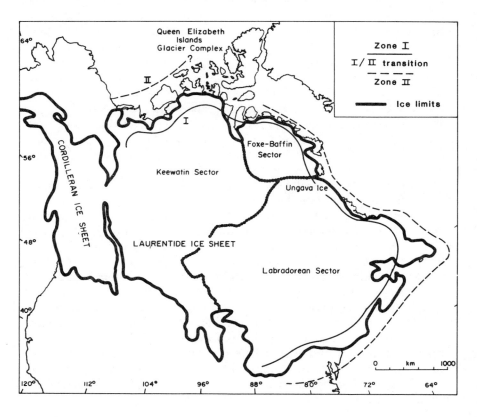

Source: After Prest (1984).

vocably into the highest marine feature — the marine limit (Løken, 1962, 1965; Andrews, 1966; Andrews *et al.*, 1970; Pheasant and Andrews, 1973; Miller, 1980). This mapping of the ice margin and the associated littoral/glacial marine facies provide a point in space and time for an associated particular sea-level height (above present) and ice margin. Figures 4.2 and 4.3 illustrate aspects of these relationships as shown by air photographs from Arctic Canada.

In Arctic Canada, and similarly in Norway, the change of relative sea level after ice withdrawal has frequently been graphed as a smoothly decelerating curve (Fig. 4.4) or, in the case of Fennoscandia, with the

98

Figure 4.2: Air photograph at the head of Cambridge Fiord, Baffin Island, NWT, showing the extensive raised outwash sediments in the southern valley and the more complex reworked fluvial deposits in the northern valley. The marine limit may be delimited by the terrace located on the bedrock shoulder. Abundant shells occur within forset beds but a radiocarbon chronology has not yet been adequately developed. Moraines of the late Foxe glacial maximum are located on the walls of the fiord (Air photo A-16295-19, Canadian Govt. Copyright)

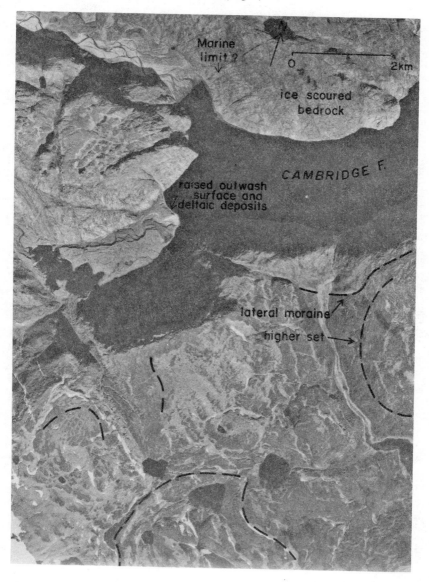

Figure 4.3: Ground photograph of lateral moraine and raised outwash/till/glacio-marine complex at the head of Omega Bay, Cambridge Fiord, Baffin Island, NWT. No specific marine deposits can be associated with the terminus of the lateral moraines. The marine limit is between 70 and 90 m asl

Figure 4.4: A series of Late Quaternary sea-level curves from Zone I (Fig. 4.1) including a curve from Norway (Lie *et al.*, 1983) and two from northern Canada

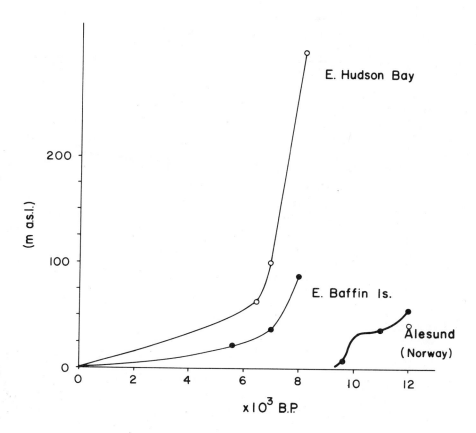

noticeable 'kink' or even marine incursion associated with a change of sea level during the later 'Litorina or Tapes transgressions' (e.g. Mörner, 1969; Lie *et al.*, 1983; see Devoy, this vol., Ch. 10 for further discussion of regional sea-level data). In this pattern of change, a question that has not been investigated in any rigorous sense in Arctic Canada, although this is not the case in Fennoscandia, is: do glacial readvances result in associated changes of relative sea level? The question is expressed graphically as Figure 4.5. In a different situation in British Columbia, Matthews *et al.* (1970) interpreted the field evidence such as to invoke a significant sequence of marine removal/inundation/removal, the middle event associated with the Sumas ice readvance. In Arctic Canada, deglaciation from the lateglacial maximum is dated at < 12,000 BP at virtually all sites.

101

Figure 4.5: Hypothetical relationship between changes of ice extent (left) and response of sea level at a site close to the ice margin. The short readvance is accompanied by a marine incursion associated with glacial loading

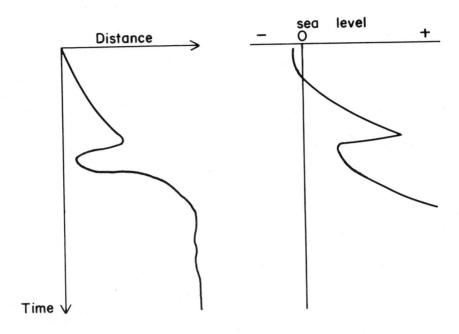

Deglaciation, as recorded by shells at or associated with the marine limit, was fairly rapid along the northwest and north margin but was significantly slower within the fiords of eastern Baffin Island (Andrews, 1972, 1982b). Evidence for major glacial readvances here is scarce and it is difficult to decide whether certain moraines were due to climatic factors and which may have been caused by a change in the style of deglaciation from marine contact to terrestrial (Andrews, 1972; Hillaire-Marcel *et al.*, 1981). In eastern Baffin Island, however, moraines formed during the Cockburn substage (Andrews and Ives, 1978) were often associated with actual readvances of the ice (Smith, 1966; Andrews *et al.*, 1970). We will return later to the question of the importance of sea level on glacial activity itself, but ... at this stage a question is simply being raised that has not been investigated in Arctic Canada (see Fig. 4.5). If we can determine (1) the extent of the glacial retreat/readvance and (2) the sea-level response, if any, we can provide geophysicists with valuable information on the response of the lithosphere to relatively short-lived events (as stated previously by Peltier, this vol., Ch. 3). Depending on regional variations in the thickness and properties of the lithosphere the local response of the crust to glacial readvances will provide constraints on lithosphere variability.

Although such detailed glacial–sea-level–lithostratigraphic investigations have not been carried out in Arctic Canada, they have been a subject of study in Norway and Sweden. In recent years a number of papers have appeared from the Norwegian research that are worth emulating in North America, Britain and other areas (Anundsen, 1978; Anundsen and Fjeldskaar, 1983; Lie *et al.*, 1983; Thomsen, 1982). The research strategy is simple, but none the less effective. Small, rock bounded basins are cored. Changes in sediment and more critically in diatom assemblages are then used to assess the type of former environment, that is: marine, brackish or freshwater. Thus a cycle of marine removal/inundation/removal (Fig. 4.6) would be represented in a lake core as a change in the diatom flora as follows: marine–brackish–freshwater–brackish–marine–brackish–freshwater. Alternatively, if the change of sea level with respect to the lake basin was slight, the change might only encompass the marine–brackish–marine–brackish–freshwater sequence. Intuitively we might expect that close to the ice margin the change in relative sea level might be slight or difficult to determine but, further from the margin, crustal depression associated with a readvance might be clearly recorded in critically located lake basins (Fig. 4.6). Thus we will return to this sort of investigation in the section of the sea-level Zone I/II transition.

Figure 4.6: A theoretical example of how coring lakes below the local marine limit might resolve subtle and/or short-lived changes in relative sea level (shown on right)

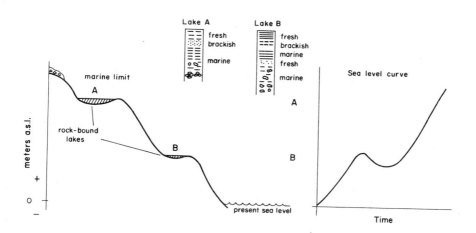

Zone I/II Transition

The area distal from the maximum position of the Late Wisconsin ice sheet for approximately 100 km is considered to represent the transition from Zone I to II sea-level response (Fig. 4.7). This transition area is one of the most fascinating to work in and, from a geophysical viewpoint, it is also an important region to gain an understanding of the relationship between glaciation and sea level. In the last decade or more, workers in Arctic Canada, Greenland and Spitsbergen have started to realise that the glaciation/sea-level response in the I/II transition is not easily interpreted (Andrews, 1978). A researcher commencing work in this zone, following a period of research in Zone I, would automatically associate the age of the marine limit with deglaciation from the lateglacial maximum. However, field studies, air photographic converage (Fig. 4.8) and a ^{14}C dating programme indicate the following: (1) the lowermost major beach/terrace is 'fresh'; (2) the sediments associated with it frequently overlie organic/soil layers' (3) above this clearly younger unit (1 and 2) are one or more older sets of marine deposits and (4) in many cases the uppermost unit, in a height sense, is associated with ice-contact deposits. Figures 4.8 and 4.9 illustrate some of these situations from eastern Baffin Island and western

Figure 4.7: Sketch of the trend of Late Quaternary sea level in the Zone I/II transition (Fig. 4.1)

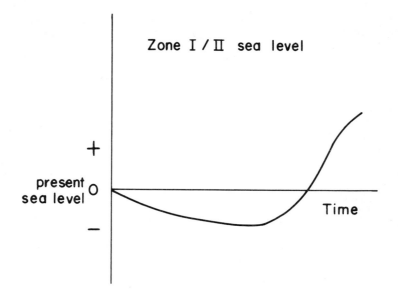

Figure 4.8: Air photograph from the outer coast of eastern Baffin Island north of Clark Fiord showing the upper limit of the early Holocene marine incursion (arrow). Above this limit (~ 15-25 m asl) are older marine and glacial marine deposits that are ^{14}C dated at >40,000 BP. (Air Photograph A-16301-30, Canadian Govt. Copyright.)

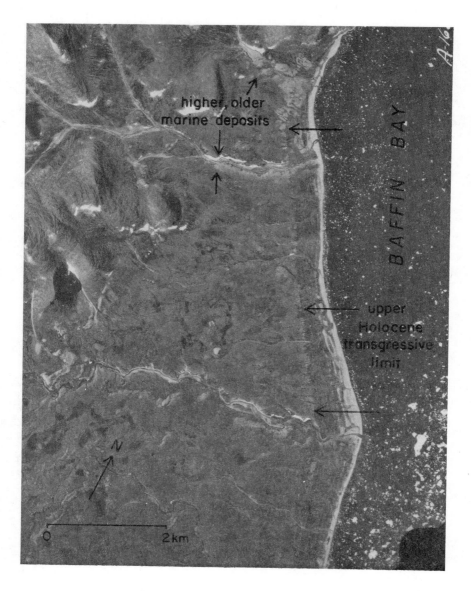

Figure 4.9: Series of raised beaches on western Spitsbergen (see Forman and Miller, 1984 for more detailed discussion). The limit of 'fresh' beaches, episode A, lies at 44 m asl, episode B deposits extend to 55 m whereas the highest marine sediments extend to the marine limit at ~ 78 m and are tentatively dated at >70,000 BP

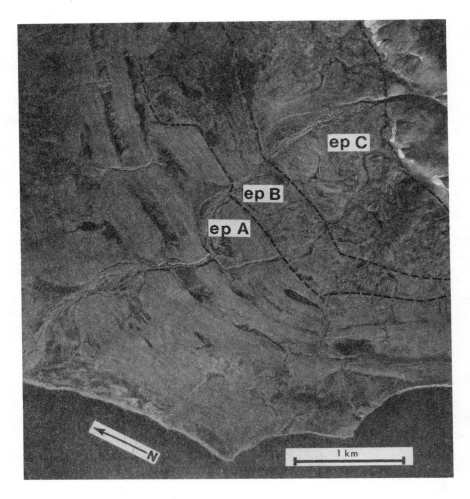

Spitsbergen (e.g. Miller *et al.*, 1977; Nelson, 1981, 1982; Brigham, 1983; Mode *et al.*, 1983; Miller, 1973; Boulton *et al.*, 1982; Forman and Miller, 1984), whereas Figure 4.10 sketches the stratigraphic relationships explicit in these areas (see references above). In the sense of glaciation/sea level relationships it is *vital* to note that the youngest major unit is a shallow-water littoral sediment that has no clear spatial link with an ice margin.

Figure 4.10: Diagram of the stratigraphic relationships seen and inferred from the areas illustrated as Figures 4.8 and 4.9. The influence of changes of relative sea level on the style of sedimentation are discussed in Andrews (1978), Boulton *et al.* (1982), and Mode *et al.* (1983)

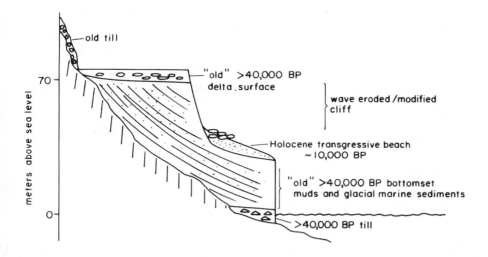

They are shallow-water distal facies, in a glacio-marine sense (Andrews and Matsch, 1983), often with a cold macro- and micro-fauna (Feyling-Hanssen, 1985; Miller, 1980). Along eastern Baffin Island, ^{14}C dates on shells from these sediments, or on the organics below the sediments, range in age from ~ 9,500 to 10,500 BP (Table 4.1). The ^{14}C age of the marine sediment that underlies this latest Pleistocene/Early Holocene sequence has been dated repeatedly at > 40,000 BP (Miller *et al.*, 1977; Miller, 1985; Brigham, 1983; Nelson, 1982). Although in places these sediments may be broadly of Mid-Wisconsin age the majority appear to be Early Wisconsin (marine isotope stage 5/4) in age (Miller, 1985; Klassen, 1982, 1985) based on amino acid ratios and other considerations.

With the exception of Frobisher Bay and Cape Hatt, northern Baffin Island (Miller, 1985; Klassen, 1985) ^{14}C dates on undoubted proximal glacial marine sediments associated with deposits of the Baffinland stage (Latest Wisconsin) are < 9,500 BP and most frequently < 9,000 BP (Falconer *et al.*, 1965; Miller and Dyke, 1974). However, to the west the marine limit over wide areas of Somerset Island and Boothia Peninsula (Dyke, 1983, 1984) date from ~ 9,300 BP.

A basic unknown in the sea-level history of Arctic Canada during the Late Quaternary was the position and trend of sea-level changes > 10,000

Table 4.1: ^{14}C dates associated with the interval of the 'Cape Adair Transgression'

Location	Site	^{14}C	Lab no.
? Loks Land	Shell	9,960 ± 230 (10,370 ± 230)	GSC-2752
Allen Island	Shell	9,230 ± 110 (9,640 ± 110)	GSC-2618
Kingnait Fiord	Shell	8,680 ± 160 (9,090 ± 190)	GSC-2478
Qivitu Foreland	Peat	9,950 ± 185	QC-453
Qivitu Foreland	Shell	9,280 ± 120 (9,690 ± 120)	
Qivitu Foreland	Peat	9,935 ± 165	QC-451
Qivitu Foreland	Peat	9,092 ± 150	QC-454
? Itiribilung Fiord	Shell	9,110 ± 160 (9,520 ± 160)	GSC-2215
Clyde Foreland	Plant	9,880 ± 200	GSC-2201
? Clyde Foreland	Plant	11,360 ± 320	SI-2614
Cape Adair Peat	Peat	9,480 ± 165	DIC-374
Broughton Island	Seaweed	9,100 ± 140	GSC-1969
Broughton Island	Shell	9,850 ± 250	GaK-2573
? Henry Kater Peninsula	Shell	10,210 ± 180	Y-1986
? Cape Kater	Shell	9,260 ± 150 (9,670 ± 150)	GSC-392
? McBean Bay	Shell	9,280 ± 150 (9,690 ± 150)	GSC-241

Notes: ? = stratigraphic context not clear.
Date in brackets is GSC date + 400 a for comparison with other laboratories.

and < 18,000 BP. At least three suggestions have been made in the literature. Andrews (1980) inferred that because of large, submerged deltas at −20 to −40 m below present sea level on extreme eastern Baffin Island (Miller, 1975), the sea level was *below* present and rose throughout the so-called 'Cape Adair transgression' which peaked <10,500 and >8,000 BP. England (1983), working on the northeast coast of Ellesmere Island, proposed that sea level remained high between ~ 12,000 BP and deglaciation at ~ 8,000 BP. Evidence for this suggestion was noted in the occurrence of shell deposits which dated elevations close to the marine limit throughout the interval noted above. Finally, Quinlan (1981, 1985) has modelled the response of the eastern Canadian Arctic to glacial loading using a development of Peltier's approach (Peltier, 1976; Peltier and Andrews, 1976; Quinlan, 1981, 1985; Quinlan and Beaumont, 1983). His results indicate that sea level remained above present throughout the Late Pleistocene. Figure 4.11 illustrates the three different suggestions (note that England's data does not come from Baffin Island).

At the moment there is no easy way to reconcile these models of sea-level variation; however, it is suggested that a programme of coring selected lakes in the Zone I/II transition would be a valuable contribution to Quaternary studies and to the geophysical meaning of changes in local sea levels (e.g. Fig. 4.6). The submerged deltas off large, former valley glaciers in easternmost Cumberland Peninsula (Miller, 1975; Andrews, 1980) cannot be adequately explained by the available geophysical model of glaciation/sea-level interaction and new work along the lines described is now required.

Changing direction, we may now turn to discuss the growing evidence that Early Holocene sea-level movement in the Zone I/II region was more

Figure 4.11: Three proposed sea-level curves from eastern northern Canada illustrating different histories of sea level during the Late Pleistocene. Note that prior to 12,000 BP there are *no* dated raised marine deposits

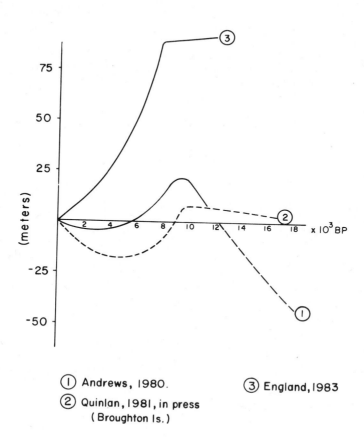

① Andrews, 1980.   ③ England, 1983

② Quinlan, 1981, in press
   (Broughton Is.)

109

complicated than a 'simple' marine inundation/removal cycle (Andrews, 1980). Even in this last noted paper (1980) the author had remarked that the data of King (1969) from the region of McBeth and Itirbilung fiords, Baffin Island, appeared to require a prolonged period when the sea remained close to the lateglacial marine limit. Since King's study, field mapping and dating have gradually brought together a body of information that *implies* significant Early Holocene oscillations of sea level (Nelson, 1978, 1982; Andrews and Miller, 1985). Figure 4.12 graphs three sea-level curves from the eastern Canadian Arctic. Here field observations and the results of ^{14}C dating apparently indicate that the trend of sea level, during the 'Inugsuin regression', was more complex than a simple relaxation and rebound from the highest Holocene shoreline (cf. Andrews, 1980). These observations have also to be incorporated into the findings of Løken (1962, 1964) along the coast of northernmost Labrador, where he noted that the S-3 strandline represented a significant marine incursion that reached an elevation of ~ 18 m asl. Radiocarbon dates on shells bracket the age of this Early Holocene rise in sea level to between 8,000 and 9,000 BP. Figure 4.13 is a simplified version of Løken's diagram (1962) and indicates how a

Figure 4.12: Three relative sea levels from the east coast of northern Canada between Labrador and Qivitu (*c.* 68 N). Note the agreement in the timing of a significant fluctuation in sea level around 8,200 BP

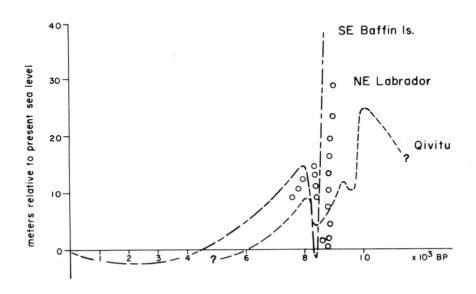

Sources: Data from Nelson (1978) (Qivitu), Andrews and Miller (1985) (SE Baffin Is.) and Løken (1962).

Figure 4.13: Simplified shoreline diagram (after Løken, 1962) for the coast of northernmost Labrador

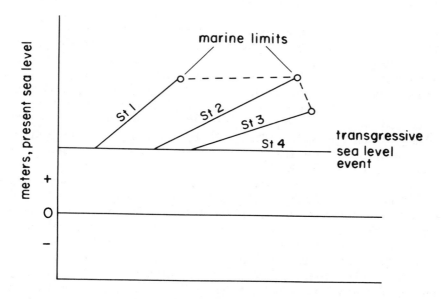

Distance normal to isobases

shoreline diagram can be used to infer a marine flooding event. To the north, along the outer coast of Meta Incognita Peninsula, Baffin Island, Miller has observed glacio-marine and deltaic sequences, whose stratigraphy indicates that after initial deposition the sediments were eroded by fluvial action and later back-filled as sea-level rose again (Andrews and Miller, 1985). The curves of sea-level response, shown in Figure 4.12, certainly imply that the Early Holocene history of sea level along the outer coast of the eastern Canadian Arctic contains complexities that are not apparent, or have not been recognised, in published relative sea-level curves from the inner fiord region (e.g. Løken, 1965; Andrews *et al.*, 1970).

What explanation is there for these short-lived but apparently major events of marine removal/incursion? Two explanations may be advanced. *First* it is worth noting that the east coast of Baffin Island and Baffin Bay is an area of major earthquakes (Basham *et al.*, 1977). There is also some folklore from the Innuits (in Andrews and Miller, 1985) that suggests a large earthquake affected the outer coast of northern Frobisher Bay. Large sections of the Baffin Island coastline are fault-controlled (MacLean, 1985)

111

and this, plus the occurrence of large magnitude earthquakes, indicates that we cannot dismiss a *non-glacial isostatic* component from consideration as a cause of sea-level variations during glacial unloading. In Britain and mainland Europe there is increasing evidence that the stresses induced by glacial loading can be released along pre-existing faults (Mörner, 1980; Sissons and Cornish, 1982). Arguments against a tectonic/unloading origin for the Early Holocene sea-level fluctuations might be the apparent similarity in timing of the major variations over ~ 1,000 km (see Fig. 4.12). A *second* explanation for the event (or events?) might be sought through an analogy with the Norwegian studies on the sea-level oscillations that are associated with the Younger Dryas (11,000-10,300 BP) glacial event (Anundsen and Fjeldskaar, 1983). In this last-mentioned paper the authors examine the possibility that the significant variations in sea level that occurred in western Norway between 13,000 and 10,000 BP were largely driven by glacial isostatic unloading and renewed loading. Anundsen and Fjeldskaar (1983) used a Brotchie and Silvester (1969) two-dimensional crustal model to investigate the effects of renewed ice advance on relative sea-level changes. They also considered the additional effects of geoid deformation and hydro-isostasy. The basic model was similar to one used to develop a first approximation of sea-level history for Baffin Island (Andrews, 1975, 1980). Although the model is a useful first approximation, the major problem with its application is that the relaxation constant has to be specified and is not derived from internal considerations of the rheological model. Despite this caveat, the approach of Anundsen and Fjeldskaar (1983) represents a first quantitative examination of the problem. Their results indicate that the sea-level variations in western Norway (e.g. Fig. 4.4) may be explained by the glacial history.

Bearing these results in mind we can now examine what is known about the Early Holocene glacial history of Baffin Island (e.g. Andrews, 1982b for review). Ever since the publication of Ives and Andrews (1963), it has been known that a system of large end and lateral moraines lie at high elevations above the fiord heads and project seaward as a series of major moraine units. Smith (1966), Andrews *et al.* (1970), Pheasant and Andrews (1973), and Dyke *et al.* (1982), among others, have studied various parts of this moraine system and provided a number of ^{14}C dates (Andrews and Ives, 1978 for review). In specific areas, such as Sam Ford Fiord (Smith, 1966) and Tingin Fiord (Andrews *et al.*, 1970), the fiord outlet glaciers readvanced in the period between 8,000 and 8,500 BP. In Frobisher Bay, Miller (1980) noted evidence for ice advances at ~ 10,700 and 10,000 BP, but thereafter the ice retreated rapidly down Frobisher Bay toward the inner bay, where a major moraine system delimits another interval of position mass balance (Blake, 1966; Miller, 1980; Lind, 1983; Squires, 1984).

Lind (1983) has argued that there was a significant readvance of the ice within inner Frobisher Bay which reached a maximum extent close to 8,500 BP, at which time the ice extended to the complex of moraines that Blake (1966) first identified and which are now referred to as the Frobisher Bay moraines (Miller, 1980). Colvill (1982), Lind (1983), and Squires (1984) have described and mapped the Frobisher Bay moraine and discussed deglaciation from this moraine to the head of Frobisher Bay. The interval of time during which the ice lay at the line of the Frobisher Bay moraines was reasonably extensive, as the marine limit changes across the moraine from an elevation of ~ 120 m on the distal side to < 50 m on the proximal side of the moraine.

Evidence that the ice margin readvanced during the latter part of the Cockburn substage on eastern Baffin Island (i.e. between 8,000 and 9,000 BP (Andrews and Ives, 1978)) has been noted above. At the same time local ice readvanced from ice caps on the south side of Frobisher Bay (Miller, 1980) and local glaciers were in contact with the sea in parts of northern and southern Cumberland Peninsula (Miller, 1973, 1975; Dyke, 1979). However, in view of our concern of the interrelationship between 'Glaciation and Sea Level' the critical question is: how extensive was the advance and was it preceded by an interval of ice recession and subsequent ice build-up? There are at present no good or firm answers to these questions. There are, however, some hints that climate may have become more conducive to glaciation during the Cockburn substage, although it is unknown whether this was the result of enhanced snow accumulation, cooler summers, or some combination of these conditions. Evidence from a variety of approaches, for example from palynology, foraminifera and molluscs studies, indicates that critical oceanographic and atmospheric changes were in progress across the eastern Canadian Arctic between 8,000 and 9,000 BP (Andrews, 1972, in press; Miller, 1980; Osterman, 1982; Short et al., 1985) but how these link into the observed changes in ice extent and sea-level history is not presently understood.

A long piston core from central Frobisher Bay (Fig. 4.14) contains some evidence that there may indeed have been a substantial change in glacial conditions between 8,500 and 8,000 BP. The core (labelled HU77-159) has been studied extensively by Osterman and others (Osterman, 1982; Osterman and Andrews, 1983; Dowdeswell et al., in press). Initial ^{14}C dates obtained from the < 2 μm fine fraction, after removal of detrital carbonates (Osterman, 1982), suggested that the base of the core dated to ~ 12,000 BP. The lithology and biostratigraphy of the core indicated that significant changes in environment occurred between 12,000 and 10,000 BP and these events were linked to the chronology of glaciation/deglaciation established by Miller (1980), in which marine molluscs from raised marine strata were used as dating control. However, with the advent of the tandem accelerator technique for dating small samples (30-50 mg), a series of small bivalves from core HU77-159 were submitted for examination

113

Figure 4.14: Graph of foraminifera numbers from core HU77-159, Frobisher Bay, Baffin Island (62 50.05 N & 67 02.04 W) (Osterman, 1982) and two models of the rate of accumulation of foraminifera. The short timescale is shown enhanced on the right

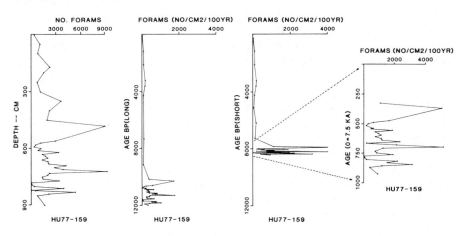

(Andrews and Jull, 1984). Radiocarbon dates on these molluscs indicated that the base of the core dated from ~ 8,500 BP rather than 12,000 BP and that the transition to a distal, ice rafting environment occurred at ~ 7,000 BP. Thus the changes in the marine environment that were initially believed to have occupied 2,000-3,000 years, starting at 12,000 BP, are now thought to have taken place in < 1,000 years immediately after 8,500 BP. Figure 4.14 graphs the numbers of foraminifera per 100 g versus depth in the first column; thereafter foraminifera numbers $cm^2 \ a^{-100}$ are shown for the two chronologies. Finally, in the fourth column a 'blow-up' of the foraminifera accumulation data is shown for the critical interval between 8,000 and 8,400 BP. Note the interval of extremely low productivity centered around ±8,200 BP.

Although the data are not substantial, in either quantity or depth of investigation, they do suggest that glaciological conditions between 8,000 and 8,500 BP favoured glacial advance. It may be a coincidence that this interval coincides with the period when sea-level appears to have been fluctuating markedly along the outer east coast of Baffin Island. It is suggested, therefore, that a vital project to investigate glaciation and sea-level interactions would involve a programme of detailed lake (Fig. 4.6) and fiord (Fig. 4.14) coring, in part after the manner already developed in Norway (e.g. Anundsen, 1978) and discussed earlier in this chapter. Frobisher Bay would appear to be an ideal location for such a research project because of the availability of marine cores, the number of rock basins below the local marine limit, the ease of access to the area (relatively speaking, of course!)

and the extent of research that has been carried out over the last decade.

In a different context the recent work by Hodgson and Vincent (1984) indicates a situation where there may be no clear relationship between glaciological events and sea-level response. In the region of Victoria, Banks, and Melville islands, Arctic Canada, the marine limit has been dated at ~ 12,000 BP. However, below the marine limit there is abundant evidence that a significant glacial event occurred. Hodgson and Vincent (1984), after a careful review of the evidence, suggested that deglaciation of the arctic channels triggered a collapse of a major ice dome (the McClintock Dome (Dyke, 1983, 1984)). The collapse resulted in a surge and the formation of a thin floating ice shelf. This event occurred close to 10,000 BP and happened at a time when sea-level was falling, or at least relative sea level was well below the local marine limit. The suggested collapse of the ice dome may have had an effect on sea-level history in the vicinity of the proposed ice centre (i.e. very rapid unloading), but the surge added no substantial load to the deglaciated channels and hence little or no isostatic response should be expected. This example of a significant glacial event being disassociated from a specific sea-level response leads naturally to the next section of this chapter.

SEA-LEVEL CONTROL OF ICE SHEETS

Denton and Hughes (1981) have made the important suggestion that the fundamental, global control on Northern and Southern Hemisphere deglaciation is the eustatic rise associated with deglaciation. They argue that it is the rise of sea level against the margins of ice sheets terminating in the sea which induces ice sheets to thin rapidly and collapse. Ruddiman and McIntyre (1981) have invoked this notion of an ice sheet collapse to explain a barren zone in North Atlantic deep sea cores. Indeed, they have argued on the basis of their data that some 50 per cent of the volume of the Northern Hemisphere ice sheets was added to the oceans as meltwater in the interval of *only 3,000 years!* These latter authors correctly noted that isochrone maps on the deglaciation of the Laurentide Ice Sheet (Prest, 1969; Bryson *et al.*, 1969) indicated that between 13,000 and 16,000 BP the *area* of the ice sheet changed relatively slowly and in a linear manner. Thus to explain the barren zones by the injection of massive amounts of meltwater, at a time when the areal extent of the ice sheet showed slow recession, Ruddiman and McIntyre (1981) invoked a collapse of the central portion of the Laurentide Ice Sheet. Previously, field studies and theory had suggested that the ice sheet had indeed thinned dramatically, but the date for the onset of this was later than the 13,000-16,000 BP interval (Andrews and Peltier, 1976) and was dated to between 8,000 and 12,000 BP.

From the viewpoint of glaciation and sea-level interactions there are two main interests in the concept of rapid ice sheet collapse. First is the idea that rapid changes in the volume of ice sheets should be reflected in associated changes of sea level at sites distant from the ice margin. Second is the concern that changes in sea level might be the major mechanism for causing globally synchronous deglaciation on geological timescales.

In the last one to two decades there is no doubt that as our knowledge of Late Quaternary sea-level variations has improved, so also has our awareness that there exist additional underlying complexities (hydro-isostasy, geoidal deformation, etc; see Mörner, 1980, and earlier discussion by Peltier, this vol., Ch. 3). Thus the cause(s) for local or regional marine removal/inundation events might not be associated with the global process of sea water subtraction and addition associated with ice sheet history. Nevertheless, at the scale suggested by Ruddiman and McIntyre (1981) an estimated 35-50 m of water was being added to the ocean level in a matter of 3,000 years. This suggests a rate of sea-level rise of 1.2 to 1.7 cm a^{-1} or > 1 m a^{-100}. One would surely think that a change of this magnitude could be verified or rejected on the basis of studies of sea-level variations on coasts that were not directly influenced by glaciation (i.e. Zones II/III/IV of Clark *et al.*, 1978). However, this belief in a rational universe is dispelled by a critical examination of the data! There are very few, if any, reliable curves (time-depth data plots) from the critical interval that we need to assess.

One recent curve is that by McManus and Creager (1984) from the region of the continental shelf of Beringia, the Bering Sea. Their analysis of the ^{14}C dates here suggests that sea-level between 16,000 and 13,000 BP rose from −57 m to −35 m msl, a change of only 22 m, far short of the catastrophic rise in sea level called for in the Ruddiman and McIntyre (1981) scenario. However, although these data would support a 'thin' ice sheet model, another paper from the area of Greece (Van Andel and Lianos, 1984) uses data from Bloom (1977) to suggest a far different sea-level history, with a change of sea level approaching 35 m in the 3,000 years between 16,000 and 13,000 BP. The important question is: can we be sure that both curves are *not right*, in the sense that they faithfully record relative sea-level events for each site? In assessing such data we need to ascertain if they contain information on the global scale. This is a necessary condition for any evaluation of the relationship between sea-level history and the causative mechanism(s) of deglaciation.

The main problem with the notion that ice sheet collapse is associated with changes of global sea level, is to extract a eustatic sea-level curve from glacio-isostatic sources of changes in the sea bed, at the margin of an ice sheet terminating in the sea. The rate of glacio-isostatic change may be several metres per 100 years, at least judging from published rebound curves (see Fig. 4.4), and thus for world synchrony of glacial response, the

116

entire process of ice build-up and attainment of pseudo-isostatic equilibrium must proceed in concert. But we know that this did not occur. The timing of the deglaciation of marine-based portions of the various Northern and Southern Hemisphere ice sheets is relatively well known, at least as a first approximation, and one of the notable features is the differences in the dates of deglaciation of the outer coasts. This suggests that the global control on ice sheet disintegration either is not valid, or is complicated by a variety of regionally varying controls (such as glacio-isostatic response to variations in ice sheet build-up).

Thomas (1977) has developed a model to examine the retreat of the Laurentide Ice Sheet in the vicinity of the St Lawrence Valley, during the Late Wisconsin deglaciation of eastern Canada. Sea level is not explicitly included in the model, but it is implicitly included in the parameter that specifies the depth of water at the grounding line. The proposed instability of marine-based ice sheets is primarily associated with the fourth-power dependence of the creep-thinning rate on ice thickness at the grounding line. Unless this can be balanced by frictional drag against the margins of the ice shelf (or by a few ice rises), it is virtually impossible to restrain a marine-based ice sheet from rapid collapse. Figure 4.15 is a sketch of an ice sheet terminating in a marine environment. The effect of a rise of sea level at the ice margin is to increase the buoyancy force on the ice tongue which,

Figure 4.15: Sketch of an ice sheet/ice shelf junction and terminology

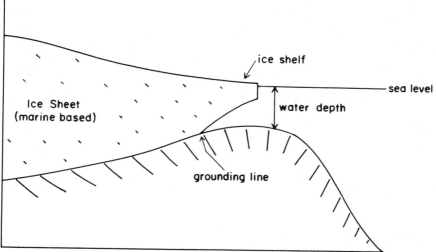

Source: After Thomas (1977).

if it is not compensated for by an increase in the rate of accumulation upstream, will lead to a retreat of the grounding line (Fig. 4.15). If the slope of the surface behind the grounding line is negative there is a positive feedback and retreat of the ice margin can proceed extremely rapidly. Thomas (1977; Fig. 4) gives an indication of the possible effect of a sea-level rise for different sized ice catchment areas and for different values of shear stress and a sliding law parameter. If we consider changes of the basal depth at the grounding line to be equivalent physically to a negative (i.e. deepening) up-glacier slope, then this graph, and the equation from which it is derived, can be used to gain some insights into the sensitivity of the recession process to a rise in sea level. What is currently missing is a physically plausible model for the calving rate of the margin. From Thomas (1977: p. 350) an estimate of the rate of retreat of the grounding line can be derived from:

$$a' = H'(a)/(1-p_w/p_i)0(a) - (a) \qquad \text{(Eq. 4.1)}$$

where a' is the retreat rate (m a^{-1}) of the grounding line, H' is the thinning rate at the grounding line, p_w and p_i are the density of water and ice respectively, and $0(a)$ and (a) are the bedrock and ice slopes. In detail the effect of a rise in sea level is a non-linear forcing function, whose effect on the collapse of the ice sheet is a function of the following variables and their interactions: (1) thinning rate at the grounding line; (2) rate of change of the surface depth behind the grounding line, (3) changes in glaciological conditions such as changes in bed substrate and controls on sliding, (4) the depth at the grounding line at the start of the sea-level rise. Thomas's calculations indicate that the rate of grounding line retreat is a non-linear function of the water depth at the grounding line. He stresses the importance of changes in relative sea level associated with glacial isostasy repeatedly, concluding that 'the retreat was triggered by some combination of sea bed depression, locally rising sea level due to lateral gravitational attraction by the ice sheet, reduction in total upstream snow accumulation, or reduction in the bed friction of the ice stream' (Thomas, 1977: p. 355). Although Thomas was talking about the Laurentian Channel, it is suggested here that these remarks (mechanism) have general validity. It might even be that the retreat of the great ice sheets is more associated with *relative* changes of sea level, driven by glacio-isostatic response, than they are by variations in the Milankovitch orbital parameters, although such remarks are probably heretical in view of the enormous pressure being exerted for the latter theory in the scientific literature.

The maximum and minimum estimates of the eustatic sea-level rise associated with the Late Wisconsin deglaciation are between 80 and 160 m. If 50 per cent by volume of this water was released by rapid calving in 3,000 years, the limits on the rates of sea-level rise would be between

13.3 and 26.6 mm a^{-1}. These are equivalent, as rates, to the fall in present sea level around the shores of Hudson Bay. The proposition of this catastrophic deglaciation is a 'chicken and egg' proposition. If ice-sheet collapse is triggered by a rise in global (eustatic) sea level, what process causes early deglaciation to produce the required rise in sea level? In addition, where does this early deglaciation occur? It certainly is not evident around the Laurentide Ice Sheet prior to 16,000 BP (e.g. Prest, 1984); some retreat was under way in Antarctica prior to 16,000 BP (Stuiver *et al.*, 1981) but can this be caused by global sea-level changes? Returning to an earlier point, it is suggested here that the feature of deglaciation that is most evident is the non-synchrony of major deglacial events at the global scale. This, one may conclude, excludes the mechanism that these events were controlled by changes in global (eustatic) sea level. Rather it is suggested that regional deglaciation of marine-based sections of ice sheets is related to changes in *relative* sea level and associated factors at each ice margin. Such changes are more likely associated with glacio-isostatic depression of the sea bed, which will *lag* the build-up of the ice sheet (Fig. 4.16) by a

Figure 4.16: Diagram illustrating the advance of an ice sheet toward the edge of a continent. Because of the viscoelastic response of the earth relative sea-level at site Z (and other sites for that matter) will continue to rise even though the ice sheet is no longer expanding. The increase in water depth will affect the stability of the ice at the grounding line and at some point grounding line retreat will commence

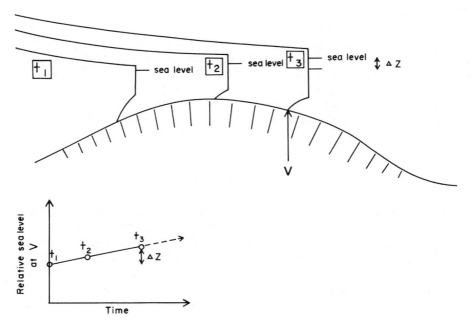

characteristic interval associated with an appropriate rheological model of the earth. Thus if a glacial maximum is attained at 18,000 BP due to the readvance of ice from some Mid-Wisconsin interstadial minima (cf. Andrews *et al.*, 1983), glacio-isostatic depression of the sea bed at the margin of the ice sheet will continue after the maximum ice-sheet configuration has been reached. Thus relative sea-level will continue to rise and, depending on glaciological and climatic factors, this relative rise in sea level will eventually destabilise the marine-based portions of ice sheets. It would thus appear that glacial isostasy may have a major control on ice-sheet stability and be a self-limiting factor in ice-sheet growth.

Ice divides and ice mass

Finally, a commentary, albeit a brief one, on the confusion that can arise if we do not clearly distinguish between ice divides based on glacial flow indicators, as opposed to areas of maximum ice thickness. The latter are often inferred from the pattern of postglacial isobases on radiocarbon-dated marine beaches. Andrews (1982) drew attention to the essential difference between these two in terms of their use in the reconstruction of ice sheets. Over the Barents Sea, the Queen Elizabeth Islands and Hudson Bay the areas of greatest postglacial rebound are located either over, or close to, former marine-based portions of ice sheets. These data have been used by the author and others to locate centres of former ice sheets. However, I now feel that the simple utilisation of this argument can be misleading if not totally wrong. An examination of the Greenland and the Antarctic ice sheets indicates that the areas of greatest ice thickness do not coincide with ice divides. If anything ice divides tend to be located over plateaux or mountain chains where the ice is relatively thin and hence cold. Thus I would strongly suggest that the relationship between the patterns of isobases and the surface configuration of the former ice sheets need not coincide. To paraphrase: maximum ice thickness does not an ice divide make!

RESEARCH NEEDS: GLACIATION AND SEA LEVEL

It is appropriate to end this chapter with a brief statement as to where future research might be directed. In reviewing the chapter and the suggestions contained therein, a conclusion reached is that the paramount need is for detailed relative sea-level curves to be constructed, specifically in the vicinity of the margins of former ice streams that link into marine-based portions of ice sheets. Examples of such areas would be in the vicinity of the outer St Lawrence Valley, outer Hudson Strait and some of the large channels within the High Arctic Canadian archipelago. By and large, sea-level studies from Canada and Northwest Europe have documented the progression of sea level *after* deglaciation. Recent work in Norway strongly suggests that a glacial event, the magnitude of the Younger Dryas, influences glacio-isostatic response and hence relative sea level, by causing depression of the sea bed. The strategy of coring small lake basins within a critical elevational range would seem to be capable of resolving some of the questions that have been posed. Certainly around the margin of outer Hudson Strait, on either southeast Baffin Island or northernmost Labrador, there was ice-free terrain exposed at the maximum of the Late Wisconsin (e.g. Osterman *et al.*, 1985; Clark, 1984) and some of this terrain existed along portions of both coasts. Seismic or acoustic stratigraphy of the lake basin sediments, combined with a serious coring effort, might give important clues on the relationship between glaciation and sea level, and on the rheological properties of the earth.

ACKNOWLEDGEMENTS

This chapter is a contribution to NSF grant EAR-84-09915 awarded to the author and G.H. Miller. Jim Walters drafted the figures for the chapter and Giff Miller supplied the air photograph from Spitsbergen.

REFERENCES

Aharon, P. (1983) '140,000-yr isotope climatic record from raised coral reefs in New Guinea', *Nature*, *304*, 720-3.

Andel, T.H. Van and Lianos, N. (1984) 'High-resolution seismic reflection profiles for the reconstruction of postglacial transgressive shorelines: an example from Greece', *Quat. Res.*, *22*, 31-45.

Andrews, J.T. (1966) 'Pattern of coastal uplift and deglacierization west Baffin Island, N.W.T', *Geogr. Bull.*, *8*, 174-93.

—— (1970) *A Geomorphological Study of Post-glacial Uplift with Particular Reference to Arctic Canada*, Institute of British Geographers Special Publication, 2.

—— (1972) 'Recent and fossil growth rates of marine bivalves, Canadian Arctic and Late Quaternary Arctic marine environments', *Palaeogeography, Palaeoclimatol., Palaeoecol.*, *11*, 157-76.

—— (1975) 'Support for a stable Late Wisconsin ice margin (14,000 to *ca* 9000 BP); a test based on glacial rebound', *Geology*, *4*, 617-20.

—— (1978) 'Sea level history of arctic coasts during the Upper Quaternary', *Prog. Phys. Geogr.*, *2*, 375-407.

—— (1980) 'Progress in relative sea level and ice sheet reconstructions Baffin Island N.W.T. for the last 125,000 years', in N.-A. Mörner (ed.), *Earth Rheology, Isostasy and Eustasy*, Wiley, Chichester and New York, pp. 175-200.

—— (1982a) 'On the reconstruction of Pleistocene Ice Sheets: A review', *Quat. Sci. Rev., 1,* 1-30.

—— (1982b) 'Holocene glacier variations in the eastern Canadian Arctic: A review', *Striae, 18,* 9-14.

—— (in press) *Reconstruction of Environmental Conditions in the Eastern Canadian Arctic During the Last 11,000 years,* National Museums of Canada, Ottawa.

—— Buckley, J.T. and England J.H. (1979) 'Late-glacial chronology and glacio-isostatic recovery, Home Bay, east Baffin Island', *Bull. Geol. Soc. Am., 81,* 1123-48.

—— and Ives, J.D. (1978) '"Cockburn" nomenclature and the Late Quaternary history of the Eastern Canadian Arctic', *Arctic and Alpine Res., 10,* 617-33.

—— and Jull, A.J.T. (1984) 'Rates of fiord and shelf sediment accumulation, Baffin Island, Canada, based on 14C accelerator dates on *in situ* molluscs', *Geol. Soc. Am.,* Abstracts, *16,* 431.

—— and Matsch, C.L. (1983) *Glacial Marine Sediments and Sedimentation. An Annotated Bibliography,* Geo-Abstracts, Norwich.

—— and Miller, G.H. (1984) 'Quaternary glacial and nonglacial correlations in the Eastern Canadian Arctic', *Geol. Surv. Can., Paper 84-10,* 101-66.

—— and Miller, C.H. (1985) 'Holocene Sea-level variation within Frobisher Bay', in J.T. Andrews (ed.), *Quaternary Environments: Eastern Canadian Arctic, Baffin Bay, and West Greenland,* Allen & Unwin, London, pp. 585-608.

—— and Peltier, W.R. (1976) 'Collapse of the Hudson Bay ice sheet', *Geology, 4,* 73-5.

—— Shilts, W.W. and Miller, G.H. (1983) 'Multiple deglaciation of the Hudson Bay lowlands since deposition of the Missinaibi (Last Interglacial?) formation', *Quat. Res., 19,* 18-37.

Anundsen, K. (1978) 'Marine transgression in Younger Dryas in Norway', *Boreas, 7,* 49-60.

—— and Fjeldskaar, W. (1983) 'Observed and theoretical Late Weischelian shore-level changes related to glacier oscillations at Yrkje, south-west Norway', in H. Schroeder-Lanz (ed.) *Late and Post-glacial Oscillation of Glaciers: Glacial and Periglacial Forms,* A.A. Balkema, Rotterdam, pp. 133-70.

Basham, P.W., Forsyth, D.A. and Wetmiller, R.J. (1977) 'The seismicity of northern Canada', *Can. J. Earth Scis., 14,* 1646-67.

Bird, J.B. (1954) 'Post-glacial marine submergence in central Arctic Canada', *Bull. Geol. Soc. Am., 65,* 457-64.

Blake, W., Jr (1966) 'End moraines and deglaciation chronology in northern Canada, with special reference to southern Baffin Island', *Geol. Surv. Can., Paper 66-21,* 1-31.

—— (1970) 'Studies of glacial history in Arctic Canada. I. Pumice, radiocarbon dates, and differential postglacial uplift in the eastern Queen Elizabeth Islands', *Can. J. Earth Scis., 7,* 634-64.

—— (1975) 'Radiocarbon age determinations and post-glacial emergence at Cape Storm, southern Ellesmere Island', *Geografiska Annaler, 62A,* 1-71.

—— (1976) 'Sea and land relations during the last 15,000 years in the Queen Elizabeth Islands, Arctic Archipelago', *Geol. Surv. Can., Paper 76-1B,* 201-7.

Bloom, A.L. (comp.) (1977) *Atlas of Sea-level Curves.* International Geological Correlation Program No. 61, Sea-level Project. Limited distribution (mimeo), various paginations.

Boulton, G.S., Baldwin, C.T., Peacock, J.D., McCabe, A.M., Miller, G., Jarvis, J., Horsfield, B., Worsley, P., Eyles, N., Chroston, P.N., Day, T.E., Gibbard, P.,

Hare, P.E. and von Brunn, V. (1982) 'A glacio-isostatic facies model and amino acid stratigraphy for late Quaternary events in Spitsbergen and the Arctic', *Nature*, *298*, 437-41.

Brigham, J.K. (1983) 'Stratigraphy, amino acid geochronology, and correlations of Quaternary sea-level and glacial events, Broughton Island, arctic Canada', *Can. J. Earth Sci.*, *20*, 577-98.

Brotchie, J.F. and Sylvester, R. (1969) 'On crustal flexure', *J. Geophys. Res.*, *74*, 5240-52.

Bryson, R.A., Wendland, W.M., Ives, J.D. and Andrews, J.T. (1969) 'Radiocarbon isochrones on the disintegration of the Laurentide Ice Sheet', *Arctic and Alpine*, *1*, 1-14.

Clark, J.A., Farrell, W.E. and Peltier, W.R. (1978) 'Global changes in post-glacial sea level: a numerical calculation', *Quat. Res.*, *9*, 265-87.

Clark, P.U. (1984) 'Glacial geology of the Kangalaksiorvik-Abloviak region, northern Labrador', Ph.D dissertation, University of Colorado, Boulder, Col.

Colvill, A., (1982) 'Glacial landforms at the head of Frobisher Bay, Baffin Island, Canada', MA dissertation, University of Colorado, Boulder, Col.

Denton, G.H. and Hughes, T.J. (eds) (1981) *The Last Great Ice Sheets*, Wiley, New York.

—— and Hughes, T.J. (1983) 'Milankovitch theory of Ice Ages: Hypothesis of ice-sheet linkage between regional insolation and global climate', *Quat. Res.*, *20*, 125-44.

Dowdeswell, J.A., Osterman, L.E. and Andrews, J.T. (in press) 'SEM and other criteria for distinguishing glacial and non-glacial events in a marine core from Frobisher Bay, N.W.T., Canada', in W.B. Whalley (ed.), *Scanning Electron Microscopy: A Symposium*, Geo-Abstracts, Norwich.

Dyke, A.S. (1974) *Deglacial Chronology and Uplift History: Northeastern Sector, Laurentide Ice Sheet*, Institute of Arctic and Alpine Research, Occasional Paper 12.

—— (1979) 'Glacial and sea-level history of the southwestern Cumberland Peninsula, Baffin Island, N.W.T., Canada', *Arctic and Alpine Res.*, *11*, 179-202.

—— (1983) 'Quaternary geology of Somerset Island, District of Franklin', *Geol. Surv. Can.*, *Memoir 404*.

—— (1984) 'Quaternary geology of Boothia Peninsula and Northern District of Keewatin, Central Arctic Canada', *Geol. Surv. Can.*, *Memoir 407*.

—— Andrews, J. T. and Miller, G.H. (1982) 'Quaternary geology of Cumberland Peninsula, Baffin Island, District of Franklin', *Geol. Surv. Can.*, *Memoir 403*.

England, J.H. (1976) 'Late Quaternary glaciation of the eastern Queen Elizabeth Islands, N.W.T., Canada: Alternative models', *Quat. Res.*, *6*, 185-203.

—— (1983) 'Isostatic adjustments in a full glacial sea', *Can. J. Earth Scis.*, *20*, 895-917.

Falconer, G., Ives, J.D., Loken, O.H. and Andrews, J.T. (1965) 'Major end moraines in eastern and central Arctic Canada', *Geogr. Bull.*, *7*, 137-53.

Feyling-Hanssen, R.W. (1985) 'Late Cerozoic marine deposits of East Baffin Island and East Greenland: microbiostratigraphy, correlation, age', in Andrews,J.T. (ed.), *Quaternary Environments: East Canadian Arctic, Baffin Bay, and West Greenland*, Allen & Unwin, London, pp. 345-72.

Flint, R.F. (1971) *Glacial and Quaternary Geology*, Wiley, New York.

Forman, S.L. and Miller, G.H. (1984) 'Time-dependent soil morphologies and pedogenic processes on raised beaches, Bröggerhalvöya, Spitsbergen, Svalbard Archipelago', *Arctic and Alpine Res.*, *16*, 381-94.

Grosswald, M. (1983) *Ice Sheets of the Continental Shelves*, Nauka, Moscow.

123

Translation I. Plam (mimeo), *North American Shelves*, Institute of Arctic and Alpine Research.

Hillaire-Marcel, C., Occhietti, S. and Vincent, J.-S. (1981) 'Sakami moraine, Quebec: a 500 km-long moraine without climatic control', *Geology*, 9, 210-14.

Hodgson, D.A., and Vincent, J.-S. (1984) 'A10,000 yr BP extensive ice shelf over Viscount Melville Sound, Arctic Canada', *Quat. Res.*, 22, 18-30.

Ives, J.D. and Andrews, J.T. (1963) 'Studies in the physical geography of north-central Baffin Island, N.W.T.', *Geogr. Bull.*, 19, 5-48.

King, C.A.M. (1969) 'Glacial geomorphology and chronology of Henry Kater Peninsula, east Baffin Island, N.W.T.', *Arctic and Alpine Res.*, 1, 195-212.

Klassen, R.A. (1982) 'Quaternary stratigraphy and glacial history of Bylot Island, N.W.T., Canada', PhD dissertation, University of Illinois, Urbana-Champaign, Ill.

—— (1985) 'An outline of glacial history of Bylot Island, District of Franklin, N.W.T', in J.T. Andrews (ed.), *Quaternary Environments: Eastern Canadian Arctic, Baffin Bay, and West Greenland*, Allen & Unwin, London, pp. 428-60.

Lie, S.E., Stabell, B. and Mangerud, J. (1983) 'Diatom stratigraphy related to Late Weischelian sea-level changes in Sunmore, Western Norway', *Nor. Geol. Unders.*, 380, 203-19.

Lind, E.K. (1983) 'Sedimentology and palaeoecology of the Cape Rammelsberg area, Baffin Island, Canada', MSc dissertation University of Colorado, Boulder, Col.

Løken, O.H. (1962) 'The lateglacial and postglacial emergence and deglaciation of northernmost Labrador', *Geogr. Bull.*, 17, 23-56.

—— (1964) 'A Study of the late and postglacial changes of sea level in northernmost Labrador', unpublished report to the Arctic Institute of North America (mimeo).

—— (1965) 'Postglacial emergence at the south end of Inugsuin Fiord, Baffin Island, N.W.T.', *Geogr. Bull.*, 7, 243-58.

—— (1966) 'Baffin Island refugia older than 54,000 years', *Science*, 153, 1378-80.

MacLean, B. (1985) 'Geology of the Baffin Island shelf', in J.T. Andrews (ed.), *Quaternary Environments: Eastern Canadian Arctic, Baffin Bay, and West Greenland*, Allen & Unwin, London, pp. 154-77.

McManus, D.A. and Creager, J.S. (1984) 'Sea-level data for parts of the Bering–Chukchi shelves of Beringia from 19,000 to 10,000 14C yr BP', *Quat. Res.*, 21, 317-25.

Mangerud, J. (1972) 'Radiocarbon dating of marine shells, including a discussion of recent shells from Norway', *Boreas*, 1, 143-72.

—— Larsen, E., Longva, O. and Sonstegaard, E. (1979) 'Glacial history of western Norway 15,000-10,000 BP', *Boreas*, 8, 179-87.

Matthews, W.H., Fyles, J.G. and Nasmith, H.W. (1970) 'Postglacial crustal movements in southwestern British Columbia and adjacent Washington State', *Can. J. Earth Sci.*, 7, 690-702.

Miller, G.H. (1973) 'Late Quaternary glacial and climatic history of northern Cumberland Peninsula, Baffin Island, N.W.T., Canada', *Quat. Res.*, 3, 561-83.

—— (1975) 'Quaternary glacial and climatic history of northern Cumberland Peninsula, Baffin Island, Canada, with particular reference to fluctuations during the last 20,000 years', PhD dissertation, University of Colorado, Boulder, Col.

—— (1980) 'Late Foxe glaciation of southern Baffin Island, N.W.T., Canada', *Bull. Geol. Soc. Am.*, Part I, 91, 399-405.

—— (1982) 'Quaternary depositional episodes, western Spitsbergen, Norway: aminostratigraphy and glacial history', *Arctic and Alpine Res.*, 14, 321-40.

—— (1985) 'Aminostratigraphy of Baffin Island shelf-bearing deposits', in J.T. Andrews (ed.), *Quaternary Environments: Eastern Canadian Arctic, Baffin Bay, and West Greenland*, Allen & Unwin, London, pp. 394-427.

—— Andrews, J.T. and Short, S.K. (1977) 'The last interfacial-glacial cycle, Clyde foreland, Baffin Island, N.W.T.: Stratigraphy, biostratigraphy and chronology', *Can. J. Earth Sci.*, *14*, 2824-57.

—— and Dyke, A.S. (1974) 'Proposed extent of Late Wisconsin Laurentide ice on eastern Baffin Island', *Geology*, *2*, 125-30.

Mode, W.N., Nelson, A.R. and Brigham, J.K. (1983) 'Sedimentologic evidence for Quaternary glaciomarine cyclic sedimentation along eastern Baffin Island, Canada', in B.F. Molnia (ed.), *Glacial-marine Sedimentation*, Plenum Press, New York, pp. 495-534.

Mörner, N.-A. (1969) 'The Late Quaternary history of the Kattegat Sea and the Swedish West Coast: Deglaciation, shoreline displacement, chronology, isostasy and eustasy', *Sveriges Geol. Unders., Arsb. 63, Serv. C, 640*.

—— (1980) 'The Fennoscandian Uplift: geological data and their geodynamical implication', in N.-A. Mörner (ed.), *Earth Rheology, Isostasy and Eustasy*, Wiley, Chichester and New York, pp. 251-84.

—— (ed.) (1980) *Earth Rheology, Isostasy and Eustasy*, Wiley, Chichester and New York.

Nelson, A.R. (1978) 'Quaternary glacial and marine stratigraphy of the Qivitu Peninsula, northern Cumberland Peninsula, Baffin Island, Canada', PhD dissertation, University of Colorado, Boulder, Col.

—— (1981) 'Quaternary glacial and marine stratigraphy of the Qivitu Peninsula, northern Cumberland Peninsula, Baffin Island', *Bull. Geol. Soc. Am., (92)1*, 512-8; *(92)2*, 1143-261.

—— (1982) 'Aminostratigraphy of Quaternary marine and glaciomarine sediments, Qivitu Peninsula, Baffin Island', *Can. J. Earth Sci.*, *19*, 945-61.

Osterman, L.E. (1982) 'Late Quaternary history of southern Baffin Island, Canada: A study of foraminifera and sediments from Frobisher Bay', PhD dissertation, University of Colorado, Boulder, Col.

—— and Andrews, J.T. (1983) 'Changes in glacial-marine sedimentation in core HU77-159, Frobisher Bay, Baffin Island, N,W.T.: a record of proximal, distal and ice-rafting glacial-marine environments', in B.J. Molnia (ed.), *Glacial-marine Sedimentation*, Plenum Press, New York, pp. 451-94.

—— Miller, G.H. and Stravers, J.A. (1985) 'Middle and Late Foxe glacial events in southern Baffin Island', in J.T. Andrews (ed.), *Quaternary Environments: Eastern Canadian Arctic, Baffin Bay, and West Greenland*, Allen & Unwin, London, pp. 520-45.

Peltier, W.R. (1976) 'Glacial isostatic adjustment — II: The inverse problem', *Geophys. J. Roy. Astron. Soc.*, *46*, 669-706.

—— and Andrews, J.T. (1976) 'Glacial-isostatic adjustment — I: The forward problem', *Geophys. J. Roy. Astron. Soc.*, *46*, 605-46.

—— and Andrews, J.T. (1983) 'Glacial geology and glacial isostasy, Hudson Bay, Canada', in E.I. Smith (ed.), *Shorelines and Isostasy*, Academic Press, London and New York, pp. 285-319.

Pheasant, D.R. and Andrews, J.T. (1973) 'Wisconsin glacial chronology and relative sea-level movements, Narpaing Fiord Broughton Island area, eastern Baffin Island N.W.T', *Can. J. Earth Sci.*, *10*, 1621-41.

Prest, V.K. (1969) 'Retreat of Wisconsin and Recent ice in North America', *Geol. Surv. Can.*, Map 1257A.

—— (1984) 'The Late Wisconsinian glacier complex', *Geol. Surv. Can., Paper 84-*

125

10, 21-38 (Map GSC 1584A in pocket).

Quinlan, G., (1981) 'Numerical models of postglacial relative sea-level change in Atlantic Canada and the eastern Canadian Arctic', unpublished PhD dissertation, Department of Oceanography, Dalhousie University, Halifax, Nova Scotia.

—— (1985) 'A numerical model of postglacial relative sea level change near Baffin Island', in J.T. Andrews (ed.), *Quaternary Environments: Eastern Canadian Arctic, Baffin Bay, and West Greenland*, Allen & Unwin, London, pp. 560-84.

—— and Beaumont, C. (1983) 'The deglaciation of Atlantic Canada as reconstructed from the postglacial relative sea-level record', *Can. J. Earth Sci.*, *19*, 2232-48.

Ruddiman, W.F. and McIntyre, A. (1981) 'The mode and mechanism of the last deglaciation: Oceanic evidence', *Quat. Res.*, *16*, 125-34.

Salvigsen, O. (1981) 'Radiocarbon dated raised beaches in Kongs Karl Land, Svalbard, and their consequences for the glacial history of the Barents Sea area', *Geografiska Annaler*, *63*, 283-91.

Short, S.K., Mode, W.N. and Davis, P.T. (1985) 'The Holocene record from Baffin Island: modern and fossil pollen studies', in J.T. Andrews, (ed.), *Quaternary Environments: Eastern Canadian Arctic, Baffin Bay, and West Greenland*, Allen & Unwin, London, pp. 608-42.

Sim, V.W. (1960) 'Maximum post-glacial marine submergence in northern Melville Peninsula', *Arctic*, *13*, 178-93.

Sissons, J.B. and Cornish, R. (1982) 'Differential glacio-isostatic uplift of crustal blocks at Glen Ray, Scotland', *Quat. Res.*, *18*, 268-88.

Smith, J.E. (1966) 'Sam Ford Fiord: a study in deglaciation', MSc thesis. McGill University, Montreal.

Squires, C. (1984) 'The Late Foxe deglaciation of the Burton Bay area, southeastern Baffin Island, N.W.T.', MA dissertation, University of Windsor, Windsor, Ont.

Stuiver, M., Denton, G.H., Hughes, T.J. and Fastook, J.L. (1981) 'History of the marine ice sheet in West Antarctica during the last glaciation: a working hypothesis', in G.H. Denton and T.J. Hughes (eds), *The Last Great Ice Sheets*, Wiley, Chichester and New York, pp. 319-39.

—— and Polach, H.A. (1977) 'Discussion. Reporting of 14C data', *Radiocarbon*, *19*, 355-63.

Thomas, R.H. (1977) 'Calving bay dynamics and ice sheet retreat up the St. Lawrence valley system', *Géographie physique et quaternaire*, *31*, 347-56.

Thomsen, H. (1982) 'Late Weischelian, shore-level displacement on Nord-Jaeren, south-west Norway', *Geol. Fören. Stockh. Förh.*, *103*, 447-68.

Vincent, J.-S. (1982) 'The Quaternary history of Banks Island, N.W.T., Canada', *Géographie physique et quaternaire*, *36*, 209-32.

5

Tectonic Processes and their Impact on the Recording of Relative Sea-level Changes

Kelvin Berryman

INTRODUCTION

Tectonic processes have both very pronounced and very subtle effects on the earth's crust, which may be measured by the deformation of datum planes at the earth's surface. The position of sea level with respect to adjacent landmasses may be regarded as a datum plane that records the interplay between climatic, geodynamic and tectonic processes through time.

Clear examples of the instability of the earth's crust include collision zones between crustal plates that result in the building of young mountain belts such as the Himalayan Ranges, the crustal rifts and associated volcanism of East Africa and Iceland and the very deep oceanic trenches that exist adjacent to island arc complexes. At many of these sites of crustal movement, features formed by successive still-stands of sea level are likely to be modified from their originally near-horizontal plane. Relationships of successive shoreline features here will no longer record absolute variations in sea-level position, but will record relative sea level with respect to present or an inferred palaeosea-level position.

The tectonic processes that may influence the position of past sea levels may vary in time and space, and feedback relationships particularly between climate and tectonism occur. Mörner (1980) has suggested the possibility of changes in mantle flow causing changes in gravity fields, and hence the isostatic response of crustal blocks and, therefore, sea level. In another way the different tectonic processes themselves affect the recording of sea-level position over differing time-spans. At the long-term scale some deep crustal and mantle processes, such as changes in mantle flow and total water budget on the earth, show little or no change over relatively short time-spans of 1-100 ka. On the other hand, rapid changes in ice volumes (causing crustal loading and flexure) in polar regions at the ends of glacial periods induce changes in sea-level position at a faster pace than any other causative factor for periods of 0.1-10 ka. Middle scale,

127

second order tectonic deformations are induced by the volume of ocean basins increasing or decreasing during glacial cycles, or by rapid continental margin sedimentation (see this vol., Chappell, Ch. 2 and Devoy, Ch. 16). At the short timescale the essentially instantaneous shifts in sea-level position, brought about by tectonic faulting in coastal areas, brings about the most rapid sea-level change, in short time-spans of a few seconds to a few years, (see Peltier, this vol., Ch. 3). In this context the chapter will discuss some of the tectonic processes that have changed the relative position of former shoreline features. Examples used here of earthquake associated (co-seismic) uplift of the land relative to sea level are particularly characteristic of circum Pacific localities and these emphasise the transient nature of former shoreline features in relation to present sea-level position. The tectonic processes and examples of tectonically affected palaeoshoreline features discussed in this chapter are restricted to Late Quaternary time, generally less than 500 ka old.

The tectonic component of flights of marine terraces is determined in a number of ways, which generally include the assumption that rates of tectonic uplift through time have been uniform. The deformation of marine terraces at the Huon Peninsula, New Guinea, support this common assumption (Bloom *et al.*, 1974). However, data on Holocene and last Interglacial uplift rates from Japan and New Zealand illustrate changes in uplift rates through time. Many localities close to plate margins, especially convergent margins, show deformation dominated by folding leading to rapid lateral variations in rates of tectonic movement (differential vertical tectonics). In these environments it is common for the axis of fold structures to migrate with time, so that palaeo-shoreline positions are not only differentially displaced in space but the rate of uplift at any one locality may also vary. The relationships of tectonics and sea-level fluctuations in such environments are thus complex.

What are the tectonic processes?

Long-term global processes, embodied in plate tectonic theory (Cox, 1973), are involved in generating crustal deformation which affects the position of sea-level relative to adjacent landmasses. The driving mechanism of plate tectonics is considered to be convective flow within the mantle (McKenzie, 1968), which causes rifting and spreading at the diverging segments of plate boundaries. Assuming that the earth is not expanding as rapidly as the plates grow, various types of plate interaction must occur to take up excess lateral movement (Fig. 5.1). The rate of interactive processes at plate margins depends on the rate at which material is intruded along spreading centres and the rate of lateral spreading (Le Pichon, 1968) (Fig. 5.2).

Figure 5.1: Block diagram illustrating schematically the configurations and roles of the lithosphere, asthenosphere and mesosphere in processes of plate tectonics. Arrows on lithosphere indicate relative movements of adjoining blocks. Arrows in asthenosphere represent compensating flow in response to downward movement of segments of lithosphere. One arc-to-arc transform fault appears at left between opposing facing zones of convergence (island arcs), two ridge-to-ridge transform faults along ocean spreading ridge at centre. Simple arc structure at right

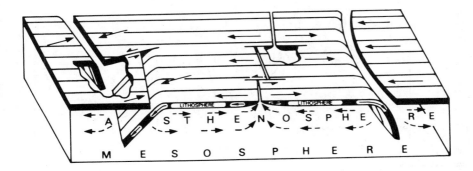

Source: Isacks *et al.* (1968).

Continents are carried by oceanic crustal material and display essentially three styles of deformation at their margins.

Convergent margins

These dynamic margins occur at a plate boundary where compressive inter-action, generally between a continental landmass and oceanic plate, occurs. Volcanic island chains adjacent to deep ocean trenches (island arc-trench system, see Fig. 5.3) are also characteristic of convergent margins. Oceanic crustal rocks are denser than continental rocks and, at large distances from the spreading centres, tend to flex downward. Driven by the gravitational sinking of old oceanic lithosphere and intrusion of new crustal material at spreading centres, slabs of oceanic crust plunge beneath the adjacent continental landmasses (subduction process). At later stages of subduction, oceanic crust is remelted and extruded through partially melted continental rocks forming volcanoes 50-300 km landward of the associated trench axis (Gill, 1981).

The locations of convergent margins are often clearly seen on bathy-metric maps by the location of deep oceanic trenches, and are further characterised by the occurrence of dipping zones of seismicity (Benioff Zone) beneath adjacent continental landmasses or island arcs. Interactions at plate boundaries vary from extension to simple shear, through oblique

129

Figure 5.2: Crustal plates and rates of movement of those plates. The lines with arrowheads pointing outward show where plates are moving apart at ridges and lines with opposed arrowheads show where plates are moving toward each other, usually at trenches. Numbers indicate the relative velocity of plates in cm a^{-1}

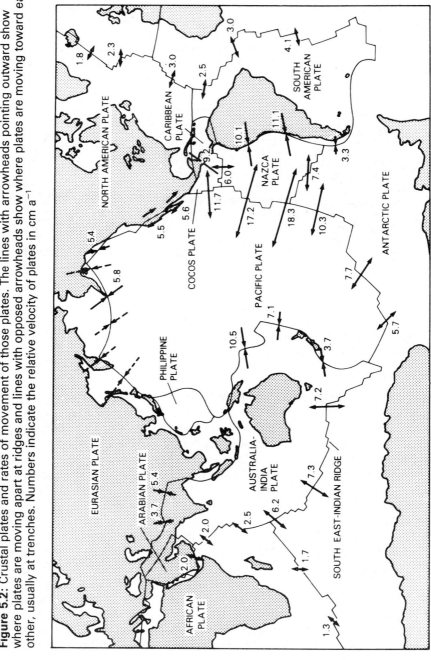

Source: After McKenzie and Richter (1976).

130

convergence to head-on convergence. Rates of convergence vary according to spreading rates and orientation of boundaries to the spreading direction but generally lie in the 10-100 mm a^{-1} range. The East Pacific Rise is an active spreading centre and associated convergence about the affected margins of the Pacific Ocean ranges from 20 to 180 mm a^{-1} (McKenzie and Richter, 1976; Chase, 1978; Minster and Jordan, 1978). These rapid rates of convergence often result in rapid thickening of the sedimentary pile accreted to the continental landmass by imbricate thrust faulting. As a result rapid uplift occurs. Horizontal datums such as shoreline features in this tectonic environment may be subject to rates of tectonic uplift of up to 10 mm a^{-1} (Yonekura, 1975; Turcotte, 1983).

Where subduction occurs beneath island arcs, rather than continental landmasses, lesser rates of uplift of islands are often observed even though convergence rates remain similar. The late Pleistocene uplift of islands in the Tonga chain in the south Pacific Ocean are examples of this style. Taylor and Bloom (1977) found that last Interglacial reef complexes data at 133 ± 12 ka are only 6.7 m above present sea level. These data indicate little if any Late Pleistocene uplift although tectonic convergence across the Tonga trench is about 70 mm a^{-1} (Chase, 1978) and there is an active, deep, earthquake zone. Similar slow rates of tectonic uplift are observed in the Mariana Islands.

If the contrast in densities between converging plates is not as marked as that between oceanic and continental crust then a variety of collision zones occur. These situations are common where plate boundaries pass between and within continental landmasses such as the Himalayan suture zone and Taiwan. In Taiwan, subduction zones terminate to the east and southwest. Between these margins a 300 km-wide zone comprising Taiwan and adjacent offshore areas is undergoing rapid, east-west convergence which began in the Late Pliocene (Wu, 1978). The fault system is complex with many thrust faults, particularly in the west of the island. At the Longitudinal Valley fault system in the east of the island reverse faulting with a sinistral shear component has been observed. Uplift rates of coastal marine terraces are notably higher than from the Ryukyu island arc to the northeast. Holocene marine features are found to 46 m above present sea level and coral from 35 m altitude has been dated at 6,132 ± 184 BP (Pirazzoli, 1978). In other parts of Taiwan, uplift rates that range between 1.8 and 9.4 mm a^{-1} have been calculated (Yonekura, 1983).

Passive margins

In many shoreline situations around the world both continent and ocean form parts of the same crustal plate and no relative motion occurs between continental crust and oceanic crust. Examples occur along the present-day continental margin on both sides of the Atlantic Ocean. The mid-Atlantic spreading centre is currently moving at 13 mm a^{-1} (Chase, 1978), but the eastern margin of the Americas is moving with the oceanic spreading, thus widening the Atlantic Ocean. The potential to document sea-level fluctuations, uncluttered by tectonic effects, may therefore exist along the eastern seaboard of the USA and Canada. However, north of about latitude 33°N the continent is affected by isostatic rebound following ice melt of the last glacial episode (glacio-isostasy) and by forebulge subsidence (Walcott 1972). Newman *et al.* (1980) have discussed crustal deformation brought about by water loading (hydro-isostasy) and sediment loading on the continental shelf areas of the eastern USA during the rise in sea level following the last glacial maximum. These factors make the resolution of eustatic sea-level (sea level relative to the geoid) position in this apparently favourable environment more difficult than on first inspection (see Chapters 3, 4 and 10 for further discussion).

Bermuda is situated off the eastern seaboard of North America at 32°N 65°W. It is a particularly important site for eustatic sea-level studies since it is situated about 1200 km eastward from the continental landmass and is on a 'stable' mid-oceanic platform. Harmon *et al.* (1983: p. 42) indicate that

> by comparison [with tectonically emergent coastal areas] the palaeo sea-level record of stable carbonate platforms, coastal areas and oceanic islands is much more complex, typically consisting of intricate and complex associations of both terrestrial and marine deposits and erosional features which are vertically compressed into a zone within a few metres of present sea level.

Bermuda is of particular importance because of the probable stability of the carbonate platform for at least the past 1 Ma. There exists here a complex sequence of marine carbonates overlapping and interfingering with aeolianites that are related to high stands of sea level. Residual soils and, in limestone caverns, extensive speleothems have developed during glacial periods. It has therefore been feasible to date and determine the possible eustatic position of high stands of sea level with respect to present sea level during the past ~ 200 ka. From the dated positions of speleothems, the heights of maximum sea-level in the intervening intervals have been determined. Similar stratigraphic sequences have also been dated from near-stable areas in Florida (Broecker and Thurber, 1965), the Bahamas (Mesolella *et al.*, 1969; Neumann and Moore, 1975) and Mexico (Szabo *et al.*, 1978). These and other studies emphasise the global synchroneity of several sea-level fluctuations and especially the '+2 to +10 m' sea-level positions that occurred about 118-130 ka ago (oxygen isotope stage 5e).

Rifting margins

Present-day rifting situations where palaeo sea-level relationships can be studied are limited to isolated localities on spreading centres such as Iceland, narrow spreading zones such as the Gulf of California and the Red Sea or to back arc and marginal seas behind island arc convergent margins (Fig. 5.3). The use of palaeo-shorelines in such rifting regions as datum horizons is little studied in relation to Quaternary vertical deformation, largely because net downdrop below the present relative high-stand of the sea has obscured most palaeo-shoreline features. For example, in a transect across the Taupo Volcanic Zone, a probable back-arc spreading centre in the North Island of New Zealand, Chappell (1975: Fig. 8) found that Late Pleistocene marine terraces increase in height both east and west of the extensional zone. The lack of discernible marine terraces within the extensional zone suggests tectonic downdrop.

Figure 5.3: Types of arc-trench systems and their nomenclature

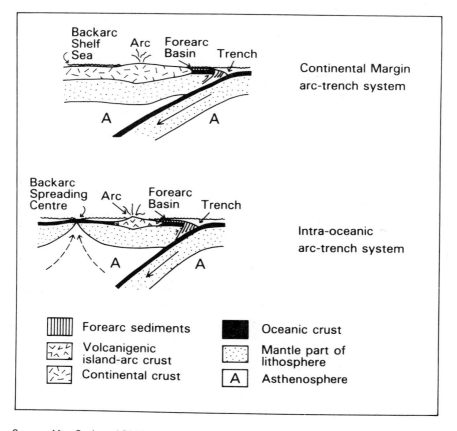

Source: After Seely and Dickinson (1977).

Reconnaissance studies of marine terraces in the rifting Gulf of California (Orme 1980) illustrate the complex vertical deformation pattern adjacent to the complexly faulted East Pacific Rise (Fig. 5.4). Within the Gulf short spreading segments of the East Pacific Rise are separated by very active strike-slip faults. Locally, sequences of marine terraces that range back to Late Pliocene or Early Pleistocene in age rise to 340 m

Figure 5.4: Structural features of Baja California and altitudes of Pleistocene and last Interglacial marine terraces. Inset map shows location in relation to major tectonic components of western North America

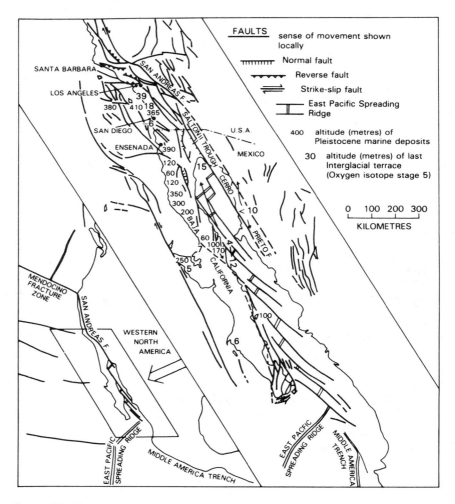

Source: After Orme (1980).

above present sea level and this uplift is closely associated with faults that occur close to the coastline. However, elsewhere, marine terraces are rarely found more than 20 m above present sea level. One or more low terraces correlated to the last interglacial period (oxygen isotope stage 5) are widespread and well preserved along both shores of the Gulf of California, with beach deposits and wave-cut terraces occurring between 4 m and 12 m above present sea level (Fig. 5.4). Thus general vertical stability with only local subsidence and uplift is indicated by the height distribution of marine terraces along the Gulf of California.

HISTORIC ACCOUNTS OF TECTONIC EVENTS AFFECTING SHORELINES

Two examples of historic tectonic events in coastal areas are introduced here to illustrate some of the early scientific observations of the effect of tectonism on shoreline elevation. Additionally the observations serve as *a priori* cases for the geological interpretation of prehistoric occurrences of coastal uplift.

Boso Peninsula, Japan, earthquakes of AD 1703 and AD 1923

The Boso Peninsula is situated on the southeast coast of Honshu facing the Japan Trench and the Sagami Trough (Fig. 5.5). Large magnitude ($M_s >$ 8) earthquakes struck the region in 1703 and 1923 and resulted in dramatic uplift of the coastal area. Imamura (1925), cited in Matsuda *et al.* (1978), reported pre-1703 documents that recorded the pre-earthquake shoreline position. These clearly demonstrated the permanent 800 m advance of land at the southern part of the Boso Peninsula that occurred at the time of the 1703 earthquake. Imamura also reported an old map of the fishing port of Aihama (Fig. 5.5b) drawn in 1654. This map clearly showed a different coastal physiography and the 6 m-high marine abrasion platform observed at Aihama today is attributed to the 1703 event. Farther east at the promontory at Nojimazaki (Fig. 5.5b) there was a sea stack about 200 m offshore prior to 1703. During the 1703 earthquake uplift connected this stack to the main landmass by a rocky terrace elevated about 5 m above present sea level. This terrace is known throughout the region as the Genroku Terrace. In 1923, the south Kanto area including the Boso Peninsula was again struck by a magnitude 8 earthquake. Although the uplift pattern for the 1923 event differs from the 1703 event, the southern part of the Boso Peninsula was uplifted by up to 1.5 m.

It was these historic events in the Boso Peninsula that led Sugimura and Naruse (1954, 1955) to conclude that the stepped topography of the Holo-

Figure 5.5: A. Map of Japan showing locations of trenches and troughs. Arrows indicate direction of convergence on subducting plate. Numbers in square boxes are localities listed in Table 5.1. B. Map of the Boso Peninsula, (locality 5 in A) showing extent of Holocene marine terraces (Numa Terraces) and localities mentioned in text

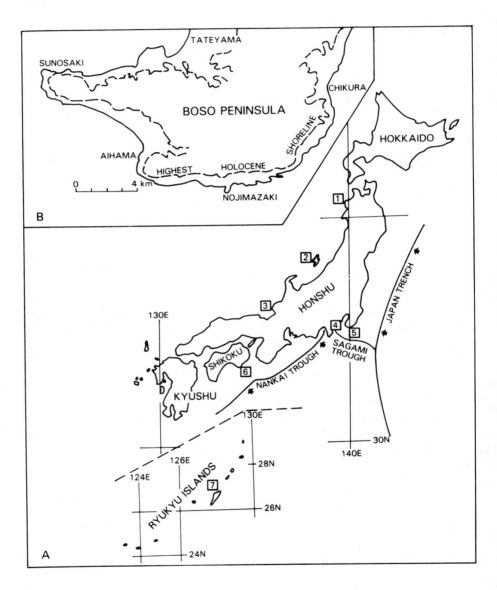

cene coastal plain in several parts of Japan is the result of episodic earth-quake events that uplift marine abrasion platforms formed in interseismic periods. Yoshikawa *et al.* (1964) extended this concept to suggest Late Pleistocene marine terraces were also uplifted by periodic tectonic events. Sugimura and Naruse (1954, 1955) recognised four Holocene terraces on the Boso Peninsula and demonstrated that the height distribution of each of the four marine terraces had a similar form. If each of the palaeo-shoreline features is representative of a tectonic uplift event, then four palaeo-shorelines have been created in the past ~ 6 ka (Yonekura, 1975) with essentially a stationary eustatic position of sea level. In other localities, where the historic and prehistoric dating control is not so rigorous, it becomes increasingly difficult to interpret stepped topography in terms of land uplift relative to a stable sea level, or to rapid eustatic sea-level falls with tectonic stability or, indeed, a combination of both processes.

Central Chile earthquake of 20 February 1835

In 1835 Charles Darwin was visiting Chile aboard the *Beagle* under the command of Captain Fitzroy. On 20 February Darwin was in Valdivia where he experienced strong earth tremors and observed a moderate seiche at the time of the earthquake. On 4 March, the *Beagle* arrived in Concepcion (Fig. 5.6) to a scene of total devastation caused largely by the effects of a large tsunami. There were eye-witness reports of ground fissures and landslides, and of people and animals being thrown to the ground during the earthquake. There was evidence that the land at Concepcion had been uplifted 0.6 to 1.0 m relative to sea level, but because the coastline was composed mostly of sandy beaches the evidence of elevated shoreline features was not well recorded. However,

> [on] the island of Santa Maria [~ 50 km away] ... the elevation [coseismic uplift] was greater; on one part Captain Fitzroy found beds of putrid mussel-shells still adhering to the rocks ten feet [~ 3 m] above the high-water ... the inhabitants had formerly dived at low-water spring tides for these shells. ... The elevation of this province is particularly interesting from its having been the theatre of several other violent earthquakes, and from the vast number of sea-shells scattered over the land up to a height of certainly 600 [feet] and I believe 1,000 feet [200-300 m]. (Darwin, 1851: p. 300)

Together these examples from Japan and Chile serve to illustrate how palaeo-shoreline indicators can be used to establish the position of a past sea level. They also illustrate how tectonic histories may be elucidated using palaeo-shoreline features.

Figure 5.6: Map of the area about Concepcion, southern South America. 1. Arrows indicate convergence direction on subducting plate. 2. Margins of central valley of Chile. 3. Amount of uplift (metres) that occurred at the time of the 1835 earthquake. 4. Contours of the amount of uplift in metres. The location of Santa Maria Island and Mocha Island are shown

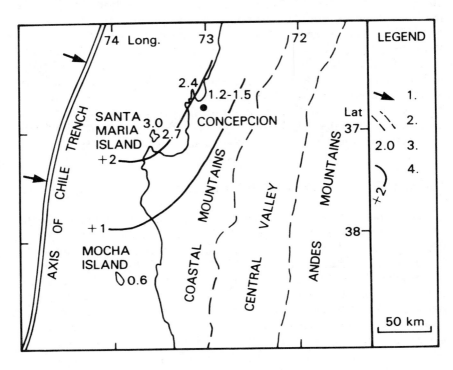

Source: After Kaizuka *et al.* (1973).

EXAMPLES OF RELATIONSHIPS BETWEEN RELATIVE SEA-LEVEL POSITION AND TECTONISM

If sea level were to oscillate between limits controlled by climatic factors then a single high-stand and single low-stand of sea level would be recorded with each successive cycle. If, however, some form of episodic or creep-wise uplift is imposed on the land/sea boundary then successive high-stands and low-stands of sea level will be recorded as a flight of marine terraces. However, because climatic oscillations through time have not been of equal magnitude the pattern is not so simple. Further, sea-level fluctuation has itself imposed smaller order deformations of shorelines by

138

loading the continental shelf with the water column (see Peltier, this vol., Ch. 3). Other complicating factors include variations in uplift rate through space and time, and the possibility of superposition and destruction of palaeo-shoreline positions.

Variable uplift rates in space and time

(1) At the Huon Peninsula, New Guinea (Fig. 5.7a), a spectacular uplifted marine terrace sequence is well documented (Bloom *et al.*, 1974; Chappell, 1974, 1983; see also Pillans, this vol., Ch. 9) and is now widely accepted as the most important sequence in establishing an absolute (eustatic) sea level curve for the Late Quaternary. This sequence of marine terraces consists of about 20 dated coral terraces and others that are beyond the range of present dating methods. The terraces can be traced for about 80 km along the northeast coast of the Huon Peninsula and rise to about 600 m above present sea level. The peninsula faces the New Britain trench and is above a zone of shallow, southwest dipping, seismicity suggestive of an active, convergent plate boundary (Denham, 1969) (Fig. 5.7). Individual terraces vary in height when traced along the 80 km of coastline. The terrace sequence is disrupted by a series of active faults (Fig. 5.8). Several terraces in the sequence have been age-dated by the ^{230}Th/^{234}U (Bloom *et al.*, 1974) and ^{14}C methods (Polach *et al.*, 1969) (see Sutherland, this vol., Ch. 6, for discussion of these techniques).

Three eustatic sea-level tie points can be applied to the New Guinea data. Chappell (1974, 1983) firstly adopted the 120 ka shoreline as a datum. This terrace occurs in many parts of the world. At more than 30 localities considered to be 'near-stable' (although opinions differ as to the identification or existence of such zones, see this vol., Chs 2, 3 and 4), it occurs in a narrow height range of 6 ± 4 m above present sea level. Additional terraces have been dated at 'near-stable' Florida and Barbados at 82 ka and 105 ka (Broecker and Thurber, 1965; Broecker *et al.*, 1968), suggesting eustatic sea-level positions at these times of −10 m and −15 m respectively. Terraces at both these ages have also been described from the Huon Peninsula at elevations ranging from 60 m to 250 m asl. Assuming constant uplift rates in the time period between dated terraces, and accepting the eustatic tie points of + 6 m at 120 ka, −10 m at 105 ka and −15 m at 82 ka, then eustatic sea-level positions for all the dated terraces at the Huon Peninsula have been determined (Chappell, 1983). However, the sequences of dated terraces, on eleven different transects (Bloom *et al.*, 1974; Chappell, 1974) across the peninsula, show differing altitude ratios between terraces in adjacent transects and illustrate non-uniform uplift in space (Fig. 5.8). The analysis requires a change in uplift rate through time in the most rapidly uplifting area,

Figure 5.7: A. Locality map showing the position of the Huon Peninsula, on northeast coast of New Guinea. B. Geophysical context of tectonism at the Huon Peninsula showing volcanic and seismic zones. Pacific and Australian crustal plates are shown, and also orientation and rate of convergence across the plate boundary

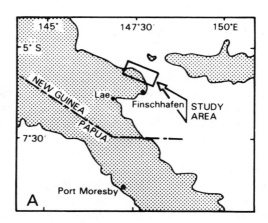

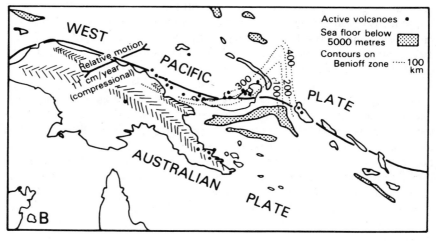

Sources: Denham (1969) (Benioff data); Chappell (1974).

southeast of Kanzarua, such that a roughly uniform rate characterises the period 0-80 ka and a different, slightly higher, rate characterises the period 80-133 ka (Bloom *et al.*, 1974). These data illustrate both the importance of well-dated uplifted terrace sequences in establishing eustatic sea-level fluctuations and also the variability of uplift rate in space in active tectonic environments.

(2) Another set of data that illustrates variable rates of tectonic uplift over

Figure 5.8: Map of the uplift pattern of coral reef marine terraces at the Huon Peninsula. The rate (m ka^{-1}) reaches a maximum about Kanzarua and Tewai. Principal faults, isochrons based on palaeo-shoreline positions and other localities discussed in the text are also shown. Tentative estimates of uplift rate (dashed lines) are based on assumed uniform uplift prior to 230 ka along a traverse inland from Sialum

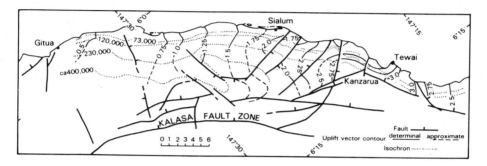

Source: Chappell (1974).

different time periods is the comparison of average uplift rates from localities where both Holocene and Late Pleistocene marine terraces are preserved. Data are presented from Japan (Fig. 5.5, Table 5.1) from a compilation by Ota and Yoshikawa (1978) and from New Zealand (Fig. 5.9, Table 5.2). The rates derived from the Holocene data have higher uncertainties than for the period back to the last Interglacial. Age dating, altitude measurement, eustatic sea-level position and position within a cycle of deformation contain errors that result in uncertainties up to 40 per cent of calculated Holocene uplift rates. Equivalent errors in calculating rates from the last Interglacial terrace are no more than 5 per cent. Therefore, if the ratio of the Holocene to the Late Pleistocene uplift rate exceeds about 1.5, then differences in rates (the Holocene rates being generally more rapid) are considered significant. Both data sets (Tables 5.1 and 5.2) illustrate consistently faster Holocene rates of tectonic uplift. Because most of the sites in Japan and New Zealand are on short wavelength folds at convergent plate margins, it may be that these structures characteristically begin their development at a slow rate and then accelerate. This situation is, however, the reverse of that documented from the Ventura Anticline in southern California by Keller *et al.*, (1982). Whatever the structural implications, there are clearly potential shortcomings to the common assumption that tectonic uplift rates at active folds and faults are constant through time.

(3) In active tectonic areas such as at convergent margins, variable uplift of palaeo-shoreline position in space and time make correlation difficult on

141

Figure 5.9: Map of New Zealand showing positions of subduction zones. Arrows on subducting plates indicate rate and orientation of subduction. Localities are those listed in Table 5.2. The Alpine Fault, the transform fault linking the two subducting margins, is also shown

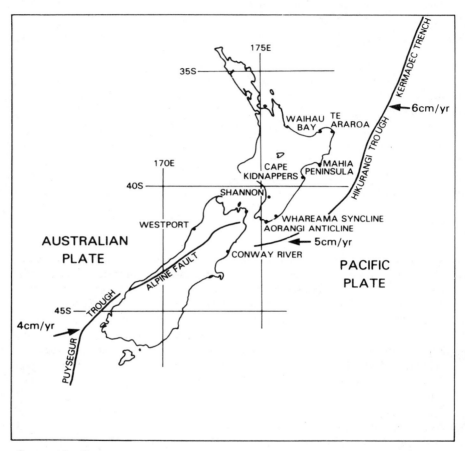

Source: After Chase (1978).

the basis of height alone. The development of forearc basins at convergent plate margins is very variable and many types are recognised (Seely and Dickinson, 1977). Most have in common (using the terminology of Seely and Dickinson, 1977) an arc massif, a shelf or slope basin, an outer structural high, a subduction complex and an outer oceanic arch (Fig. 5.10). It is common for the subduction complex to migrate toward the trench as subduction proceeds and for new structures, generally folds and thrust faults, to develop closer to the trench. Deformation also continues in the structural high and earlier formed structures of the subduction complex.

142

Table 5.1: Average rate of tectonic uplift estimated from heights of palaeo-shorelines — Japanese data. Locations are shown in Figure 5.5

Locality	Uplift rate – Holocene mm a^{-1}	Uplift rate – Last Interglacial mm a^{-1}	Ratio Holocene/ Last Interglacial
1. Shirakami Mts	1.3	0.7	1.9
2. Sado Island	1.2	0.9	1.3
3. Nyu Mts	1.0	0.9	1.1
4. Oiso Hills	3.8	1.2	3.2
5. Boso Peninsula*	1.5	1.0	1.5
6. Muroto Peninsula	4.0	1.5	2.7
7. Kikai Island	1.8	1.5	1.2

Note: *Not maximum rates but altitudes measured from equivalent position on structure. Ratio will therefore be applicable.
Source: Ota and Yoshikawa, 1978.

Table 5.2: Average rate of tectonic uplift estimated from heights of palaeo-shorelines — New Zealand data. Locations are shown in Figures 5.9 and 5.11. Data sources are listed

Locality	Uplift rate – Holocene mm a^{-1}	Uplift rate – Last Interglacial mm a^{-1}	Ratio Holocene/ Last Interglacial	Reference
1. Waihau Bay	1.0	0.6	1.7	Sumosusatro (1984)
2. Te Araroa	2.9	2.4	1.2	Yoshikawa et al. (1980)
3. Mahia Peninsula	3.1	1.3	2.4	Berryman unpub. data
4. Cape Kidnappers*	2.6	0.8	3.3	Hull (1985)
5. Whareama Syncline	1.3	0.5	2.6	Ghani (1978)
6. Aorangi Anticline	2.7	2.0	1.3	Ghani (1978)
7. Conway River	2.0	0.8	2.5	Ota et al. (1984)
8. Westport	0.8	0.4	2.0	Nathan (1975)
9. Shannon	0.4	0.2	2.0	Hesp and Shepherd (1978)

Note: *Not maximum rates but altitudes from equivalent position on structure. Ratio will therefore be applicable.

Deformation occurs characteristically by reverse faulting and folding overlying decollement zones within shallow, landward dipping sediments above the descending oceanic lithosphere. Folding plays an important role in accommodating horizontal shortening by crustal thickening. As deformation proceeds, older thrust faults are tilted and imbricated and smaller basins develop on the flanks of folds. The potential for rapid changes in the locus of maximum deformation and the development of parasitic faulting and folding on larger structures is apparent. Shoreline features are affected in similar fashion.

143

Figure 5.10: Generalised forearc model showing typical structural, morphological and volcanic elements. Generalised motion relative to sea level is shown at lower left

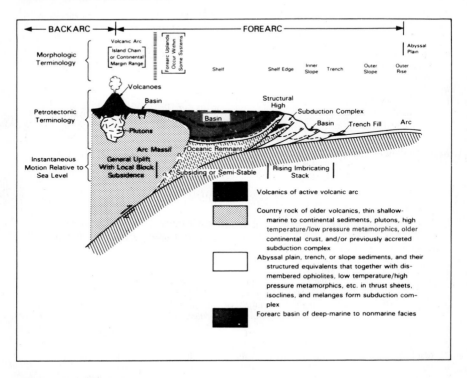

Source: Seely and Dickinson (1977).

Many of the marine terrace sequences documented on emergent coast-lines occur within this structural framework, for example along the southeast coast of Japan, the south coast of the Indonesian islands, the Gulf of Alaska and the east coast of the North Island of New Zealand. At Cape Kidnappers (northeast New Zealand — Fig. 5.9) deformation of the last Interglacial marine terrace defines a young, active anticline (Fig. 5.11) that itself occurs on the west flank of a more major active anticlinal structure. The last Interglacial shoreline is cut in coarse clastic marine sediments dated at less than 1 Ma. Figure 5.12 shows the northwest dipping Mid Pleistocene strata truncated by the last Interglacial shoreline that has subsequently been deformed into an anticline. The geological history illustrates the rapid lateral variability in locus of deformation. Since the Mid Pleistocene, a rapidly subsiding, shallow marine and terrestrial basin has filled, then been uplifted and deformed. Shoreline features have then been cut

144

Figure 5.11: Structural elements of the Cape Kidnappers areas of northwest New Zealand (see Fig. 5.9 for location). Altitudes of marine terrace remnants, broken by normal faults, define a parasitic anticline on the flank of the older structure with an axis to the east

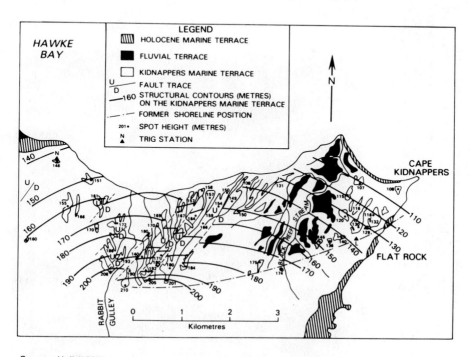

Source: Hull (1985).

across the folded sediments and have been subsequently folded into an anticline with an amplitude of about 100 m and uplifted by ~ 200 m at the axis of the anticline (Hull, 1985). An uplifted Holocene marine terrace also occurs at this locality; its deformation is accordant with the last Interglacial terrace but it is being uplifted at a faster rate (Table 5.2).

Superposition or erosion of palaeo-shoreline features in tectonic environments

The Huon Peninsula, New Guinea, Late Pleistocene sea-level curve is now based on 12 dated coral terraces formed in the past 133 ka (Chappell, 1983) and 20 over the past 400 ka. If land uplift here has been uniform through time (but as we have seen this assumption is not always acceptable) a unique altitudinal spacing of terraces would exist for any particular

145

Figure 5.12: Oblique air photograph of Cape Kidnappers area looking southwest onto the nose of a north-trending anticline defined by deformation of dissected terrace remnants of last Interglacial age. Marine terrace is cut with angular unconformity in Mid-Pleistocene marine and terrestrial gravels/sands which have been tilted on the northwest flank of an older anticline. Numerous normal faults occur near the axis of the young parasitic fold and can be seen disrupting the terrace surface and the older strata along cliffs in the foreground. (Photograph: D.L. Homer, New Zealand Geological Survey, Reference Number A9198b.)

uplift rate. Altitudes of terrace formation since 133 ka record an overall decline in sea level to an ~ 18 ka low point of at least −120 m (Bloom *et al.*, 1974; Veeh and Veevers, 1970). This progression of successively lower sea-level maxima since the last Interglacial, when coupled with uplift, is the reason for the preservation of flights of marine terraces with age decreasing with elevation. Some terraces may have been destroyed by subsequent erosion, or have been inundated by rising sea level if uplift rates were less than about 1 mm a^{-1}.

To give an example (using the terrace numbering system of Chappell (1983), modified by Bull (1984)), the 6A terrace (176 ka) would not have been preserved unless an average uplift rate of about 1.0 mm a^{-1} was attained (Table 5.3) and inundation would have destroyed the 4A (76 ka) and 5B (94 ka) terraces unless an average uplift rate of about 2 mm a^{-1} was attained. From Table 5.3 it can be seen that at low average uplift rates (say 0.3 mm a^{-1}) the only marine terraces emergent are restricted to Holocene bench(es) and the commonly prominent Late Pleistocene terraces 5A (83 ka), 5C (103 ka) and merged 5D/5E (120-133 ka).

Table 5.3: Altitudes of global marine terraces along a hypothetical coast with four different rates of uniform uplift during the last 176 ka. Negative signs for present altitude indicate terrace levels below present sea level. Brackets indicate those terraces that are unlikely to be preserved because of submergence subsequent to their formation. Asterisks indicate those terraces that are unlikely to be separated in field situations because of their similar present-day altitudes. Column 1 are marine isotope stages, and the capital letters designate separate global high stands of sea level. Column 2 are marine terrace ages. Column 3 are formation heights of global marine terraces. Columns 4-7 are predicted present-day altitudes at various constant uplift rates

1	2	3	4	5	6	7
No.	Age (ka)	Formation ht (m)	0.3 mm a^{-1}	0.6 mm a^{-1}	1.0 mm a^{-1}	2.0 mm a^{-1}
1A	6	0	+2	+4	+6*	+12
2A	30	−42	(−33)	(−24)	(−12)	+18
3A	40	−37	(−25)	(−13)	(+3)	+43
3B	46	−37	(−23)	(−9)	+9*	+55
3C	58	−29	(−12)	(+6)	+29	+87
3D	62	−26	(−7)	+12	+36	+98
4A	76	−46	(−23)	(0)	(+30)	+106
5A	83	−13	+12	+37	+70*	+153
5B	94	−20	(+8)	(+36)	+74*	+168
5C	103	−10	+21	+52	+93	+196
5D	120	+6	+42*	+78	+114	+246
5E	133	+5	+45*	+85*	+128	+271
6A	176	−21	(+32)	+85*	+155	+331

Sources: After Shackleton and Opdyke (1973) and Chappell (1983).

147

However, at a moderate to rapid average uplift rate (say 2 mm a^{-1}) a complete flight of terraces recording all the high still-stands of sea level in the period since 176 ka may be preserved with steps of 10 to 60 m altitude between successive terraces (Table 5.3). Regression analyses based on altitude ratios have been used to assign ages to undated marine terrace sequences, or to test how well the fit is if uniform uplift is assumed (Ota *et al.*, 1968; Miyoshi, 1983; Bull, 1984). The assigning of ages based on height relationships is critically dependent on uniform land uplift rate and without independent dating control age assessments of terraces in these ways must remain speculative.

Palaeo-shoreline features arranged in stepped fashion

Sequences of marine terraces in many parts of the world display a distinctive morphology of near horizontal surfaces separated by steep cliffs. The mode of formation of four of these sequences from the circum-Pacific region is now discussed.

The Late Quaternary sea-level curves that have been developed indicate that high-stands of sea level have not exceeded +10 m or so in the past ~ 400 ka (Chappell 1974, Fig. 19) and provide clear evidence that flights of Late Quaternary marine terraces at elevations of tens to hundreds of metres above present sea level result from tectonic uplift. Flights of Holocene terraces that range in height up to 41 m above present sea level (Middleton Island, Alaska) in regions removed from the effects of isostatic rebound are certainly tectonically uplifted.

Middleton Island, Alaska

(See Fig. 5.13) The island is situated near the outer margin of the Alaskan continental margin facing the Aleutian Trench. Pleistocene marine sediments dipping at 22-30° NW (landward tilted away from the trench) underlie six Holocene marine terraces that constitute much of the geomorphology of the island. The six terraces are separated by steps of 3.5 to 9 m high and have been dated by wood in marine cover deposits and by terrestrial peats growing on the exposed abrasion platforms, or above the marine cover deposits (Fig. 5.14). A recurrence interval of uplift of 500 to 1350 years is calculated, with the last uplift of 3.5 m occurring at the time of the 1964 (M$_s$ 8.4) Alaskan Earthquake (Plafker and Rubin, 1978). It is interesting to note that this uplift was only half the magnitude or less of prior uplift events (maximum 9.0 m). Plafker and Rubin (1978) noted that, although it is possible the 1964 earthquake released all of the accumulated crustal elastic strain in this area, it is more likely that a large future earthquake will occur in a period that is short, compared with the average recurrence interval, to produce land uplift similar to the average recorded

Figure 5.13: Locality map of the Gulf of Alaska showing the position of the Aleutian Trench, the orientation and rate of convergence with arrows on the subducting plate, and the epicentre and deformation contours for the 1964 earthquake. The outer edge of the continental shelf at −2000 m is indicated by a dotted line and active or dormant volcanoes are shown by stars. Middleton Island is shown midway between the zone of maximum uplift and the trench axis

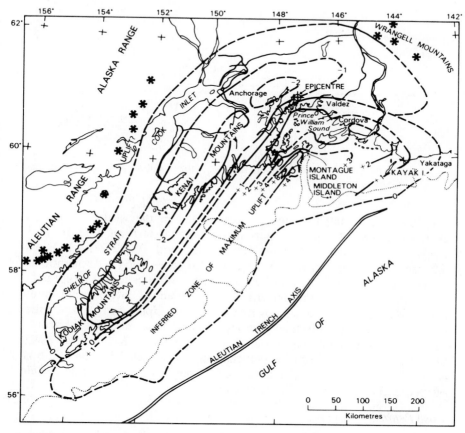

Source: Plafker and Rubin (1978).

by older terraces. The two closely spaced uplift events will then appear in the geological record as a single event, since insufficient time would have passed probably to have cut an abrasion platform after the 1964 earthquake and before the next uplift episode. It is possible that older, large steps between terraces are also the result of several closely spaced uplifts.

There has been a monotonic decrease in uplift rate here since the first terrace became emergent about 4.3 ka BP. The average uplift rate of

149

Figure 5.14: Height/age plots of uplifted Holocene marine terraces at Middleton Island (A), Boso Peninsula (B), Mocha Island (C) and Mahia Peninsula (D)

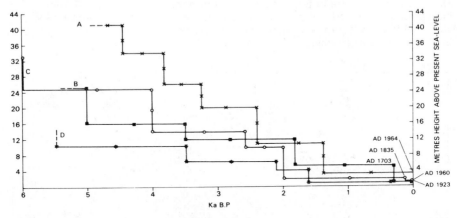

Sources: After Plafker and Rubin (1978); Matsuda et al. (1978); Kaizuka et al. (1973); Berryman, unpublished data.

9.4 mm a^{-1} is among the highest documented in the world. In detail it averages about 12.5 mm a^{-1} during the time of terraces I to III (4.3-3.2 ka), 8.7 mm a^{-1} between terraces III and V (3.2-1.4 ka) and 5.9 mm a^{-1} between terraces V and VI (1.4 ka — 1964). As mentioned above, the short-term rate may be low because of a postulated future uplift event and it is possible that uncertainties in the actual date of uplift events may smooth the data. Other possibilities are that uplift on the local structure is decreasing with time although maintaining constant horizontal shortening (as documented for the Ventura anticline — Keller et al., 1982) or that the assumption of near-constant relative sea level for the past 6 ka is significantly in error; a rise in sea level along with tectonic uplift would be recorded as a reduction in the rate of uplift.

Boso Peninsula, Japan

(See Fig. 5.5.) In addition to the historic earthquakes and uplift events recorded in 1703 and 1923, mapping and dating of the Holocene deposits on the Boso Peninsula reveal a flight of terraces with dated marine cover deposits, suggesting very rapid relative changes in sea level between terraces (Sugimura and Naruse 1954, 1955; Yonekura, 1975; Shimazaki and Nakata, 1980). The plot of uplift against age is similar to that obtained from Middleton Island (Fig. 5.14) although the average rate of uplift is slower. The occurrence of two uplift events closely spaced in historic time indicates the possibility that large discrete uplift events of 5 to 7 m, spaced at least 1 ka apart as suggested by the geological record, are also composite. The composite Genroku Terrace was constructed largely by the 1703 uplift of the Boso Peninsula and was only extended a small amount by the 1923 uplift event. It is probable that on a steep rocky coastline, such as at the Boso Peninsula, very detailed resolution of the prehistoric formation of marine terraces is difficult to determine. The geological record will indicate only a minimum number of uplift events but overestimate individual uplift amplitude.

Mocha Island, Chile

(See Fig. 5.6.) This island is situated near the outer edge of the continental shelf about 90 km eastward of the axis of the Chile Trench. Both Mocha Island and Santa Maria Island, where Darwin observed uplifted shoreline features immediately after the 1835 earthquake, have Late Quaternary marine terraces that show landward tilting (Kaizuka *et al.*, 1973). Mocha Island has higher- and middle-level terraces that range in altitude from 200 to 400 m asl. The lower terrace reaches 33 m above present sea level and is divided into several sub-levels. Each sub-level is overlain by a thin veneer of beach gravel, sand and shell fragments, all partly covered by sand dunes. The lower sub-levels are not all dated and the 33 m terrace is correlated, by extending average uplift rates, to the culmination of the postglacial rise in sea level, that is, about 6 ka BP in this area. The data of Kaizuka *et al.* (1973) are plotted as land uplift against age and again the pattern resembles other flights of terraces formed in regions facing seismically active subduction margins (see Fig. 5.14).

Mahia Peninsula, New Zealand

(See Fig. 5.9 for location.) This peninsula is situated on the east coast of the North Island of New Zealand, about 120 km west of the Hikurangi Trough. The region is subjected to oblique convergence as the Pacific Plate subducts at the Hikurangi Trough beneath the continental landmass of New Zealand (Australian Plate). Convergence of the Pacific Plate relative to the Australian Plate is calculated at 5.5 cm a^{-1} (Chase, 1978). Landward tilting of Late Pleistocene terraces occurs here. A flight of five Holocene terraces, which attain 15 m elevation above present sea level about the peninsula (Fig. 5.15), record episodic shifts in relative sea level (Berryman, 1983). The Holocene terraces have been dated by tephrochronology and radiocarbon methods and are interpreted as indicating that co-seismic uplift of 1 to 4 m occurs with a recurrence interval of 400 to 1500 years (Fig. 5.14).

The Mahia Peninsula data fit the time predictive model of co-seismic uplift proposed by Shimazaki and Nakata (1980) who tested the Boso Peninsula data described earlier. Other examples, such as Middleton Island, do not fit a time predictive model because of the observed change in uplift rate and

Figure 5.15: Oblique air photograph of Mahia Peninsula on the northeast coast of New Zealand. Looking northwest down the western flank of an anticline that has uplifted and tilted a sequence of Late Pleistocene marine terraces (left). In the foreground there is a sequence of uplifted Holocene marine terraces ranging in age from ~ 300 to ~ 3,500 a BP (see Fig. 5.9 for location and Fig. 5.14 for plot of Holocene terraces). (Photograph: D.L. Homer, New Zealand Geological Survey, Reference Number A9138b)

at Mocha Island there are uncertainties in the height distribution and age of some of the terraces. The time predictive model infers that uplift and large magnitude earthquakes occur when the upper stress state of a region is constant with time while the initial stress state varies. It follows, therefore, that the smaller the uplift event in flights of marine terraces, the shorter the time interval to the next uplift event. These qualitative observations have some support in the independent assessment of the likely future uplift at Middleton Island, since the 1964 uplift was small compared with the long-term average. The more recent events affecting Mocha Island (1835, 1960) are small compared to previous steps and large earthquake activity is likely in the future. At Mahia Peninsula the 1.1 m uplift event ~ 300 years ago is predicted to be followed within the next 250 years by an uplift event and accompanying large magnitude earthquake.

The general observations that can be made from these four cited examples of uplifted Holocene marine terraces are: the similarity in geomorphic form, the uniformity of tectonic process affecting coastal areas facing convergent plate boundaries and the usefulness of palaeo-shoreline positions as datums in tectonic studies.

Tectonism affecting shoreline position on mid-ocean islands

Several recent studies in both passive continental and mid-ocean regions have sought to use the height distribution of dated Holocene and Late Pleistocene marine terraces to define deformation created by lithospheric loading and thereby obtain an estimate of lithospheric rigidity and mantle viscosity. At the eastern continental margin of Australia, relative uplift and downwarp of Holocene reefs appear to be related to loading of the wide continental shelf by ocean water during the postglacial rise in sea level and to a lesser extent by tectonism (Chappell *et al.*, 1982; Hopley, 1983). In a different situation the possible existence and concept of a 'Higher-than-present eustatic sea level' at mid-ocean islands during the Late Holocene (Curray *et al.*, 1970), has only recently been rationalised to the view that there may have been different relative sea-level histories on islands that had once been considered tectonically stable (see Hopley, this vol., Ch. 12). Whereas workers in the 1960s and early 1970s developed global sea-level curves, more recently they have recognised both local and regional variations in eustatic sea-level position (Bloom, 1980; Clark *et al.*, 1978). Differences in relative sea-level histories have been used to elucidate the melting history of past ice sheets. Studies of geodynamic vertical motions involving the viscosity of the mantle have also used these data (Nakiboglu *et al.*, 1983) (see also Chapters 3, 4 and others in this volume).

A number of mid-ocean volcanic islands and atolls occur within the South Pacific Ocean with Pleistocene coral reefs up to at least 70 m asl.

Clearly these islands do not fit the pattern of submergence initiated by contraction of the oceanic lithosphere. Fairbridge (1961) and Veeh (1966) attempted to explain their elevation as the result of Pleistocene high-stands of sea level. Such high-stands are now generally discounted. McNutt and Menard (1978) noted that many of these raised volcanic islands and atolls were close to active or recently active basaltic shield volcanoes on the ocean floor. Menard (1964) has shown that the bathymetry across the oceanic volcanoes reveals moats and arches around each volcano, indicating that the oceanic crustal support for the excess weight of these volcanoes involves an annulus of deformation of the sea floor (Fig. 5.16).

Figure 5.16: Model of apparent sea-level change on coral atolls caused by volcanic loading on the elastic lithosphere. A. Atolls developed on two extinct volcanoes are in equilibrium with sea level on an oceanic lithosphere surface that is slowly sinking as it cools (right to left). B. Active volcanism loads the lithosphere inducing subsidence in a near-field 'moat' and emergence at greater distance (typically about 200 km). In the moat region new reef formation is induced by the subsidence and reefs become emerged at greater distance. C. Continued volcanic activity (volcano has greater volume and weight) causes further subsidence and uplift

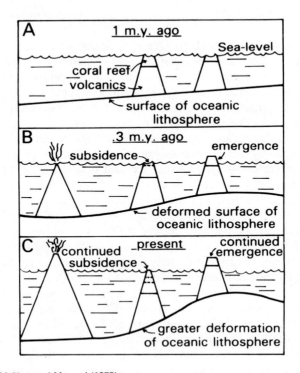

Source: After McNutt and Menard (1978).

154

The islands of the Southern Cook group (Fig. 5.17) at about 160°W, 20°S illustrate diverse relative sea-level histories over relatively short distances of about 200 km. Mangaia is a volcanic island reaching 169 m asl and is surrounded by raised, apparently Pleistocene, reef limestone up to at least 70 m asl. The volcanic rocks are oceanic alkali basalts and dated to 16.6-21.4 Ma (Dalrymple *et al.*, 1975; Jarrard and Turner, 1979). Two Th/U ages of 90 ± 20 ka and 110 ± 20 ka and a uranium series date of 110 ± 50 ka (Veeh, 1966) have been obtained from reef coral elevated about 2 m above modern equivalents. Recent dating of raised microatolls and cored coral reef on Mangaia (Yonekura *et al.*, 1983) have shown that sea level relative to the present rose from + 1.2 m to + 1.8 m between 5,020 ± 190 and 3,410 ± 170 BP, after which an erosion notch was

Figure 5.17: Southern Cook Islands, south Pacific Ocean, showing bathymetry (interval is 1000 m), ages of volcanic core rocks and noting evidence of relative emergence in Pleistocene, last Interglacial and Holocene time

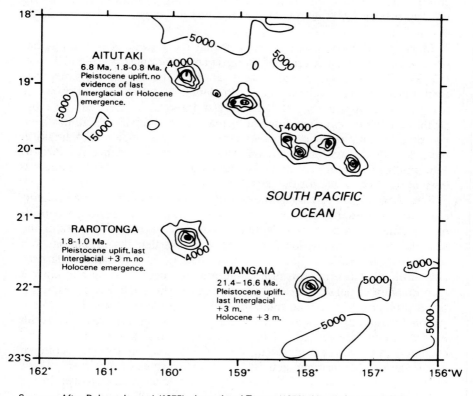

Sources: After Dalrymple *et al.* (1975); Jarrard and Turner (1979); Yonekura *et al.* (1983).

155

formed (~ 3,000 BP) and then sea level fell about 2 m to its present level.

Rarotonga, approximately 200 km northwest of Mangaia (Fig. 5.17) is a young volcanic island, formed by volcanism at 1.83-1.04 Ma. It reaches 653 m elevation and is surrounded by a fringing reef. Marine terrace deposits may extend to 30 m asl and ^{14}C dates of 28.2 ka and > 43.4 ka (Schofield, 1970) on shell within reef limestone up to 3 m asl are suggestive of last Interglacial age considering the degree of recrystallisation of the samples. No evidence of a higher than present relative sea level in the Late Holocene at Rarotonga was found by Yonekura et al. (1983). Aitutaki is about 250 km north of Rarotonga (Fig. 5.17) at the northwest end of a chain of volcanic islands and atolls. The island is volcanic in origin, with a 124 m hill at the north of the 'almost-atoll' exposing volcanic rocks. The bulk of the island formed about 6.8 Ma ago (Jarrard and Turner, 1979) with a younger phase at 1.8-0.8 Ma (Dalrymple et al., 1975). Elsewhere twelve small reef islands (motus; cf. Pirazolli et al., 1985) occur on the eastern, windward, reef and a further two small motus occur on the southern reef. Again no evidence of Late Holocene emergence has been found, and dating of coral from drillholes on the reef suggests sea level was above −0.7 m msl by 4,690 ± 100 BP and reached its present level about 3,000 BP.

McNutt and Menard (1978) suggest that the growth of Rarotonga has been responsible for the uplift of Mangaia and although Jarrard and Turner (1979) criticised some of the quantification of the deformation response they generally accepted the thesis. In detail, however, palaeo-shoreline features on Mangaia suggest that the last Interglacial reef may be close (within a few metres) to its original formation height, while there has been probable tectonic uplift of 2 m compared with Rarotonga and Aitutaki in the past 3,000 years. Similarly, on Aitutaki there is evidence for Pleistocene uplift of a young volcano that should generally be subsiding and has been apparently stable for the past 4,000 years.

These examples illustrate some of the complexities in Holocene relative sea-level histories that are now recognised on what were previously considered stable mid-ocean islands that served as 'dipsticks' in evaluating eustatic sea-level fluctuations. It is also apparent that the resolution, to a few metres, of sea-level fluctuation from tectonic deformation is particularly difficult. Uncertainties in altitude measurements, dating and correct identification of palaeo sea-level position in the shoreline feature are significant problems in resolving small relative changes in sea level. In addition, regional variations in the form and timing of sea-level rise occur as a result of: (a) sea-surface altitude change caused by atmospheric pressure and currents, (2) hydro-isostasy, (c) lithospheric loading due to sediment accumulation, (d) volcanic activity and (e) slow rates of tectonism. These may all have impacts on the recording of sea-level change if it is in the order of a few metres per thousand years.

156

CONCLUSIONS

(1) Tectonic processes resulting in vertical deformation have a visible impact on recording palaeo sea-level position. The examples of stepped coastal topography in tectonically active environments where former sea-level positions are now widely separated in altitude are some of the most striking.

(2) Rates of vertical tectonic movement are most rapid at convergent plate margins. Passive and rifting tectonic margins show, generally, tectonic stability with rates not exceeding 1 mm a^{-1}.

(3) In tectonic environments where rapid uplift occurs, palaeo sea-level position serves as an excellent, originally near-horizontal datum. Processes such as isostatic movement, induced by water loading on continental shelves, and other second order movements affecting palaeo sea-level position are small uncertainties in calculating rates of tectonic movement.

(4) Rapidly uplifting coastal environments, providing there is close age control of the former sea-level position, provide excellent sites in establishing a general or 'global' eustatic sea-level curve. The data from the Huon Peninsula, New Guinea are most important in this regard. It is necessary that several, dated palaeo sea-level positions in stable tectonic environments are known in order to provide tie-points with the tectonically affected data.

(5) Palaeo sea-level position can be estimated in stable environments, although superposition of successive sea-level still-stands makes the unravelling of the position of individual sea-level positions extremely complex. Age control of individual sea-level positions is again crucial.

(6) Second-order tectonic and isostatically related tectonic movements are best evaluated on stable or slowly rising coasts since their amplitude is small compared with tectonic events.

(7) Assumptions of uniformity in tectonic processes may not be justified in attempts to establish the former position of sea level. Deformation rates are demonstrated to vary on time scales of 1-100 ka years (Middleton Island, Alaska; Kanzarua area of the Huon Peninsula; Ventura Anticline, southern California). The locus of maximum deformation is seen to vary with time so that rapidly submerging basins are everted and the rates of movement of palaeo-shoreline features have no uniformity. At convergent margins the mechanism of uplift is often episodic and related to large magnitude earthquakes. Recurrence intervals of these events at affected coastal sites are commonly seen to vary by about 100 per cent.

(8) Past sea-level positions are not readily documented from mid-ocean island sites as once believed, especially if recent volcanic activity has occurred in the area. Recent work has documented regional variations in sea-level histories because of glacio-isostasy and lithospheric loading has induced recent and ongoing deformation about active mid-ocean volcanoes.

ACKNOWLEDGEMENTS

Firstly to Robert Devoy for an invitation to contribute to a very worthwhile volume. Several colleagues including Graham Bishop, Graham Mansergh, and Alan Hull (NZ Geological Survey), Brad Pillans (Victoria University of Wellington) and Yoko Ota (Yokohama National University) commented on various versions of this chapter. I appreciate their considered comments which have greatly improved it. Martin Heatherington drafted the figures and Alison Lee typed several versions of the text before its final form; their contribution is gratefully acknowledged.

REFERENCES

Berryman, K.R. (1983) 'Tectonic implications of the Mid-Late Holocene geology of Mahia Peninsula, East Coast, North Island, New Zealand' (abstract), *International Symposium on Coastal Evolution in the Holocene*, Komazawa University, Tokyo. A contribution to IGCP Project 200.

Bloom, A.L. (1980) 'Late Quaternary sea-level change on South Pacific coasts: A study in tectonic diversity', in Mörner, N.-A. (ed.), *Earth Rheology, Isostasy and Eustasy*, Wiley, Chichester and New York, pp. 505-16.

—— Broecker, W.S., Chappell, J., Matthews, R.K. and Mesolella, K.J. (1974) 'New uranium-series dates from the emerged reef terraces on Huon Peninsula, New Guinea', *Quat. Res.*, 4, 185-205.

Broecker, W.S. and Thurber, D.L. (1965) 'Uranium series dating of corals and oolites from Bahaman and Florida key limestones', *Science, 149*, 55.

—— Thurber, D.L., Goddard, J., Ker, T.L., Matthews, R.K. and Mesolella, K.J. (1968) 'Milankovitch hypothesis supported by precise dating of coral reefs and deep-sea sediments', *Science, 159*, 297-300.

Bull, W.B. (1984) 'Correlation of flights of global marine terraces', in M. Morisawa and J. Hack (eds), *Tectonic Geomorphology. Proceedings of the 15th Annual Geomorphology Symposium, State University of New York at Binghamton, Boston*, Allen & Unwin, London, pp. 129-54.

Chappell, J. (1974) 'Geology of coral terraces, Huon Peninsula, New Guinea: A study of Quaternary tectonic movements and sea-level changes', *Bull. Geol. Soc. Am., 85*, 553-70.

—— 'Upper Quaternary warping and uplift rates in the Bay of Plenty and west coast, North Island, New Zealand', *NZ J. Geol. Geophys., 18*, 129-53.

—— 'A revised sea-level record for the last 300,000 years from Papua New Guinea', *Search 14*, 99-101.

—— Rhodes, E.G., Thom, B.G., and Wallensky, E. (1982) 'Hydroisostasy and the sea level isobase of 5500 years BP in north Queensland, Australia', *Mar. Geol., 49*, 81-90.

Chase, C.G. (1978), Plate kinematics: the Americas, East Africa and the rest of the world. *EPSL, 37*, 353-68.

Clark, J.A., Farrell, W.E., and Peltier, W.R. (1978) 'Global changes in postglacial sea level: A numerical calculation', *Quat. Res., 9*, 265-87.

Cox, A. (1973) *Plate Tectonics and Geomagnetic Reversals*, W.H. Freeman, San Francisco.

Curray, J.R., Shepard, F.P., Veeh, H.H. (1970) 'Late Quaternary sea level studies in

Microneisa-Carmasel Expedition', *Bull. Geol. Soc. Am.*, *81*, 1865-80.

Dalrymple, G.B., Jarrard, R.D., and Clague, D.A. (1975) 'K-Ar ages of some volcanic rocks from the Cook and Austral Islands', *Bull. Geol. Soc. Am.*, *86*, 1463-7.

Darwin, C.A. (1851) *Geological Observations on Coral Reefs, Volcanic Islands and on South America; being the geology of the Voyage of the Beagle, under the Command of Captain Fitzroy, R.N. during the Years 1832 to 1836*, London.

Denham, D. (1969) 'Distribution of earthquakes in the New Guinea–Solomon Islands region', *J. Geophys. Res.*, *74*, 4290-9.

Fairbridge, R.W. (1961) 'Eustatic changes in sea level', in L.H. Ahrens, F. Press, K. Rankama and S.K. Runcorn (eds), *Physics and Chemistry of the Earth*, Pergamon Press, London, vol. 4, pp. 99-185.

Ghani, M.A. (1978) 'Late Cenozoic vertical crustal movements in the southern North Island, New Zealand', *NZ J. Geol. Geophys.*, *21*, 117-26.

Gill, J.B. (1981) *Orogenic Andesites and Plate Tectonics*, Springer-Verlag, Heidelberg.

Harmon, R.S., Mitteier, R.M., Kriausakul, N., Land, L.S., Schwarcz, H.P., Garrett, P., Lason, G.J., Vacher, H.L. and Rowe, M. (1983) 'U-Series and amino-acid racemization geochronology of Bermuda: implications for eustatic sea-level fluctuations over the past 250,000 years', *Paleogeography, Paleoclimatol., Paleoecol.*, *44*, 41-70.

Hesp, P.A. and Shepherd, M.J. (1978) 'Some aspects of the Late Quaternary geomorphology of the lower Manawatu Valley, New Zealand', *NZ J. Geol. Gephys.*, *21*, 402-12.

Hopley, D. (1983) 'Evidence for 15,000 years of sea-level change in tropical Queensland', in D. Hopley (ed.), *Australian Sea Levels in the Last 15,000 Years: A Review*, Monogr. Ser. Occ. Paper No. 3, Dept. of Geography, James Cook University of North Queensland, pp. 93-104.

Hull, A.G. (1985) 'Late Quaternary geology of the Cape Kidnappers Region, Hawke's Bay, New Zealand', unpublished MSc thesis, Victoria University of Wellington, New Zealand.

Imamura, A. (1925) 'Change of the coastline in Boso Peninsula', in *Report of the Imperial Earthquake Investigation Committee*, *100B*, pp. 91-4. (in Japanese.)

Isacks, B., Oliver, J. and Sykes, L.R. (1968) 'Seismology and the new global tectonics', *J. Geophys. Res.*, *73*, 5855-99.

Jarrard, R.D. and Turner, D.L. (1979) 'Comments on "Lithospheric flexure and uplifted atolls" by M. McNutt and H.W. Menard', *J. Geophys. Res.*, *84*, 5691-7.

Kaizuka, S., Matsuda, T., Nogami, M. and Yonekura, N. (1973) 'Quaternary tectonic and recent seismic crustal movements in the Arauco Peninsula and its environs, central Chile', *Geographical Reports of Tokyo Metropolitan University*, *8*, 1-50.

Keller, E.A., Rockwell, T.K., Clark, M.N., Dembroff, G.R. and Johnson, D.L. (1982) 'Tectonic geomorphology of the Ventura, Ojai and Santa Paula areas, Western Transverse Ranges, California', in *Neotectonics in Southern California*, Guidebook prepared for the 78th annual meeting of the Cordilleran section of the Geological Society of America, Anaheim, California, pp. 25-42.

Le Pichon, X. (1968) 'Sea-floor spreading the continental drift', *J. Geophys. Res.*, *76*, 3661-97.

McKenzie, D.P. (1968) 'The influence of the boundary conditions and rotation on convection in the earth's mantle', *Geophys. J. Roy. Astron. Soc.*, *15*, 457-500.

—— and Richter, F. (1976) 'Convection currents in the earth's mantle', *Sci. Am.*, *235*, 72-89.

McNutt, M. and Menard, H.W. (1978) 'Lithospheric flexure and uplifted atolls', *J.*

Geophys. Res., *83*, 1206-12.

:suda, T., Ota, Y., Ando, M., Yonekura, N. (1978) 'Fault mechanism and recurrence interval of major earthquakes in southern Kanto district, Japan, as deduced from coastal terrace data', *Bull. Geol. Soc. Am.*, *89*, 1610-18.

Menard, H.W. (1964), *Marine Geology of the Pacific.* McGraw-Hill, New York.

Mesolella, K.J., Matthews, R.K., Broecker, W.S. and Thurber, D.L. (1969) 'The astronomical theory of climatic change: Barbados data', *J. Geol.*, *77*, 250-74.

Minster, J.B. and Jordan, T.H. (1978) 'Present-day plate motions', *J. Geophys. Res.*, *83*, 5331-54.

Miyoshi, M. (1983) 'Estimated ages of Late Pleistocene marine terraces in Japan, deduced from uplift rate', *Geogr. Rev. Japan*, *56*, 819-34.

Mörner, N.-A. (1980) 'Eustasy and geoidal changes as a function of core mantle change', in N.-A. Mörner, (ed.), *Earth Rheology, Isostasy and Eustasy*, Wiley, Chichester and New York, pp. 535-54.

Nakiboglu, S.M., Lambeck, K. and Aharon, P. (1983) 'Postglacial sealevels in the Pacific: implications with respect to deglaciation regime and local tectonics', *Tectonophys.*, *91*, 335-58.

Nathan, S. (1975), *Sheets S23 and S30, Foulwind and Charleston (1st edn.), Geological Map of New Zealand 1:63,360 Map (1 sheet) and Notes*, New Zealand Department of Scientific and Industrial Research, Wellington, New Zealand.

Neumann, A.C. and Moore, W.S. (1975) 'Sea-level events and Pleistocene coral ages in the northern Bahamas', *Quat. Res.*, *5*, 215-24.

Newman, W.S., Ciquemani, L.J., Pardi, R.R. and Marcus, L.F. (1980) 'Holocene delevelling of the United States east coast', in N.-A. Mörner (ed.), *Earth Rheology, Isostasy and Eustasy*, Wiley, Chichester and New York, pp. 449-63.

Orme, A.R. (1980) 'Marine terraces and Quaternary tectonism; Northwest Baja, California, Mexico', *Physical Geography*, *1 & 2*, 138-61.

Ota, Y., Kaizuka, S., Kiluchi, T. and Naito, H. (1968) 'Correlation between heights of younger and older shorelines for estimating rates and regional differences of crustal movements', *Quat. Res.*, *7*, 171-81.

—— Yoshikawa, T. (1978) 'Regional characteristics and their geodynamic implications of Late Quaternary tectonic movement deduced from deformed former shorelines in Japan', *J. Phys. Earth*, *26* (Suppl.), 379-89.

—— Yoshikawa, T., Iso, N., Okada, A. and Yonekura, N. (1984) 'Marine terraces of the Conway Coast, South Island, New Zealand', *NZ J. Geol. Geophys.*, *27*, 313-26.

Pirazzoli, P.A. (1978) 'High stands of Holocene sea levels in the Northwest Pacific', *Quat. Res.*, *10*, 1-29

—— Brousse, R., Delibrias, G., Montaggione, L.F., Sachet, M.H. Salvat, B. and Sinoto, Y.H. (1985) 'Leeward islands (Maupiti, Tupai, Bora Bora, Huahine), Society archipelago', in Delesalla, B., Galzin, R. and Salvat, B. (eds), *French Polynesian Coral Reefs, Reef Knowledge and Field Guides*, pp. 17-72, Fifth Int. Coral Reef Congr., Tahiti, 1985, vol. 1.

Plafker, G. and Rubin, M. (1978) 'Uplift history and earthquake recurrence as deduced from marine terraces on Middleton Island, Alaska', in J.F. Evernden (convenor), *Proceedings of Conference VI. Methodology for Identifying Seismic Gaps and Soon-to-Break Gaps*, USGS open file report 78-943. pp. 687-721

Polach, H.A., Chappell, J.M.A. and Lovering, F. (1969) 'ANU radiocarbon date list', *Radiocarbon*, *11*, 245-52.

Schofield, J.C. (1970) Notes on Late Quaternary sea levels, Fiji and Rarotonga. *NZ. J. Geol. Geophys.*, *13*, 199-206.

Seely, D.R. and Dickinson, W.R. (1977) 'Stratigraphy and structure of compressional margins', in *Geology of Continental Margins*, AAPC Course No. 5, pp. C1-23.

Shackleton, N.J. and Opdyke, N.D. (1973) 'Oxygen isotope and palaeoma stratigraphy of equatorial Pacific core V28-238: oxygen isotope temper and ice volumes on a 10^5 and 10^6 year scale', *Quat. Res., 3*, 39-55.

Shimazaki, K. and Nakata, T. (1980) 'Time-predictable recurrence model for large earthquakes', *Geophys. Res. Lett., 7*, 279-82.

Sugimura, A. and Naruse, Y. (1954) 'Changes in sea-level, seismic upheavals and coastal terraces in Southern Kanto region, Japan (I)', *Jap. J. Geol. Geophys., 24*, 101-13.

—— and Naruse, Y. (1955) 'Changes in sea-level, seismic upheavals and coastal terraces in Southern Kanto region, Japan (II)', *Jap. J. Geol. Geophys., 26*, 165-76.

Sumosusastro, P.A. (1984) 'Late Quaternary geology of Whangaparoa area, East Cape, New Zealand', unpublished MSc (Hons) dissertation, Victoria University of Wellington.

Szabo, B.J., Ward, W.C., Weidie, A.E. and Brady, M.J. (1978) 'Age and magnitude of the Late Pleistocene sea-level rise on the eastern Yucatan Peninsula. *Geology, 6*, 713-5.

Taylor, F.W. and Bloom A.L. (1977) 'Coral reefs on tectonic blocks, Tonga Island arc', in *Proc. Third Int. Coral Reef Symp.*, pp. 275-81.

Turcotte, D.L. (1983) 'Mechanisms of crustal deformation', *J. Geol. Soc. Lond., 140*, 701-24.

Veeh, H.H. (1966) 'Th^{230}/U^{238} and U^{234}/U^{238} ages of Pleistocene high sea-level stand', *J. Geophys. Res., 71*, 3379-86.

—— and Veevers, J.J. (1970) 'Sea-level at −175 m off the Great Barrier Reef 13,600-17,000 years ago', *Nature, 226*, 536-7.

Walcott, R.I. (1972) 'Past sea levels, eustasy and deformation of the earth', *Quat. Res., 2*, 1-14.

Wu, F.T. (1978) 'Recent tectonics of Taiwan', in S. Uyeda, R.W. Murphy and K. Kobayashi (eds), *Geodynamics of the Western Pacific, J. Phys. Earth (Suppl.)*, pp. 265-99.

Yonekura, N. (1975) 'Quaternary tectonic movements in the outer arc of S.W. Japan with special reference to seismic crustal deformations', *Bulletin Dept. of Geology, University of Tokyo, 7*, 19-71.

—— (1983) 'Late Quaternary vertical crustal movements in and around the Pacific as deduced from former shoreline data', in W.C. Hilde and S. Uyeda (eds), *Geodynamics of the Western Pacific—Indonesian Region*, Geodynamic Series 11, AGU-CSA, pp. 41-50.

——, Matsushima, Y., Maeda, Y., Matsumoto, E., Togashi, S., Sugimura, A., Ida, Y. and Ishii, T. (1983) 'Holocene sea-level changes in the southern Cook Islands', Abstract, *Int. Symp. on Coastal Evolution in the Holocene Komazawa University, Tokyo*, pp. 151-4. Contribution to IGCP Project 200.

Yoshikawa, T., Kaizuka, S. and Ota, Y. (1964) 'Crustal movement in the Late Quaternary revealed with coastal terraces on the southeast coast of Shikoku, southwest Japan', *J. Geod. Soc. Japan, 10*, 116-22.

—— Ota, Y., Yokekura, N., Okada, A. and Iso, N. (1980) 'Marine terraces and their tectonic deformation on the northeast coast of the North Island, New Zealand', *Geogr. Rev. Japan, 53*, 238-62 (Japanese with English abstract).

Part Two

The Evidence and Interpretation of Sea-surface (Sea-level) Movements

6

Dating and Associated Methodological Problems in the Study of Quaternary Sea-level Changes

Donald G. Sutherland

INTRODUCTION

In the last forty years there has been a remarkable growth in the number of techniques available for dating Quaternary sea-level change. Radiometric methods such as radiocarbon and uranium series dating, biochemical methods such as amino acid diagenesis and the application of palaeo-magnetism when allied to the more traditional forms of relative dating such as pollen analysis have permitted the determination of complex histories of sea level throughout the Quaternary. The degree of resolution possible in studies of former sea level is primarily a function of the degree of preservation and accessibility of the field evidence. It is therefore inevitable that most recent work on sea-level change has concentrated on the period following the last glacial maximum (at ~18,000 BP) when widespread sequences of coastal sediments were deposited. It is also during this period that the most accurate and widely applicable of the 'absolute' dating methods, radiocarbon (^{14}C), can be used with most confidence.

This chapter is therefore divided into two broad sections, the first relating to this most recent phase of sea-level change during which radiocarbon is the dominant technique applied, and the second relating to earlier events in the Quaternary for which radiocarbon is either of more limited accuracy or inapplicable. It should be emphasised, however, that whatever the merits of any one dating technique, the objective of dating a particular sea-level event is more reliably achieved if more than one technique is applied. This permits cross-checks on the age of the event. Moreover, cross-checks involving techniques that are independent of each other are preferable to re-applications of the same method, as this may only compound a systematic error.

A further general point is that it is of fundamental importance to relate the material sampled directly to the event being dated. As will be discussed, a number of techniques used to date sea-level change are also useful in identifying the appropriate horizons to be dated. Frequently, however, the

165

means of associating a sample with a particular sea-level event is independent of the dating method.

DATING OF SEA-LEVEL CHANGE SINCE THE LAST GLACIAL MAXIMUM

Sea-level change since the last glacial maximum has been extremely complex, in some regions resulting in a net rise of over 100 m whilst other regions have experienced a net fall of over 100 m. The causes of this diversity are the interactions of glacio-isostatic, glacio-eustatic, hydro-isostatic and geoidal-eustatic effects (Clark *et al.*, 1978; Mörner, 1976), but the consequences in coastal sedimentary sequences around the world have broad stratigraphic similarities: at a variety of altitudes terrestrial sediments are overlain or underlain, or both, by marine sediments. Such widespread and well-preserved intercalations of terrestrial and marine sediments offer abundant opportunities to trace the changes of sea level that produced the stratigraphic sequences. In discussing the application of dating techniques to these sediments, a distinction can usefully be drawn between terrestrial or freshwater substances and those of marine origin. This distinction relates both to the field contexts and the type of information that can be extracted from the different sediments as well as, importantly, the technicalities of radiocarbon dating and the differences between the marine and the terrestrial parts of the carbon cycle.

Terrestrial and freshwater samples

A schematic sequence of sediments that may result from sea-level change and that is typical of many lower energy coastal areas such as estuaries is shown in Figure 6.1a. Stratigraphies such as these are widely reported (Jelgersma, 1961; Sissons *et al.*, 1966; Tooley, 1978; Devoy, 1979) and dating opportunities arise at the various organic/minerogenic contacts. For instance, where local relief is low and the underlying deposits are free-draining, sea-level rise results in a corresponding rise in the water table and, when the water table intersects the ground surface, fen peat development (Jelgersma, 1961; Kidson and Heyworth, 1973). Secondly, where seral development from estuarine to saltmarsh to freshwater conditions (or vice versa) can be demonstrated by pollen or diatom analyses across the minerogenic/organic contacts, then these horizons clearly relate to advance or retreat of the shoreline (Scholl, 1964, 1965; Sissons and Brooks, 1971; Tooley, 1978). Thirdly, soils directly overlain by estuarine or marine sediments also imply landward movement of the shoreline (Mörner, 1969). Two major problems exist in using the above types of field evidence to date sea level: (1) the accurate establishment of the age of the sedimentary contacts and (2) the relationship of the samples dated to sea level, or more precisely, to a particular part of the tidal cycle.

Radiocarbon dating is the principal method applied in such situations, whilst the independent relative dating can be derived from pollen studies where regional pollen assemblage zones are known and from the integration of the local sedimentary sequence into regional stratigraphies, or shoreline systems. The relative dating relationships may serve as a direct check on the accuracy of the local radiocarbon dates if they are themselves independently dated at other localities by radiocarbon or, in certain areas, varves.

Sample treatment

The radiocarbon (^{14}C) in living terrestrial plants exchanges rapidly and is in equilibrium with that of the atmosphere. If no contamination is present, the measured radiocarbon activity of a sample can be simply related to the time that has lapsed since the death of the plant when the equilibrium with atmospheric radiocarbon was disrupted (Libby, 1955). Peats and woody materials are thus good for radiocarbon dating, but the assumption that there may be insignificant contamination is frequently erroneous. Potential contaminants in the situations being discussed are organic (humic) acids circulating in ground water, penetration by younger rootlets, the presence of detrital 'old' carbon and secondary carbonate deposition.

Samples for radiocarbon dating should, therefore, be treated to identify and remove such contaminants, otherwise significant inaccuracies may result. Various studies have recorded deep rootlet penetration (Kaye and Barghoorn, 1964; Scholl and Stuiver, 1967). Olsson (1979) has suggested that rootlets may contribute a disproportionately small amount of carbon with respect to their total mass in certain sediments but the empirical observations of Streif (1972), indicating systematic contamination by rootlets up to a maximum error of 845 years, demonstrate the presence and possible magnitude of rootlet contamination. Critical field examination of all samples for rootlets followed by either removal under a microscope or dating of only a fine size fraction may be expected to minimise this problem (cf. van de Plassche, 1980).

Minor amounts of secondary carbonates are common in many samples and radiocarbon laboratories normally treat all relevant samples with acid to remove this material. The presence of dissolved 'old' carbon in ground water may result in a 'hard water effect', particularly if a component of the organic material being dated is either derived from aquatic plants that synthesise their carbon from water rather than air, or is derived from contemporaneous carbonate deposition. For instance, Scholl and Stuiver (1967) showed that a groundwater induced carbonate-rich sediment that was succeeded by seral mangrove development consequent upon sea-level

167

rise along the Florida coast had an apparent age of ~ 400 years.

Organic acids circulating in ground water may be precipitated if they move into an environment with appropriate pH conditions. The separation and dating of alkali-soluble and insoluble fractions, with organic acids being mainly in the former, can allow the assessment of this form of contamination.

Soils in particular contain complex organic compounds of diverse origins and they present considerable problems for radiocarbon dating (Scharpenseel, 1971). In addition to the need to separate and date particular organic fractions from soils, marked age-depth gradients have been demonstrated in soil profiles (Matthews, 1981; Matthews and Dresser, 1983). Where marine deposits bury soil horizons then there should be no erosion of the latter if a date on the upper layer of the soil is to provide an accurate estimate of the time of soil burial.

The dated sample as a sea-level indicator: problems in collection and interpretation

The relationship of the dated samples to sea level introduces a different series of problems (see also Devoy, this vol., Ch. 1). First is the relationship to a part of the tidal cycle. The present-day altitudes with respect to the tidal cycle of various of the relevant sedimentary boundaries have been reviewed by Tooley (1978) who concluded that local factors could seriously distort any general relationships. Despite this caution, seral development from saltmarsh to freshwater peats is broadly assumed to occur near High Water Mark (HWM). Ground-water induced fen peat formation was originally suggested to relate to HWM (Jelgersma, 1961) but this interpretation has since been revised and such peats are now considered to relate to mean tide level (Jelgersma, 1980; van de Plassche, 1982). It is normal, on the basis of altitudes of dated samples, to infer variations in sea level but the possibility that tidal range has also changed over the same time period, perhaps in part because of the sea-level change itself produces uncertainty in the actual value for sea-level change that may be inferred.

A second problem in relating dated samples from such stratigraphic sequences, as illustrated in Figure 6.1a, to the level of the sea at the time of their formation is the autocompaction of peat (Kaye and Barghoorn, 1964) and the compaction and consolidation of the associated sediments (Heyworth and Kidson, 1982). These factors may vary locally (Skempton, 1970; van de Plassche, 1980, 1986) as well as being of different magnitude at different positions in the sedimentary sequence. Thus where a sample is from near the base of a peat resting on, for example, sand and gravel (lower core A, Fig. 6.1a) little alteration of this nature is likely. The upper sample of core A (Figure 6.1a), however, may be expected to have been influenced to some extent in this manner. Some studies have attempted to correct for these effects (Kidson and Heyworth, 1973) whilst others, although noting the problem, have made no corrections (Tooley, 1978; Devoy, 1979). In the lagoonal areas of the Netherlands these effects have been so irregular that data from this area cannot be used for sea-level change studies (Jelgersma, 1980).

Finally, in order to interpret sea-level change accurately, the full stratigraphy of the area being studied should be established and tied, where possible, to a shoreline sequence (Sutherland, 1983). An example of how critical knowledge of the overall stratigraphy may be is illustrated in Figure 6.1b, which is based on the work of Sissons and Smith (1965) in the upper Forth Valley, central Scotland. The stratigraphic sequence shown resulted from an initial fall of sea level which exposed a now buried estuarine flat on which peat accumulated. Subsequently sea level rose and minerogenic estuarine sediments were deposited on top of the peat. In two areas of the Forth Valley, however, such as around A (Fig. 6.1b), peat growth was rapid

Figure 6.1: (a) Schematic representation of intercalated estuarine and terrestrial deposits. (b) Interbedded estuarine and terrestrial deposits in part of the upper Forth valley, Scotland. *Key*: 1. Peat; 2. Silt and clay; 3. Sand; 4. Gravel; 5. Soil horizon; 6. Boreholes and sample positions

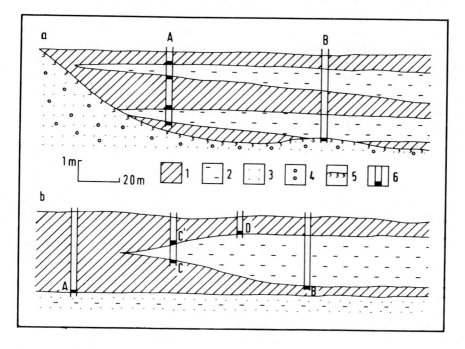

Source: After Sissons and Smith (1965).

enough to match sea-level rise and hence a wedge of estuarine sediments was deposited in the margin of the peat bog. Sea level thereafter fell and peat accumulated over the abandoned estuarine surface.

Seral vegetational development can be identified (Sissons and Brooks, 1971) at the sedimentary contacts labelled A, B, C, C' and D (Fig. 6.1b). Two points of interpretation should be noted. First, if samples had only been recovered from around A, then only an early fall in sea level would have been dated and a significant part of the sea-level history missed. Second, a core taken at C would have identified the lower (C) and upper (C') contacts of the upper estuarine sediments which may have been interpreted as relating to rising and falling sea levels respectively. Both contacts, however, date the period of sea-level rise, the local stratigraphy simply reflecting a change in the relative rates of peat accumulation and estuarine sedimentation. The maximum sea level is dated by sample D which is on the surface of the estuarine flats and accords with a prominent shoreline (Main Postglacial Shoreline, Sissons, 1983). In both these instances major errors of interpretation of the dated horizons could result from an incomplete survey of the local stratigraphy.

Away from estuaries, coastal sediments may be more sporadically developed and in high latitudes driftwood on raised shorelines has been dated (Blake, 1961, 1970; Marthinussen, 1962; Salvigsen, 1981; Stewart and England, 1983). Wood, and in particular cellulose which can be extracted from it, is a relatively stable organic compound (Olsen and Broecker, 1958) and has proved reliable for radiocarbon dating. In all wood dates, however, the portion of the tree dated is important as heartwood may have an initial age up to several hundred years greater than the sapwood (cf. Campbell and Baxter, 1979). Tree ring counting may allow the dated sample to be related to the time of death of the tree (Heyworth and Kidson, 1982). It is also important to establish that the sample being dated relates closely in time to the formation of the shoreline. There is the possibility of reworking from older deposits, whilst branches or tree trunks that are found at the base of peat bogs may have sunk in the peat and hence be considerably younger than the surface on which they rest (Kaye and Barghoorn, 1964; Tooley, 1978). Moreover, tree trunks resting on surfaces overlain by marine sediments may be significantly older than the period of marine invasion of the site.

In ice-scoured areas that have experienced strong glacio-isostatic uplift, coastal depositional sequences are often of limited extent. A widely applied method of reconstruction of sea-level change in such areas is to utilise the many enclosed rock basins that were at one time below sea level and which have been raised above and hence isolated from the sea (Eronen, 1974; Hafsten, 1983; Kaland, 1984; Krzywinski and Stabell, 1984). A schematic example of a series of such 'isolation basins' is illustrated in Figure 6.2. On the right of the figure is a hypothetical sea-level history (S_1, S_2, S_3 and S_4)

Figure 6.2: Isolation basin sedimentation resulting from the sea-level history depicted by curve S_1–S_4. In stratigraphic columns diagonal shading represents marine and stipple freshwater sediments

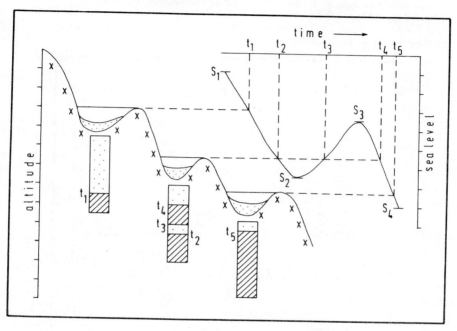

and on the left the changes between marine and freshwater lacustrine sediments in three enclosed basins. Dating of the contracts between the marine and freshwater sediments provides ages for the times when sea level fell below or rose above the thresholds of the basins. The altitude of the threshold is therefore critical in reconstructing a sea-level history. Where possible, basins whose threshold is in drift should be avoided, for these are likely to have suffered erosion at the time of, or subsequent to, the marine invasion.

The response of basin sedimentation to either marine isolation or invasion will depend, for example, on the size of the basin, its freshwater intake and the nature of the surrounding bedrock, soils and vegetation. As in areas of estuarine sediments many lines of evidence have been used to identify sampling positions and Figure 6.3 illustrates schematically the response of a variety of indicators to changes in conditions from marine to freshwater or terrestrial. The sequence of sediments in Figure 6.3 accords with the sea-level events illustrated in the middle basin in Figure 6.2, or the estuarine sediments in Figure 6.1b. As shown, marine horizons are typically characterised by high percentages of marine diatoms, the presence

171

Figure 6.3: Response of various indicators to marine-freshwater changes in sedimentation. (a) Lithostratigraphy; (b) diatoms (cross-hatching, marine; diagonal shading, brackish; blank, freshwater); (c) Freshwater phytoplankton (e.g. *Pediastrum* sp.); (d) Dynoflagellate cysts; (e) Secondary pollen; (f) Aquatic pollen; (g) Saltmarsh pollen (e.g. Cheonopodiaceae); (h) Marine macro- and microfauna; (i) Magnetic susceptibility or intensity

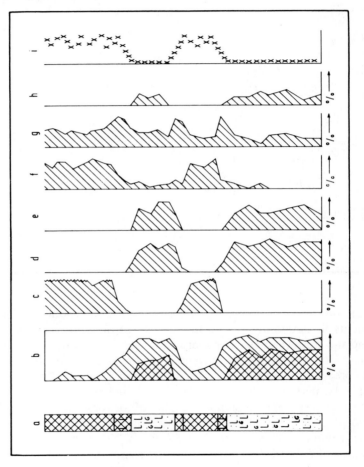

of dynoflagellate cysts (hystrichosphaerids), often abundant secondary and/or degraded pollen, a generally higher level of minerogenic sediment and marked changes in magnetic susceptibility or intensity. In contrast, terrestrial or freshwater lacustrine sediments contain freshwater diatoms, phytoplankton such as *Pediastrum sp.* or *Botryococcus sp.*, generally higher levels of pollen from aquatic vegetation and increased organic content. The transitional layers are typified by increased percentages of brackish water diatoms, pollen of saltmarsh plants such as *Armeria maritima* and certain of the Chenopodiaceae and, in areas of former reed swamp, abundant *Phragmites* macrofossils which may also be reflected in increased percentages of Gramineae pollen grains. As sedimentation varies locally, as well as between sites, it is advisable that a number of these variables should be studied when reconstructing the former environment of deposition.

The principal objective of the above analyses is the identification of samples for radiocarbon dating of the marine-freshwater contacts. The use of pollen analysis, as already described, or palaeomagnetic measurements allow inferences to be drawn as to the age of the sediments. The magnetic intensity and susceptibility of the sediments reflect the changing depositional conditions (Oldfield, 1981), whilst the systematic variations in the inclination and declination, at least for the last 10,000 years when comparison can be made with regional 'master curves', allow the sediments to be dated (Thompson, 1977; Thompson and Turner, 1979).

For radiocarbon dating, samples should be selected from above, or below, rather than across the sedimentary contacts. Terrestrial (peat) or freshwater sediments normally have higher organic contents than the associated marine or estuarine sediments. They also normally have less derived (redeposited) organic material and do not contain sediment with carbon that has a significant 'apparent age' because of its assimilation from seawater (see later discussion). It is therefore normal practice to sample the terrestrial or freshwater sediments in direct contact with the marine or estuarine sediments. Such samples therefore give maximum age estimates for the onset and minimum age estimates for the end of a marine event. Samples dated generally vary between 1 and 10 cm in thickness and hence relate to a period of time the length of which is a function of the rate of sedimentation. Errors in dating marine events of up to 250 years have been suggested because of this sampling factor (Kaland *et al.*, 1984). In a sequence of sediments interpreted in terms of sea-level rises and falls this factor will, therefore, overemphasise the periods assigned to the rises and, by an equivalent amount, reduce the periods assigned to the falls.

The nature of sea-level change is such that the shoreline is constantly migrating, frequently over large distances in relatively short periods. This means that particular localities only have stratigraphic sequences that relate to a portion of the sea-level history of the last 18,000 years and the information from a variety of such localities must be integrated to reconstruct the complete history of sea-level change. In many areas time-stratigraphic markers such as distinct shorelines, pumice horizons (Rose, 1978), or particular faunal elements have been used as independent bases for correlation between widely spaced sites. Such stratigraphic markers are themselves forms of relative chronology (Sutherland, 1983) and hence serve the dual purpose of aiding correlation and cross-checking the dating of individual study areas.

173

Plotting of data

The most common method of integrating radiocarbon-dated sea-level data is in the form of time-depth curves and many of these have been published (Bloom, 1977). Whilst there is broad agreement that the major form of these curves reflects the melting of the northern hemispheric ice sheets, and hence is broadly climatically controlled (see Peltier, this vol., Ch. 3), there is considerable disagreement as to the detailed history of sea-level change, even within the same regions (Kidson, 1982) (see also Shennan, this vol., Ch. 7 and chapters in Part Three). This disagreement mainly relates to the reality of minor phases of sea-level rise or fall that certain workers interpret as being superimposed on the overall deglacial trend. The uncertainties that have been discussed above in the dating methods and in relating the dated samples to a particular part of the tidal cycle, have meant that small-scale events in sea-level change have been difficult to verify. As it is an important geological problem to establish whether there have been fluctuations superimposed on the deglacial trend, a separate approach to the problem has been made through histogram analysis (Geyh, 1971).

In constructing histograms of radiocarbon dates from studies of sea-level change, Geyh (1971) argued that, excepting basal ground-water induced peats, landward migration of the shoreline would limit opportunities for peat growth whilst seaward migration would favour peat growth. Hence regional peaks and troughs on radiocarbon histograms would reflect real falls and rises in sea level. However, it has been shown (Shennan, 1979) that peaks and troughs can be the result of random events, particularly when the sample size is small as is normal with radiocarbon dates related to sea-level change in any given region. On statistical grounds, therefore, Geyh (1980) has suggested that reliable histograms should have at least 25 dates to each class interval (typically 50-100 years) and unreliable histograms would have 4 or less dates to each class interval. This is clearly a severe restriction on histogram analysis. It is also of note, as has been pointed out for the construction of U-series date histograms (Gordon and Smart, 1984), that the statistical procedures commonly adopted are strictly incorrect and tend to give undue weight to samples with large standard deviations, that is those which are most uncertain. Geyh (1980) has also noted that the peaks and troughs of his radiocarbon histograms coincide with the peaks and troughs of natural radiocarbon production (de Jong *et al.*, 1979). Berendsen (1984), however, found no such correlation using dates from a much more limited geographical area. Such a coincidence, if observed widely, could either mean that the histogram was simply a function of radiocarbon production, or that the sea-level changes purportedly represented by the histogram and the variations in natural radiocarbon production were the result of a common cause.

A modified approach to the construction of histograms is the separation of dated samples into those related to landwards and those related to

174

seawards movements of the shoreline. Separate histograms are then constructed for the two sets of samples (Roeleveld, 1974; Tooley, 1982; Shennan, 1982; Shennan *et al.*, 1983, see also the sea-level tendency concept, Shennan, this vol., Ch. 7). Such histograms as published to date suffer severely from the small numbers of radiocarbon dates in each class interval and, for most areas and much of the time period studied, fall within Geyh's (1980) 'unreliable' class of histograms. In such circumstances, histograms cannot be regarded as analytical devices but have only illustrative value. Very large numbers of radiocarbon dates are needed to produce statistically reliable histograms: estimated, for the Holocene in an area with common sea-level history, as between 400 (Tooley, 1982) and 1000 (Shennan, 1979) evenly distributed in time. Such numbers of radiocarbon dates would be more than adequate to date the stratigraphically defined field evidence on which all reconstructions are ultimately dependent.

Marine samples

Sediments containing fossils of marine origin are very widely distributed. In addition to aiding in the interpretation of the origin of such sediments, the contained marine organisms are of particular utility in dating the sea-level changes implied by the sediment distributions. The use of distinctive fossil assemblages (e.g. diatoms, foraminifera, ostracods), or of the occurrence of particular marker species can aid in the establishment of a relative chronology, but for deposits formed during the last ~ 18,000 years radiocarbon dating is the dominant method in studies of chronology.

The sea water effect in radiocarbon dating

A basic assumption of radiocarbon dating is that the radiocarbon activity of the material being dated was, at the time of death of an organism or precipitation of a sediment, the same as that of the laboratory standard used in age calculation. The normal laboratory figure used for age calculation is equivalent to the radiocarbon activity of wood during the latter half of last century (i.e. prior to significant industrial influence). The above assumption can be invalidated if the level of natural radiocarbon production has varied through time (see later discussion), or if the carbon cycle operates in such a manner as to produce variations in the radiocarbon activity of contemporaneous materials. Studies of the distribution of radiocarbon introduced into the atmosphere as a result of atom bomb testing have shown that radiocarbon is rapidly (in < 10 years) mixed throughout the atmosphere, and hence all terrestrial materials that synthesise their carbon from and are in equilibrium with the atmosphere have the same level of radiocarbon activity at death. In the oceans, however, the mixing process of radiocarbon is very much slower and materials that derive their carbon

175

from the marine environment have a lower radiocarbon activity than do contemporaneous terrestrial materials. In radiocarbon dating, therefore, marine samples have a 'reservoir' or 'apparent' age, the magnitude of which varies depending upon the rate of mixing of the surface layers of the ocean, the presence of upwelling of 'old' deep water and dilution effects near the coast (Broecker et al., 1960; Mangerud, 1972; Mangerud and Gulliksen, 1975). In addition to this effect it may be noted that estuaries receiving drainage containing ancient carbon from bedrock may produce anomalously high ages (Broecker and Bender, 1972; Mangerud, 1972).

The magnitude of the apparent age of sea water has been estimated by measuring the radiocarbon activity of shells collected during the last ~ 100-150 years, prior to atom bomb testing. A summary of these results is given in Table 6.1 which indicates that for many parts of the world the apparent age of sea water is 350-450 years. Towards the polar regions considerably higher apparent ages have been recorded because of greater upwelling in these areas. As Table 6.1 also indicates, apparent ages of the same magnitudes also apply to marine vegetation, marine mammals and to land animals (including humans) whose diet is dominantly marine products.

A further factor that distinguishes marine from terrestrial samples is isotopic fractionation, the process whereby different isotopes are preferred or discriminated against because of their different atomic masses. Isotopic fractionation has resulted in marine carbonates having, on average, 5 per cent greater radiocarbon activity than does terrestrial vegetation. This is equivalent to an age difference ('younging'), on average, of ~ 410 years.

The adjustments for apparent age of sea water and isotopic fractionation that are necessary to make a radiocarbon date for a marine carbonate directly comparable to that for a contemporaneous sample of terrestrial vegetation are in opposition (arithmetically) and, for many samples, cancel each other out. It has, therefore, been argued (Nydal et al., 1972; Krog and Tauber, 1974) that neither adjustment need be made. This is unsatisfactory, however, as the apparent age of sea water may be significantly different from the adjustment for isotopic fractionation. The two factors should thus be independently assessed and the sample age adjusted accordingly.

Marine carbonates: types and interpretational problems

As with terrestrial samples, it is important to relate any dated sample of marine carbonate or other material to a part of the tidal cycle. Marine shells or corals occur either *in situ* or as detrital accumulations and it is necessary to distinguish between them in the field prior to inferring a relationship to a particular sea level (cf. van de Plassche, 1986).

In situ specimens are of greatest utility in sea-level studies, for there is a marked zonation related to the tidal cycle around all coasts and many samples can be closely related, given knowledge of their constituent

Table 6.1: Apparent radiocarbon ages of marine organisms

Locality	Apparent age ± σ	Reference
(a) Marine molluscs		
British Isles	405 ± 40	Harkness (1983)
Swedish W. coast	340 ± 30	Olsson (1980)
S. Norway	370 ± 25	Mangerud and Gulliksen (1975)
		Olsson (1980)
N. Norway	445 ± 35	Mangerud and Gulliksen (1975)
		Olsson (1980)
Spitsbergen	425 ± 25	Mangerud and Gulliksen (1975)
		Olsson (1980)
Iceland	365 ± 20	Hakansson (1983)
Faroes	380 ± 75	Krog and Tauber (1974)
		Hakansson (1983)
E. Greenland	515 ± 25	Hjort (1973)
		Hakansson (1983)
Canadian Arctic	750 ± 40	Mangerud and Gulliksen (1975)
S. Georgia	650 ± 45	Harkness (1979)
E. Australia	450 ± 35	Polach et al. (1978)
(b) Marine plants		
S. Sweden	245 ± 30	Olsson (1980)
(c) Marine mammals		
W. Sweden	320 ± 35	Olsson (1980)
E. Sweden	310 ± 50	Olsson (1980)
S. Georgia	770 ± 30	Harkness (1979)
(d) Mammals with marine food sources		
Greenland	480 ± 25	Olsson (1980)
Spitsbergen	420 ± 75	Olsson (1980)
Denmark	400 ± 55	Olsson (1980)

ecology, to their position in such a zonation. The nature of the occurrence of the material being sampled together with its overall sedimentary context should be assessed in considering whether it is in place. Coral reefs provide abundant opportunities for relating both organic and inorganic marine carbonates to different parts of the tidal cycle. Reef-top corals, especially in the form of microatolls, are related to MLWST (McLean et al., 1978; Scoffin and Stoddart, 1978) whilst beach rock can be tied to MHWST (Hopley, 1974; McLean et al., 1978) (see Hopley, this vol., Ch. 12). A variety of other carbonate-cemented sediments which occur on coral reefs are of utility in studies of sea-level change (McLean et al., 1978; Scoffin and McLean, 1978) whilst facies models for both fringing and barrier atolls, relating both biostratigraphy and sediment distribution to sea level, have been proposed (Chappell, 1974).

Molluscs have been sampled widely as indicators of former sea level.

177

Gasteropods, such as certain types of barnacles that grow in the upper part of the intertidal zone, may be found still attached to rock surfaces (Ten Brink, 1974) whilst certain of the Vermitid gasteropods construct ramparts that can be related to LWM (Labordel, 1980). *In situ* bivalve molluscs should have both valves together and infaunal species should be in their position of growth. In addition, the associated sediment should be that which the species normally inhabits. Where a number of species, particularly of associated microfauna or microflora, can be identified, a coherent life population should be recognisable. Frequently, molluscs have a considerable depth range of occurrence and although this may be narrowed by analysis of the associated species, it may only be possible to bracket the sea level at the time of their death by a range of ~ 10 m (Peacock *et al.*, 1977, 1978).

In the littoral zone, detrital accumulations are more often encountered than are *in situ* occurrences. They can be recognised by their sedimentary context, by the occurrence of abundant broken and abraded fragments and by the lack of a life population, with certain species being over-represented or derived from distinct parts of the littoral or nearshore zones. Death assemblages such as these frequently are found in landforms or sediments such as shingle ridges, beach gravels, delta foresets or in channels on mud or sand flats that may themselves be related closely to a particular sea level. The principal uncertainty involved in using such material to date the associated sea level is that it cannot be established whether the material dated is contemporaneous with deposition of the sediments. Such contemporaneity is usually assumed rather than questioned until an 'anomalous' date is recorded.

Radiocarbon dating, as with techniques such as U-series and amino acid dating discussed later, is dependent upon the 'closed system' assumption, that is that there has been no introduction of radiocarbon into the sample after its death or precipitation. Recrystallisation, with uptake of extraneous carbon, is the principal form by which marine carbonates can infringe the closed system assumption and considerable attention has been directed to identifying samples that have not suffered recrystallisation.

Corals have an aragonitic mineralogy as do certain molluscs whilst other molluscs have a calcitic or part-calcitic, part-aragonitic mineralogy, the actual proportions of the two being a function of species and the temperature and salinity of the sea they inhabit (Milliman, 1974). Recrystallisation normally produces calcite and carbonate cements that may have enclosed mollusc or coral samples are also typically calcitic. For corals and aragonitic molluscs the problem of contamination therefore can be approached by establishing the presence of calcite. For molluscs of a mixed or calcitic mineralogy, this approach cannot be followed.

During field sampling, specimens can be broken and examined for calcitic cleavage (Bloom *et al.*, 1974) and the degree of preservation of molluscs can often be assessed by the surface character. Shells with the periostracum still attached are likely to be well preserved and approximate degrees of alteration may be identified along the scale of slight surface pitting, loss of detail such as growth lines, a cover of dusty calcite that may be removed on touch, to disintegration on removal from the surrounding sediment. Large, thick-walled shells or massive corals should be sampled where possible, as these provide sufficient material for sample sub-division and cross-checking in the laboratory. Shells that have complex structures with a large surface area to weight ratio and a large number of voids should be avoided where possible.

A considerable range of laboratory techniques have been applied to identify the degree of alteration in samples that were originally composed of aragonite. Visual inspection under a binocular microscope or in thin section (Bloom et al., 1974; Polach et al., 1978) or of acetate peals (Vita-Finzi, 1980) allow detailed examination for calcite, or for void-filling carbonate cement. Quantitative analysis has been carried out by infrared spectrophotometry (Polach et al., 1978) and X-ray diffraction (Chappell and Polach, 1972; Bloom et al., 1974; Polach et al., 1978; Vita-Finzi, 1980).

The detailed approach outlined above has undoubtedly improved the quality of samples analysed. It is not possible, however, to quantify the effect that a given degree of recrystallisation will have on a radiocarbon age. Chappell and Polach (1972) have demonstrated two types of recrystallisation, a 'sparry calcite' type and a 'subtle coarsening' type, the former being interpreted as evidence of open-system conditions and the latter being related to a closed system. Reliable radiocarbon ages were often achieved on those samples that had only experienced the subtle coarsening type of alteration.

It is normal practice in radiocarbon laboratories to discard the outermost 15-40 per cent of carbonate by acid leaching, as it is argued that recrystallisation is most likely to proceed from the outer surfaces inward. Where the sample is of sufficient size, many laboratories produce two further fractions by progressive leaching, normally termed the 'inner' and 'outer' fractions, and date both of these. Agreement between the ages of these two fractions is regarded as a check on the likelihood of contamination. This test is not rigorous (Polach et al., 1978) and in certain situations may be counter-productive (Vita-Finzi and Roberts, 1984), but for the majority of samples analysed it does allow greater confidence to be placed in the age of the sample if both inner and outer fractions are in agreement to less than one standard deviation (Mangerud, 1972; Sutherland, 1986).

Occasionally the reliability of radiocarbon dates on marine carbonates has been questioned (Shotton, 1967). However, a large number of such samples have yielded results in conformity with their stratigraphical or geomorphological contexts and in accord with the ages of equivalent samples from terrestrial or freshwater environments. This indicates that marine carbonates, subject to the cross-checks and adjustments for apparent age of sea water and isotopic fractionation as discussed above, provide reliable radiocarbon dates.

Conclusion

The principal method used in the dating of sea-level changes that have occurred in the last ~ 18,000 years has been radiocarbon dating and the various other relative dating techniques that have been applied, such as pollen analysis or palaeomagnetic secular variation, have themselves been calibrated against the radiocarbon timescale. It has already been noted that the application of radiocarbon dating is dependent upon the assumption that the radiocarbon activity of the sample was, at the time of death or precipitation, the same as that of modern standards. The validity of this assumption has been discussed with respect to certain parts of the carbon cycle, but it has been established (see papers in Olsson, 1970) that the natural production of radiocarbon has not been constant over the time periods of concern and hence the original radiocarbon activity will have varied. The radiocarbon timescale is therefore, not in accord with sidereal time and this should be considered when rates of sea-level rise or fall are being calculated.

The deviation of radiocarbon from sidereal years can be monitored by radiocarbon dating material that can be related to annual growth markers, such as tree rings, or annual sediment layers, such as varves. Most work has been done on tree rings and a radiocarbon calibration curve extending to ~ 8,000 years is now available (Klein *et al.*, 1982). These measurements have indicated a maximum deviation of ~ 900 years between radiocarbon and sidereal years around 6,000-7,000 radiocarbon years ago, the radiocarbon timescale being too 'young' at this period. Varve measurements have not proved as precise as those on tree rings although they have been extended over a longer period (Tauber, 1970; Stuiver, 1971). Together with comparisons of radiocarbon dates and other dating methods (Stuiver, 1978) these studies suggest that the radiocarbon timescale, as distinct from individual radiocarbon dates, may not deviate by more than 2,000 years from sidereal time.

DATING OF SEA-LEVEL CHANGES PRIOR TO THE LAST GLACIAL MAXIMUM

Evidence for sea-level change prior to the last glacial maximum is much less abundantly preserved in the geological record than similar evidence for the last 18,000 years. The type of evidence used for sea-level change studies is the same as that discussed above. However, the accuracy obtainable by dating techniques is rather less, partly because the quantitative dating methods used are less precise, partly because samples have had much greater opportunity for contamination and partly because of uncertainties in the application of biostratigraphical relative dating schemes (cf. Bowen, 1978). It follows, therefore, that the application of more than one dating method to any particular problem is desirable.

Radiometric techniques

Application of radiocarbon dating

Radiocarbon can, in principle, be applied to material to at least ~ 45,000 years old, and older, using enrichment techniques, Grootes (1978). Serious reservations have, however, been expressed (Broecker and Bender, 1972) as to the reliability of many such ages, particularly on marine carbonates (e.g. Olsson, 1968) $>\sim$ 18,000 years old. The principal reason for this is that only a few per cent contamination by recent material, which may be extremely difficult to identify or remove, can give a very old sample a radiocarbon age in the range 20,000-45,000 BP (Olsson, 1968). It can, therefore, be very difficult to distinguish 'true' radiocarbon ages from this time period from those that are the product of contamination. In the context of sea-level change studies this is most acutely demonstrated by the problem of high interstadial sea levels during the period 35,000±10,000 years ago.

During the 1960s a number of studies reported radiocarbon-dated samples that suggested that during the interstadial prior to the last glacial maximum, sea level came close to its present level (Shepard, 1963; Curray, 1965; Milliman and Emery, 1968). Such an idea contrasted with other interpretations of sea level during this period (Fairbridge, 1961; Mesolella et al., 1969) and in particular conflicted with both direct and indirect evidence for the extent of northern hemispheric ice sheets, the principal control on world sea level (Broecker and van Donk, 1970; Mörner, 1971). This 'dilemma' was reviewed in detail by Thom (1973), who indicated the inadequacy of many of the dates used to support the high interstadial sea level concept. Application of more rigorous criteria as to the suitability of the samples for dating (Thom, 1973; Bloom, 1983), together with the use of U-series dating methods (particularly on corals, see following sections), has indicated that certain deposits previously ascribed to this interstadial are in fact of last Interglacial age (Marshall and Thom, 1976). Other studies utilising U-series methods indicate that whilst there may have been relatively high sea levels during this period, they are likely to have been no higher than −30 to −40 m msl (Bloom et al., 1974; Moore, 1982).

The U-series technique

The above discussion has implied that the U-series dating method may be more reliable than radiocarbon dating for certain substances that are older than 18,000 years old. The principles of U-series dating have been described in detail by Ivanovich (1982). Briefly, the two naturally occurring isotopes of uranium (^{235}U and ^{238}U) decay to give rise to a series of daughter elements of which lead is the final stable form. In an ideal or mature natural system, the rate of decay of a daughter element is in equilibrium with its rate of production. However, disruptions to this ideal can occur by leaching, preferential deposition or during growth of organisms such that there is produced either an excess or a deficiency of a particular parent relative to its daughter. Upon removal of the material from the process producing the disequilibrium, the system will progress towards a new equilibrium. If the degree of excess or deficiency is known, or can be reasonably assumed, and the present degree of disequilibrium is measured, then with knowledge of the half-lives of the various elements involved (see Table 6.2), the time that has lapsed since the excess or deficiency occurred

Table 6.2: Decay series of ^{238}U and ^{235}U

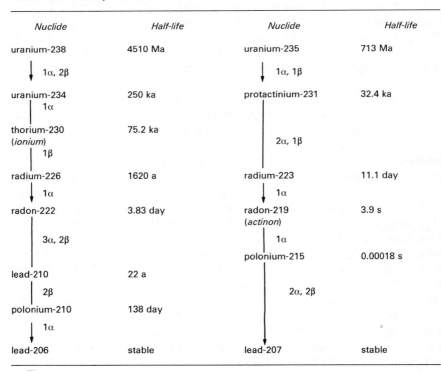

Nuclide	Half-life	Nuclide	Half-life
uranium-238	4510 Ma	uranium-235	713 Ma
↓ 1α, 2β		↓ 1α, 1β	
uranium-234	250 ka	protactinium-231	32.4 ka
↓ 1α			
thorium-230 (*ionium*)	75.2 ka		
↓ 1β		2α, 1β	
radium-226	1620 a	radium-223	11.1 day
↓ 1α		↓ 1α	
radon-222	3.83 day	radon-219 (*actinon*)	3.9 s
3α, 2β		1α	
		polonium-215	0.00018 s
lead-210	22 a		
↓ 2β		2α, 2β	
polonium-210	138 day		
↓ 1α			
lead-206	stable	lead-207	stable

Source: After Bowen (1978).

can be calculated (Fig. 6.4). A further relationship that has been used for dating results from the formation of helium by alpha-particles produced during radioactive decay. This helium is retained in certain substances, such as corals, and if it is assumed that there was initially no helium in the system, the ratio of ^{4}He/U is a measure of the time lapse since death or precipitation (Fanale and Schaeffer, 1965).

There are numerous ratios between various parents and daughters that may be used to establish the age of substances (Ivanovich, 1982) but in practice ^{230}Th/^{234}U and ^{231}Pa/^{235}U are the most commonly used ratios (Table 6.1). The ratios ^{234}U/^{238}U, ^{226}Ra/^{230}Th, ^{230}Th/^{232}Th and ^{231}Pa/^{230}Th may also be used for dating but these are often considered as cross-checks on the validity of ^{230}Th/^{234}U and ^{231}Pa/^{235}U ages. The half lives of thorium (Th) and protactinium (Pa) are such that ^{230}Th/^{234}U is suitable for dating substances $\leqslant$ 350,000 years old and ^{231}Pa/^{235}U for substances $\leqslant$200,000 years old. The accumulation of helium is progressive through time and may in principle be used to date substances up to 100 Ma, but in

Figure 6.4: Changes over time in the ratio of U-series isotapes ^{238}U and ^{235}U. The curves shown are based on experimental data and the age of a sample may be read from the graph. The equilibrium value here is represented by 1.00 and with the exception of the ^{234}U/^{238}U ratio, which declines from 1.15 (its value in sea water) to zero, the ratios increase from zero to 1.00

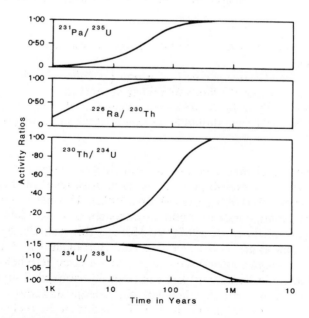

Source: After Broecker and Bender (1972).

practice helium leakage and sample recrystallisation set an upper limit of 1-10 Ma (Bender, 1973). Uncertainties as to the initial presence of helium place a lower dating limit to this method of ~ 200 ka.

Application of U-series to sea-level indicators

Although the greater time range covered by U-series dating gives it a considerable advantage over radiocarbon, for sea-level change studies the method has only been applied reliably to corals (Thurber *et al.*, 1965; Mesolella *et al.*, 1969; Bloom *et al.*, 1974) and, for indirect inferences on sea level, speleothems (Harmon *et al.*, 1978). U-series dates on molluscs have been much more controversial, some authors maintaining that such dates are inherently unreliable (Kaufman *et al.*, 1971; Broecker and Bender, 1972; Moore, 1982) whilst others have claimed that the results obtained have been of use in solving dating problems (Stearns and Thurber, 1965; Butzer, 1975; Andrews and Miller, 1980). U-series dating, as with radiocarbon dating, is dependent on the assumption that the sample has been a closed system since death or precipitation. The field and laboratory criteria discussed previously to ascertain whether a marine carbonate accords with this assumption should be applied to samples for U-series dating. In addition, a series of other relationships internal to the U-series decay products can be measured to assess the validity of the closed system assumption. The variable results achieved for mollusc dating are due to their frequent lack of accordance with these relationships (Kaufman *et al.*, 1971) which are:

(1) samples should have a uranium concentration equivalent to that of living species;
(2) the $^{234}U/^{238}U$ ratio, corrected for age (as calculated by $^{230}Th/^{234}U$) should be 1.15, within the limits of analytical precision;
(3) the $^{226}Ra/^{230}Th$ ratio should be consistent with the age of the sample;
(4) the $^{230}Th/^{232}Th$ ratio should be in excess of 20, indicating no contaminating ^{232}Th.

For sea-level change studies, the principal achievement of U-series dating has been to establish in detail the periods of relatively high sea level during the last ~ 140,000 years (Moore, 1982). This has principally been achieved by dating corals on uplifted sequences of coral reefs such as on Barbados (Mesolella *et al.*, 1969; see Mörner, this vol., Ch. 8) or in New Guinea (Bloom *et al.*, 1974) (see this vol., Berryman, Ch. 5 and Pillans, Ch. 9). In such areas, littoral sequences that are elsewhere below sea level have been tectonically uplifted and can be easily sampled. Estimates of the long-term rates of uplift together with the present altitudes of the dated coral reefs, allow the original altitude of formation of the reefs to be calculated. These studies have confirmed the close correspondence between worldwide fluctuations of sea level and the variations in the oxygen isotope record in deep sea cores, itself dominantly a reflection of northern hemispheric ice volume, and have been taken as confirmation that variations in the earth's orbit around the sun have been the principal control on climatic fluctuations during the Quaternary (Broecker *et al.*, 1968; Broecker and van Donk, 1970).

Helium dating has been much less widely applied but its potential for dating long sequences of uplifted coral reefs has been demonstrated in Barbados, where eleven marine terraces have been dated between 180 k BP and 640 k BP (Bender *et al.*, 1973; 1979).

Alternative techniques

In the context of sea-level change studies, the application of U-series dating has been limited, because reliable dates have only been achieved on corals and speleothems. Other techniques have therefore been developed that are of much wider potential application to either marine organisms (amino acid dating) or to sediments (thermoluminescence dating and palaeomagnetism).

Amino acid dating

Amino acid diagenesis has been shown to proceed by a series of complex reactions such as racemisation (conversion of L: to D: isomers in amino acids), epimerisation (e.g. formation of D-alloisoleucine from L-isoleucine), hydrolysis (the breaking of the peptide bonds to produce 'free' amino acids) and a variety of decomposition reactions (e.g. the decomposition of threonine and serine) (Hare and Mitterer, 1967; 1969; Miller and Hare, 1980). The progress of these reactions is broadly time and temperature dependent, as well as being related to the type of substance containing the amino acids and other environmental conditions such as the pH of groundwater. Molluscs and foraminifera have been shown to preserve amino acids particularly well (Miller and Hare, 1980 Miller *et al.*, 1979; Davies, 1983), both because of the tight crystal structure and because the carbonate provides a buffer against changing groundwater pH. The principal variable that has to be controlled in using amino acid diagenesis in molluscs and foraminifera as a dating technique is therefore temperature.

The main diagenetic reactions that have been used in chronological studies are the racemisation reaction in the more stable amino acids such as aspartic acid, glutamic acid or leucine and the epimerisation reaction of L-isoleucine to D-alloisoleucine. These reactions are typically expressed as enantiomeric (D:L) ratios, the valves for which range between 0 (immediately on death) to 1:1 for racemisation 1:1.3 for epimerisation when the reversible reaction has achieved equilibrium. The time necessary to achieve

185

equilibrium is mainly a function of the thermal environment. In polar regions, samples of Late Tertiary age may not have achieved equilibrium, whilst in the tropics the limit of the technique may only be a few hundred thousand years. It is important to note that the rate of progress of the reactions is also genus-dependent (King and Hare, 1972; Lajoie *et al.*, 1980; Wehmiller, 1980; Andrews *et al.*, 1979).

The temperature control on amino acid diagenesis can be partly minimised if samples from a relatively small geographical area are considered together. It can then be argued that contemporaneous samples will have experienced a common temperature history and hence should have the same enantiomeric ratios. Older samples will also have experienced that temperature history, together with an earlier one, and hence will have higher enantiomeric ratios. Hence a relative chronology, an 'aminostratigraphy' of the deposits from which the samples are derived (Miller and Hare, 1980), can be established irrespective of the absolute temperatures experienced. This approach has been widely used (Miller *et al.*, 1977; Miller *et al.*, 1979; Mangerud *et al.*, 1981; Boulton *et al.*, 1982; Wehmiller and Belknap, 1982). In distinguishing between different horizons, graphs of relative rates of racemisation in different amino acids or in the 'free' and the 'total' amino acid fractions for the same amino acid can be of considerable utility (Lajoie *et al.*, 1980).

A second approach to this problem is the use of calibration graphs, in which the enantiomeric ratios from horizons within a relatively small region of known age are plotted against those ages. The ratios of samples of unknown age can then be plotted and their ages interpolated or extrapolated (Fig. 6.5). The accuracy of calibration graphs is clearly a function both of the accuracy of the independent dating methods and that of the amino acid technique. A refinement to calibration graphs has been introduced where the known age racemisation ratios are combined with laboratory information on the temperature sensitivity of the relevant reactions and a model of the racemisation kinetics developed (Wehmiller, 1982). Such models are in principle capable of finer resolution than simple linear interpolation or extrapolation (Miller *et al.*, 1983), but the form of the kinetic model is not always certain (cf. Mitterer, 1975; Wehmiller and Belknap, 1978).

Amino acid dating is being increasingly applied and it has proved particularly useful as a relative dating technique within limited regions where contemporaneous samples have experienced similar thermal histories. It has the advantages of being applicable to a wide range of commonly occurring marine fossils, of only needing small samples for individual analyses and of being a relatively quick and cheap form of analysis. The ability to do multiple analyses from single samples, as well as from different samples from the same horizon, together with the development of diagnostic ratios of amino acids that allow identification of species (Andrews *et al.*, 1985) or

Figure 6.5: 'Absolute' age interpolation of groupings in allo/isoleucine ratios derived from Pliocene–Pleistocene shell and foraminifera data in southeast–eastern England. The sites used represent former freshwater-terrestrial as well as coastal- marine environments, the latter having been used in shoreline reconstruction studies. Curve 1 gives the possible ages of Groups 3–5 on the assumption/U-series dating that shells in Group 2 date from ~ 124,000 BP. Curve 1 also assumes that the isoleucine epimerisation reaction in samples used follows first order reversible kinetics exactly, Curve 2 that Group 3 dates from ~ 210,000 BP, Curve 3 that Group 3 is from ~ 310,000 BP and Curve 4 that Group 3 dates from ~ 410,000 BP. Ages in Curves 2-4 are based on oxygen isotape data (see Miller *et al.*, 1979 for further discussion)

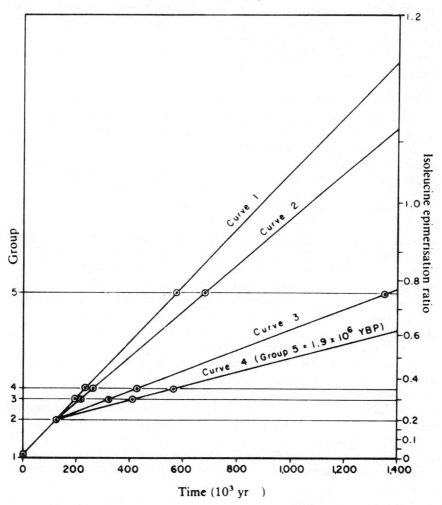

Source: Miller *et al.* (1979).

the presence of contaminating amino acids, form strong internal checks on the method. More research is necessary, however, to refine the technique. Published inter-laboratory studies suggest more standardisation in sample handling is necessary (Kvenvolden, 1980; Wehmiller, 1984). Amino acid dating has been applied in conjunction with other techniques, such as U-series dating of molluscs (Szabo *et al.*, 1981), thermoluminesence dating (Boulton *et al.*, 1982) or radiocarbon, towards the margin of its reliability (Mangerud *et al.*, 1981); all these independent dating methods having uncertainties attached to them. Relatively few studies have been published in which detailed comparisons have been possible with well-dated independent chronologies. Where such studies have been carried out (McCartan *et al.*, 1982; Miller *et al.*, 1983) the broad time dependence of amino acid racemisation or epimerisation has been confirmed, although certain anomalous results have been reported, suggesting that local effects may occasionally mask the time factor.

Thermoluminescence dating

The desire for a dating technique that is widely applicable and capable of dating material throughout the whole of Quaternary time has led to the application of thermoluminescence (TL) dating to sediments. Certain glassy or crystalline materials, such as quartz, that have been subject to ionising radiation emit light when heated, hence the term 'thermolumine-scence'. The amount of light emitted depends on the temperature to which the material is heated and this variation in light intensity with different temperatures when graphed is called a glow curve. Glow curves vary with material and radiation history and the intensity of the glow curve is broadly a function of the time lapsed since the time of the radiation exposure (Drei-manis *et al.*, 1978; Wintle and Huntley, 1982). It follows that if the radiation dose to which the material has been subjected is known, or can be reconstructed, and assuming that there was a 'zero' time when the sample was devoid of TL, then, knowing from laboratory studies the equivalent radiation dose that gives rise to a similar glow curve in the same material, the time lapsed since 'zero' year can be calculated.

A particular problem in applying TL dating to sediments is that the 'zero' year accords to the geological event being dated. The exposure of sediment to sunlight for at least a day appears to be sufficient to zero the TL in certain materials (Wintle and Huntley, 1982). Most TL dates of natural sediments have been carried out on aeolian material, where stratigraphic-ally concordant dates in broad agreement with independent dating techniques have been recorded. Other sediments which have been dated are tills, colluvial horizons, marine and deep-sea deposits. Of particular interest to sea-level change studies have been the applications to marine sediments, as in Spitsbergen (Troitsky *et al.*, 1979) and western Norway (Hutt *et al.*, 1983). These dates are thought by the relevant authors to give minimum

188

estimates of the ages of the horizons sampled and the associated error terms are quite large. The results, however, are sufficiently encouraging to anticipate a much larger contribution by TL to sea-level dating programmes in the future.

Relative dating frameworks

The dating methods described up to this point have all been laboratory techniques. Yet their value is entirely dependent upon the field samples being placed in a rigorous geomorphological or stratigraphical relative dating framework. Biostratigraphy, in particular, is a traditional form of relative dating and many detailed studies of Quaternary sediments have been published in which sea-level changes have been inferred and dated approximately on the basis of palynomorphs (pollen and spores) (see Devoy, this vol., Ch. 16) or marine fossils. The Quaternary is of sufficiently brief duration that no evolutionary trends modify the species studied and biostratigraphies can be interpreted in terms of changing environments, the principal control on which is climate. The cyclic nature of Quaternary climatic change, however, introduces the problem that similar biostratigraphical assemblages may have been produced at different times (Bowen, 1978) and this may introduce a serious error in any biostratigraphical correlation scheme. Certain species have become extinct at various times during the Quaternary and these may, therefore, be used as broad time markers.

One of the most detailed records of Quaternary sea-level change that is primarily based on biostratigraphical work is that developed around the southern margin of the North Sea Basin. This record is principally based on pollen studies, although marine faunas have also been studied in detail (West, 1972; 1980; Norton, 1977; Zagwijn, 1979). One of the most detailed such studies that illustrates the potential of the method is that of Zagwijn (1983) who, on the basis of pollen work, has traced the progressive sea-level changes throughout the last (Eemian) Interglacial. Figure 6.6 shows this sea-level curve and illustrates the similarity in form between the curves depicting the rising phase of sea level during the Eemian and that, based on radiocarbon dating, for the rising sea level during the Holocene.

Biostratigraphy is essentially a relative dating method and must be calibrated by other techniques in order to infer absolute ages. For the Northwest European biostratigraphic sequences beyond the range of radiocarbon, the principal method of independent dating that has been used is palaeomagnetism (van Montfrans, 1971). This approach utilises the established sequence of magnetic reversals and 'excursions' (in which the virtual geomagnetic pole deviates by more than 45° of latitude from the pole and then returns to the original polarity). The palaeomagnetic sequence is itself a relative dating technique but it has been calibrated by other methods, principally K–Ar dating of volcanic rocks (Mankinen and Dalrymple,

Figure 6.6: Comparison of sea-level curves for the last (Eemian) Interglacial and the Holocene in the Netherlands. Eemian chronology based on pollen zonation (Zagwijn, 1983) and Holocene chronology on radiocarbon dating (Jelgersma, 1961). The altitudinal difference between the curves is dominantly the result of long-term tectonic subsidence of the southern North Sea Basin

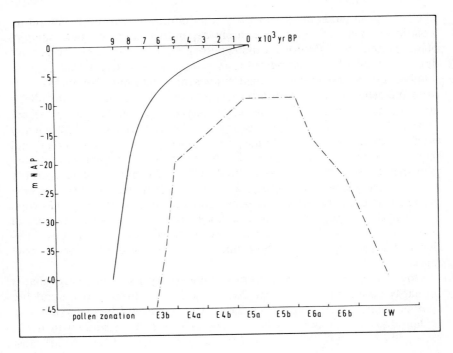

1979). Ultimately, it may be hoped that the Northwest European sequence will be cross-checked with other techniques such as amino acid and TL dating.

CONCLUSION

This chapter has reviewed the principal techniques that are currently available for studying Quaternary sea-level change. It is apparent that a wide variety of techniques that are applicable to a range of naturally occurring substances have been developed to cover the whole of Quaternary time. The accurate application of some of the techniques is still subject to further research and refinement, but all methods have inherent uncertainties that must be appreciated by the user. Such uncertainties relate both to the basic assumptions of the methods and to the problems of field occurrence and an attempt has been made to discuss these issues. The period 1955-86 has witnessed remarkable changes in the understanding of the variations in Quaternary sea level. The basic chronology is better understood than ever before. It is to be hoped that the next thirty years will be just as successful and stimulating.

REFERENCES

Andrews, J.T., Bowen, D.Q. and Kidson, C. (1979) 'Amino acid ratios and the correlation of raised beach deposits in southwest England and Wales', *Nature*, 281, 556-8.

—— and Miller, G.H. (1980) 'Dating Quaternary deposits more than 10,000 years old', in R.A. Cullingford, D.A. Davidson and J. Lewin (eds), *Timescales in Geomorphology*. Wiley, Chichester and New York, pp. 263-87.

—— Miller, G.H., Davies, D.C. and Davies, K.H. (1985) 'Generic identification of fragmentary Quaternary molluscs by amino-acid chromatography: a tool for Quaternary and palaeontological research', *Geol. J.*, 20, 1-20.

Bender, M.L. (1973) 'Helium-uranium dating of corals', *Geochim. Cosmochim. Acta*, 37, 1229-47.

—— Taylor, F.T. and Matthews, R.K. (1973) 'Helium-uranium dating of corals from Middle Pleistocene Barbados reef tracts', *Quat. Res.*, 3, 142-6.

—— Fairbanks, R.G., Taylor, F.W., Matthews, R.K., Goddard, J.G. and Broecker, W.S. (1979) 'Uranium-series dating of the Pleistocene reef tracts of Barbados, West Indies', *Bull. Geol. Soc. Am.*, 90, 577-94.

Berendsen, H.J.A. (1984) 'Quantitative analysis of radiocarbon dates of the perimarine area in the Netherlands', *Geol. Mijnb.*, 63, 343-50.

Blake, W. (1961) 'Radiocarbon dating of raised beaches in Nordaustlandet, Spitsbergen', in G.O. Raasch, (ed.), *The Geology of the Arctic*, University of Toronto Press, Toronto, pp.133-45.

—— (1970) 'Studies of glacial history in Arctic Canada: I. Pumice, radiocarbon dates, and differential postglacial uplift in the eastern Queen Elizabeth Islands', *Can. J. Earth Sci.*, 7, 634-64.

Bloom, A.L. (1977) *Atlas of Sea-level Curves*, IGCP Project 61, Cornell University, Ithaca, NY.

—— (1983) 'Sea level and coastal morphology of the United States through the Late Wisconsin glacial maximum', in S.C. Porter (ed.), *Late Quaternary Environments of the United States. Volume 1: The Late Pleistocene*, Longman, London, pp. 215-29.

—— Broecker, W.S., Chappell, J.M.A., Matthews, R.K. and Mesolella, K.J. (1974) 'Quaternary sea level fluctuations on a tectonic coast: new ^{230}Th/^{234}U dates from the Huon Peninsula, New Guinea', *Quat. Res.*, 4, 185-205.

Boulton, G.S., Baldwin, C.T., Peacock, J.D., McCabe, A.M., Miller, G., Jarvis, J., Horsfield, B., Worsley, P., Eyles, N., Chroston, P.N., Day, T.E., Gibbard, P., Hare, P.E. and von Brunn, V. (1982) 'A glacio-isostatic facies model and amino acid stratigraphy for late Quaternary events in Spitsbergen and the Arctic', *Nature*, 298, 437-41.

Bowen, D.Q. (1978) *Quaternary Geology: A Stratigraphic Framework for Multidisciplinary Work*, Pergamon Press, Oxford.

Broecker, W.S. and Bender, M.L. (1972) 'Age determinations on marine strandlines', in W.W. Bishop, and J.A. Miller, (eds), *Calibration of Hominoid Evolu-*

tion, Oliver & Boyd, Edinburgh, pp. 19-38.
—— and van Donk, J. (1970) 'Insolation changes, ice volumes and the 0^{18} record in deep-sea cores', *Rev. Geophys. Space Phys.*, 8. 169-98.
—— Gerard, R., Ewing, M. and Heezen, B.C. (1960) 'Natural radiocarbon in the Atlantic Ocean', *J. Geophys. Res.*, 65, 2903-31.
—— Thurber, D.L., Goddard, J.T., Ku, T.-L., Matthews, R.K. and Mesolella, K.J. (1968) 'Milankovitch hypothesis supported by precise dating of coral reefs and deep-sea sediments', *Science*, 159, 297-300.
Butzer, K.W. (1975) 'Pleistocene littoral-sedimentary cycles of the Mediterranean Basin: a Mallorquin view', in K.W. Butzer and G.L. Isaac (eds), *After the Australopithecines*, Mouton, The Hague, pp. 25-71.
Campbell, C.A. and Baxter, M.S. (1979) 'Radiocarbon measurements on submerged forest floating chronologies', *Nature*, 278, 409-13.
Chappell, J. (1974) 'Geology of coral terraces, Huon Peninsula, New Guinea: A study of Quaternary tectonic movements and sea-level changes', *Bull. Geol. Soc. Am.*, 85, 553-70.
—— and Polach, H.A. (1972) 'Some effects of partial recrystallization on ^{14}C dating Late Pleistocene corals and molluscs', *Quat. Res.*, 2, 244-52.
Clark, J.A., Farrell, W.E. and Peltier, W.R. (1978) 'Global changes in postglacial sea level: a numerical calculation', *Quat. Res.*, 9, 265-87.
Curray, J.R. (1965) 'Late Quaternary history, continental shelves of the United States', in H.E. Wright, and D.G. Frey, (eds), *The Quaternary of the United States*, Princeton University Press, Princeton, NJ, pp 723-35.
Davies, K.H. (1983) 'Amino acid analysis of Pleistocene marine molluscs from the Gower peninsula', *Nature*, 302, 137-9.
Devoy, R.J.N. (1979) 'Flandrian sea level changes and vegetational history of the lower Thames estuary', *Phil. Trans. Roy. Soc. Lond. B.*, 285, 355-407.
—— (1982) 'Analysis of the geological evidence for Holocene sea-level movements in southeast England', *Proc. Geol. Ass.*, 93, 65-90.
Dreimanis, A., Hutt, G., Raukas, A. and Whippey, P.W. (1978) 'Dating methods of Pleistocene deposits and their problems: I. Thermoluminescence dating', *Geoscience Canada*, 5, 55-60.
Eronen, M. (1974) 'The history of the Litorina Sea and associated Holocene events', *Comment. Phys.-Math.*, 44, 79-195.
Fairbridge, R.W. (1961) 'Eustatic changes in sea level', *Phys. Chem. Earth*, 4, 99-185.
Fanale, F.P. and Schaeffer, O.A. (1965) 'Helium-uranium ratios for Pleistocene and Tertiary fossil aragonites', *Science*, 149, 312-17.
Geyh, M.A. (1971) 'Middle and young Holocene sea-level changes as global contemporary events', *Geol. Fören. Stockh. Förh.*, 93, 679-91.
—— (1980) 'Holocene sea-level history: case study of the statistical evaluation of ^{14}C dates', *Radiocarbon*, 22, 695-704.
Gordon, D. and Smart, P.L. (1984) 'Comments on "Speleothems, travertines, and palaeoclimates" by G.J. Hennig, R. Grun, and K. Brunnacker', *Quat. Res.*, 22, 144-7.
Grootes, P.M. (1978) 'Carbon-14 timescale extended: comparison of chronologies', *Science*, 200, 11-15.
Hafsten, U. (1983) 'Biostratigraphical evidence for Late Weichselian and Holocene sea-level changes in southern Norway', in D.E. Smith and A.G. Dawson (eds), *Shorelines and Isostasy*, Academic Press, London and New York, 161-81.
Hakansson, S. (1983) 'A reservoir age for the coastal waters of Iceland', *Geol. Fören. Stockh. Förh.*, 105, 64-7.
Hare, P.E. and Mitterer, R.M. (1967) 'Nonprotein amino acids in fossil shells', *Carnegie Institute Washington Year Book*, 65, 362-4.

——— and Mitterer, R.M. (1969) 'Laboratory simulation of amino-acid diagenesis in fossils', *Carnegie Institute Washington Year Book, 67*, 205-8.

Harkness, D.D. (1979) 'Radiocarbon dates from Antarctica', *Br. Antarct. Surv. Bull., 47*, 43-59.

——— (1983) 'The extent of natural ^{14}C deficiency in the coastal environment of the United Kingdom', *Pact, 8*, 351-64.

Harmon, R.S., Schwarcz, H.P. and Ford, D.C. (1978) 'Late Pleistocene sea level history of Bermuda', *Quat. Res., 9*, 205-18.

Heyworth, A. and Kidson, C.(1982) 'Sea-level changes in southwest England and Wales', *Proc. Geol. Ass., 93*, 91-111.

Hjort, C. (1973) 'A sea correction for East Greenland', *Geol. Fören. Stockh. Förh., 95*, 132-4.

Hopley, D. (1974) 'Investigations of sea level changes along the coast of the Great Barrier Reef', *Proc. Second Int. Coral Reef Symp., 2*, 551-62.

Hutt, G., Punning, J.-M. and Mangerud, J. (1983) 'Thermoluminescence dating of the Eemian–early Weichselian sequence at Fjøsanger, western Norway', *Boreas, 12*, 227-31.

Ivanovich, M. (1982) 'Uranium series disequilibrium applications in geochronology', in M. Ivanovich and R.S. Harmon (eds), *Uranium Series Disequilibrium: Applications to Environmental Problems*, Clarendon Press, Oxford, pp. 56-78.

Jelgersma, S. (1961) 'Holocene sea level changes in the Netherlands', *Meded. Geol. Sticht. Nederland, Ser, C VI.7*, 1-100.

——— (1980)'Late Cenozoic sea level changes in the Netherlands and the adjacent North Sea Basin', in N.-A. Mörner (ed.), *Earth Rheology, Isostasy and Eustasy*, Wiley, Chichester and New York, pp. 435-47.

Jong, A.F.M. de, Mok, W.G. and Becker, B. (1979) 'Confirmation of Suess wiggles: 3,200–3,700 BC', *Nature, 280*, 48-9.

Kaland, P.E. (1984) 'Holocene shore displacement and shorelines in Hordaland, western Norway', *Boreas, 13*, 203-42.

——— Krzywinski, K. and Stabell, B. (1984) 'Radiocarbon dating of transitions between marine and lacustrine sediments and their relation to the development of lakes', *Boreas*, 13, 243-58.

Kaufman, A., Broecker, W.S., Ku, T.-L. and Thurber, D.L. (1971) 'The status of U-series methods of mollusk dating', *Geochim. Cosmochim. Acta*. 35, 1155-83.

Kaye, C.A. and Barghoorn, E.S. (1964) 'Late Quaternary sea-level change and crustal rise at Boston, Massachusetts, with notes on the auto-compaction of peat', *Bull. Geol. Soc. Am., 75*, 63-80.

Kidson, C. (1982) 'Sea level changes in the Holocene', *Quat. Sci. Rev., 1*, 121-51.

——— and Heyworth, A. (1973) 'The Flandrian sea-level rise in the Bristol Channel', *Proc. Ussher Soc., 2*, 565-84.

King, K. and Hare, P.E. (1972) 'Species effects in the epimerization of L-isoleucine in fossil planktonic foraminifera', *Carnegie Institute Washington Year Book, 71*, 596-8.

Klein, J., Lerman, J.C., Damon, P.E. and Ralph, E.K. (1982) 'Calibration of radiocarbon dates: tables based on the concensus data of the Workshop on Calibrating the Radiocarbon Timescale', *Radiocarbon, 24*, 103-50.

Krog, H. and Tauber, H. (1974) 'C-14 chronology of Late- and Post-glacial marine deposits in North Jutland', *Danm. Geol. Unders. Arbog*, 1973, 93-105.

Krzywinski, K. and Stabell, B. (1984) 'Late Weichselian sea level changes at Sotra, Hordaland, western Norway', *Boreas, 13*, 159-202.

Kvenvolden, K.A. (1980) 'Interlaboratory comparison of amino acid racemization in a Pleistocene mollusc, *Saxidomus giganteus*', in P.E. Hare, T.C. Hoering and K. King, (eds), *Biogeochemistry of Amino Acids*. Wiley, New York, pp. 223-32.

Labordel, J. (1980) 'Les Gastéropodes vermitidés: leur utilisation comme marqueurs biologiques de rivages fossiles', *Oceanis*, *5*, 221-39.

Lajoie, K.R., Wehmiller, J.F. and Kennedy, G.L. (1980) 'Inter- and intrageneric trends in apparent racemization kinetics of amino acids in Quaternary mollusks', in P.E. Hare, T.C. Hoering and K. King (eds), *Biogeochemistry of Amino Acids*, Wiley, Chichester and New York, pp. 305-40.

Libby, W.F. (1955) *Radiocarbon Dating*, University of Chicago Press, Chicago.

McCartan, L., Owens, J.P., Blackwelder, B.W., Szabo, B.J., Belknap, D.F., Kriausakul, N., Mitterer, R.M. and Wehmiller, J.F. (1982) 'Comparison of amino acid racemization geochronometry with lithostratigraphy, biostratigraphy, uranium-series coral dating, and magnetostratigraphy in the Atlantic Coastal Plain of the southeastern United States', *Quat. Res.*, *18*, 337-59.

McLean, R. F., Stoddart, D.R., Hopley, D. and Polach, H. (1978) 'Sea level change in the Holocene on the northern Great Barrier Reef', *Phil. Trans. Roy. Soc. Lond. A.*, *291*, 167-86.

Mangerud, J. (1972) 'Radiocarbon dating of marine shells, including a discussion of apparent age of recent shells from Norway', *Boreas*, *1*, 143-72.

—— and Gulliksen, S. (1975) 'Apparent radiocarbon ages of recent marine shells from Norway, Spitsbergen and Arctic Canada', *Quat. Res.*, *5*, 263-73.

—— Gulliksen, S., Larsen, E., Longva, O., Miller, G.H., Sejrup, H.P. and Sonstegaards, E. (1981) 'A Middle Weichselian ice-free period in western Norway: the Alesund Interstadial', *Boreas*, *10*, 447-62.

Mankinen, E.A. and Dalrymple, G.B. (1979) 'Revised geomagnetic polarity time scale for the interval 0-5 m y BP', *J. Geophys. Res.*, *84*, 615-26.

Marshall, J.F. and Thom, B.G. (1976) 'The sea level in the last interglacial', *Nature*, *263*, 120-1.

Marthinussen, M. (1962) 'C-14 datings referring to shorelines, transgressions, and glacial substages in northern Norway', *Nor. Geol. Unders.*, *215*, 37-67.

Matthews, J.A. (1981) 'Natural ^{14}C age/depth gradient in a buried soil', *Naturwissenschaften*, *68*, 472-4.

—— and Dresser, P.Q. (1983) 'Intensive ^{14}C dating of a buried palaeosol horizon', *Geol. Fören. Stockh Förh.*, *105*, 59-63.

Mesolella, K.J., Matthews, R.K., Broecker, W.S. and Thurber, D.L. (1969) 'The astronomical theory of climatic change: Barbados data', *J. Geol.*, *77*, 250-74.

Miller, G.H. and Hare, P.E. (1980) 'Amino acid geochronology: integrity of the carbonate matrix and potential of molluscan fossils', in P.E. Hare, T.C. Hoering, and K. King (eds), *Biogeochemistry of Amino Acids*, Wiley, Chichester and New York, pp. 415-43.

—— Andrews, J.T. and Short, S.K. (1977) 'The last interglacial–glacial cycle, Clyde Foreland, Baffin Island, N.W.T.: stratigraphy, biostratigraphy and chronology', *Can. J. Earth Sci.*, *14*, 2824-57.

—— Hollin, J.T. and Andrews, J.T. (1979) 'Aminostratigraphy of UK Pleistocene deposits', *Nature*, *281*, 539-43.

——, Sejrup, H.P., Mangerud, J. and Andersen, B.G. (1983) 'Amino acid ratios in Quaternary molluscs and foraminifera from western Norway: correlation, geochronology and paleotemperature estimates', *Boreas*, *12*, 107-24.

Milliman, J.D. (1974) *Marine Carbonates*, Springer Verlag, Berlin.

—— and Emery, K.O. (1968) 'Sea levels during the past 35,000 years', *Science*, *162*, 1121-3.

Mitterer, R.M. (1975) 'Ages and diagenetic temperatures of Pleistocene deposits in Florida based on isoleucine epimerization in *Mercenaria*', *Earth Plant. Sci. Lett.*, *28*, 275-82.

Montfrans, H.M. van (1971) 'Palaeomagnetic dating in the North Sea Basin', *Earth*

Planet. Sci. Lett., 11, 226-35.

Moore, W.S. (1982) 'Late Pleistocene sea-level history', in M. Ivanovich, and R.S. Harmon, (eds) *Uranium Series Disequilibrium: Applications to Environmental Problems*, Clarendon Press, Oxford, pp. 481-96.

Mörner, N.-A. (1969) 'The Late Quaternary history of the Kattegatt Sea and the Swedish west coast: deglaciation, shorelevel displacement, chronology, isostasy and eustasy', *Sveriges Geol. Unders., Ser. C, 640*, 1-487.

—— (1971) 'The position of the ocean level during the interstadial at about 30,000 BP — a discussion from a climatic–glaciologic point of view', *Can. J. Earth Sci., 8*, 132-43.

—— (1976) 'Eustasy and geoid changes', *J. Geol., 84*, 123-51.

Norton, P.E.P. (1977) 'Marine mollusca in the East Anglian Preglacial Pleistocene', in F.W. Shotton (ed.), *British Quaternary Studies: Recent Advances*, Oxford University Press, pp. 43-53.

Nydal, R., Gulliksen, S. and Lovseth, K. (1972) 'Trondheim natural radiocarbon measurements VI', *Radiocarbon, 14*, 418-51.

Oldfield, F. (1981) 'Peats and lake sediments: formation, stratigraphy, description and nomenclature', in A. Goudie (ed.), *Geomorphological Techniques*, Allen & Unwin, London, 306-26.

Olsen, E.A. and Broecker, W.S. (1958) 'Sample contamination and reliability of radiocarbon dates', *Trans. New York Acad. Sci., Ser. II., 20 B*, 593-604.

Olsson, I.U. (1968) 'Modern aspects of radiocarbon datings', *Earth Sci. Rev., 4*, 203-18.

—— (ed.) (1970) *Radiocarbon Variations and Absolute Chronology. Twelfth Nobel Symposium, Uppsala, 1969*, Wiley, New York.

—— (1979) 'A warning against radiocarbon dating of samples containing little carbon', *Boreas, 8*, 203-7.

—— (1980) 'Content of ^{14}C in marine mammals from northern Europe', *Radiocarbon, 22*, 662-75.

Peacock, J.D., Graham, D.K., Robinson, J.E. and Wilkinson, I. (1977) 'Evolution and chronology of Lateglacial marine environments at Lochgilphead, Scotland', in J.M. Gray and J.J. Lowe (eds), *Studies in the Scottish Lateglacial Environment*, Pergamon Press, Oxford, pp. 89-100.

—— Graham, D.K. and Wilkinson, I.P. (1978) 'Late-Glacial and post-Glacial marine environments at Ardyne, Scotland, and their significance in the interpretation of the history of the Clyde sea area', *Rept. Inst. Geol. Sci., 78/17*.

Plassche, O. van de (1980) 'Compaction and other sources of error in obtaining sea-level data: some results and consequences', *Eiszeit. Gegenw., 30*, 171-81.

—— (1982) 'Sea-level change and water-level movements in the Netherlands during the Holocene', *Meded. Rijks. Geol. Dienst., 36(1)*, 1-93.

—— (1986, in press) *Sea-level Research: A Manual for the Collection and Evaluation of Data*, Geo-Abstracts, Norwich.

Polach, H.A., McLean, R.F., Caldwell, J.R. and Thom, B.G. (1978) 'Radiocarbon ages from the northern Great Barrier Reef', *Phil. Trans. Roy. Soc. Lond., A., 291*, 139-58.

Roeleveld, W. (1974) 'The Holocene evolution of the Groningen marine-clay district', *Bericht. Rijks. Oudheid. Bodem., 24*, 1-132.

Rose, J. (1978) 'Glaciation and sea-level change at Bugöyfjord, south Varangenfjord, north Norway', *Norsk. Geogr. Tidsskr., 32*, 121-35.

Salvigsen, O. (1981), 'Radiocarbon dated raised beaches in Kong Karls Land, Svalbard, and their consequences for the glacial history of the Barents Sea area', *Geogr. Ann., 63A*, 283-91.

Scharpenseel, H.W. (1971) 'Radiocarbon dating of soils — problems, troubles, hopes', in D.H. Yaalon (ed.), *Paleopedology: Origin, Nature and Dating of Palaeosols*, International Society of Soil Science and Israel Universities Press, Jerusalem, pp. 77-88.

Scholl, D.W. (1964) 'Recent sedimentary record in mangrove swamps and rise in sea level over the southwestern coast of Florida: Part 1', *Mar. Geol.*, *1*, 344-66.

—— (1965) 'Recent sedimentary record in mangrove swamps and rise in sea level over the southwestern coast of Florida: Part 2', *Mar. Geol.*, *2*, 343-64.

—— and Stuiver, M. (1967) 'Recent submergence of southern Florida: a comparison with adjacent coasts and other eustatic data', *Bull. Geol. Soc. Am.*, *78*, 437-54.

Scoffin, T.P. and McLean, R.F. (1978) 'Exposed limestones of the Northern province of the Great Barrier Reef', *Phil. Trans. Roy. Soc. Lond.*, A, *291*, 119-38.

—— and Stoddart, D.R. (1978) 'The nature and significance of microatolls', *Phil. Trans. Roy. Soc. Lond.*, B, *284*, 99-122.

Shennan, I. (1979) 'Statistical evaluation of sea-level data', *Sea-Level*, 1, 6-11.

—— (1982) 'Interpretation of Flandrian sea-level data from the Fenland, England', *Proc. Geol. Ass.*, *83*, 53-63.

—— Tooley, M.J., Davis, M.J. and Haggart, B.A. (1983) 'Analysis and interpretation of Holocene sea-level data', *Nature*, *302*, 404-6.

Shepard, F.P. (1963) 'Thirty-five thousand years of sea level', in T. Clements (ed.), *Essays in Marine Geology in Honor of K.O. Emery*. University of California Press, Los Angeles, Calif., pp. 1-10.

Shotton, F.W. (1967) 'The problems and contributions of methods of absolute dating within the Pleistocene period', *Quart. J. Geol. Soc. Lond.*, *122*, 357-83.

Sissons, J.B. (1983) 'Shorelines and isostasy in Scotland', in D.E. Smith and A.G. Dawson (eds) *Shorelines and Isostasy*, Academic Press, London and New York, pp. 209-25.

—— and Brooks, C.L. (1971) 'Dating of early postglacial land and sea level changes in the western Forth valley', *Nature Phys. Sci.*, *234*, 124-7.

—— and Smith, D.E. (1965), 'Peat bogs in a post-glacial sea and a buried raised beach in the western part of the Carse of Stirling', *Scott. J. Geol.*, *1*, 247-55.

—— Smith, D.E. and Cullingford, R.A. (1966), 'Late-glacial and post-glacial shorelines in South-East Scotland', *Trans. Inst. Br. Geogr.*, *39*, 9-18.

Skempton, A.W. (1970) 'The consolidation of clays by gravitational compaction', *Quart. J. Geol. Soc. Lond.*, *125*, 373-412.

Stearns, C.E. and Thurber, D.L. (1965) 'Th^{230}/U^{234} dates of late Pleistocene marine fossils from the Mediterranean and Moroccan littorals', *Progress in Oceanography*, *4*, 293-305.

Stewart, T.G. and England, J. (1983) 'Holocene sea-ice variations and palaeoenvironmental change, northernmost Ellesmere Island, N.W.T., Canada', *Arctic and Alpine Res.*, 15, 1-17.

Streif, H. (1972) 'The results of stratigraphical and facial investigations in the coastal Holocene of Woltzeten/Ostfriesland, Germany', *Geol. Fören. Stockh. Förh.*, *94*, 281-99.

Stuiver, M. (1971) 'Evidence for the variation of atmospheric C^{14} content in the late Quaternary', in K.K. Turekian (ed.), *Late Cenozoic Glacial Ages*, Yale University Press, New Haven, pp. 57-70.

—— (1978) 'Radiocarbon timescale tested against magnetic and other dating methods', *Nature*, *273*, 271-4.

Sutherland, D.G. (1983) 'The dating of former shorelines', in D.E. Smith and A.G. Dawson (eds),*Shorelines and Isostasy*, Academic Press, London, pp. 129-57.

—— (1986) 'A review of Scottish marine shell radiocarbon dates, their standardization and interpretation', *Scott. J. Geol.*, *22*, 145-64.

Szabo, B. J., Miller, G. H., Andrews, J.T. Stuiver, M. (1981) 'Comparison of uranium-series, radiocarbon, and amino acid data from marine molluscs, Baffin Island, Arctic Canada', *Geology, 9,* 451-7.

Tauber, H. (1970) 'The Scandinavian varve chronology and C-14 dating', in I.U. Olsson (ed.), *Radiocarbon Variations and Absolute Chronology. Twelfth Nobel Symposium, Uppsala, 1969,* Wiley, New York, pp. 173-96.

Ten Brink, N.W. (1974) 'Glacio-isostasy: new data from West Greenland and geophysical implications', *Bull. Geol. Soc. Am., 85,* 219-28.

Thom, B.G. (1973) 'The dilemma of high interstadial sea levels during the last glaciation', *Prog. Geogr., 5,* 167-246.

Thompson, R (1977) 'Stratigraphic consequences of palaeomagnetic studies of Pleistocene and Recent sediments', *J. Geol. Soc. Lond., 133,* 51-9.

—— and Turner, G.M. (1979) 'British Geomagnetic Master Curve 10000-0 yr BP for dating European sediments', *Geophys. Res Lett., 6,* 249-52.

Thurber, D.L., Broecker, W.S., Blanchard, R.L. and Potratz, H.A. (1965) 'Uranium-series ages of Pacific atoll corals', *Science, 149,* 55-8.

Tooley, M.J. (1978) *Sea-level Changes in North-West England during the Flandrian Stage,* Clarendon Press, Oxford.

—— (1982) 'Sea-level changes in northern England', *Proc. Geol. Ass., 93,* 43-51.

Troitsky, L., Punning, J.-M., Hutt, G. and Rajanae, R. (1979) 'Pleistocene glaciation chronology in Spitsbergen', *Boreas, 8,* 401-7.

Vita-Finzi, C. (1980) '^{14}C dating of recent crustal movements in the Persian Gulf and Iranian Makran', *Radiocarbon, 22,* 763-73.

—— and Roberts, N. (1984) 'Selective leaching of shells for ^{14}C dating', *Radiocarbon, 26,* 54-8.

Wehmiller, J.F. (1980) 'Intergeneric differences in apparent racemization kinetics in mollusks and foraminifera: implications for models of diagenetic racemization', in P.E. Hare, T.C. Hoering and K. King (eds), *Biogeochemistry of Amino Acids.* Wiley, Chichester and New York, pp. 341-55.

—— (1982) 'A review of amino acid racemization studies in Quaternary mollusks: stratigraphic and chronologic applications in coastal and interglacial sites, Pacific and Atlantic coasts, United States, United Kingdom, Baffin Island, and tropical islands', *Quat. Sci. Rev., 1,* 83-120.

—— (1984) 'Interlaboratory comparison of amino acid enantiomeric ratios in fossil Pleistocene mollusks', *Quat. Res., 22,* 109-20.

—— and Belknap, D.F. (1978) 'Alternative kinetic models for the interpretation of amino acid enantiomeric ratios in Pleistocene mollusks: examples from California, Washington and Florida', *Quat. Res., 9,* 330-48.

—— and Belknap, D.F. (1982) 'Amino acid age estimates, Quaternary Atlantic Coastal Plain: comparison with U-series dates, biostratigraphy, and paleomagnetic control', *Quat. Res., 18,* 311-36.

West, R.G. (1972) 'Relative land — sea-level changes in south-eastern England during the Pleistocene', *Phil. Trans. Roy. Soc. Lond., A, 272,* 87-98.

—— (1980) *The Pre-glacial Pleistocene of the Norfolk and Suffolk Coasts,* Cambridge University Press, Cambridge.

Wintle, A.G. and Huntley, D.J. (1982) 'Thermoluminescence dating of sediments', *Quat. Sci. Rev., 1,* 31-53.

Zagwijn, W.H. (1979) 'Early and Middle Pleistocene coastlines in the southern North Sea Basin', in E. Oele, R.T.E. Schuttenhelm and A.J. Wiggers (eds). *The Quaternary History of the North Sea,* Acta Univ. Ups. Symp. Univ. Ups. Annum Quingentesimum Celebrantis: 2, Uppsala, pp. 31-42.

—— (1983) 'Sea-level changes in the Netherlands during the Eemian', *Geol. Mijnb., 62,* 437-50.

7

Global Analysis and Correlation of Sea-level Data

Ian Shennan

INTRODUCTION

Current research in sea-level changes varies tremendously in the range of both temporal and spatial scales studied and in the techniques of analysis used. Yet there should always be an implicit aim that the results of each piece of research should be able to draw from, and contribute to, other investigations. To what extent this is possible varies. Given the correct research design the results should be able to be compared and correlated at any chosen scale, to be used as the input for, or to provide the data to test, another line of research. The difficulties facing anyone attempting to correlate sea-level data from different research works at any scale from a single estuary to a global analysis are numerous and may be most clearly seen by limiting the discussion primarily to the Holocene since the uncertainties in correlating pre-Holocene sea-level data are increased due to a restricted data base.

The decade starting in 1970 was a period of increasing interest in sea-level changes. A coarse indication of this can be seen in Table 7.1.

In 1974 the International Geological Correlation Programme (IGCP) Board approved the proposal for a sea-level project. Thus IGCP Project 61

Table 7.1: Year of publication of sea-level curves according to the *Atlas of Sea-Level Curves* bibliography (Bloom 1977, 1982)

Year	Number of curves
1940-4	6
1945-9	3
1950-4	2
1955-9	4
1960-4	20
1965-9	34
1970-4	78
1975-9	107

'Sea-Level Movements During the Last Deglacial Hemicycle (About 15,000 Years)' was formalised and ran until 1982. The primary objective of Project 61 'was to establish a graph of the trend of mean sea-level during the period of deglaciation and continuing to the present day, based on compilations of sea-level index points from all over the world' (Tooley 1982a; p. 3). For the majority of participants in Project 61 research efforts were concentrated on the production of local sea-level studies, either in areas which had not been studied before or in increasing the detail of knowledge of sea-level and palaeoenvironmental change in previously studied areas. Comparison and correlation between different areas, usually involving the work of more than one research worker or group, was attempted in order to reveal local and regional factors affecting the registration of the relative sea-level history in a specific field study area. In an attempt to aid the compilation, comparison and correlation of sea-level data a computer-based data bank of radiocarbon dated sea-level index points was set up as part of Project 61.

Project 61 ended in 1982 and has been followed by a new project: Project 200 (1983-7) 'Late Quaternary sea-level changes: measurement, correlation and future applications'. Many of the problems that existed at the beginning of Project 61 remain for the participants in the new project to tackle although some firm conclusions were reached between 1974 and 1982. Kidson (1982) has described three main conclusions. Firstly, the search for a universal eustatic curve must be regarded as over. This stems from the general acceptance of the concept of geoidal-eustasy since the important paper by Mörner (1976a) and Clark *et al.* (1978). From this point it follows that, secondly, regional differences in response to changes in the geoid mean that eustatic sea-level curves can have only regional and not global validity. Thirdly, it is now accepted that no part of the earth's crust can be regarded as wholly stable. Further, although it is not listed by Kidson as one of the firm conclusions, it became obvious by the end of Project 61 that there was a need for increased rigour in data collection, assessment of errors and the definition of terms.

IGCP Project 200, like its predecessor, is firmly based on both careful, locally based sea-level studies and correlation from the local to regional to continental to global scale. Ultimate aims, whether or not they are recognised by 1987, are focused on the prediction of near-future changes for application to a variety of coastal problems. The degree with which existing problems of correlating between different areas at different temporal and spatial scales can be overcome will influence the success of achieving these aims. These problems are described in this chapter.

METHODOLOGICAL APPROACHES

Studies of local sea-level change lack any accepted formal methodology and therefore scientific laws and theories. The treatment of data remains essentially inductive. This, the development of explanation via inductive models, is the normal route followed by a science during the period of data collection (Harvey, 1969). Problems arise when a common language is required for classification, since without adequately rigid operational definitions, and strict adherence to them, even statistically significant features are not capable of comparison. This problem has been discussed at length by Shennan and Tooley (Shennan 1982a, b, 1983; Shennan et al., 1983; Tooley, 1982b) and a methodology based on the concept of sea-level tendencies has been proposed which allows for a variety of approaches, at different scales, to be integrated.

In addition to the dependence on inductive inferences it is worthwhile reiterating that the philosophical principles of *local* sea-level studies are the same as those outlined by Birks and Birks (1980) for geology and palaeo-ecology. These include (1) the employment of the methods of multiple working hypotheses, (2) methodological uniformitarianism, and (3) to let the simplest explanation suffice until more evidence is available which necessitates more complicated explanations.

Analyses of *global* sea-level data, which can be viewed as the other extreme of the range of analyses from local sea-level histories, employ quite different methodologies. *A priori* knowledge of deglaciation chrono-logies, the mantle viscosity profile and observed relative sea-level data for different areas allow a model testing methodology to be used (e.g. Peltier, 1982), quite different to the inductive approach of local sea-level studies, but ultimately dependent on the latter for verification. The problem of model robustness consequent upon a change in precision of the relative sea-level data requires investigation. The same can be said for palaeo geoidal modelling (e.g. Newman et al., 1980).

Inherent in the different approaches, ranging from local to global studies, is the requirement to identify the relevant temporal and spatial scale and thus the relative importance of each variable. There is a clear causal relationship between relative sea-level change and the sediments and forms produced at the coastline (Fig. 7.1). This figure only holds true at a spatial scale larger than the individual site since there will be significant feedback relationships between form and the registration of sea level at each site. Thus, even at this conceptual level it is necessary to delimit the scales of study and define the terms: e.g. mean sea-level at the open coast, mean height of similar sea-level index points along the section of coast, a timescale of perhaps tens of years. Assuming that the scale of study is large enough to ignore the feedback mechanisms, Figure 7.1 indicates that the inductive route to explanation reverses the path, sometimes missing out

Figure 7.1: Cause, effect and explanation of relative sea-level changes

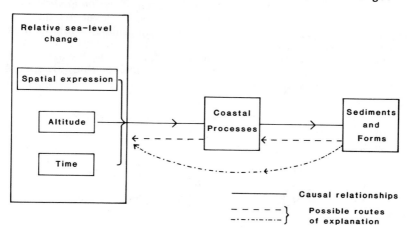

stages, of the causal relationships. This may cause problems with the reliability of the explanation reached if the correct temporal and spatial scales are not chosen and specified, if the variables are inadequately defined, if the feedback relationships are not considered, and finally if the processes themselves are not identified. In a different field of study, in fluvial geomorphology, these relationships between time, variables, processes and form were discussed in an important paper by Schumm and Lichty (1965) and their arguments are pertinent to all aspects of sea-level research. The link between present process-orientated studies and Holocene sea-level studies is weak and it needs strengthening. 'A study of process must attempt to relate causality to the evolution of the system' (Schumm and Lichty, 1965: p. 110); equally geomorphic processes cannot be ignored if sea-level studies are to offer adequate explanations of local sediment suites, through to global changes.

Schumm and Lichty argue that apparently disparate approaches, process studies and study of the origin and history of a system, are not mutually exclusive but can be linked according to temporal and spatial scales. The nature of sea-level rise during the Holocene has remained a point of discussion, now with proponents of both the smooth and spasmodic schools of thought perhaps showing a degree of agnosticism (see Kidson, 1982). The two schools of thought are not mutually exclusive if the problem is compared by analogy to the drainage basin example of Schumm and Lichty (1965, Table 1 and Figure 1). Pethick (1984) uses the same analogy in considering deposition rates in the coastal environment (Fig. 7.2), as do Sugden and John (1976) for glacier systems. Thus, as the time span to be considered becomes smaller then the spatial scale of the dependent variables to be considered also decreases. Other variables change from

201

Figure 7.2: (a) The relationship between time and landform scale for fluvial landscapes. (b) The concept applied to the coastal estuarine environment. Estuaries react slowly to causal processes such as sea-level changes, whereas at smaller spatial scales their bounding mudflats or saltmarshes reach a steady state between process and form relatively quickly. This relationship can be applied to other variables, for example the axis could be labelled mean annual sea level, in place of disposition rate; the spatial and temporal scales would remain the same

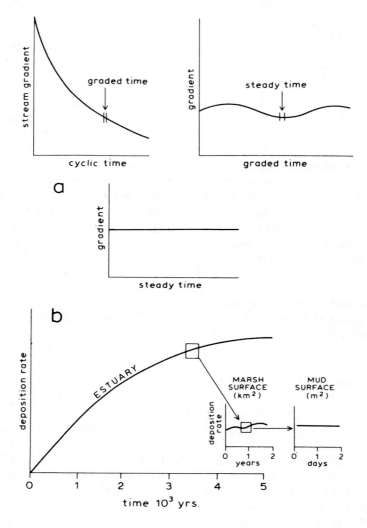

Source: Pethick (1984). After Schumm and Lichty (1965).

being an independent variable to becoming a dependent variable. For example, a specified reference water level at the coast (e.g. MSL or MHWST) would be a dependent variable where the timescale of study is in the order of thousands of years and may be represented by a sea-level band, but at a timescale of hundreds of years the same variable is now classified as independent. At this timescale, analogous to the graded condition of Schumm and Lichty, only components of the previous system under study can be considered. Indeed only components of the system can be described as 'graded' since the whole system, known from the previous scale of study, reveals a progressive change. Furthermore, while 'the whole system may be undergoing a progressive change of very small magnitude, some of the components of the system will show no progressive change' (Schumm and Lichty, 1965: p. 114). The sea-level analogy is that of a fluctuating sea level of low amplitude superimposed on a smooth curve. Alternatively, in sea-level tendency terms, some components, for example the developing transgressive or regressive overlap, will show a progressive change, whereas other components, for example coastal peat bog stratigraphy or offshore subtidal sediment suites, may appear in dynamic equilibrium with the same parameters of change. Equally, the components in dynamic equilibrium under one given set of conditions would reveal progressive change under different conditions.

Schumm and Lichty also warn that during brief spans of time 'cause and effect' can seem to be reversed, due to feedback from the dependent to the independent variables. Thus short-term changes in sediment availability or tidal velocities, due for example to man's interference on the coastline, may give misleading results if the nature and rate of presently monitored processes are uncritically extrapolated back through time.

Sea-level studies range from the development of local sea-level histories, dealing with relatively short time spans, small areas and specific details, to global analyses, which tend to generalise. At each scale the status of each variable must be assessed relative to the chosen timescale. If this is done then the apparently disparate approaches are reconciled, including the important contribution from present-day process studies.

Shennan and others (1982a, 1983; Shennan *et al.*, 1983) have proposed a method by which individual sea-level index points are assessed in a consistent manner. An individual sea-level index point is unlikely to show unequivocally a regionally significant process such as a rise or fall in sea-level. Such processes, which affect a larger spatial scale than those site specific processes which ultimately control the registration of any single sea-level index point, should be interpreted in terms of the dominant tendency of sea-level movement. A positive tendency of sea-level move-ment is defined as an apparent increase in the marine influence and a negative tendency of sea-level movement is the apparent decrease of the marine influence. It is a further, and often very difficult, stage to show unequivocally that tendencies of sea-level movement indicate changes in sea-level through time. Transgressive overlap and regressive overlap are just two types of sea-level index points (Shennan, 1982a; Tooley, 1982b); others include palaeobotanical evidence, morphological and archaeological data, palaeosols and other lithostratigraphic changes. Following the analysis of these data and the assessment of errors relating to the age, altitude and meaning of numerous sea-level index points, the accumulated evidence, at the local scale, can be interpreted in terms of the dominant local sea-level tendency. By adopting rigorous operational definitions disparate data can be assimilated at increasing scales to investigate different phenomena in a step-wise fashion to reach the required spatial and temporal resolution for reliable conclusions to be made. The application of a consistent methodology, based on unambiguous operational definitions, is required for correlation at all stages. The methodology has been most formalised for data from the UK but it clearly originated, in part, from West Germany (see especially Streif, 1979a). Given the required approach, a range of investigations can use the method (Fig. 7.3) (see also Geyh *et al.*, 1979; Streif, 1979b). The application of the method to the analysis of UK sea-level data is discussed later in the chapter.

LOCAL, REGIONAL AND GLOBAL SEA-LEVEL ANALYSES

Because of ... basic inconsistencies most sea-level data cannot be used in their published form. Re-classification and re-evaluation of the data are required before meaningful correlations are possible. It can be argued that even if the sea-level data currently used for correlations between different areas ... were all defined in a suitable form, the techniques of correlation are rarely developed beyond the visual similarities of two or more curves upon a graph. More reliable techniques of correlation are required. (Shennan, 1982b: p. 53)

This statement remains valid. Methodological weakness still hinders the comparison and correlation of sea-level data at all scales. The validity of global sea-level analyses ultimately depends on reliable local sea-level analyses and their correlations. Agreed operational definitions and the rules of stratigraphic correlation need to be rigorously applied in the methodological approach taken. A number of case studies will now be presented to highlight methods of approach at the various scales, their limitations and advantages.

Figure 7.3: The application of the sea-level tendency methodology to local sea-level analysis, regional sea-level tendency analysis and the analysis of crustal movements

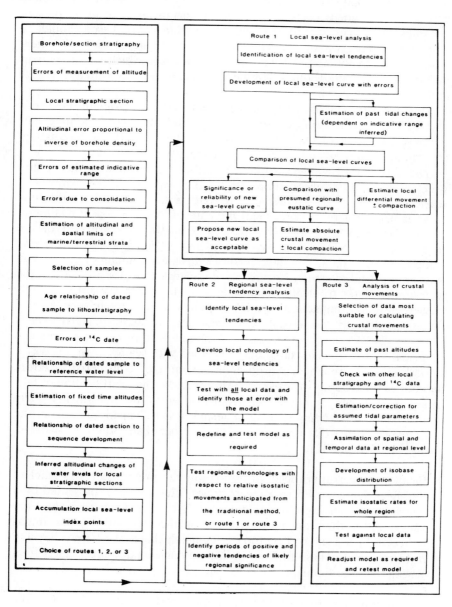

Source: Shennan (1983).

Local and regional sea-level histories: the inductive method

Local and regional sea-level studies depend largely on inductive inferences and reasoning where extrapolations are made in an attempt to present generalisations. The method of multiple working hypotheses should be employed where as many explanations as possible are considered, otherwise a single hypothesis may lead to a 'theory into which all subsequent observations are fitted, often without due regard for the evidence' (Birks and Birks, 1980: p. 7).

This criticism is valid when the argument that took place in the 1960s and 1970s between the different 'schools of thought' about the nature of the post glacial rise of sea level is considered; the 'smooth' or 'fluctuating' (spasmodic) sea-level controversy (Jelgersma, 1966).

The adoption of a single working hypothesis as a tentative theory into which all subsequent observations are fitted is apparent in the synthesis of sea-level data from the Netherlands by Jelgersma (1979), where the final curve is based solely on dates from the base of the lowest peat resting on Pleistocene sands (see Devoy, this vol., Ch. 10 for discussion of the regional pattern of Holocene sea-level change) (Fig. 7.4). These provide compaction-free data, but their relationships to a contemporaneous tide level are difficult to assess. Regardless of the arguments for and against the inclusion of dates from intercalated peat layers, which are susceptible to post-deposition compaction, the index points plotted by Jelgersma on a time–altitude graph do not fall on a single smooth curve. Jelgersma states (1979: p. 243), 'only the points located in the lowest places for a given age represent a ground water table that coincided with sea level. In this way a smooth curve is obtained. All observations must be considered as errors and not as fluctuations of sea-level.' This excludes the identification of falls of sea level in her methodological stance whereas the data quality really demand alternative hypotheses to be tested. It is salient to quote from the same paper, in the discussions of Eemian rather than Holocene sea-level changes (1979: p. 237):

> Finally, the maximum level of 8 m below the present one was reached during pollen zone E5 (hornbeam zone). This level is represented by a marsh clay overlying older Eemian sandy marine deposits in the Eem valley. After that time, a regression of the Eemian sea sets in, as indicated by a peat layer on top of the marsh clay in the Eem valley and in Friesland ... These findings indicate a considerable drop in sea level during the Late Eemian.

The temporal and spatial scales of study barely differ yet for one time period, i.e. the last Interglacial, a fall of sea level is interpreted from a regressive overlap, yet similar evidence for Holocene sequences is attri-

Figure 7.4: Relative sea-level curve and sea-level index points from the Dutch coastal plain. The cross-section (E–W) shows diagrammatically the main lithostratigraphic pattern of Holocene sediments above the sloping sand gravel surface in the Netherlands

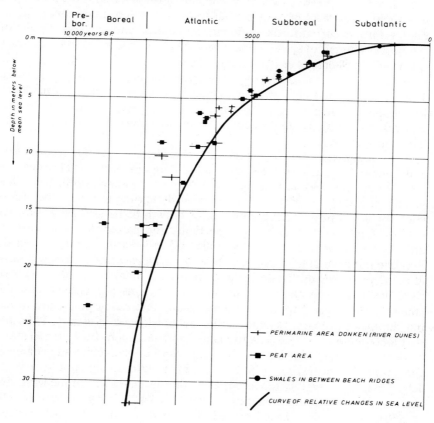

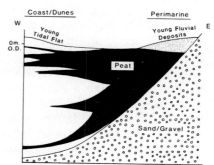

Sources: Jelgersma (1979); Jelgersma (1961).

207

buted to other factors. Of course, there are much more data available for the latter and more variables have to be considered. Until proven not to be a valid hypothesis then the possibility of fluctuations in sea level should be considered. A regressive overlap is one line of evidence showing a negative tendency of sea-level movement, but to be able to conclude that a fall in sea level has occurred further evidence is required and the exact timing of the start of the fall of sea level may be difficult to assess (see the examples discussed by Shennan *et al.*, 1983). A clear change in altitude since the time indicated by the dated regressive overlap (e.g. Sissons and Brooks, 1971; Geyh *et al.*, 1979) gives unambiguous evidence of a fall in sea level. In the case of Jelgersma (1979) the fall in sea level at the end of the Eemian is essentially confirmed on *a priori* knowledge of the glacial/interglacial cycle.

The comparison of the interpretation of regressive overlaps of Holocene and Eemian age illustrates the point about temporal and spatial scales and the different variables to be considered. Van de Plassche (1980, 1981) has greatly extended the work of Jelgersma by attempting to isolate the different variables controlling the initiation of growth of basal (basis) peat. By detailed field analyses, using open sections, grid pattern borehole surveys and the usual methods of stratigraphic and radiocarbon analyses van de Plassche has been able to propose how various local scale factors will have affected peat growth. These include changes in tidal levels, the effect of river discharge, the nature of the substrate and its topography, and the changing palaeogeographic conditions. Many interesting points are raised but the major problem remains, and is acknowledged by van de Plassche, that there are not enough data to make firm conclusions because too many of the assumptions that have to be made can be relatively easily changed. For example, he writes (1980a: p. 343), 'Accepting [the curve of Jelgersma, 1979, slightly modified] as a fair approximation of the MSL rise and assuming a constant tidal range during the last 7,200 years' and (1980: p. 350), 'her graph is based on only a limited number of widely spaced time-depth points, additional data may in due course reveal the presence of fluctuations in the sea-level rise.' Such a modification would change the nature of the gradient-effect reduction curve shown later by van de Plassche (Fig. 7.5b).

Notwithstanding the limitations posed by the current data base a number of the points made by van de Plassche deserve greater attention. In particular, a difficult problem of operational definition is highlighted by the two papers (van de Plassche 1980, 1981). In the first paper he considers that the differences between Jelgersma's curve and sea-level index points (sea-level/shoreline indicators used to define sea-level position(s)) from a series of sites in the Rhine–Meuse delta reveal a change in tidal range and river gradient effect through time (Fig. 7.5a, b). This is dependent on a number of assumptions, noted by van de Plassche, and the results are

Figure 7.5: (a) Time-depth diagram showing the slightly corrected MSL curve by Jelgersma (1979) with the curves based on data from the 'donken' of Brandwijk and Barendrecht, and the Hazendonk data (boxes) in the area of the Rhine-Meuse distributary. (b) Plot of the vertical distance above the MSL curve of the graphs and time-depth data shown in (a). Van de Plassche suggests that convergence of the Brandwijk and Barendrecht curves upon MSL and of the Hazendonk data on coastal MHW reflects the decrease in river gradient effect. It is argued that a line connecting the oldest Brandwijk datum and Hazendonk datum 4 in a slightly concave fashion produces a gradient-effect reduction curve for the Brandwijk-Hazendonk area

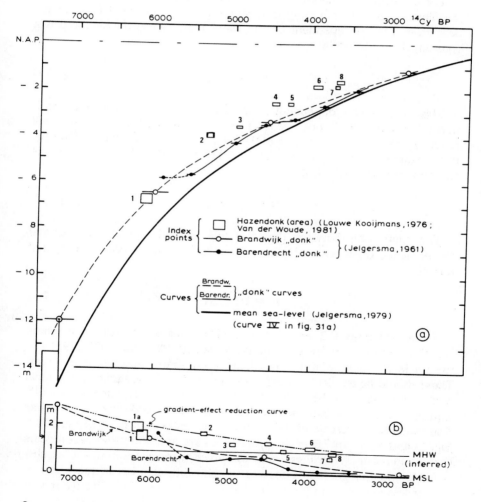

Source: van de Plassche (1982).

explicitly stated as provisional, including the assumption that all the time–depth data indicate former local MHW or local mean ground-water table. In the second paper, also making the assumption about Jelgersma's curve, deviations from the base curve are explained by peat growth commencing at different altitudes, MSL or MHW, due to local changes in palaeo-geographic conditions. This difficulty of relating the growth of basal (basis) peat to a particular tide level has also been discussed, quite independently from van de Plassche, by Godwin (1940), Behre *et al.* (1979) and Shennan (1980, 1986).

The problem of definition in this situation is as follows. Firstly, assuming the fenwood peat forms away from depressions in the basal topography, peat growth (p) is dependent on the water table (w) at the site, and is a constant (k) relationship:

$$p = kw \qquad \text{(Eq. 7.1)}$$

This water table is dependent on local sedimentary and drainage conditions; as the intertidal zone becomes closer to the site the water table will reflect a level closer to the local MHWST rather than local MTL. Thus the relationship to local tide level, MTL or MHWST (m), is dependent on time (t) and spatial (s) variables:

$$w = f(m,t,s) \qquad \text{(Eq. 7.2)}$$

The relationship between the local tidal conditions (m) and the tidal regime at the open coast (M) is also dependent on temporal and spatial variables:

$$m = f(M,t,s) \qquad \text{(Eq. 7.3)}$$

From equations 7.2 and 7.3 it is clear that the indicative meaning (relationship to a sea-level position; see van de Plassche, 1977, 1986) of the sea-level index point is dependent on the spatial and temporal scales. These should be explicitly stated since the status of one variable, the local tidal condition, can be either an independent or dependent variable. Equation 7.2 is relevant on a spatial scale in the order of 50 m linear distance and a timescale of 400 radiocarbon years (van de Plassche 1981: Fig. 6) and equation 7.3 in the order of 25 km and 400 radiocarbon years (van de Plassche 1980: Figs 1 and 2). Given these relationships it is questionable whether gradient-effect curves showing changes in the order of 1 m or less can be clearly separated from the changes in indicative meaning, which show a similar change in altitude.

Indeed, due to the various errors associated with altitude reconstruction

210

(see Kidson, 1982; Shennan, 1982a; Devoy, 1982, 1983 for example), precise sea-level curves appear misleadingly accurate and some indication of both altitudinal and age errors should be shown (see Devoy, this vol., Ch. 1 and Fig. 1.5). A further point to be considered when comparing the altitudes of different sea-level curves is that when the comparison is made, by subtracting the altitudes at fixed points on the time axis, the altitudinal difference calculated is extremely influenced by small shifts of the curve on the time axis, when the curve is steeply rising. Therefore, a minor adjustment of one point (cf. van de Plassche, 1980) may appear more significant than it should be if errors boxes of the dimensions shown by Devoy (1982), Heyworth and Kidson (1982), Preuss (1979) and Shennan (1982a) were given.

There are many factors which affect the local registration of sea-level change. Of particular importance are changes in coastal geomorphology, sediment supply and river discharge (see Orford, this vol., Ch. 13 for a discussion of some of these). Changes in any of these could be very important at the local scale but it is necessary to assess, as well as possible, whether the cause of such changes can be related to a regional phenomenon. For example, the build-up and break-down of coastal barriers is an attractive, and often used explanation for the alternation of freshwater peats and marine/brackish clastic sediments in the coastal zone without involving fluctuations in sea level (Jelgersma, 1961; Kidson and Heyworth, 1978; Heyworth and Kidson, 1982). But alternative hypotheses should be proposed and tested. What causes the alternation of barrier build-up and break-down? Why are there similar sediments found on coasts presently without barriers and with no barrier sediments recorded in the stratigraphic record, e.g. the Fenlands, England, and on coastlines presently with a dune coast, e.g. northwest Lancashire, England? The chronology of positive and negative tendencies, based in part on the analysis of transgressive and regressive overlaps, shows a high degree of synchroneity, inviting an explanation based on non-local factors (Shennan et al., 1983). Increasingly local sea-level studies should explicitly state the alternative hypotheses that require testing and what analyses are required to reject each hypothesis. There are sufficient data for well-studied areas, for example Northwest Europe, the USA and Canada, for a more rigorous hypothesis-testing methodology to be adopted (e.g. Shennan, 1986). Only in this way can cause and effect be identified.

Numerical analysis of sea-level data

The increasing amount of quantitative sea-level data that is available is slowly allowing numerical analyses to be performed. These allow existing observations to be formulated into testable hypotheses. The analyses so far

211

performed vary considerably in terms of the number of variables considered.

Perhaps the simplest numerical methods that can be used to aid comparison and correlation of sea-level data are the analysis of time series using cumulative histograms (e.g. Geyh 1969, 1971, 1980; Geyh and Streif 1970; Morrison 1976; Shennan 1980, 1982b; Shennan *et al.*, 1983). This technique is essentially exploratory and provides a good visual impression of the data base. It is possible to use quantitative measures to summarise the changes in the chronology but these depend on the data set being large enough to nullify the effect of randomness. By simulating a large data set, using random numbers, Shennan (1979) has shown that an average density of dates of at least 40 per 1,000 years (Fig. 7.6) would be required to achieve sufficient accuracy (see Sutherland, this vol., Ch. 6). This assumes a rather simple hypothesis is being tested, yet is a considerably larger data set than most currently available for local sea-level analyses. The methods assume that each date is carefully checked and classified correctly according to the hypotheses under test. Morrison (1976) has indicated how different sets of dates can be combined to test different hypotheses. The limitations to these methods being more widely used are those of requiring a large reliable data set. This problem is discussed further later in this chapter.

To date, histogram analysis has been shown to be useful with real data sets, although the theory has been advanced further than the applications. It has also been integrated with the sea-level tendency method of analysis (Fig. 7.7) (Shennan *et al.*, 1983). Shennan (1982b) has also developed the theoretical framework of how local sea-level chronologies can be compared assuming *a priori* knowledge of the relative crustal movement histories of each area. Visual and statistical correlations, using the chi-square cross-association test, can be used to identify periods of both positive and negative tendencies of sea level which appear to be of more than local significance. So far these analyses have only been used to identify similarities between the chronologies but no explanation of the cause of the similarities has been attempted.

Many explanations could be offered, but each one needs to be presented in such a way that it is a testable hypothesis. It is impossible to test a hypothesis such as one that the chronologies are simply the result of the random build-up and break-down of offshore barriers. Indeed the randomness factor may be rejected since it has already been shown that the sequence correlations are not random (Shennan, 1982b). The field evidence does not allow the detailed reconstruction of individual barrier systems. Even if this could be shown, a process operating at a regional or continental scale would still be required to explain the synchroneity (see also Devoy, 1982 for further discussion of this point). One method of hypothesis testing is as follows.

Figure 7.6: (a) Combined frequency histograms of 100 random dates within a 5000 year period. Each date is represented by a histogram approximation of a normal distribution curve of base width ± 2 standard deviations for each date. (b) Combined frequency histograms of 115 random dates within a 5000 year period with the constraint that no dates may occur in the periods shown in black on the time axis. (c) Combined frequency histograms of 154 random dates within a 5000 year period with the constraint that the chance of a date being obtained during the periods shown in black is only 20 per cent that of the rest of the time. (d) Summary of a series of combined frequency histograms using coefficient of variation as a summary statistic of the fluctuations of the data sets. The curves and the θ symbol represent a series of simulations similar to (a), the ☉ symbols refer to the assumptions made for (b), and symbols □ and ■ for (c). Such histograms may be used to explore a number of hypotheses. For example, that peat growth will be inhibited during major ingressions of the sea (e.g. b or c). If the data set is too small fluctuations of similar magnitude can be produced where peat growth is not limited (a). At least 200 dates per 5,000 years would be required to show the difference (d) between these two situations given this broad hypothesis. More specific hypotheses would require alternative tests to show departure from randomly produced distributions

(a)

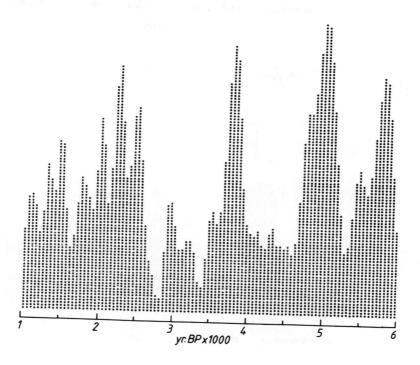

yr. BP x 1000

Figure 7.6 *continued*

(b)

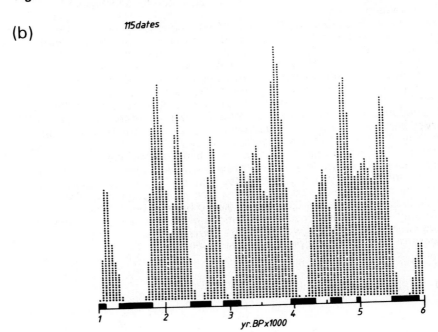

115 dates

yr. BP x 1000

(c)

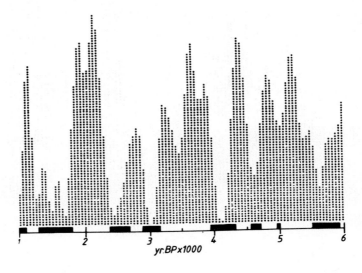

154 dates

yr. BP x 1000

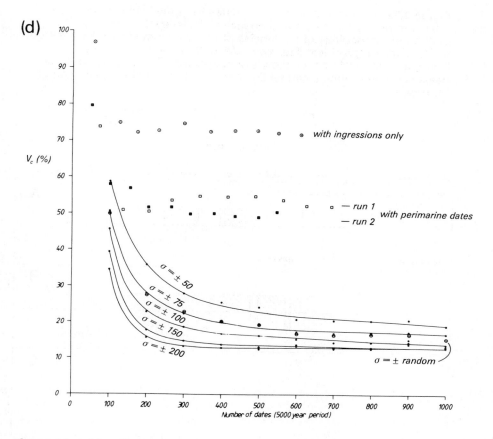

Source: Adapted from Shennan (1979).

The background

The identification of periods of positive and negative tendencies of sea level of more than local significance in the sea-level chronologies from the Fenland and northwest England (Shennan *et al.*, 1983) could be due to absolute rises and falls of the regional eustatic sea level. Fluctuations of proposed 'eustatic' curves (see Mörner, this vol., Ch. 8) in excess of 2 m have been published since 1960, the most widely quoted being those by Fairbridge (1961) and Mörner (1969). These were criticised on a number of counts: the 'smooth vs spasmodic schools' argument again, including the lack of evidence for synchroneity of fluctuations (e.g. numerical measures of goodness of fit were never made); further, the fluctuations in ice sheets that would have been required to explain such altitudinal variations were

215

Figure 7.7a: A. Distribution of 47 ^{14}C dates from the Fenland which are related to positive or negative tendencies of sea-level movement. B. Continuous chronology of tendencies of sea-level movement derived from all ^{14}C and stratigraphical data available. C. Combined frequency histograms of the 47 dates: the y-axis represents a probability scale for dates occurring within a certain 50-year interval given the distribution of ^{14}C data in A

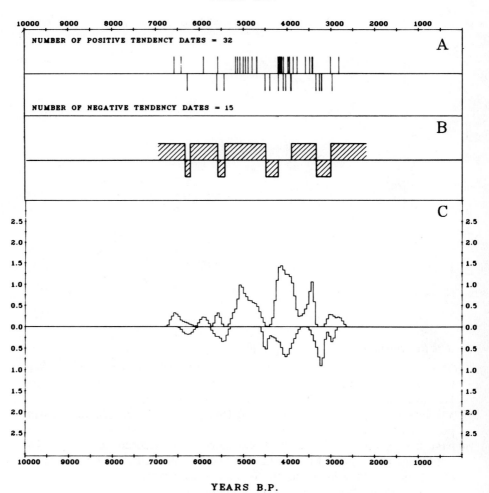

Source: Shennan *et al.* (1983).

216

Figure 7.7b: A. Distribution of 70 ^{14}C dates on transgressive or regressive overlaps from northwest England. B. Distribution of 74 ^{14}C dates related to positive or negative tendencies of sea-level movement. C. Combined frequency histograms of the 74 dates. D. Partial chronology of tendencies of sea-level movement in northwest England

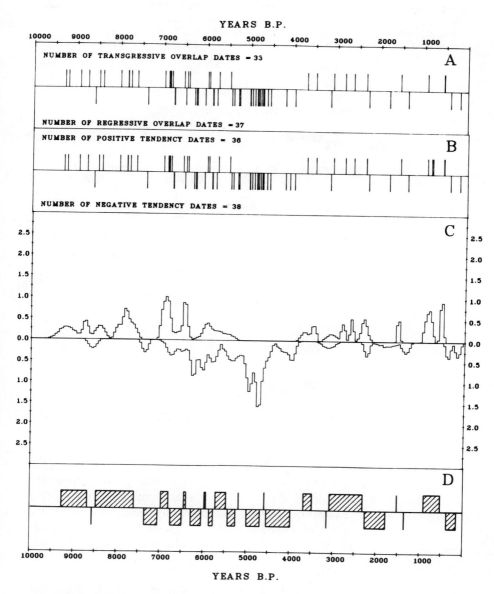

Source: Shennan *et al.* (1983).

Figure 7.7c: A. Chronologies of tendencies of sea-level movement from the Tay Estuary area (obtained via analysis similar to that shown in Figures 7.7a and b), northwest England and the Fenlands. B. Partial chronologies from northwest England and the Fenland for tendencies of more than local significance. C. Partial chronologies from the Tay Estuary area and northwest England for tendencies of more than local significance. This application of the concept of sea-level tendency to multifarious data from different areas collected and published by different authors has permitted meaningful correlations between rising and subsiding areas based on an objective method

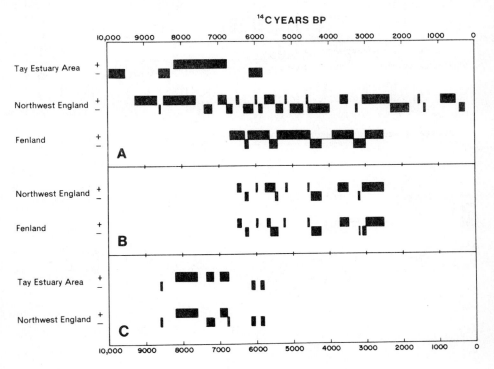

Source: Shennan *et al.* (1983).

not apparent from any analysis of known ice sheet limits or by climate and ice sheet simulation models. Some regional synchroneity of ocean-level changes is claimed by Mörner (1980) and this would be supported by the findings of Shennan *et al.* (1983). Mörner suggests that the altitudinal changes are palaeogeoid changes due to small palaeomagnetic fluctuations.

Testing

The sea-level tendency methods can be extended to give some indication of the possible fluctuations in sea level that could explain the observed sea-level chronologies. Shennan (1982b) developed a model to illustrate the interaction of regional eustasy and isostatic movement at different locations and to show how the regional similarities could be obtained from two or more local chronologies. Regional eustasy (E) was modelled as a function of time (T).

$$E = 0.5 \sin (T\pi/500) \text{ if } T < 3100 \tag{Eq. 7.4}$$

$$E = 0.5 \sin (T\pi/500) + 50 (\cos ((T-3100)\pi/2000)-1) \text{ if } T > 3100 \tag{Eq. 7.5}$$

These equations give a reasonable model of the Holocene rise of sea level. The resultant curve is a smoothly rising curve reaching present sea level at 3,100 BP, modelled by the sine function, with superimposed oscillations of 1 m ampltitude and 1000-year wavelength. The periods of regionally significant sea-level tendencies from the Fenland and northwest England data approximate to one period of positive sea-level tendency and one period of negative sea-level tendency per thousand years between 6,500 and 2,500 BP (Shennan, 1982b).

The model can, therefore, be hypothesised as a reasonably good fit to the observed data. It was further shown (Shennan, 1982b) that the effect of crustal subsidence, the I-factor, was to enhance the periods of sea-level rise. Thus falls in sea level would be observed later than any real eustatic falls and would end earlier. Neglecting standstills of sea level, rises of sea level would, obviously, start earlier and finish later. In an area affected by crustal uplift the converse holds true. The time lag observed, in the model, is a function of the rate of uplift/subsidence and the rate of sea-level rise/ fall. In order to conform with the observed frequency of 1 ka the model, Eq. 7.4 and 7.5, was rerun for varying amplitudes between 0.1 m to 2 m. The results are summarised in Figure 7.8. Where a specific solution was required the actual parameters could be entered into the programme. The boundaries used are those for the periods of negative sea-level tendencies (6,300 to 6,200, 5,600 to 5,400, 4,500 to 4,200/3,900 and 3,300 to 3,000 BP) defined by Shennan (1982a).

The graph, and specific solutions, can be used to estimate the minimum amplitude of oscillation in the regional eustatic curve to give the observed advance/delay of sea-level change. The results presented below should be taken as preliminary since both the Fenland and northwest England chronologies are lacking specific details in a number of crucial locations. Given these limitations, though, the results are summarised in Table 7.2.

The values given in Table 7.2 should be considered with the following points in mind. Firstly a number of the boundaries are defined by only one radiocarbon date and no allowance has been made for the standard errors. The 'boundary ages reversed' comment indicates that the advance or delay

219

Figure 7.8: Relationship between the amplitude of sea-level oscillation (0.1 to 2.0 m), crustal uplift or subsidence (+2.0 m ka^{-1} to −2.0 m ka^{-1}) and the time lag of initiation of a fall in sea level (years). The latter axis can be used, with the obvious adjustment of sign as required, for the end of a fall in sea level and the initiation or the end of a rise in sea level. The graph is based on the assumptions explained in the text and the relationship given in equation 7.4

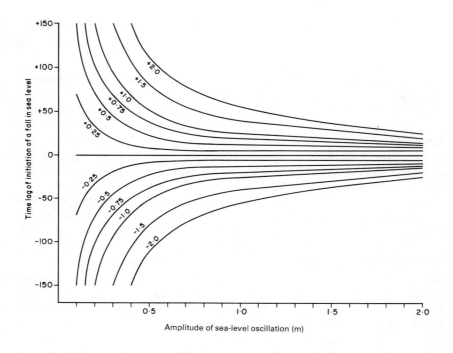

Table 7.2: Modelled minimum amplitude oscillations of regional eustatic sea-level fluctuations around the UK, 2,500-6,500 BP

Boundary: a BP (defined from Fenland chronology)	Maximum advance/delay of equivalent boundary from NW England (a)	Minimum amplitude of oscillation (m)
6,300	120	0.50
6,200	215	0.30
5,600	boundary ages reversed	—
5,400	150	0.20
4,500	45	0.55
4,200	505	0.10
3,900	205	0.17
3,300	190	0.17
3,000	boundary ages reversed	—

of the boundary is in the opposite direction to that expected, but, as noted in Shennan (1982b), within the confidence limits of the radiocarbon ages. Calculation of the amplitude of oscillation allowing for the range of values indicated by the quoted errors of the radiocarbon data would serve little purpose at present. The data base is not yet large enough to put any great confidence in quantifying rates of change. Table 7.2 is just to illustrate the order of magnitude indicated. The values are obtained using existing knowledge of the crustal histories of the two areas, approximated by comparing each area with Mörner's (1976b) regional eustatic curve. The Fenland is estimated as undergoing a net subsidence of 0.9 m ka^{-1} for the whole period (Shennan, 1986) and northwest England as being essentially stable 0-5,500 BP but with an average uplift of 1.3 m ka^{-1} around 6,200-6,300 BP. The method only gives a minimum estimate for the oscillation since the chronology for northwest England contains gaps; thus there is the possibility that each boundary, except those at 6,200, 5,400 and 4,500 BP, was essentially synchronous. If this were so then the rate of change of sea level must be so great, either due to a large amplitude oscillation or a change not analagous to a sine curve, that no time lag caused by the crustal effect is observed. Equally, those boundaries which are shown to be reversed or synchronous, when more data are available to make the chronologies more reliable, may indicate the effects of local factors on the timing of change.

The method just outlined can be used to define the limits of altitudinal change indicated by the available data, namely sea-level chronologies and estimated crustal uplift/subsidence rates. It does not prove fluctuations in the regional eustatic sea level and does not indicate the cause, for example, geoidal-eustatic or glacio-eustatic. However, quantitative measures are provided which can be used to test subsequent hypotheses. This represents an advance from just listing the possible causes of synchroneity or dia-chroneity. 'Possible Causes' represent multiple hypotheses but they are only useful when they are in a form that allows verification or rejection.

Statistical treatment of sea-level data have been carried out by a number of authors including Segota (1973), Devoy (1982), Flemming (1982), Aubrey and Emery (1983) and Newman and others (Newman et al., 1980, 1981; Marcus and Newman, 1983). The work of Devoy and of Flemming is focused at a similar scale to Shennan et al. (1983), i.e. the United Kingdom, and represents further interesting lines for future analysis, but it is at the moment acknowledged, by both authors, to be essentially pre-liminary and limited by the available data base. However, given the better data and improved theoretical bases, they should provide interesting results.

The papers by Newman and others represent one of the recent attempts to synthesise sea-level data on a global scale (see also Pirazzoli, 1977). Numerous criticisms can be levelled at the work of Newman and his

221

colleagues. There are problems of definition of terms such as palaeo-geodesy (see Mörner's comment in Newman *et al.*, 1980: p. 567) and isobase (Marcus and Newman, 1983). The data base only includes the variable latitude, longitude, laboratory number, date and standard deviation, elevation with respect to the local datum and the nature of the dated material. Research workers familiar with local sea-level studies would point out the limitations of using the entries in the journal *Radiocarbon* as the major data source and the omission, for example, of any correction for indicative meaning and local tide levels. They also use numerical methods not well suited to the nature of the data. For example, the packages SYMAP and SYMVU (Newman *et al.*, 1981) are notorious when used with unevenly distributed data (Liebenberg, 1976).

While these criticisms are quite valid, and their importance will only be quantified when the global scale analyses can be adequately tested against the local data at the local scale, the state of sea-level research and the transfer of data between research workers effectively places these constraints on those attempting global analyses. It is impossible for one research team to accumulate the detailed information about thousands of individual sea-level index points so that they can be analysed on a global scale in the detail that would satisfy each of those workers studying local scale changes. If IGCP Project 200 has inherited one major difficult aim from IGCP Project 61 it is to accumulate the detailed data bank of radiocarbon-dated sea-level index points. There are few groups of research workers attempting to work with regional or global data sets but the completion of the data bank is dependent on many other researchers spending a lot of time supplying data for which they may have no direct use and for which they may see little advantage to themselves. The papers by Newman and his colleagues are very important. They illustrate the temporal and spatial variability of Holocene sea levels and the methods used for analysis will become more important in the future to test various models of palaeo-geodesy, sea-level change and crustal deformation. By way of example, two of the maps produced by Marcus and Newman (1983) (Fig. 7.9) illustrate the spatial variability of relative sea levels at 7,000-8,000 BP on a world-wide and regional scale. A number of questions arise from such distributions but it is not possible to use the maps for separating the causative factors, ranging from those operating on a local scale, e.g. sediment compaction, local tidal range, etc., to those worldwide factors, e.g. ocean volume and ocean water distribution changes. It is clearly seen that the options taken within the computer programs used have greatly differing effects as the scale of study is changed, along with the size of the data set. Weighting of individual data points and smoothing algorithms should be followed by an analysis of the residuals from the computed surface. At present the data base and methods of analysis can be too easily criticised but the problems and the possibilities should be appreciated. In a related

Figure 7.9: (a) Contours of elevation with respect to the local datum for *c.* 380 ^{14}C dates, worldwide, 7,500 ± 500 BP. (b) Contours of elevation with respect to the local datum for *c.* 50 ^{14}C dates, for eastern North America, 7,500 ± 500 BP. Both maps are drawn using a computer based plotting and contouring package and the raw data are subjected to smoothing

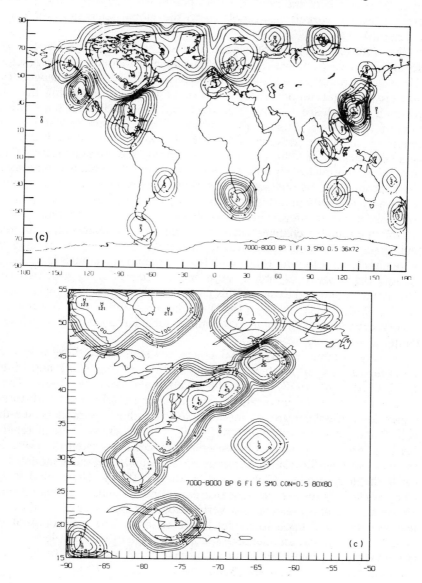

Source: Masters and Flemming (1983).

223

context the available radiocarbon sea-level index points for the UK have yet to be fully checked to the IGCP Project 61 standard (cf. van de Plassche and Preuss, 1978). This has involved considerable work by Tooley, Shennan and co-workers at Durham (Tooley, 1982a) since less than 20 per cent of the potential data points were actually submitted by other research workers. Extrapolated to the global scale the problem is enormous!

The future, therefore, of global analyses will ultimately depend on acquiring a large, reliable data base. Theory and modelling have reached the stage of requiring testing, particularly in the fields of palaeogeodesy and crustal deformation. These are two of the most appealing lines of research but both are dependent on the local sea-level data for the verification of hypotheses. Mörner (1976a and numerous subsequent papers) totally changed the paradigm of sea-level studies with his discussion of geoid changes. Prior to his paper much effort was spent on correlating fluctuations in worldwide eustatic sea level or trying to explain variations in local chronologies by untestable hypotheses of local changes. At present it is too easy to attribute differences to geoid changes. This line of argument soon becomes as infertile as the previous sea-level controversy unless reliable quantifiable changes can be identified for the temporal, altitudinal and spatial coordinates (see Devoy, this vol., Ch. 1). Patterns may be recognised (e.g. Mörner, 1980, Fig. 10) but the advance must be made to identify the process, variables, cause and effect producing such patterns. The methods outlined earlier in this paper and those discussed by Newman et al. (1980, 1981), Devoy (1982), Flemming (1982) and Aubrey and Emery (1983) can all be used to make such advances at the relevant temporal and spatial scales.

Equally as appealing as the palaeogeodesy studies are those modelling experiments by Clark, Peltier, Andrews and others (e.g. Clark and Lingle 1977; Clark et al., 1978; Peltier, 1982; Peltier and Andrews, 1983). These have been extremely important in integrating sea-level studies with three-dimensional ice-sheet growth models and earth rheology models (see this vol., Peltier, Ch. 3 and Andrews, Ch. 4). The goodness of fit of relative sea-level predictions at sites at varying distances away from the polar ice sheets have been crucial in determining acceptable ice-sheet models and earth rheology models (Peltier and Andrews, 1983). These models are being constantly revised (at least two papers per year have appeared since 1976) and accurate testing and verification will be required as they are increasingly called upon to model future changes. The robustness of the models are such that an adequate spatial coverage of local sea-level curves is perhaps more important than altitudinal detail, as would be expected from a global-based approach (Schumm and Lichty, 1965).

PREDICTION OF NEAR-FUTURE CHANGES FOR APPLICATION TO COASTAL PROBLEMS

The title of this final section is taken from the aims of IGCP Project 200 since there are strong arguments for increasing the effort in directing palaeo sea-level research towards an applied science. (Further aspects of this theme are discussed in later chapters in Parts Four and Five.) It is possible, for example, that local sea-level studies can be used to supply information about the development and stability of the coastline. Lincolnshire County Council, England (1982) have discussed the problems of continuing man-made reclamation in the estuary of the Wash, particularly with regard to the debate about the nature of sediment supply, but also the effect on sea defences. Studies of past sea levels and sedimentary history here have been able to contribute to the debate, illustrating where some of the assumptions about sediment supply are wrong and how linear trends should not be used due to the transient nature of sedimentary sequences in the past and the variations possible in local sea levels over relatively short timescales (Shennan, in press, 1986).

At the global scale studies of sea-level change also have an important role to play. The US Environmental Protection Agency (EPA) have recently initiated a project to assess the effect of higher concentrations of atmospheric carbon dioxide and other greenhouse gases, via a global warming, on sea level due to thermal expansion of the oceans and the transfer of water from the polar ice sheets to the oceans (Hoffman et al., 1983; Barth and Titus, 1984). The project has, so far, been based on a model of future energy supply and demands and a model of climate–ice sheet–sea-level response to the various atmospheric parameters. Mid-range estimates of average global sea-level rise range between 144 and 217 cm by 2100 AD, although high and low assumptions for the various atmospheric variables give estimates of 345 and 56 cm respectively (see Titus, this vol., Ch.15). An extreme case of a high level of atmospheric retention of carbon dioxide and very high levels of trace gases coinciding with an increase in solar luminosity by 0.5 per cent would produce a rise of 429 cm. The model is constructed in such a way that the final figures are calculated for the effect of the rise in global temperatures on the thermal expansion of the oceans. The figures for sea-level rise given above are obtained by multiplying the thermal expansion figure by two or three. The rationale for this is that 4 to 5 cm of the observed sea-level rise, a total of 10 to 15 cm during the last 100 years, can be explained by thermal expansion. This represents a major weakness in the model since this relationship clearly could not be applied uniformly to previous time periods, i.e. throughout the Holocene.

Another factor not considered in the model includes the uneven redistribution of water added to the ocean (Clarke and Lingle, 1977). In addition local factors, such as changes in sediment supply, coastal geo-morphology, crustal subsidence, changes in storm systems, would be required for realistic predictions of local conditions.

The EPA project is ongoing but at present clearly lacks an adequate input from palaeo sea-level studies. Such studies can contribute by indica-ting the scale of change that has occurred in the past and what responses are most likely in the future. Areas that have shown similar responses in the past can be identified, as can areas of variable or out of phase sea-level or crustal movement histories. The problem of geoidal eustasy and tidal change may be important. Indeed, if such changes in sea level do occur then satellite monitoring of geoid changes will provide very valuable evidence which could be used to calibrate models of past situations. In the first instance, however, sea-level data are the only way to test the climate–ice sheet–ocean volume–ocean-level models which are at the centre of the EPA progamme. The models should be able to give acceptable results in modelling the deglaciation of the present Interglacial on a timescale of 100 and 1000 years and to model the variability of sea-level response, the observed fluctuations in sea-level tendencies. In particular the relationship between the two factors, thermal expansion and the addition of water mass to the oceans, should be studied and tested more carefully. The existence of thresholds in the system should be considered and it is suggested that thresholds in the glacial response system, which have not been recognised, are indicated by points of greatest divergence between the existing models and the known sea-level history. For example, 7,500 years ago sea level rose relative to the coast of northwest England by some 7 m in only about 200 years and this is attributed to the disintegration of the Laurentide ice sheet (Tooley, 1978). A similar effect of ice surge on sea-level rise is suggested by Hollin (1977) for earlier interglacial episodes. Mercer (1978) has suggested that the present West Antarctic ice sheet may be in a similar threshold situation, which may not be indicated by a climate model, but the effects of such an event have been clearly demonstrated by sea-level research.

CONCLUSIONS

Sea-level studies remain dependent on high-quality fieldwork and labora-tory analyses. There are many new and attractive methods of data analysis, including modelling, which can be used to correlate the various data and provide stimulating results. The attractiveness of these new lines of enquiry must not, however, be at the cost of basic data collection and related research.

At all scales of inquiry in sea-level research more effort should be concentrated on adhering to accepted philosophical principles. This includes the explicit statement of multiple working hypotheses and the methods and data required to verify or reject each hypothesis. Clearly formulated testable hypotheses have been noticeable by their absence for too long. Implicit in such an approach are the use of unambiguous operation definitions, specific reference to the relevant temporal and spatial scales, a clear understanding of the effects of data smoothing, the necessity to explain the distribution of residuals, and the limitations of correlation at a scale which removes the research worker from the detailed knowledge of the individual site factors.

IGCP Project 200 aims to initiate a move towards global analyses and applied science, both based in the first instance on high-quality local sea-level data. Whether or not the 'C' stands for correlation or co-operation, a number of aims and advances will be thwarted without the unselfish contribution of every sea-level research worker contributing to the exchange of data, especially the compilation of the data banks.

REFERENCES

Aubrey, D.G. and Emery, K.O. (1983) 'Eigenanalysis of recent United States sea levels', *Continental Shelf Res.* 2, 21-33.

Barth, M.C. and Titus, J.G. (1984) *Greenhouse Effect and Sea-Level Rise*, Van Nostrand Reinhold, New York.

Behre, K.-E., Menke, B. and Streif, H. (1979) 'The Quaternary geological development of the German part of the North Sea', in E. Oele, R.T.E. Schüttenhelm and A.J. Wiggers, (eds), *The Quaternary History of the North Sea*, Acta Univ. Ups. Symp. Univ. Ups. Annum Quingentesimum Celebrantis: 2, Uppsala, pp. 85-113.

Birks, H.J.B. and Birks, H.H. (1980) *Quaternary Palaeoecology*, Edward Arnold, London.

Bloom, A.L. (1977), *Atlas of Sea-Level Curves*, IGCP Project 61, Cornell University, New York.

—— (1982), *Atlas of Sea-Level Curves (Supplement)*, Draft version, Cornell University, New York.

Clark, J.A., Farrell, W.E. and Peltier, W.R. (1978), 'Global Changes in Postglacial Sea Level: A Numerical Calculation', *Quat. Res.* 9, 265-87.

—— and Lingle, C.S. (1977) 'Future sea-level changes due to West Antarctic ice sheet fluctuations', *Nature*, 269, 206-9.

Devoy, R.J.N. (1982) 'Analysis of the geological evidence for Holocene sea-level movements in south-east England', *Proc. Geol. Ass.*, 93, 65-90.

—— (1983) 'Late Quaternary shorelines in Ireland: an assessment of their implications for isostatic land movement and relative sea-level changes', in D.E. Smith and A. Dawson (eds), *Shorelines and Isostasy*, Academic Press, London and New York, pp. 227-54.

Fairbridge, R.W. (1961) 'Eustatic changes in sea-level', in L.H. Ahrens *et al.* (eds), *Physics and Chemistry of the Earth*, Pergamon Press, vol. 4, pp. 91-187.

Flemming, N.C. (1982) 'Multiple regression analysis of earth movements and eustatic sea-level changes in the United Kingdom in the past 9000 years', *Proc. Geol. Ass.*, 93, 113-25.

Geyh, M.A. (1969) 'Versuch einer chronologischen Gliederung des marinen

Holozans an der Nordseekuste mit Hilfe der statistischen Auswertung von 14C-Daten', *Z. dt. geol. Ges.*, *118*, 356-60.
—— (1971) 'Middle and young Holocene sea-level changes as global contemporary events', *Geol. För. Stockh. Förh.*, *93*, 679-92.
—— (1980) 'Holocene sea level history: case study of the statistical evaluation of 14C dates', *Radiocarbon*, *22*, 695-704.
—— and Streif, H. (1970) 'Studies on coastal movements and sea-level changes by means of the statistical evaluation of 14C-data', in *Proceedings of the Symposium on Goedesy (Munich)*, pp. 599-611.
Geyh, M.A., Kurdrass, H.-R., and Streif, H. (1979) 'Sea-level changes during the Late Pleistocene and Holocene in the Strait of Malacca', *Nature*, *278*, 441-3.
Godwin, H. (1940) 'Studies of the Post Glacial History of British Vegetation. III. Fenland Pollen Diagrams. IV. Post-Glacial Changes of Relative Land and Sea-Level in the English Fenland', *Phil. Trans. Roy. Soc., Lond., B*, *230*, 239-303.
Harvey, D. (1969) *Explanation in Geography*, Edward Arnold, London.
Heyworth, A and Kidson, C. (1982) 'Sea-level changes in south-west England and Wales', *Proc. Geol. Ass.*, *93*, 91-111.
Hoffman, J.S., Keyes, D. and Titus, J.G. (1983) *Projecting Future Sea Level Rise. Methodology, Estimates to the Year 2100, and Research Needs.* US Environment Protection Agency, Washington, D.C.
Hollin, J.T. (1977) 'Thames interglacial sites, Ipswichian sea levels and Antarctic ice surges', *Boreas*, 6, 33-52.
Jelgersma, S. (1961) 'Holocene sea-level changes in the Netherlands', *Meded. Geol. Sticht. (Ser. C6)*, 7, 1-100.
—— (1966) 'Sea-level changes during the last 10,000 years', in J.S. Sayer (ed.), *World Climate 8000 to 0 BC*, Roy. Met. Soc., London, pp. 54-69.
—— (1979) 'Sea-level changes in the North Sea basin', in E. Oele, R.T.E. Schüttenhelm and A.J. Wiggers (eds), *The Quaternary History of the North Sea*, Acta Univ. Ups. Symp. Univ. Ups. Annum Quingentesimum Celebrantis: 2, Uppsala, pp. 233-48.
Kidson, C. (1982) 'Sea level changes in the Holocene', *Quat. Sci. Rev.*, *1*, 121-51.
—— and Heyworth, A. (1978) 'Holocene eustatic sea-level change', *Nature*, *273*, 748-50.
Liebenberg, E. (1976) 'SYMAP: it uses and abuses', *Cartogr. J.*, *13*, 26-36.
Lincolnshire County Council (1982) *Development on the Lincolnshire Coast: Subject Plan*, Lincoln.
Marcus, L.F. and Newman, W.S. (1983) 'Hominid migrations and the eustatic sea level paradigm: a critique', in P.M. Masters and N.C. Flemming (eds) *Quaternary Coastlines and Marine Archaeology*, Academic Press, London and New York, pp. 63-85.
Mercer, J.H. (1978) 'West Antarctic ice sheet and CO_2 greenhouse effect: a threat of disaster', *Nature*, *271*, 321-5.
Mörner, N.-A. (1969) 'The Late Quaternary History of the Kattegatt Sea and the Swedish West Coast', *Sveriges. Geol. Unders., Series C.*, *640*, 1-487.
—— (1976a), 'Eustasy and geoid changes', *J. Geol.*, *84*, 123-51.
—— (1976b), 'Eustasy changes during the last 8000 years in view of radiocarbon calibration and new information from the Kattegatt region and other North-Western European coastal areas', *Palaeogeography, Palaeoclimatol., Palaeoecol.*, *19*, 63-85.
—— (1980) 'Eustasy and geoid changes as a function of core/mantle change', in N.-A. Mörner (ed.), *Earth Rheology, Isostasy and Eustasy*, Wiley, Chichester and New York, pp. 535-53.
Morrison, I.A. (1976) 'Comparative stratigraphy and radiocarbon chronology of

Holocene marine changes on the western seaboard of Europe', in D.A. Davidson and M.L. Shackley (eds), *Geoarchaeology*, Duckworth, London, pp. 159-75.

Newman, W.S., Marcus, L.F., Pardi, R.R., Paccione, J.A. and Tomecek, S.M. (1980) 'Eustasy and deformation of the geoid: 1000-6000 radiocarbon years BP', in N.-A. Mörner (ed.), *Earth Rheology, Isostasy and Eustasy*, Wiley, Chichester and New York, pp. 555-67.

—— Marcus, L.F. and Pardi, R.R. (1981) 'Palaeogeodesy, Late Quaternary geoidal configurations as determined by ancient sea levels', in I. Allison (ed.), *Sea Level, Ice and Climatic Change*, IAHS Publ. No. 131, IAHS, Washington, pp 263-75.

Peltier, W.R., (1982) 'Dynamics of the Ice Age Earth', *Adv. Geophys.*, 24, 1-146.

—— and Andrews, J.T. (1983) 'Glacial geology and glacial isostasy of the Hudson Bay region', in D.E. Smith and A.G. Dawson (eds), *Shorelines and Isostasy*, Academic Press, London and New York, pp. 285-319.

Pethick, J.S. (1984) *An Introduction to Coastal Geomorphology*, Edward Arnold, London.

Pirazzoli, P.A. (1977) 'Sea level relative variations in the world during the last 2000 years', *Zeit. Geomorph, N.F.*, 21, 284-96.

Plassche, O. van de (1977) *Sea-level Changes During the Last Deglacial Hemicycle (ca 15,000 y): A Manual for Sample Collection and Evaluation of Sea-Level Data*, Free University, Amsterdam.

—— (1980) 'Holocene water-level changes in the Rhine–Meuse delta as a function of changes in relative sea level, local tidal range, and river gradient', *Geol. Mijnb.*, 59, 343-51.

—— (1981) 'Sea level, groundwater and basal peat growth — a reassessment from The Netherlands', *Geol. Mijnb.*, 60, 401-8.

—— (1982) 'Sea-level changes and water-level movements in the Netherlands during the Holocene', *Meded. Rijks. Geol. Dienst.*, 36 (1), 1-93.

—— (1986), *Sea-level Research: A Manual for the Collection and Evaluation of Data*, Geo-Abstracts, Norwich.

—— and Preuss, H. (1978) 'IGCP Project 61 sea-level movements during the last deglacial hemicycle (ca. 15000 y). Explanatory guidelines for completion of the computerform for sample documentation', unpublished typescript, Free University, Amsterdam.

Preuss, H (1979) 'Progress in computer evaluation of sea-level data within the IGCP Project No. 61', in *Proceedings of the 1978 International Symposium on Coastal Evolution in the Quaternary. Sao Paulo, Brazil (1979)*, pp. 104-34.

Schumm, S.A. and Lichty, R.W. (1965) 'Time, space and causality in geomorphology', *Am. J. Sci.*, 263, 110-19.

Segota, T. (1973) 'Radiocarbon measurements and the Holocene and Late Würm sea level rise', *Eiszeit. Gegenw.*, 23/24, 107-15.

Shennan, I. (1979) 'Statistical evaluation of sea-level data', *Sea Level. Information Bull. of IGCP Project No. 61.*, 1, 6-11.

—— (1980) 'Flandrian sea-level changes in the Fenland', unpublished PhD dissertation, University of Durham.

—— (1982a) 'Interpretation of Flandrian sea-level data from the Fenland, England', *Proc. Geol. Ass.*, 93, 53-63.

—— (1982b) 'Problems of correlating Flandrian sea-level changes and climate', in A.F. Harding (ed.), *Climatic Change in Later Prehistory*, Edinburgh University Press, Edinburgh, pp. 52-67.

—— (1983) 'Flandrian and Late Devensian sea-level changes and crustal movements in England and Wales', in D.E. Smith and A.G. Dawson (eds) *Shorelines and Isostasy*, Academic Press, London and New York, pp. 255-83.

—— (in press) 'Impact of man on the coast of eastern England', in H.J. Walker

229

(ed.), *Artificial Structures and the Shoreline.*
—— (1986) 'Flandrian sea-level changes in the Fenland', *J. Quat. Sci., 1,* 119-79.
—— Tooley, M.J., Davis, M.J. and Haggart, B.A. (1983) 'Analysis and interpretation of Holocene sea-level data', *Nature, 302,* 404-6.
Sissons, J.B. and Brooks, C.L. (1971) 'Dating of early post-glacial land and sea-level changes in the western Forth Valley', *Nature, 234,* 124-7.
Streif, H. (1979a) 'Cyclic formation of coastal deposits and their indications of vertical sea-level changes', *Oceanis, 5,* 303-6.
—— (1979b) 'Holocene sea level changes in the Strait of Malacca', in *Proceedings of the 1978 International Symposium on Coastal Evolution in the Quaternary. Sao Paulo, Brazil (1979),* pp. 552-72.
Sugden, D.E. and John, B.S. (1976) *Glaciers and Landscape,* Edward Arnold, London.
Tooley, M.J. (1978) *Sea-Level Changes in North-West England during the Flandrian Stage,* Clarendon Press, Oxford.
—— (1982a) 'Introduction: IGCP project No. 61 in the U.K', *Proc. Geol. Ass., 93,* 3-6.
—— (1982b), 'Sea-level changes in northern England', *Proc. Geol. Ass., 93,* 43-51.

Part Three

The State of Current Knowledge: Regional and Global Views

8a

Pre-Quaternary Long-Term Changes in Sea Level

Nils-Axel Mörner

INTRODUCTION

Long-term changes in sea level are dominated by tectono-eustasy due to alternations in the volume of the ocean basins (i.e. the hypsographic land/ sea distribution), by geoidal eustasy due to redistribution of mass within the Earth and rotational changes, and by local or regional crustal movements. Glacio-eustasy had, of course, only effect during periods of glaciations; namely the Upper Cenozoic, the Permian, the Ordovician, the Upper Precambrian and at around 2.3 Ga. Mörner (1983a, 1985a) estimated the rates and amplitudes possible for the different eustatic factors. Tectono-eustasy is a slow process (operating at maximum rates of 0.5-0.7 mm a^{-1}) whilst geoidal eustasy (as well as glacial eustasy) also includes very rapid processes (some 10-30 mm a^{-1}).

It is important to realise that earth movements giving rise to hypso-graphic deformations of the ocean basin volume leading to tectono-eustatic sea-level changes involve redistributions of mass that simultaneously must lead to deformations of the geoid configuration as illustrated in Fig. 8a.1.

Much has been written about pre-Quaternary sea-level changes (e.g. Bond, 1978; Donovan and Jones, 1979, Hallam, 1977; Hancock and Kauffman, 1979; Hardenbol et al., 1981; Hays and Pitman, 1973; Loutit and Kennett, 1981; Mörner, 1976, 1980a, 1983a, 1985a; Pitman, 1978; Pitman and Golovchenko, 1983; Sleep, 1976; Steckler, 1984; Thorne and Watts, 1984; Vail et al., 1977, 1980; Vail and Hardenbol, 1979; Vail and Mitchum, 1980; Watts, 1982; Watts and Steckler, 1979; Wise, 1974; Worsley et al., 1984; Yanshin, 1973), and we can here only discuss some of the main results and sea-level trends.

One striking observation concerning the analysis of pre-Quaternary sea-level changes is that most of the scientists dealing with these problems are not familiar with the Late Pleistocene, Holocene and instrumental records of short-term changes in sea-level and often make statements about the rates and amplitudes of sea-level changes that do not concur with these

Figure 8a.1: Causal connection between tectono-eustatic changes in sea level and simultaneous geoidal changes in sea level (cf. Mörner, 1985a)

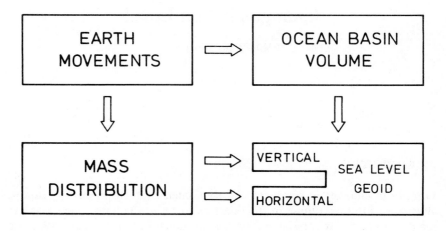

basic observational data. It is therefore quite clear that the pre-Quaternary geologists have very much to learn from the studies of the Quaternary sea-level changes. At the same time, however, there has previously been a general oversimplification amongst Quaternary geoscientists in interpreting all the changes in terms of glacio-eustasy. In order to understand the long-term changes due to tectono-eustasy and the effects and interpretation of geoidal deformations, it is important for Quaternary geoscientists to be familiar also with the longer-term pre-Quaternary trend.

BASE DATA AND SOURCES OF INFORMATION

The pre-Quaternary sea-level data are partly similar and partly different to those generally used in Quaternary sea-level analyses. The data are rather qualitative than quantitative (i.e., for example, changes between marine and terrestrial conditions without information about the total amplitude involved, bathymetric changes with wide margins of error, stratigraphic information with gaps of uncertain length). All data represent relative sea-level changes; often with a dominance from vertical tectonic movements. Because of the long periods of time involved, long-term tectonic processes and deformations of the ocean basin volume (i.e. tectono-eustasy) play a significant role.

The analyses of pre-Quaternary sea-level changes may therefore be grouped as follows:

(1) Establishment of local sea-level records where the eustatic and tectonic components cannot be separated and where large portions of the curves are usually based on qualitative estimates.

(2) Comparisons of different relative sea-level records (i.e. those considered to be most reliable) and establishment of a combined curve claimed to represent approximate eustatic changes (e.g. Hancock and Kauffmann, 1979).

(3) Comparisons of numerous sea-level records illustrating the complexity of sea-level changes (e.g. Matsumoto, 1977) and, maybe even, the absence of recognisable eustatic similarities (e.g. Jeletsky, 1978).

(4) Analysis of all available sea-level records for a longer period of time (such as the Cretaceous) with respect to their geographical location in order to elucidate possible latitudinal and/or longitudinal waves (e.g. Mörner, 1980a).

(5) Estimations of the percentage of larger continents that, with changing time, was covered by the sea (e.g. Hallam, 1977).

(6) Analysis of marine seismic stratigraphy in terms of 'coastal onlap' records and transformation of this into corresponding sea-level changes (e.g. Vail et al., 1977). Even deep-sea hiatus became integrated in this analysis (e.g. Moore et al., 1978).

The pre-Quaternary sea-level changes are, therefore, usually quite different to those of the Quaternary, with respect both to type and quality of base data and to their dominant driving mechanisms.

DISCUSSION AND INTERPRETATION

The available sea-level records will be discussed in terms of their interpretation and the possible causative mechanisms. The main emphasis here will be put on the records of the last 200 Ma, for which there exists a wealth of data and fundamentally different interpretations.

1. The Upper Cenozoic

Continental Ice Ages began ~2.5 Ma ago and had a glacio-eustatic power of lowering the sea level by about 130 ± 40 m (B in Fig. 8a.2). The Antarctic ice cap, which stores water equivalent to a glacio-eustatic rise of ~60 m had grown to a full ice cap by at least the end of Miocene (around 5.5 Ma) and probably even as early as by ~ 13 Ma (A in Fig. 8a.2). Prior to 13 Ma, the Antarctic glaciers were of small to insignificant glacio-eustatic capacity (probably less than 10 m glacio-eustatic power).

2. The last 200 Ma

On the basis of marine seismic data ('coastal onlap' records), Vail *et al.* (1977) constructed a sea-level graph for the last 200 Ma (Fig. 8a.2); the well-known and famous so-called 'Exxon Eustatic Curve' (also see: Vail and Hardenbol, 1979; Vail and Mitchum, 1980; Vail *et al.*, 1980; Kerr, 1980; Hardenbol *et al.*, 1981, Loutit and Kennett, 1981).

This curve (Fig. 8a.2) is characterised by numerous, very rapid, not to say 'instantaneous', regressions of high amplitude, and by a long-term trend with sea level 300 m above the present in late Cretaceous time.

The curve has unfortunately gained a wide acceptance as a eustatic standard (e.g. Kerr, 1980), although several authors have drawn attention to serious problems (e.g. Mörner, 1983a; Pitman and Golovchenko, 1983; Thorne and Watts, 1984).

I have discussed the Exxon 'Eustatic' Curve in detail before (Mörner, 1985a, 1985b, 1983a, 1982a, 1980a) and shown that it is impossible to combine the rates and amplitudes with a global ('eustatic' in the old sense) validity. Something must be changed.

One can question the rates, i.e. the time control, and suggest less steep sea-level regressions. This is not enough, however. One can question the amplitudes, i.e. the transformation of the seismic structures into sea-level changes, and suggest smaller regressional amplitudes. This is not enough either. One can question the global validity and explain the recorded changes in terms of tectonics (e.g. Thorne and Watts, 1984), local interplay between shelf subsidence and sea-level changes (e.g. Pitman and Golovchenko, 1983), or palaeogeoidal changes (Mörner, e.g. 1983a). One can

Figure 8a.2: The so-called 'Exxon Eustatic Curve' of Vail *et al.* (1977) with vertical scale in metres. Line A gives time and amplitude of the Antarctic ice cap and line B the same for the Ice Ages of the last 2.5 ma. The curve includes 33 'sudden' falls in relative sea level of high amplitude

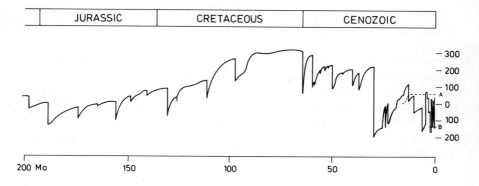

also question the origin, and claim that the recorded structures do not represent sea-level changes. This is especially attractive when it concerns the explanation of the deep-sea hiatus (e.g. Mörner, 1984a).

With respect to the old concept of eustasy, Pitman and Golovchenko (1983) quite correctly concluded 'that either there is a sufficient, but as yet unknown, mechanism or that the interpretation of the sedimentary patterns in terms of sea-level changes is incorrect'. The 'unknown' mechanism is proposed to be palaeogeoid changes (Mörner, 1982a, 1983a, 1985a, 1985b).

The only process that can create such rapid changes is palaeogeoidal change. But this implies that the changes are irregular and even of different sign (compensational) over the globe; i.e. that they are not globally valid. Tectono-eustasy is too slow a process to be able to be used to explain the rapid regressions; only some of the minor falls in sea-level can be achieved by this process within the time-scale of 1 Ma (Mörner 1985a, Fig. 4).

There are even larger problems with the original data upon which the Exxon 'Eustatic' Curve was first constructed (see Vail et al., 1977, Fig. 4.5; Mörner, 1985a, Fig. 2). The tremendous 365 m regression in the Mid Oligocene was based upon a 365 m regression in the North Sea, a stratigraphic gap (no information) off northwest Africa, a 20 m regression in Gippsland (Australia), and a 120 m regression in California, to which we may now add a 200 m transgression along the North American east coast (Olsson et al., 1980). Together these data seem to give evidence rather of high amplitude, irregular sea-level changes of compensational nature over the globe; namely geoidal eustasy or palaeogeoid deformations (Mörner, 1982a, 1983a, 1985a, 1985b).

There are even larger problems connected with the Late Miocene regression; for example an inverted lead-lag relationship (Mörner, 1985a, Fig. 3).

The sudden regressions in the Exxon 'Eustatic' Curve have been claimed often to correlate with deep-sea hiatus, in some obscure causal relationship where the sea-level fall should have created the deep-sea hiatus (e.g. Moore et al., 1978; Thiede, 1981; Keller and Barron, 1983). This cannot be the case, however. (What does a surface some 4000-6000 m below sea level care about a 100-200 m change in level at the surface?) Instead, Mörner (1985a) suggested that the primary factor for the formation of deep-sea hiatus was changes in the ocean bottom currents and bottom circulation, due to rotational changes and transfer of angular momentum between the 'solid' Earth and the hydrosphere (i.e. a similar mechanism to that which seems to have caused major short-term climatic changes; Mörner 1984a, 1985c).

A detailed analysis of the available information about the Cretaceous sea-level changes over the globe (Mörner, 1980a, 1981a) gave a quite different picture as to that of Vail *et al.* (1977). Besides tectono-eustatic effects (dominant during the Albian and Cenomanian periods of very rapid plate tectonics), local tectonic effects and possible irregular palaeogeoid changes, a novel factor was established; namely wave-like deformations of the geoid that travelled up and down the globe; a kind of 'gravitational drop motions' (Fig. 8a.3). Similar changes have also been found in the Miocene, Pliocene and probably also in the Holocene (Mörner, 1981a).

The long-term trend with a very high level (+300 m) in the Late Cretaceous (Vail *et al.*, 1977; Pitman, 1978; and the lower high level of Watts and Steckler, 1979) cannot be of global validity. It has to do with the fact that the ocean water masses must have been drastically and differently distributed over the globe in response to the quite different distributions of lithospheric and asthenospheric mass (and, maybe, even the core centration) before, during and after the main destruction of the Gondwana continent, the intense sea-floor spreading and opening of the Atlantic during the last 200 Ma (Mörner, 1980a, 1983b, 1983c, 1983d).

3. The Phanerozoic

Vail *et al.* (1977) extended their curve back into the top of the Precambrian and established a main, '1st-order cycle', sea-level curve for the Phanerozoic. Hallam (1977) presented curves of the percentage inundation through time of the USA and USSR, respectively.

In Fig. 8a.4, these curves are compared. They all show high values in the Lower Palaeozoic, low values in the Upper Palaeozoic (and early Mesozoic, too, according to Vail *et al.*), high values in the Late Mesozoic, and falling values from the end of the Cretaceous. A major cyclic pattern with some 300-400 Ma per 'cycle' is suggested (cf. Fischer, 1984; Worsley *et al.*, 1984). These changes must be seen and analysed in the light of other geodynamic variations like the Earth's spin rate, preferential polarity, plate tectonics, and asthenospheric changes, which seem to exhibit similar 'cyclic' changes (Mörner, 1983b, 1983c, 1983d).

On the long-term basis, sea-level changes are controlled mainly by the deformations of the hypsographic land/sea distribution (i.e. tectono-eustasy), changes in mass distribution and shape of the Earth, variations in the rate of rotation or the axis of tilting, and finally, of course, local and regional vertical displacements of former sea-level marks.

During the Permian and Ordovician glaciations, glacio-eustasy was, of course, in operation.

There are many records of cyclic changes that must be caused by sea-level changes. These cycles have the normal orbital geometry frequencies of 23,000, 41,000 around 100,000 a and 2.0-2.5 Ma. Because of the absence of glaciations (i.e. glacio-eustasy), these sea-level changes must be

238

Figure 8a.3: 'Gravitational drop motions' or waves of geoid deformations travelling up and down the globe as identified by Mörner (1980a, 1981a) for the Cretaceous (also found in the Miocene, Pliocene and probably Holocene). The rate for the waves change from ~ 15 Ma to 2–4 Ma in connection with the very rapid sea-floor spreading and opening of the Atlantic at around 100 Ma

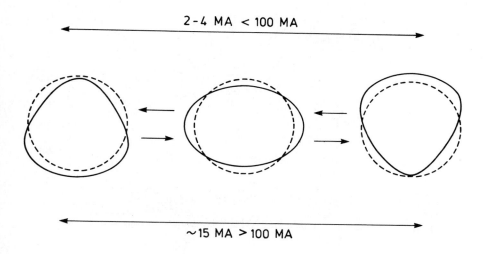

caused by geoidal eustasy (Mörner, 1980a). This is taken to indicate that the so-called 'Milankovitch Variables' are effective in deforming the geoid configuration via endogene processes as suggested by Mörner (1978, 1984a). This has a bearing on the interpretation of cyclic Pleistocene sea-level records; not necessarily always representing glacial volume changes.

4. The Precambrian

Sea-level changes, must, of course, have occurred also during the Precambrian. Usually, we do not know so much about them. There are several exceptions, however, where we have quite detailed records; usually from a limited region and for a limited time unit. Grotzinger (1984) was, for example, able to trace about 150 sea-level cycles (a rapid transgression followed by a slow regression) with an amplitude in the order of 10 m and a frequency in the order of 43,000 a in ~ 1.1 Ga old rocks in northeastern Canada.

Glacio-eustasy was in operation in association with the ice ages in the Upper Precambrian and at around 2.3 Ga.

239

Figure 8a.4: Main sea-level changes during the last 600 Ma (A) according to Vail *et al.* (1977) in metres (vertical scale to the right), (B) per cent inundation of USSR and (C) North America according to Hallam (1977)

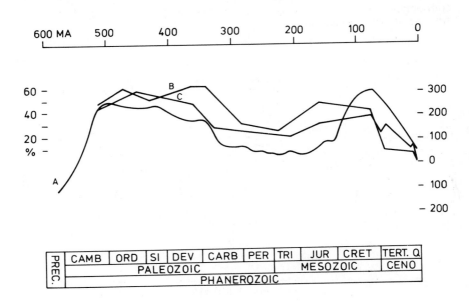

CONCLUSIONS

(1) The long-term changes in sea level are dominated by tectono-eustasy, palaeogeoid deformations due to mass redistribution, density changes and rotational changes, and by vertical Earth movements. During the periods of glaciation (uppermost Cenozoic, Permian, Ordovician, uppermost Precambrian and at around 2.3 Ga) glacio-eustasy was effective.

(2) The so-called 'Exxon Eustatic Curve' of Vail *et al.* (1977) is in need of an extensive revision (Mörner, 1985a) because the proposed origin, the rates and amplitudes given and the global validity claimed are, in fact, not compatible.

(3) Cyclic changes of frequencies similar to the so-called 'Milankovitch Variables' seem to be identified during most of the geologic records. This can only be explained in terms of cyclic deformations of the geoid surface due to endogene responses to the planetary beat (Mörner, 1984a). It indicates that our planetary system has been very conservative and does not change significantly.

240

(4) Sea-level changes are a sensitive tool for the recording and understanding of several fundamental geodynamic and geophysical processes on the Earth. It also sheds light on the uniformity of our planetary system.

8b

Quaternary Sea-level Changes: Northern Hemisphere Data

Nils-Axel Mörner

INTRODUCTION

Analyses of Holocene and Late Pleistocene sea-level data show (Mörner, 1976, 1981b, 1983a, 1983e) the combined effects of (1) glacio-eustasy, (2) geoidal eustasy, (3) local tectonism and isostasy, and (4) perhaps some effects of tectono-eustasy. We will here analyse the long-term Quaternary sea-level records. There is no scientific reason, however, to separate the records from the Northern and Southern Hemispheres (this was simply a practical matter); glacio-eustasy should (in the old sense) give rise to 'worldwide simultaneous changes in sea level', and geoidal eustasy should give rise to irregular compensational sea-level changes over the globe (Mörner, e.g. 1976, 1987).

The characteristic alternations of the Quaternary climate between periods of continental glaciations and periods of interglacial climatic conditions began about 2.2-2.5 Ma ago (e.g. Mörner, 1978, 1981b). The exact position of the Pliocene/Pleistocene boundary has been much debated, however, and a majority seem now to place it about 1.4-1.6 Ma.

During the 1.5 or 2.5 Ma of the Quaternary period, sea level has changed drastically. Whilst elevated marine shorelines or marine deposits are usually easily identified in the field, submarine shorelines and continental deposits off the coast must be recorded by special oceanographic instruments generally not in operation until well after the War.

Already Ovidius (45 BC to 17 AD) mentioned evidence of former higher sea levels (Mörner, 1979a). This is not surprising as Italy is a classical area of elevated Quaternary shorelines; Calabrian, Emilian, Sicilian, Milazzian, Tyrrhenian, (Monastirian), Versilian. In other parts of the world the records are quite different, however.

The Quaternary sea-level changes have always been considered to have been dominated by glacio-eustatic changes in sea level (e.g. Maclaren, 1842; Penck, 1882; Daly, 1910; Fairbridge, 1961; Bloom et al., 1974). Two major problems directly emerge, however (Fig. 8b.1). One concerns

242

Figure 8b.1: The problem of the number (A) of glacial/interglacial cycles and the shape (B) of the major sea-level changes during the Quaternary

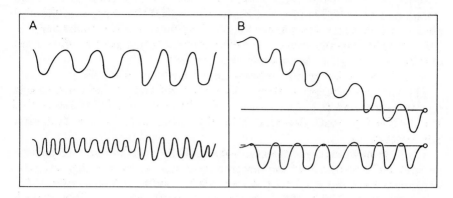

the actual number (Fig. 8b.1A) of glacial/interglacial cycles (some 6-8 major glaciations or numerous shorter glacials and interglacials). The other concerns the shape (Fig. 8b.1B) of the sea-level changes (a stepwise lowering of the interglacial high stands or oscillations with all the interglacial levels at around or slightly above the present sea level).

Another complication is the recently realised fact that two ocean levels are never parallel to each other due to deformations of the geoid configuration and that such 'geoidal eustatic' changes may provide significant and sometimes even dominant eustatic changes in sea level on a regional scale (Mörner, 1976, 1980b, 1981b, 1983e, 1985a, 1987).

THE NATURE OF SEA-LEVEL CHANGES

The relative sea-level changes (i.e. the evidence we find in the field) are the combined effects of all factors that may have affected the level of the oceans (i.e. eustasy) and the crustal (land) level (Mörner, 1987, 1983a, 1980b, 1976).

The absolute sea-level changes (i.e. the changes of the ocean level itself) are the same as eustatic sea-level changes.

Eustatic sea-level changes may be brought about by (1) glacio-eustasy, (2) tectono-eustasy, (3) geoidal eustasy, and (4) dynamic sea-surface changes (Mörner, 1987, 1985a, 1983a, 1980b, 1976).

The relative sea-level changes are directly determined in the field. The absolute (eustatic) sea-level changes can only be determined or estimated from the relative sea-level data (Mörner, 1969; Chappell, 1974).

In the case of glacio-eustasy, these changes are claimed to be able to be

243

determined also by oxygen isotope analysis of deep-sea cores (Shackleton and Opdyke, 1973). A straightforward correlation between the $\delta^{18}O$-content, the palaeoglaciation volume and the sea-level changes has been questioned by Mörner (1981b, 1983f). The remarkable thing with the $\delta^{18}O$ records are that they provide continuous long-term records, whilst our sea-level records generally only cover minor parts and often are very difficult to date or even assign a fair tentative age-estimate.

The theory of Quaternary eustasy can be formulated as follows:

(1) The old concept: eustasy is globally valid and can be expressed in one eustatic curve (e.g. Fairbridge, 1961; Bloom et al., 1974; the goal of the IGCP-61 project). Sea level is locally differentiated due to tectonism and isostasy.

(2) The new concept: eustasy is not globally valid because of palaeo-geoidal redistributions of the water masses and corresponding irregular deformations of the sea level (Mörner, 1976, 1980b, 1983a, 1984b, 1987; Newman et al., 1980, 1981). Each region must define its own eustatic curve (Mörner, 1976, 1980b); the goal of the IGCP project 200). Gravitational changes will lead to irregular deformations of the geoid, whilst rotational changes (rate or tilt) will lead to latitudinal deformations.

During the Quaternary, we may — a priori — expect to find the following types of sea-level changes (cf. Mörner, 1985a, 1987, 1983a).

(1) Dominant glacio-eustatic changes with some differentiation over the globe due to corresponding geoid deformations.
(2) Significant and sometimes dominant geoidal eustatic changes due to deformations of the gravitational and/or rotational potentials controlling the geoid surface.
(3) Significant and often dominant local differentiation due to tectonic and isostatic processes (uplift or subsidence).
(4) Minor local differentiation due to changes in the dynamic sea surface due to various meteorological, hydrological and oceanographic processes.
(5) Some long-term tectono-eustatic effects in response to the Earth's geodynamic processes.

We will now turn to the examination of the available field evidence. This will be presented in three steps; (1) the long-term fluctuations during the last 2.5 Ma, (2) the special problems connected with the last 140,000 years' records and especially the last interglacial levels, and (3) the main trends from the regression maximum in connection with the last glaciation maximum at around 20,000 BP. When this is finished we will return to the question about the origin and character of the Quaternary sea-level changes.

THE LONG-TERM CHANGES DURING THE LAST 2.5 MA

Despite the vast area and the long history of investigation, there is, in fact, not very much data on the northern hemisphere that is of such quality, and is so well dated, that it can be used for extensive sea-level analyses of the type that we are able to undertake when it concerns the Holocene and the Late Pleistocene. The records are often 'patchy' and connected with serious dating problems. Furthermore, it is mainly the elevated shorelines that are recognisable. The sea-level oscillations that reached similar levels obscured each other and are generally only traceable via extensive coring. Areas like Barbados (e.g. Bender *et al.*, 1979) and New Guinea (e.g. Bloom *et al.*, 1974) do provide well-dated sequences of uplifted marine terraces for the last 0.7-0.6 Ma, or so. From southern Australia, Cook *et al.* (1977) were able to fix the Brunhes/Matuyama polarity boundary in a long sequence of elevated beaches, and to provide a nice record of the number of elevated shorelines during the last 0.7 Ma. In coastal areas of continental glaciations, one might expect to be able to relate the glacial/interglacial changes directly to the corresponding glacio-eustatic regressions/transgressions. As yet we do not have such a record, however (only 'patches' or 'possible' correlations). In high latitudes, datable material has been a problem. The amino acid method has opened an important new means of establishing the relative ages both vertically, in cores and sections, and also horizontally along coasts (e.g. Wehmiller and Belknap, 1982; McCartan *et al.*, 1982; Mangerud and Miller, 1984; Brigham-Grette and Sejrup, 1984). Areas of high tectonic activity (such as Japan, parts of the American west coast, Central America and parts of the Mediterranean) are likely to provide quite local records that may conflict very much even within one and the same region and hence become hard to evaluate. Areas of more continuous uplift may, however, provide very useful records, such as in Barbados (Mesolella *et al.*, 1969) and in parts of Baja California (Ortlieb, 1981).

1. The Bering Strait region

One of the most important and complete sequences of Quaternary sea-level changes comes from the Bering Strait region (e.g. Hopkins, 1967, 1973). Six pre-Holocene marine transgressions were identified during the last 2.2 Ma (Fig. 8b.2, 1). The chronology is supported by several absolute age determinations and by magnetostratigraphy. The Woronzofian transgression is of Mid-Wisconsin age (Hopkins, 1967) or of early Holocene age (Mörner, 1971; Schmoll *et al.*, 1972). The Pelukian (with two maxima), Kotzebuan and Einahnuhtan transgressions all fall within the last 320,000 years. The Anvilian transgression, which reached the highest elevation, is

245

of Early Pleistocene age. The Beringian transgression (which includes two episodes) represents much warmer conditions than those of today and is dated to ~ 2.2 Ma.

In summary, (1) there was a low number of marine transgressions, (2) there was a long interval (~ 1 Ma) of non-marine conditions, (3) the sea-levels do not show a stepwise lowering, and (4) all transgressions (except for the Woronzofian) reached about 10-20 m or more (less than 100 m) above the present sea level (Fig. 8b.2, 1).

2. The American west coast

Lots of elevated shorelines have been reported from along this coastline. South of San Francisco, the marine terraces reach some 300-400 m above the present sea level (Wahrkraftig and Birman, 1965, Fig. 8). Tectonism and dating problems make the records difficult to evaluate and use. Isochrones established by amino acid analysis along the coast have recently provided fundamental lateral correlations and relative datings (e.g. Wehmiller *et al.*, 1977).

From Baja California, Ortlieb (1981, 1984) has reported a beautiful sequence of elevated terraces; 9 shorelines reaching from +9 m to +190 m psl (given an age of 100,000 or 120,000 a for the +12 m terrace, the +190 m terrace would, with a linear extrapolation, date to 1.6 or 1.9 Ma, respectively). The stepwise arrangement of shorelines found here is clearly the effect of a more or less continuous tectonic uplift.

In summary, (1) there were 9 marine transgression maxima within the last 1.6-1.9 Ma or so, and (2) the stepwise occurrence of beaches from +190 m to +12 m here is the function of crustal uplift (Fig. 8b.2, 2).

3. The Barbados Island, The West Indies

Barbados is a tectonically rising island that has an important sequence of elevated marine terraces (Broecker *et al.*, 1968; Mesolella *et al.*, 1969; Bender *et al.*, 1979). Numerous absolute age determinations have given the record a better dating control than most other records (Fig. 8b2, 3). A sequence of 15 terraces have been identified up to an elevation of ~ +200 m psl. They date back to ~ 650,000 BP and seem to represent 6 or 7 complex interglacial high sea-level stands (Bender *et al.*, 1979). The dates obtained, in general, support a correlation with the marine oxygene isotope stages 5-17 (Shackleton and Opdyke, 1973; Shackleton, 1975). Tectonic-ally, the area possesses a non-linear uplift (of up to 0.5 mm a^{-1}) and a local tilt.

In summary, (1) this rapidly rising island has numerous terraces, (2) they seem to represent 6 or 7 interglacial high stands, (3) these 'interglacials' were all complex and represented by 2-3 separate terraces, and (4) the 'interglacial' high stands seem to have reached up to around present sea level and, by no means, can be taken as evidence of a stepwise lowering of absolute sea level (Fig. 8b.2, 3).

4. The American east coast

Several shoreline terraces occur along the Atlantic coast (e.g. Richards and Judson, 1965). They are 'discontinuous and patchy', poorly dated and their marine origin has even been questioned (e.g. Flint, 1940). The terraces reach up to ~ +70 m psl and form a stepwise succession of levels. Cooke (1931) identified 7 and Puri and Veron (1959) 5 separate shorelines. Isochrones based on amino acid analyses have provided interesting relative datings and lateral correlations along the coast (Wehmiller et al., 1982; McCartan et al., 1982).

In summary, there may have been some 5-7 high sea-level stands, that form a stepwise falling-off sequence (Fig. 8b.2, 4).

5. The North Sea region

In the North Sea region, the classical glacial-interglacial changes can be directly correlated with changes between non-marine and marine conditions. The problem, however, is that there are so few marine beds in the coastal records (Zagwijn, 1984). There are only the Eemian, Dömnitzian (Wacken), Holsteinian, pre-Elsterian (top-Cromerian) and Tiglian marine transgressional beds recorded. Between the Tiglian and pre-Elsterian marine deposits there is a gap of ~ 1.3 Ma without any marine deposits. These marine transgressions all reached up to or just above the present sea-level position.

In summary, there were only 4 marine transgressions (at least recorded), there was a non-marine interval of about 1.3 Ma, and all transgressions reached about the same level (Fig. 8b.2, 5).

6. The Mediterranean

The Mediterranean is the classical area for the so-called 'Zeuner system' of stepwise lowering interglacial high sea-level stands (Zeuner, 1958; Fairbridge, 1961, 1971). Some 5-7 marine terraces, covering the last 2.0-2.5 Ma, are usually found up to elevations in the order of ~ 150-200 m asl. Hey (1971) tried to bring some order and correlation to all the recorded

247

terraces around the Mediterranean but had more or less to give it up and concluded (1978) that, 'the Mediterranean shorelines can no longer be regarded as a means of obtaining precise values for the altitudes of Quaternary eustatic stillstands', but, instead, 'can be used as evidence for recent crustal movements'. From Mallorca, Butzer (1975) reported the alternation between 5 complex high sea-level stands and 5 generations of aeolian deposits, representing interglacials and glacials, respectively. The last interglacial is represented by 3 high sea levels dating from 125,000 to 80,000 BP. The second last interglacial is represented by 2 high sea levels dated to ~ 200,000 BP. The other interglacials (consisting of 4, 2 and 4 separate high stands) are, as yet, undated. From Turkey, Erol et al., (1981) reported 7 separate erosional surfaces/terraces during the last 2.0-2.5 Ma. From the Macedonia coast in Greece, however, Kirittopoulos (1984) reports a high (+700 m) marine terrace of Miocene-Pliocene age that was succeeded by a rapid uplift at around 2.5 Ma, and then followed only by one marine terrace, at a position slightly above the present sea level.

In summary, (1) some 5-7 separate terraces can usually be identified, (2) they usually form a stepwise sequence, and (3) tectonism has played a significant role for the present position of the relative sea levels recorded (Fig. 8b.2, 6).

7. The deep-sea oxygen isotope records

Dansgaard and Tauber (1969) claimed that the deep-sea oxygen isotope records (e.g. Emiliani, 1955) were primarily palaeoglaciation records. Shackleton and Opdyke (1973), therefore, claimed that there also was a direct correlation between the $\delta^{18}0$ variations and the sea-level changes (i.e. the glacial/interglacial volumes of water) and presented a continuous record of the last ~ 2.0 Ma. This record (Fig. 8b.2, 7) includes about 25 major peaks in the $\delta^{18}0$ curve that would represent interglacial marine transgressions. Mörner (1978, 1981b) noted that the frequencies and amplitudes varied considerably through the record and claimed that factors other than sea-level changes and palaeoglaciation must also be involved. In addition he pointed out that a much lower number of 'interglacial' peaks were recorded in the very long core offshore from West Africa, described by Shackleton and Cita (1979).

According to Kukla (1977), the classical European glacial/interglacial scheme was seriously in error, and could be re-arranged so that it fitted the deep-sea records. He claimed that each of the classical marine interglacials should be divided into several separate interglacials. Amino acid analysis (Mangerud and Miller, 1984) has shown that this is not possible, however. The Eemian, Holsteinian and top-Cromerian marine beds represent 3 single interglacial events just as originally proposed.

In summary, the oxygen isotope records from the deep-sea suggest a very high number of interglacial marine transgressions during the last 2.0 Ma (Fig. 8b.2, 7).

8. Summary of the long-term trends

The data presented (like other material not mentioned due to space limitation) can, of course, not be combined into a coherent eustatic curve;

Figure 8b.2: Graphic summary of the data discussed: (1) the Bering Strait region, (2) Baja California, (3) Barbados, (4) the east coast of the USA, (5) the North Sea region, (6) the Mediterranean with onset of the main stades according to Zazo and Bonadonna (1985), (7) oxygen isotope record from core V28-238 (Shackleton and Opdyke, 1973) and corresponding magnetostratigraphy

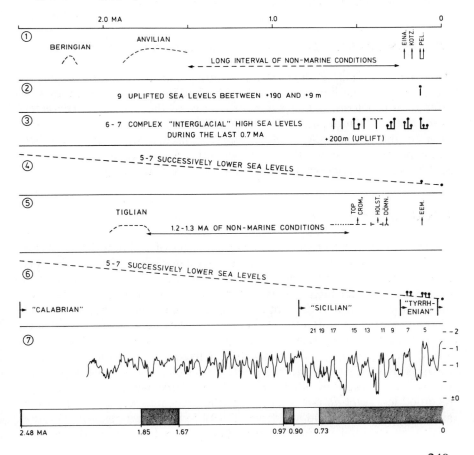

instead they are highly contradictory (Fig. 8b.2). At the same time, there are a few important points that can be made:

(1) In no case do we find any coastal record of regression/transgression cycles that would match the high number of supposed glacial expansions/contractions suggested by the oxygen isotope records from the deep-sea for the last 2.5 Ma. The records from Alaska and the North Sea region are surprisingly similar and both exhibit a very long period in the Early and Middle Pleistocene of non-marine conditions. The Barbados record (last 0.7 Ma), on the other hand, includes numerous terraces and gives an independent dating control providing a reasonable correlation with isotope stages 5, 7, 9, 13, and 15-17.

(2) The available sea-level data cannot be combined into something even approaching a globally valid eustatic curve.

(3) The available data clearly contradict a general validity of the 'Zeuner system' of Quaternary sea-level changes. This is even more obvious when southern hemisphere data are included (Mörner, 1981a).

THE MAIN CHANGES DURING THE LAST 140,000 YEARS

Although the main sea-level changes during the last interglacial/glacial cycle are of special importance and are backed up by more data, they still offer great problems. A major difficulty concerns the nature and character of the last interglacial and the beginning of the expansion of continental ice caps of the last Ice Age (Wisconsin-Weichselian-Würmian). Another major problem concerns the character and sea-level changes during the Mid-Wisconsin-Weichselian interstadial complex. (Fig. 8b.3).

1. Barbados

Broecker *et al.* (1968) reported the absolute ages of 125,000, 105,000 and 82,000 BP for three elevated terraces on Barbados (Fig. 8b.3, 1). This was a major breakthrough for the dating and identification of the character of the last interglacial high sea-level stand. Mesolella *et al.* (1969) supplemented the record with dates of two younger high stands at 65,000 BP and 42,000 BP. Steinen *et al.* (1973) claimed that they had been able to determine a major (74 m) regression between the 125,000 and 105,000 BP levels. The recorded low stand was not well fixed from a stratigraphic viewpoint as pointed out by Mörner (1972), who claimed that the regression was small or absent in other records. Later work demonstrated that the recorded low level, in fact, represented the preceding second-last Ice Age maximum (Fairbanks and Matthews, 1978). In the meantime, however, the regression reported by Steinen *et al.* (1973) was taken as evidence of a major glacial expansion at around 115,000 BP. A very important step was taken when Shackleton and Matthews (1977) and Fairbanks and Matthews (1978) were able to establish an independent and direct means of correlation based on the oxygen isotope contents in molluscs and corals respectively, between the Barbados I-III terraces and the deep-sea substages 5a, 5c, and 5e.

Matthews (1973) calculated the 'eustatic' position of the Barbados I-III terraces at −15±3 m (I), −16±3 m (II) and +6±2 m (III). Based on the oxygen isotope composition of the corals in the terraces, Fairbanks and Matthews (1978) estimated the same 'eustatic' positions at −45 m, −43 m and +5 m. It should be noted, however, that there is not a simple and straightforward relationship between isotopic composition and sea-level positions, and that the crustal movements, implied by the suggested 'eustatic' levels, are complicated and seem unlikely. We can, therefore, certainly not agree with the statement of Fairbanks and Matthews (1978) that 'the isotopic method for establishing palaeo-sea levels is a more reliable way'. The eustatic levels of the Barbados I and II terraces remain problematic. The possible regressions between terraces I and II, and II and III remain undefined.

2. Bermuda

Harmon *et al.* (1981) gave a detailed sea-level graph based on several absolute age determinations for the period 135,000-75,000 BP. Three major high stands were recorded; a transgression from somewhere below −25 m up to about +5 m at ~ 125,000 BP, a regression down to below −20 m at ~ 110,000 BP, a transgression up to about −15-20 m at ~ 105,000 a. BP, a regression to below −20 m at ~ 95,000 a. BP, a transgression up to about −15-20 m at ~ 85,000 a. BP, and finally a regression to below −20 m at ~ 75,000 BP (Fig. 8b.3, 2). The Bermuda platform is thought to be 'not significantly affected' by vertical crustal movements (Harmon *et al.*, 1981).

3. Hawaii

Stearns (1974) investigated coral terraces both on and off the coast of Oahu. He found traces of four regressional low stands and four transgressional high stands during the last 130,000 years. The dating remains a major problem, however. Ku *et al.* (1974) reported dates from the four beaches at +7.6, +3.7, +1.5 and +0.6 m psl. They found that 'all the ^{230}Th ages center about 120,000 years'. This suggests that there was a single, but complex, high sea-level stand at ~ 120,000 BP and that it was followed at

251

an unknown age, by the Kawela low, the Leahi II high (at ~ 25,000 BP) and the Mamala low (Fig. 8b.3, 3); i.e. a picture quite different from that of Stearns (1974) when it concerns the period 120,000-40,000 BP.

4. Alaska

Hopkins (1973) found that the last interglacial Pelukian transgression was double (Fig. 8b.3, 4) consisting of a first level at around +10.5 m and a second level at about +7 m that were separated by a 'minor regression' of 'unknown extent but of considerable duration', and with a climate that was 'as severe as at present' (which means interglacial climate conditions).

Hopkins (1973) also found evidence of two transgressions within the Mid-Wisconsin at about 35,000 BP and 25,000 BP which both reached up to levels of about −20 m psl.

5. The North Sea

The last interglacial in Europe, the Eemian, corresponds to a single marine transgression peak. The interglacial ends with a regression of 32 m according to Zagwijn (1983), more than 40 m according to Jelgersma (1979) and more than 50 m according to Knudsen (1984). In a drilling in northern Denmark, Knudsen (1984) found that between the interglacial bed (50 m of marine clay) and the lowermost glacial bed (20 m of marine clay with an arctic fauna), there was a transitional zone (27 m of marine clay) that represented shallower and cooler conditions than that of the interglacial sequence (Fig. 8b.3, 5).

The two long and continuous records from Grande Pile (Woillard, 1978) and Les Echets (de Beaulieu and Reille, 1984a, 1984b) provide fundamental information on the palaeoclimatic changes in Europe (Mörner, 1982b). Between the Saalian (Riss) and Weichselian (Würm) glacial beds, there is a true interglacial bed (including two geomagnetic 'excursions'), a thin but sharp cooling event (Melisey I), a second warm (not to say 'interglacial') interval (the St. Germain I interstadial), a new thin but sharp cooling event (Melisey II), and a third warm (not to say 'interglacial') interval (the St. Germain II interstadial) which includes the 'Vålbacken geomagnetic excursions' (Mörner, 1979b, 1981c). It is, of course, tempting to correlate these three warm episodes with the three high sea levels recorded in, for example, Barbados, Bermuda and Almeria. At the same time, however, the cooling events in between these seem rather to represent very short and severe climatic deteriorations than periods of real building up of continental ice caps (Mörner, 1981c). Another problem is the recording of a double geomagnetic excursion (H-I) within the true Eemian pollen zone

(Mörner, 1979b, 1981c), because the Blake double geomagnetic excursion is reported at around isotopic substage 5d (Smith and Foster, 1969) or the entire zone 5 (Denham, 1976).

6. Almeria, Morocco and Mallorca

In Morocco, high sea levels have been dated at around 140,000-120,000, 95,000, 85,000, 70,000 and 60,000 BP. The picture is still problematic, however.

In Mallorca, Butzer and Cuerda (1962) reported stratigraphical evidence of a small to insignificant regression between the Tyrrhenian IIa and IIb interglacial high stands. Butzer (1975) improved the record and was able to separate three main levels (Y1-Y3) of the last interglacial period and three generations of eolian deposition (B1-B3) of the last glacial period (Fig. 8b.3, 6). The regression between Y1 and Y2 is very small to insignificant. The regression between Y2 and Y3 is represented by a layer of terra rossa.

A remarkably detailed and conclusive record has recently been described from Almeria in southern Spain (Goy and Zazo, 1983; Zazo *et al.*, 1984). The Mediterranean guide fossil for the last interglacial, *Strombus bubonius*, occurs in 4 elevated beaches, viz. at +15 m, +11 m, +8 m and +1 m psl (in the area of east Almeria). The first 3 beaches are separated by little or no regression, whilst the +8 m and +1 m beaches are separated by a large accumulation of continental deposits representing a major regression and colder climate conditions (Fig. 8b.3, 7). Obviously, this regression represents the first major glacial expansion of the last Ice Age (i.e. the Early Weichselian maximum). This would mean that the +1 m beach is of Mid-Weichselian age which the available dates seem to indicate (Zazo *et al.*, 1984). The remarkable fact now emerges that *Strombus bubonius* must, therefore, have survived the first climatic deterioration and glacial expansion of the last Ice Age.

7. The deep-sea oxygen isotope records

If it is true that a change in $\delta^{18}O$ composition in deep-sea cores of 0.1‰ corresponds to 10 m glacio-eustatic changes in sea level, it would be easy to convert the oxygen isotope records into glacio-eustatic curves (Fig. 8b.3, 8). The problem, is, however, that the oxygen isotope records differ quite considerably (e.g. Rosholt *et al.*, 1961, Fig. 4). Mörner (1981b, 1982b), therefore, suggested that the oxygen isotope records were significantly affected by geoidal redistribution of the water masses and corresponding changes in circulation especially the Arctic/Equatorial interchange of water (cf. Mörner, 1984a).

The ocean temperature record of Sancetta *et al.* (1973) differs significantly from the isotope record and shows generally interglacial climatic conditions from about 130,000 to 75,000 BP and generally glacial climatic conditions from about 75,000 to 13,000 BP.

8. Glacial volume estimates

The major glacio-eustatic changes could, of course, also be approximately estimated from the recorded ice marginal fluctuations of the Laurentide and Fennoscandian ice caps. Mörner (1971) undertook such an estimate and presented a curve for the last 130,000 years (Fig. 8b.3, 9). From a glaciologic viewpoint one could, according to this curve, expect: no major regression at ~ 110,000 BP, a major regression followed by an interstadial transgression up to about the present sea level, a major regression of the Early Wisconsin-Weichselian maximum, a Mid-Wisconsin-Weichselian interstadial complex with the highest interstadial sea-level not exceeding −40 m psl and finally a major Late Wisconsin-Weichselian regression followed by an oscillatory rise in sea level.

9. Summary of the last 140,000 years' trend

The available data (Fig. 8b.3) are so controversial and unprecise in chronology and/or stratigraphy that they do not permit anything approaching 'final conclusions'. Too much subjective evaluation must still be put into the interpretations. We note the following, however:

(1) The data cannot be combined into anything approaching a globally valid eustatic curve. On the contrary, they often show quite different patterns of changes, suggesting that palaeogeoidal deformations may have played a significant role.

(2) The best records come from Barbados, Bermuda and Almeria. These records all show 3 separate 'interglacial' high sea-level stands. This cannot be a coincidence. At the same time, the interjacent possible regressions (isotopic substages 5d and 5b) differ considerably in character and amplitude between these records.

(3) The Almeria, Mallorca and Alaska records clearly contradict any major glacial expansion in relation to the oxygen isotope signal of substage 5d. Furthermore, there is in Alaska evidence of present day climatic conditions (i.e. full interglacial climatic conditions) during the 'minor regression' between the two interglacial high stands.

(4) The discrepancies between the sea-level records and the deep-sea oxygen isotope records (and between the various deep-sea records, too)

254

Figure 8b.3: Summary of recorded sea-level changes during the last 150,000 years: (1) Barbados, (2) Bermuda, (3) Hawaii, (4) Alaska, (5) the North Sea, (6) Mallorca, (7) Almeria in southeastern Spain, (8) the oxygen isotope record of core V19-29 converted into a generalised sea-level curve assuming a 125 m sea-level amplitude between stages 1 and 2, and (9) glacial eustatic changes as estimated from the ice marginal fluctuations of the Fennoscandian and Laurentide ice caps (Mörner, 1971)

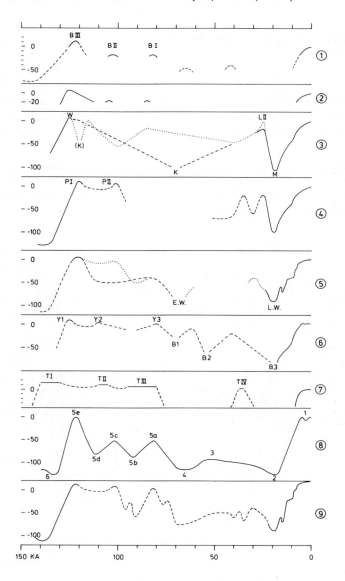

255

indicate that there is not a straightforward correlation between the $\delta^{18}0$ variations and the glacio-eustatic (i.e. palaeoglacial) changes. Geoidal redistribution of the water masses and rotational changes of the oceanic circulation, in response to the orbital geometry changes, are proposed to have affected the oxygen isotope records to a significant, and sometimes dominant, degree (Mörner, 1984a).

THE LAST 20,000 YEARS' RECORDS

At around 20,000 BP, the Weichselian ice cap in Europe and the Wisconsinan ice cap in North America both reached a final maximum position. At the same time, sea-level stood at a maximum regression level. This level differed considerably over the globe (Mörner, 1981b, 1983a). Fig. 8b.4 gives 4 different levels in a set of 20 representative localities from

Figure 8b.4: The present geoid (with respect to the rotational ellipsoid), the Holocene maximum sea-level (when above the present), the 10,000 BP level and the 20,000 BP (low stand minimum) level in 20 representative localities from around the globe

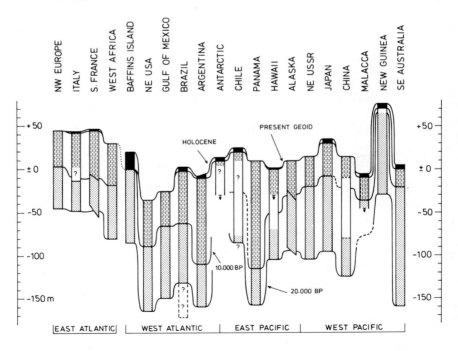

Source: Redrawn from Mörner (1983a).

all around the world; viz. the 20,000 BP level, the 10,000 BP (Pleistocene/ Holocene boundary) level, the present geoid (with respect to the rotational ellipsoid), and the Holocene optimum level when above the present level. The general rise in levels gives evidence of the glacio-eustatic rise and the melting of the ice caps. The differences between the records give evidence of significant (i.e. many tens of meters) deformation of the palaeogeoidal relief. (Local tectonism and errors in the recording and interpretation, of course, also occur, but these do not upset the conclusion about significant palaeogeoidal changes.)

CONCLUSIONS

(1) Quaternary sea-level changes, in the long-term as well as in the short-term, cannot be combined into something even approaching a eustatic curve in the old sense of eustasy (i.e. globally valid).

(2) It seems quite clear that, besides the characteristic glacio-eustatic changes in sea level, there have been significant, and sometimes even dominant, changes in sea level due to deformation of the palaeogeoidal relief, giving rise to irregular and compensational sea-level changes over the globe.

(3) The oxygen isotope records (on both a long-term as well as on a short-term basis), do not agree with the available sea-level records (the long-term records not even in general shape, and the short-term changes, at least, not in detail). This indicates that the oxygen isotope records to a significant, and often dominant, degree are controlled by another factor than glacio-eustasy (i.e. palaeoglaciation). This factor is proposed here as being that of palaeo-circulation and interchange of Arctic water. This is brought about by palaeogeoid changes of the distribution of the water masses and rotational changes affecting the circulation in the hydrosphere (Mörner, 1984a, 1985a, 1985c).

REFERENCES (combined for Chapters 8a and 8b)

Beaulieu, J.-L., de, and Reille, M. (1984a), A long Upper Pleistocene pollen record from Les Echets, near Lyon, France. *Boreas, 13*, 111-32.
Beaulieu, J.-L., de, and Reille, M. (1984b), The pollen sequence of Les Echets (France): A new element for chronology of the Upper Pleistocene. *Geogr. phys. Quatern., 38*, 3-9.
Bender, M.L., Fairbanks, R.G., Taylor, F.W., Mathews, R.K. Goddard, J.G. and Broecker, W.S. (1979), Uranium series dating of the Pleistocene reef tracts of Barbados, West Indies. *Bull. Geol. Soc. Am., 90*, 577-94.
Bloom, A.L., Broecker, W.S., Chappell, J.M.A., Matthews, R.K. and Mesolella, K.J. (1974), Quaternary sea level fluctuations on a tectonic coast: New ^{230}Th/^{234}U

dates from the Huon Peninsula, New Guinea. *Quaternary Res.*, *4*, 185-205.

Bond, G. (1978). Speculations on real sea-level changes and vertical motions of continents at selected times in the Cretaceous and Tertiary. *Geology*, *6*, 247-50.

Brigham-Grette, J. and Sejrup, H.P. (eds) *Quaternary Stratigraphy of the North Sea*, Symp. Univ. Bergen, Abstracts, pp. 30-2.

Broecker, W.S., Thurber, D.L., Goodard, J., Ku, T.L. Matthews, R.K., and Mesolella, K.J. (1968), Milankovitch hypothesis supported by precise dating of coral reefs and deep-sea sediments. *Science*, *159*, 297-300.

Butzer, K.W. (1975), Pleistocene, littoral-sedimentary cycles of the Mediterranean Basin: A. Mallorquin view. In: Butzer, K.W. and Isaac, G.L. (eds.) *After the Australopithecines*, pp. 25-71, Mouton.

Butzer, K.W. and Cuerda, J. (1962), Coastal stratigraphy of southern Mallorca and its implication for the Pleistocene chronology of the Mediterranean Sea. *J. Geol.*, *70*, 398-416.

Chappell, J. (1974), Geology of coral terraces, Huon Peninsula, New Guinea: A study of Quaternary tectonic movements and sea-level changes. *Geol. Soc. America Bull.*, *85*, 553-70.

Cook, P.J., Colwell, J.B., Firman, J.B., Linsay, J.M., Schwebel, D.A. and von der Borch, C.C. (1977) The late Cenozoic sequence of southeast South Australia and Pleistocene sea-level changes. *BMR J. Australian Geol. Geophys.*, *2*, 81-8.

Cooke, C.W. (1931), Seven coastal terraces in the southeastern states. *Washington Acad, Sci. J.*, *21*, 503-13.

Daly, R.A. (1910), Pleistocene glaciation and the coral reef problem. *Amer. J. Sci.*, *30*, 297-308.

Dansgaad, W. and Tauber, H., 1969, Glacier oxygen-18 content and Pleistocene ocean temperatures. *Science*, *166*, 499-502.

Denham, C.R. (1976), Blake polarity episode in two cores from the Greater Antilles Outer Ridge. *Earth Planet. Sci. Lett.*, *29*, 422-34.

Donovan, D.T. and Jones, E.J.W. (1979), Causes of worldwide changes in sea level. *J. Geol.*, *136*, 87-192.

Emiliani, C. (1955), Pleistocene temperatures. *J. Geol. 63*, 149-58.

Erol, O. and 11 others (1981). Morphotectonic results of the geomorphological study of the Biga Peninsula, northwestern Turkey. *Bull. INQUA Teotectonics Comm.*, *4*, 31-42.

Fairbanks, R.G. and Matthews, R.K. (1978). The marine oxygen isotope record in Pleistocene corals, Barbados, West Indies. *Quatern. Res.*, *10*, 181-96.

Fairbridge, R.W. (1961), Eustatic changes in sea level. *Physics and Chemistry Earth*, *4*, 99-185.

Fairbridge, R.W. (1971), Quaternary shoreline problems at INQUA. *Quaternania*, *15*, 1-17.

Fischer, A.G. (1984), The two Phanerozoic supercycles. In: '*The new uniformitarianism*' (W.A. Berggren & J.A. van Couvering, eds.), Princeton Univ. Press, pp. 86-105.

Flint, R.F. (1940), Pleistocene features of the Atlantic Coastal Plain. *Am. J. Sci.*, 238, 757-787.

Flint, R.F. (1971), *Glacial and Quaternary Geology*. New York, Wiley & Sons.

Goy, J.L. and Zazo, C. (1983), Pleistocene tectonics and shorelines in Almeria (Spain). *Bull. INQUA Neotectonics Comm.*, *6*, 9-13.

Grotzinger, J.P. (1984), Lecture at 'A climatic symposium in honor of Prof. R.W. Fairbridge', Barnard Coll., Columbia Univ., May 1984. Published 1986 in *Geol. Soc. America Bull.*, *97*, 1208-31.

Hallam, A. (1977), Secular changes in marine inundation of USSR and North America through the Phanerozoic. *Nature*, *269*, 769-72.

Hancock, J.M. and Kauffman, E.G. (1979), The great transgressions of the Late Cretaceous. *J. Geol. Soc. Lond.*, *126*, 175-86.
Hardenbol, J., Vail, P.R. and Farrer, P. (1981). Interpolating paleoenvironments, subsidence history and sea-level changes of passive margins from seismic and biostratigraphy. *Oceanologica Acta 1981*, 33-44.
Harmon, R.S., Land, L.S., Mitterer, R.M., Garrett, P., Schwarcz, H.P. and Larson, G.J. (1981), Bermuda sea level during the last interglacial. *Nature*, *289*, 481-3.
Hays, J.D. and Pitman III, W.C. (1973), Lithosphere plate motion, sedimentation rates and the hypsographic curve. *Earth Planet, Sci. Lett.*, *54*, 1-16.
Hey, R.W. (1971), Quaternary shorelines of the Mediterranean and Black Sea. *Quaternaria*, *15*, 273-84.
Hey, R.W. (1978), Horizontal Quaternary shorelines of the Mediterranean. *Quatern. Res.*, *10*, 197-203.
Hopkins, D.M. (1967), Quaternary marine transgressions in Alaska. In: Hopkins, D.M. (ed.) *The Bering Land Bridge*, pp. 47-90. Stanford Univ. Press.
Hopkins, D.M. (1973), Sea level history in Beringia during the past 250,000 years. *Quatern. Res.*, *3*, 520-40.
Jeletsky, J.A. (1978), Causes of Cretaceous oscillations of sea level in Arctic Canada and some general geotecnic implications. *Geol. Survey Canada*, Pap. 77-18, 1-44.
Jelgersma, S. (1979), Sea-level changes in the North Sea basin. In: Oele, E., Schüttenhelm, R.T.E. and Wiggers, A.J. (eds.) *The Quaternary history of the North Sea*, pp. 233-48. Acta Univ. Uppsala.
Keller, G. and Barron, J.A. (1983), Paleoceanographic implications of Miocene deep-sea hiatuses. *Geol. Soc. America Bull.*, *94*, 590-613.
Kerr, R.A. (1980), Changing global sea levels as a geologic index. *Science*, *209*, 483-6.
Kirittopoulos, P. (1984), Neogene lacustrine deposits in the inland basins of northern Greece; stratigraphy, paleoenvironment, paleomagnetism, and industrial usefulness. *Medd. Stockholm Univ. Geol. Inst.*, *261*, 1-96.
Knudsen, K.L. (1984). Foraminiferal stratigraphy in a marine Eemian-Weichselian sequence at Apholm, north Jutland. *Bull. geol. Soc. Denmark*, *32*, 169-80.
Ku, T.-L., Kimmel, M.A., Easton, W.H. and O'Neil, T.J. (1974), Eustatic sea level 120,000 years ago on Oahu, Hawaii. *Science*, *183*, 959-62.
Kukla, G.J. (1977), Pleistocene land-sea correlations: I. Europe. *Earth-Sci. Rev.*, *13*, 307-74.
Loutit, T.S. and Kennett, J.P. (1981), New Zealand and Australian Cenozoic sedimentary cycles and global sea-level changes. *AAPGH*, *65*, 1586-1601.
Maclaren, C. (1842), The glacial theory of Professor Agassiz. *Am. J. Sci.*, *42*, 346-65.
Mangerud, J. and Miller, G. (1984), Aminostratigraphy of marine interglacial deposits around the North Sea. In: Aarseth, I. and Sejrup, H.P. (eds.) *Quaternary Stratigraphy of the North Sea*, Symp. Univ. Bergen, Abstracts, p. 43.
Matsumoto, T. (1977), On the so-called Cretaceous transgressions. *Spec. Pap. Paleontol. Soc. Japan*, *21*, 75-84.
McCartan, L., Owens, J.P., Blockwelder, B.W., Szabo, B.J., Belknap, D.F., Kriansakul, N., Mitterer, R.M. and Wehmiller, J.F. (1982), Comparison of amino acid racemization geochronometry with lithstratigraphy, biostratigraphy, uranium-series coral dating, and magnetostratigraphy on the Atlantic Coastal Plain of the southeastern United States. *Quatern. Res.*, *18*, 337-59.
Mesolella, K.J., Matthews, R.K., Broecker, W.S. and Thurber, R.L. (1969), The astronomical theory of climatic changes: Barbados data. *J. Geol.*, *77*, 250-74.

Moore, T.C. Jr., Van Andel, Tj.H., Sancetta, C. and Psias, N. (1978), Cenozoic hiatuses in marine sediments. *Micropaleontol.*, *24*, 113-38.

Mörner, N.-A. (1969), The Late Quaternary history of the Kattegatt Sea and the Swedish West Coast; deglaciation, shorelevel displacement, chronology, isostasy and eustasy. *Sveriges Geol. Undersökn.*, *640*, 1-487.

—— (1971), The position of the ocean level during the interstadial at around 30,000 BP — A discussion from a climatologic-glaciologic point of view. *Canadian J. Earth Sci.*, *8*, 132-43.

—— (1972), When will the Present Interglacial end? *Quatern. Res.*, *2*, 341-9.

—— (1976), Eustasy and geoid changes. *J. Geol.*, *84*, 123-51.

—— (1978), Paleoclimatic, paleomagnetic and palaeogeoidal changes: Interaction and complexity. In: *Evolution of Planetary Atmospheres and Climatology of the Earth*, pp. 221-232. Toulouse, France, CNES Colloque Intern (Nice 1978).

—— (1979a), The Fennoscandian uplift and Late Cenozoic geodynamics: Geological evidence. *GeoJournal*, *3*, 287-318.

—— (1979b), The Grande Pile paleomagnetic/paleoclimatic record and the European glacial history of the last 130,000 years. *Intern. Proj. Paleolimnology Late Cenozoic Climate*, *2*, 19-24.

—— (1980a), Relative sea-level changes, tectono-eustasy, geoidal eustasy and geodynamics during the Cretaceous. *Cretaceous Res.*, *1*, 329-40.

—— (1980b), Eustasy and geoid changes as a function of core/mantle changes. In: Mörner, N.-A. (ed.) *Earth Rheology, Isostasy and Eustasy*, pp. 535-53. New York Wiley & Sons.

—— (1981a), Revolution in Cretaceous sea-level analysis. *Geology*, *9*, 344-6.

—— (1981b), Eustasy, palaeoglaciation and palaeoclimatology. *Geol. Rundschau*, *70*, 691-702.

—— (1981c), Weichselian chronostratigraphy and correlations. *Boreas*, *10*, 463-70.

—— (1982a), Sea-level changes as an illusive 'geological level'. *Bull. INQUA Neotectonics Comm.*, *5*, 55-64.

—— (1982b), Grande Pile and Les Echets; two long records of the paleoclimatic and paleomagnetic evolution in Europe during the last 130,000 years. *Intern. Proj. Paleolimnology Late Cenozoic Climate*, *3*, 17-18.

—— (1983a), 'Sea Levels'. In: Gardner, R. and Scoging, H. (eds.), *Mega-Geomorphology*, pp. 73-91. Oxford, Oxford Univ. Press.

—— (1983b), Geoid deformation, asthenospheric changes and plate motions. *Terra Cognita*, *3*, p. 116.

—— (1983c), Geophysical changes at the Paleozoic/Mesozoic boundary. Abstracts, Lunar Planetary Sci. XIV (Houston 1983).

—— (1983d), Time consistency of terrestrial processes, properties and rates. Abstracts, 18th IUGG, Hamburg 1983.

—— (1983e), Differential Holocene sea level changes over the globe: Evidence for glacial eustasy, geoidal eustasy and crustal movements. *Coastal evolution in the Holocene*. Symp., pp. 93-6. Tokyo.

—— (1983f), Illusions and problems in water-budget synthesis. In: Street-Perrott, A., Beran, M. and Ratcliffe, R. (eds.) *Variations in the global water budget*, pp. 419-23. Dordrecht, Reidel.

—— (1984a), Planetary, solar, atmospheric, hydrospheric and endogene processes as origin of climatic changes on the Earth. In Mörner, N.-A. and Karlén, W. (eds) *Climatic Changes on a Yearly to Millennial Basis*, pp. 483-507. Dordrecht, Reidel.

—— (1984b), Geoidal topography: Origin and time consistency. *Marine Geophys. Res.*, *7*, 205-8.

—— (1985a), Eustasy, unconformities and a revision of the 'Exxon Eustatic Curve'. Pres. at Symp. *North Atlantic Palaeoceanography, London 1984*, and Symp. *Sea Level — An Integrated Approach*, Houston, 1985. Submitted.

—— (1985b), Sea-level changes: Upsetting facts. *SEPM Spec. Publ.*,

—— (1985c), Short-term paleoclimatic changes. Observational data and a novel causation model. In: *A climatic symposium in honour of Professor R.W. Fairbridge*, Bernard College, New York 1984.

—— (1987), Models of global sea level changes. In: Tooley, M.J., Sheman, I. (eds) *Sea Level Changes*, pp. 332-55. Oxford, Basil Blackwell.

Newman, W.S., Marcus, L.F., Pardi, R.R., Paccione, J.A. and Tomecek, S.M. (1980), Eustasy and deformation of the geoid: 1,000-6,000 radiocarbon years BP. In: Mörner, N.-A. (ed) *Earth Rheology, Isostasy and Eustasy*, pp. 555-67. Wiley & Sons.

Newman, W.S., Marcus, L.F. and Pardi, R.R. (1981), Paleogeodesy: Late Quaternary geoidal configurations as determined by ancient sea levels. *IAHS, Publ. 131*, 263-75.

Olsson, R.K., Miller, K.G. and Ungrody, R.E. (1980), Late Oligocene transgression of middle Atlantic coastal plain. *Geology*, 8, 549-54.

Ortlieb, L. (1981), Sequence of Pleistocene marine terraces in the Santa Rosalia area, Baja California Sur, Mexico. In: Ortlieb, L. and Roldan, J. (eds.) *Geology of northwestern Mexico and southern Arizona; Fieldsguides and papers*, pp. 275-93. Univ. Nac. Auton. Mexico, Inst. Geol. Hermosillo.

Ortlieb, L. (1984), *Neotectonics and sea level variations in the Gulf of California area: Field-trip guidebook*. Geol. Inst. Hermosillo (also see: Bull: INQUA Neotectonics Comm., 7, 14-17).

Penck, A. (1882), Schwankungen des Meeresspiegel. *Geogr. Ges. München*, 7, 1-70.

Pitman III, W.C. (1978), The relationship between eustasy and stratigraphic sequences of passive margins. *Geol. Soc. America Bull.*, 89, 1389-1402.

Pitman III, W.C. and Golovchenko, X. (1983), The effect of sea-level change on the shelf edge and slope of passive margins. *SEPM Spec. Publ.*, 33, 41-58.

Puri, H. and Veron, R.O. (1959), Summary of the geology of Florida and guidebook to the classical exposures. *Florida Geol. Surv., Spec. Publ.*, 5, 1-255.

Richards, H.G. and Judson, S. (1965), The Atlantic Coastal Plain and the Appalachian Highlands in the Quaternary. In: Wright, H.E. and Frey, D.G. (eds.) *The Quaternary of the United States*, pp. 129-36. Princeton.

Rosholt, J.N., Emiliani, C., Geiss, J., Koczy, F.F. and Wangersky, P.J. (1961), Absolute dating of deep-sea cores by the Pa^{231}/Th^{230} method. *J. Geol.*, 69, 162-85.

Sancetta, C., Imbrie, J. and Kipp, N.G. (1973), Climatic record of the past 130,000 years in North Atlantic deep-sea core V23-82: Correlation with the terrestrial record. *Quatern. Res.*, 3, 110-16.

Schmoll, H.R., Szabo, B.J., Rubin, M. and Dobrovolny, E. (1972). Radiometric dating of maine shells from the Bootlegger Cove Clay, Anchorage area, Alaska. *Geol. Soc. America Bull.*, 83, 1107-14.

Shackleton, N.J. (1975). The stratigraphic record of deep-sea cores and its implications for the assessment of glacials, interglacials, stadials and interstadials in the Mid-Pleistocene. In: Butzer, K.W. and Isaac, G.L. (eds.) *After the Astralopethecines*, pp. 1-24. Mouton.

Shackleton, N.J. and Cita, M.B. (1979), Oxygen and carbon isotope stratigraphy of bentic foraminifers at Site 397: detailed history of climatic changes during the late Neogene. *Initial Rep. DSDP*, 47, 433-45.

Shackleton, N.J. and Matthews, R.K. (1977), Oxygen isotope stratigraphy of Late

Pleistocene coral terraces in Barbados. *Nature, 268,* 618-20.

Shackleton, N.J. and Opdyke, N.D. (1973), Oxygen isotope and paleomagnetic stratigraphy of Equatorial Pacific core V28-238: Oxygen isotope temperatures and ice volumes on a 10^5 to 10^6 year scale. *Quatern. Res., 3,* 39-55.

Sleep, N.H. (1976), Platform subsidence and eustatic sea-level changes. *Tectonophys., 36,* 45-56.

Smith, D.J. and Foster, J.H. (1969), Geomagnetic reversal in Brunhes Normal Polarity Epoch. *Science, 163,* 565-7.

Stearns, H.T. (1974), Submerged shorelines and shelves in the Hawaiian Islands and a revision of some of the eustatic emerged shorelines. *Geol. Soc. America Bull., 85,* 795-804.

Steckler, M. (1984), Changes in sea level. In: Holland, H.D. and Trendall, A.F. (eds.), *Patterns of changes in Earth Revolution,* pp. 103-32. NY, Springer.

Steinen, R.P., Harrison, R.S. and Matthews, R.K. (1973), Eustatic low stand of sea level between 125,000 and 105,000 BP: Evidence from the subsurface of Barbados, West Indies. *Geol. Soc. America Bull., 84,* 63-70.

Thiede, J. (1981), Reworked neritic fossils in upper Mesozoic and Cenozoic central Pacific deep-sea sediments monitor sea-level changes. *Science, 211,* 1422-4.

Thorne, J. and Watts, A.B. (1984), Seismic reflectors and unconformities at passive continental margins. *Nature, 311,* 365-8.

Vail, P.R., Mitchum, K.G., Thompson III, S., Todd, R.G., Sangree, J.B., Widmier, J.M., Bubb, J.N. and Hatlelid, W.G. (1977), Seismic stratigraphy and global changes in sea level. *AAGP Memoir, 26,* 49-212.

Vail, P.R. and Hardenbol, J. (1979), Sea-level changes during the Tertiary. *Oceanus, 22,* 71-80.

Vail, P.R. and Mitchum, R.M. (1980), Global cycles of sea-level change and their role in exploration. *Proc. 10th World Petroleum Congr., Bukarest 1979, vol. 2,* 95-104, Heyden & Sons (London).

Vail, P.R., Mitchum, R.M., Shipley, T.H. and Buffler, R.T. (1980), Unconformities of the North Atlantic. *Phil. Trans. R. Soc. London, A 294,* 137-55.

Vedder, J.G. and Wright, R.W. (1977), Correlation and chronology of Pacific coast marine terraces of continental United States by amino acid stereochemistry — technique evaluation, relative ages, kinetic model ages, and geological implications. *U.S. Geol. Surv., Open-file Rep. 77-680,* 1-196.

Wahrkraftig, C. and Birman, J.H. (1965), The Quaternary of the Pacific mountain system in California. In: Wright, H.E. and Frey, D.G. (eds.) *The Quaternary of the United States,* pp. 299-340. Princeton.

Watts, A.B. (1982), Tectonic subsidence, flexure and global changes of sea level. *Nature, 297,* 469-74.

Watts, A.B. and Steckler, M.S. (1979), Subsidence and eustasy at the continental margin of eastern North America. In: *Deep drilling results in the Atlantic Ocean: Continental margins and paleoenvironments* (M. Talwani, W. Hay, & W.B.F. Ryan, eds), AGU, Washington D.C., pp. 218—34.

Wehmiller, J.R. and Belknap, D.F. (1982), Amino acid age estimates, Quaternary Atlantic Coastal Plain: Comparison with uranium-series dates, biostratigraphy, and paleomagnetic control. *Quatern. Res., 18,* 311-36.

Wehmiller, J.R., Lajoie, K.R., Kvenvalder, K.A., Peterson, E., Belknap, D.F., Kennedy, G.L., Adicott, W.O., Vedder, J.G. and Wright, R.W. (1977), Correlation and chronology of Pacific coast marine terraces of continental United States by amino acid stereochemistry-technique evaluation, relative ages, kinetic model ages, and geological implications. *U.S. Geol. Surv., Open-file Rep. 77-680,* 1-196.

Wise, D.U. (1974), Freeboard and the volumes of continents and oceans through time. In: 'The geology of continental margins' (C.A. Burk & C.L. Drake, eds.), Springer, pp. 45-58.

Woillard, G. (1978), Grande Pile peat bog: A continuous pollen record for the last 140,000 years. Quatern. Res., 9, 1-21.

Worsley, T.R., Nance, D. and Moody, J.B. (1984), Global tectonics and eustasy for the past two billion years. Marine Geol., 58, 373-400.

Yanshin, A.L. (1973), About the so-called global transgressions and regressions. Byull. Mosk. Obshch. Ispyt. Prir., 48, 9-45.

Zagwijn, W.H. (1983), Sea-level changes in the Netherlands during the Eemian. Geol. Mijnbouw, 62, k437-50.

Zagwijn, W.H. (1984), Outline of Quaternary stratigraphy in the southern North Sea Basin. In: Aarseth, I. and Sejrup, H.P. (eds.) Quaternary Stratigraphy of the North Sea, Symp Univ. Bergen, Abstracts, pp. 68-73.

Zazo, B. and Bonadonna, F. (1985), Mediterranean and Black Sea Quaternary shorelines subcommission. INQUA Sub-Comm. Mediterranean and Black Sea Shorelines, Newsletter, 7, 1-3.

Zazo, C., Goy, J.L. and Aguirre, E. (1984), Did Strombus survive the Last Interglacial in the western Mediterranean Sea? Mediterranean Ser. Geol., 3, 131-7.

Zeuner, F.E. (1958) Dating The Past. An Introduction To Geochronology (4th ed.) London, Methuen.

EDITORIAL NOTE

Readers should note that the opinions expressed in this chapter, in particular the stress laid upon changes in the geoid as an explanation of sea-surface changes, are not necessarily supported unreservedly by other earth scientists and those working in the field of sea-level studies. An indication of the views held by other authors is given in the Introduction chapter to this book, and in chapters from Parts 1 and 2.

Readers should also note that the terms *transgression* and *regression*, as used in Chapter 8, may provide a source of confusion. These terms have been used formerly by different authors in sea level and related geological literature in a variety of contexts, and the terms may infer different interpretations of sea-level behaviour. A full discussion of the problems that may arise in the use of these terms is given by Tooley, M.J. (1982), Introduction. Proc. Geol. Ass., 93, 3-6; Tooley, M.J. (1982), Sea-level changes in northern England. Proc. Geol. Ass., 93, 43-51; Shennan, I. (1982), Interpretation of Flandrian sea-level data from the Fenland, England. Proc. Geol. Ass., 93, 53-63 and Shennan, I., Tooley, M.J., Davis, M.J. and Haggart, B.A. (1983), Analysis and interpretation of Holocene Sea-level data. Nature, 302, 404-406. Further information which may have a bearing upon issues raised in this chapter may be found elsewhere in the book in Chapters 1, 2, 3, 6, 7, 10 and 16, and also in Appendices I and II.

9

Quaternary Sea-level Changes: Southern Hemisphere Data

B. Pillans

INTRODUCTION

Quaternary sea-level changes have commonly been reconstructed from analysis of on-land sequences of coastal terraces and emergent shoreline deposits of various kinds. Early studies of this type (Cooke, 1930; Baulig, 1935) concentrated on mid-latitude shore platform/cliff type terraces. Many subsequent workers proposed altimetric correlations of terraces in such areas to the type Mediterranean sequence (e.g. Brothers, 1954; Zeuner, 1959; Ward, 1965). However, with the successful application of U/Th dating methods to fossil corals, rather than molluscs, and increasing appreciation of the effects of tectonic and isostatic deformation, the last twenty years have seen a shift in emphasis to coral reef terraces in tropical regions (Thurber *et al.*, 1965; Veeh, 1966; Mesolella *et al.*, 1969). Some of the most important on-land sequences of coral terraces are in the Southern Hemisphere, with the majority in the Australasian region (Fig. 9.1). Consequently, this chapter will focus on this area, particularly for the analysis of sea-level variations in the last 160,000 years.

In a recent review of Quaternary sea-level studies, Butzer (1983) has pointed out that coral terraces, by virtue of their development in areas of low terrigenous sediment input, lack the lithostratigraphic resolution of interdigitated marine and continental sediments. Important stratigraphic records of the latter type exist in the Southern Hemisphere, for example in Africa (Butzer and Helgren, 1972) and South America (Paskoff, 1977), but their chronology is still poorly known. One notable exception is the North Island of New Zealand, where chronology is better developed (Iso *et al.*, 1982; Pillans, 1983) through fission track dating of interbedded tephras (Seward, 1974, 1976, 1979), as well as floral (McGlone *et al.*, 1984) and faunal (Fleming, 1953; Beu and Edwards, 1984) biostratigraphy. Transgressive/regressive sediment facies relationships within emergent marginal basins in New Zealand provide a detailed record of sea-level change spanning the entire Quaternary, analysis of which is attempted here.

Figure 9.1: Locations of major on-land sites and deep-sea cores mentioned in the text

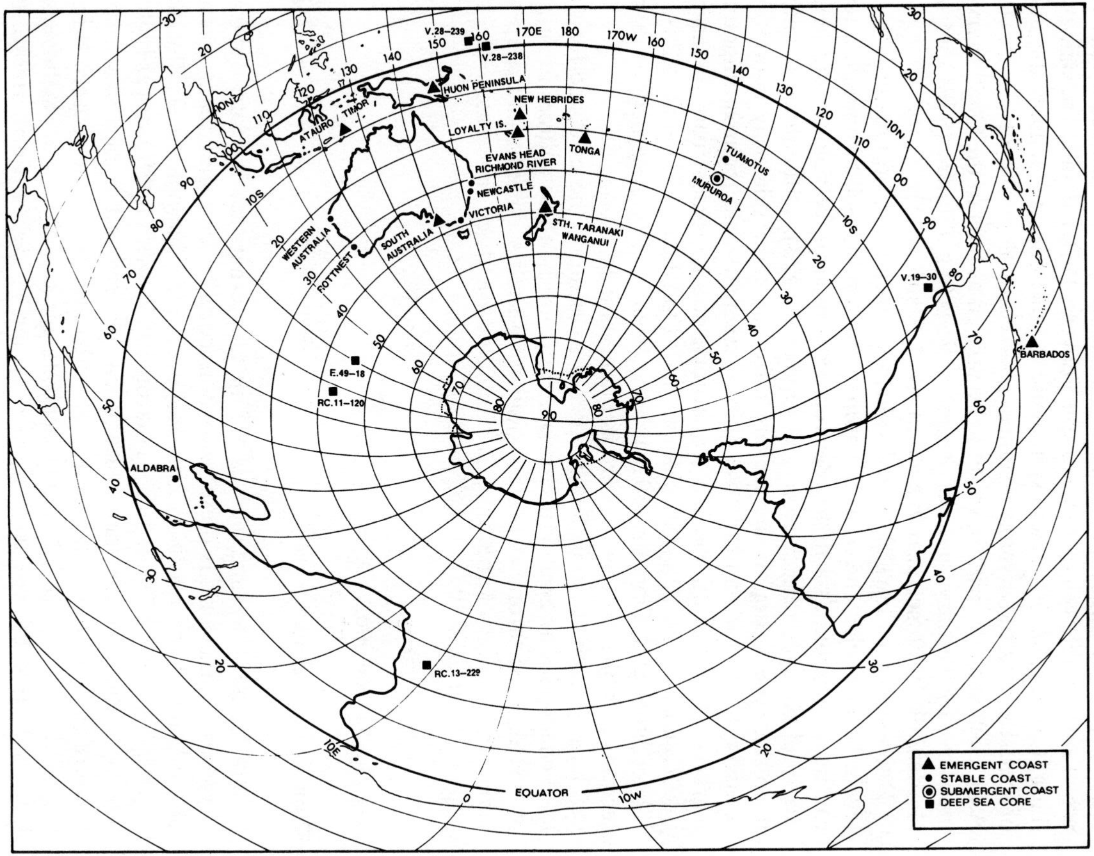

265

Onlap/offlap (transgressive/regressive) sediment sequences, similar to those found on land in New Zealand, are found below present sea level on continental shelves around the world. Although less studied than on-land sequences, their facies relationships and the presence of submerged terraces provide important data particularly for relative low sea levels. The use of submersible vehicles (Veeh and Veevers, 1970), detailed seismic profiling (Carter *et al.*, 1986) and systematic sediment coring facilitate the study of shelf deposits, but in many instances good chronological control is lacking. As a result shelf deposits have yet to be utilised as fully in the Quaternary as they have for Tertiary sea-level reconstructions (e.g. Pitman, 1978; Vail and Hardenbol, 1979; see also this vol., Chs. 2, 8 and 16).

Deep-sea cores have provided valuable information on Quaternary sea levels. In particular, the oxygen isotope variation in fossil foraminifera tests can be interpreted as a first-order representation of ice volume and hence sea-level changes (Shackleton and Opdyke, 1973). Difficulties in dating deep-sea core materials are largely offset by their relatively continuous stratigraphic records and the use of bio-, magneto- and tephrostratigraphic markers. As a result, the oxygen isotope stratigraphies from equatorial Pacific cores V28-238 (Shackleton and Opdyke, 1973) and V28-239 (Shackleton and Opdyke, 1976) have been widely regarded as 'type' sections to which correlations are frequently made.

The approach taken in this chapter is to examine the evidence for sea-level change in the Southern Hemisphere from the three sources identified above: on-land littoral deposits, continental shelf sediments and deep-sea cores. Integration of the three is attempted on three timescales: the last 160,000 years, 0-750,000 years and 0-2 million years.

THE LAST 160,000 YEARS

New Guinea data

One of the most detailed records of Late Quaternary sea-level change from anywhere in the world is preserved within coral terraces and deltaic gravels of the Huon Peninsula, New Guinea (Veeh and Chappell 1970; Bloom *et al.*, 1974; Chappell 1974, 1983; Aharon 1983). With uplift rates as high as 4 m ka^{-1}, the potential for recording and preserving a record of sea-level changes at the Huon Peninsula is higher than on most other coasts. Where the uplift rate is high, shoreline deposits are rapidly elevated above sea level and are protected from marine erosion processes. Furthermore, the stratigraphic record of relative sea-level change is 'stretched' vertically, so that shorelines of only slightly different age and palaeo sea level are clearly separated (see Berryman, this vol., Ch. 5 for discussion).

The emergent reefs at the Huon Peninsula (Figs. 9.2 to 9.4) are of two

Figure 9.2: Aerial view of the Huon Peninsula coral terraces near Sialum village (right foreground), looking south to Cromwell Range (skyline). Reef VIIb (indicated) at an elevation of ~ 200 metres has a U/Th age of ~ 120,000 BP and represents the outer barrier of a palaeo-lagoon which was similar to the present-day Sialum lagoon (foreground). A representative cross-section of the terraces near Sialum is shown in Figure 9.3. Uplift rates, estimated from the height and age of reef VIIb, increase to the southwest and decrease to the northwest. (Photograph by B.G. Thom.)

types: (a) lagoon/barrier complexes and (b) fringing reefs. Both have modern-day equivalents on the present Huon coast, with lagoon/barrier complexes dominating at the northwestern (low uplift rate) end of the coast and fringing reefs dominating at the southeastern (high uplift rate) end. A similar distribution pattern is evident in the emergent reefs. Facies geometries within the modern and fossil reefs are used to interpret relative sea-level changes. Interpretation generally follows the classical principles of reef morphology and growth identified by Darwin (1842) and summarised as follows:

(1) A rising relative sea level, slowing progressively with time, is inferred from the presence of barrier/lagoon complexes. At these times, the reefs build upward and seaward, with upwards coral growth more or less keeping pace with relative sea-level rise. (Chappell and Polach (1976) report a

267

Figure 9.3: Shore-normal (NE–SW) cross-sections at Sialum (Fig. 9.2) and Tewai Delta (Fig. 9.5), the Huon Peninsula

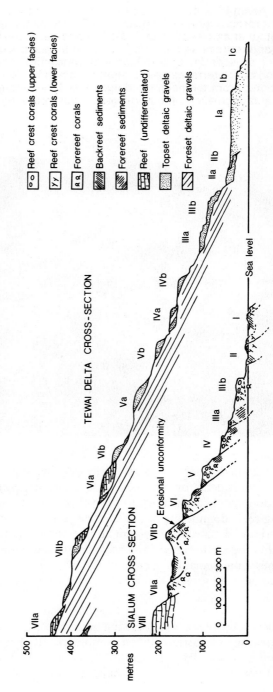

Source: After Chappell (1974, 1983).

Figure 9.4: Last Interglacial lagoon/barrier complex, ~ 20 km northwest of Sialum village, the Huon Peninsula. Patch reef (p) on palaeo-lagoon floor (L) preserved behind VIIb barrier (~ 120,000 BP), unconformably overlying reef VIIa (not seen here). Reef VIII is ~ 180,000 years old. Oxygen isotope analyses of the giant clam *Tridacna gigas* collected from patch reefs and reef VIIb at this site, and others from the Huon Peninsula, were reported by Aharon *et al.* (1980) and Aharon (1983) — see Figure 9.9. (Photograph by B. Pillans.)

mean upward growth rate of 4.7 mm a^{-1} for Holocene reef which kept pace with relative sea-level rise at the Huon Peninsula between 6,000 and 8,000 BP.) Typically, boundaries between various facies within the reefs dip landwards (Matthews, 1972).

(2) Times of approximately stable relative sea level are marked by fringing reefs, in which dominantly lateral reef growth occurs. As a result, internal facies boundaries are typically subhorizontal.

(3) During times of rapid sea-level rise, little or no reef development occurs because upwards coral growth is unable to keep pace with sea-level rise. Hopley (1982: p. 224) has summarised coral growth rates based on ^{14}C dating of Holocene reefs, which indicate maximum rates as high as 20 mm a^{-1} although typically they are less than 5 mm a^{-1}. Also, little or no reef development occurs during times of sea-level fall. Such events are generally only recorded by erosional unconformities within the reef structures, or by erosional notches produced during short stillstands (Fig. 7 in Chappell, 1974).

269

Because the Huon Peninsula is tectonically rising, and assuming that the uplift has been continuous (see Berryman, this vol., Ch. 5), the major times of reef development (both lagoon/barrier and fringing types) must represent times of absolute (eustatic) sea-level rise.

Relative low sea-level positions are indicated (Chappell, 1974, 1983) by facies relationships between foreset and topset gravels of the Tewai Delta, Huon Peninsula (Figs. 9.3, 9.5 and 9.6). The topset gravels are interpreted as beach gravels deposited at or near sea level and migration of the topset/foreset facies boundary is indicative of a relative sea-level change. Cliffs are interpreted as indicative of rapid emergence by an amount approximately equal to the cliff height. Unlike coral reef growth, the topset/foreset facies transition cannot be outstripped by rapidly rising sea level. (Thus the Tewai Delta contains an even more detailed record of sea-level change than that contained within the adjacent coral reef terraces.) As a result several topset gravel terraces have no preserved lateral equivalent amongst the coral terraces (Fig. 9.3). The topset gravel terraces cannot be dated directly, but the ages of some can be determined by correlation with U/Th and ^{14}C dated, laterally equivalent coral terraces. The ages of other topset gravel terraces and associated deposits are estimated assuming constant sediment

Figure 9.5: Aerial view of Tewai Delta, the Huon Peninsula, looking south. Height of reef VIIa (indicated) is ~ 440 metres. Gravel stratigraphy at p is detailed in Figure 9.6. A cross-section of Tewai Delta is shown in Figure 9.3. (Photograph by B. Pillans.)

Figure 9.6: Foreset and topset gravel stratigraphy, Tewai Delta, the Huon Peninsula. Terraces IVb and IVa are ~ 60,000 years old. See also Figures 9.3 and 9.7. Topset gravels occur as thin, sub-horizontal units immediately beneath terrace treads. (Photograph by B. Pillans.)

accumulation rates between points of known age (Chappell, 1983).

Overall, tectonic and eustatic factors at the Huon Peninsula can be separated by recognition of a high sea-level event which culminated ~ 120,000 BP, with a palaeo sea level (low water datum of 5 to 8 m relative to present levels (Chappell and Veeh, 1978a). Deposits relating to this event are widely identified between 1 and 10 m asl on coasts remote from plate boundaries, i.e. in supposedly stable sites (see Fig. 9.7). Terrace VIIb at the Huon Peninsula was formed during this event and a similar age terrace at Barbados is correlated with oxygen isotope stage 5e in deep-sea cores (Shackleton and Matthews, 1977). A supposed eustatic sea-level curve from the Huon Peninsula data (Fig. 9.7) is derived by assuming constant uplift along any shore-normal terrace traverse, although as Berryman has indicated (this vol., Ch. 5) this assumption cannot be accepted unequivocally. Justification for it (Bloom *et al.*, 1974) comes from: (a) the shortness of traverses, (b) the lack of increasing tilt of older terraces and (c) comparison with Barbados terrace data. Primary chronology is provided by multiple U/Th dated coral samples, mostly from reef crest facies (Bloom *et al.*, 1974; Chappell, 1974; Chappell and Veeh, 1978b), supplemented by ^{14}C dated corals and giant clams from younger terraces (Table 9.1).

Table 9.1: Summary of U/Th determinations, Huon Peninsula

Reef no.	Mean age (ka)	Number of dates	SD[2]	Maximum	Minimum	Range
I	8.2	6	1.7	9.7	5.5	4.2
II	31	1	—	31	31	—
IIIb[1]	40	5	4	46	34	12
IIIa	51	2	2	53	49	4
IV[1]	60	5	4	66	57	9
V[1]	81	3	5	86	74	12
VI	107	2	—	107	107	—
VIIb	118	2[3]	2	119	116	3
VIIa	138	3[3]	4	142	133	9
VIII[1]	185	2[3]	5	190	180	10
IXb	218	3	8	230	210	20
IXa	250	2	—	250	250	—
Xa	>250	2	—	—	—	—
XII	>250	1	—	—	—	—

Notes: [1] Samples which yielded markedly young ages have been disregarded.
[2] Standard deviation calculated as population SD.
[3] One sample run twice.
Sources: After Chappell (1974); Bloom *et al.* (1974); Chappell and Veeh (1978b).

Other on-land sites

All the dated high sea-level events recorded by terraces I-VII at the Huon Peninsula have been confirmed by radiometric dating (principally U/Th) of coral reefs elsewhere in the world. Figure 9.7 is a compilation of dated sequences in the Southern Hemisphere. Although U/Th dating of corals is commonly regarded as more reliable than that of molluscs (cf. Ku, 1976), concordant coral and mollusc dates have been recorded at several places, for example, Richmond River (Drury and Roman, 1982), Rottnest (Szabo, 1979), Huon Peninsula reefs II and IV (Chappell 1974; Chappell and Veeh, 1978b). In Victoria, U/Th dating of molluscs at two sites (Gill and Amin, 1975; Schornick, 1973) has yielded ages between 125,000 and 70,000 BP; however, no corals are present to validate these. At three localities, South Africa (Bada and Deems, 1975), South Australia (von der Borch *et al.*, 1980) and New Zealand (Pillans, 1983), amino acid racemisation dates provide chronological control. In the South Island of New Zealand, Williams (1982) used U/Th dating of speleothems to place constraints on terrace ages. Only on emergent coasts have high sea-level events younger than ~ 120,000 BP been consistently preserved above present sea level (Fig. 9.7). Furthermore, only where uplift rates exceed ~ 1.5 mm a⁻¹ will all major high sea-level events identified at the Huon Peninsula be potentially represented by on-land deposits. Thom (1973) has

Figure 9.7: Compilation of high sea-level events identified in dated on-land sequences in the Southern Hemisphere. Close circles: U/Th dates from discrete stratigraphic units, with error bars (±1σ) and number of samples indicated for multiple dates. Open circles: New Zealand terrace ages based on fission track and amino acid geochronology. Asterisk: low sea-level event identified from terraces 175 m below present sea-level (Veeh and Veevers, 1970). Amplitude of sea-level changes at the Huon Peninsula shown on a dimensionless scale 0–1 = sea-level range 0 to −150 m. Reef numbering after Chappell (1974, 1983). Times of major high sea levels identified at the Huon Peninsula are shaded to emphasise possible correlations. Arrows indicate range of ages (number of dates shown if known) where either contamination (Victoria, Mururoa) or uncertain stratigraphic position (Great Barrier Reef) of dated samples are likely

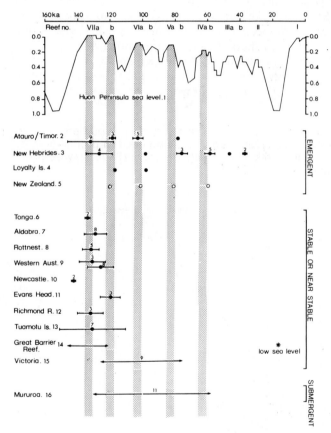

Sources: 1. Chappell (1983). 2. Chappell and Veeh (1978a). 3. Neef and Veeh (1977; Jouannic *et al.* (1980). 4. Dubois *et al.* (1977); Marshall and Launay (1978). 5. Pillans (1983). 6. Taylor and Bloom (1977). 7. Thomson and Walton (1972). 8. Szabo (1979). 9. Veeh *et al.* (1979). 10. Marshall and Thom (1976). 11. Marshall and Thom (1976). 12. Drury and Roman (1982). 13. Veeh (1966). 14. Marshall (1983). 15. Schornick (1973); Gill and Amin (1975). 16. Trichet *et al.* (1984).

discussed at length the evidence for a ^{14}C dated high sea-level event near present sea level some 30,000 BP. He concluded that, almost without exception, the data suffer from ^{14}C dating problems, stratigraphic errors and incorrect tectonic assumptions. Concordant ^{14}C determinations and a single U/Th date from Huon Peninsula reef II (Chappell and Veeh, 1978b) support a palaeo sea level of ~ −40 m at 30,000 BP.

A relation diagram (Fig. 9.8), based on the Huon Peninsula sea-level curve of Chappell (1983), can be used to predict terrace heights for various mean uplift rates. In the absence of any useful chronological control, Chappell (1975), Ghani (1978) and Bull (1984) all adopted variations of this approach to produce 'best-fit' height-age models for terraces in several areas of New Zealand. Using somewhat different approaches, Kaizuka *et al.* (1973) extrapolated Holocene uplift rates to speculate on the ages of older marine terraces in Chile, while Bowden and Colhoun (1984) reasoned that in northern Tasmania, the youngest pre-Holocene shoreline probably correlates with the 120,000 BP high sea-level event. These studies represent attempts to introduce some chronological control to areas that are currently lacking in such control.

Figure 9.8: A relation diagram based on the Huon Peninsula sea-level curve of Chappell (1983), which may be used to predict heights of terraces at various uplift rates. Conversely height and age may be used to estimate uplift rate using the diagram

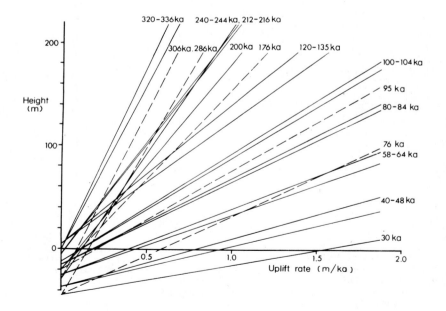

On Mururoa Atoll, U/Th ages of between 60,000 and 130,000 BP (Trichet *et al.*, 1984) have been obtained for corals from boreholes at depths between 5 and 20 m below present sea level. Unfortunately, some of the dated coral samples are recrystallised from primary to secondary aragonite and not all ages increase progressively with depth. Furthermore, K/Ar dating of basaltic basement rocks at Mururoa suggests a long-term mean subsidence rate of ~ 8 mm ka^{-1} (Trichet *et al.*, 1984). Therefore, interpretation of the age structure as evidence for relative high sea-level events younger than 120,000 BP is somewhat questionable. Finally, although the major high stands of the Huon Peninsula sea-level curve are generally confirmed at other Southern Hemisphere locations (Fig. 9.7), some dating errors are sufficiently large to allow miscorrelation of closely spaced events identified in Figure 9.8 (cf. Butzer, 1983). This is particularly true of sites which lack a sequence of age-height determinations suitable for further analysis using relation diagrams.

Isotopic records

Evidence for low sea-level stands is almost entirely lacking in the stratigraphic record of on-land deposits, except at Tewai Delta on the Huon Peninsula. Furthermore, through lack of chronological control, the full potential of continental shelf records has generally yet to be realised (cf. the work of Jongsma (1970) and Veeh and Veevers (1970) who reported ^{14}C and U/Th dates on submerged terraces and coral reefs from Northern Australia). In contrast, the record of oxygen isotope variations in deep-sea cores already has much to offer in respect of low sea-level events (see Mörner, this vol., Ch. 8). Shackleton and Opdyke (1973) summarised the evidence for interpreting $\delta^{18}O$ variations in calcareous benthonic foraminifera tests as largely the result of ice volume changes on land and the consequent effects on ocean water volume. However, whereas deep-sea cores contain a near-continuous stratigraphic record, they are less well dated than emergent terrace sequences. In addition, their $\delta^{18}O$ records are affected by diagenetic changes, bioturbation and the mixing effect consequent upon slow sedimentation rates. Thus, while $\delta^{18}O$ stratigraphy in deep-sea cores provides a useful picture of general sea-level changes through the Quaternary (cf. Shackleton and Opdyke, 1976) some workers have noted limitations for detailed analysis of sea-level variation on time scales of <10,000 a (Chappell and Veeh, 1978b; Chappell, 1981; Aharon, 1983; but see Chappell and Shackleton, 1986).

Aharon (1983) measured $^{18}O/^{16}O$ ratios on the giant clam *Tridacna gigas* for each of the reefs I-VII at the Huon Peninsula (see also Aharon *et al.*, 1980). Similar studies using molluscs (Shackleton and Matthews, 1977) and corals (Fairbanks and Matthews, 1978) have also been carried out for

Barbados coral reefs. In this way a direct comparison between the strati-graphic record of sea-level change and isotopic variation in contempor-aneous marine organisms has been made possible. Aharon (1983) demon-strated close agreement between the stratigraphic record of sea-level change at the Huon Peninsula and that inferred from isotopic data from *Tridacna*, by assuming that $\delta^{18}O$ in *Tridacna* is principally controlled by continental ice volume, after due allowance for (a) likely water temperature changes at the Huon Peninsula and (b) changes in the mean isotopic signa-ture of continental ice sheets. Aharon (1983), however, described one interesting anomaly in comparing the isotopic and stratigraphic records of sea-level history at the Huon Peninsula. Samples from terrace VIIb are consistently enriched in $\delta^{18}O$ by ~ 0.5% with respect to modern samples and the crest of reef VIIa. This corresponds to a palaeo sea level of ~ −50 m with respect to present levels if this enrichment is interpreted as solely an ice volume effect. Since the palaeo sea level for terrace VIIb is interpreted to be +5 to +8 m with respect to present sea level (Chappell and Veeh, 1978a), then either: (a) the stratigraphic estimate of palaeo sea level is in error, or (b) ocean temperatures at the Huon Peninsula were some 3°C cooler than present, or (c) there was a large floating ice cap at this time, or (d) a combination of two or more of the above. Aharon *et al.* (1980) have argued for option (d) above and interpreted the data as evidence for an Antarctic surge at ~ 120,000 BP.

While both the stratigraphic and isotopic records of sea-level change at the Huon Peninsula can be adequately reconciled, the same cannot be said for comparison between the Huon Peninsula records and the isotopic record in deep-sea cores. In Figure 9.9 the isotopic data from the Huon Peninsula are plotted beside the isotopic record from benthic foraminifera in eastern Pacific core V19-30 (Shackleton *et al.*, 1983) which, owing to its relatively high sedimentation rate, has a fairly detailed record of $\delta^{18}O$ variation spanning the last ~ 150,000 years. Of particular interest are the obvious discrepancies in the time range 120-150,000 BP and these are elaborated in the following discussion.

The last interglacial dilemma

In view of the critical importance attached to the timing and amplitude of climatic and eustatic events in the period 120-150,000 BP, particularly in relation to the Milankovitch hypothesis (Broecker *et al.*, 1968; Hays *et al.*, 1976), it seems appropriate to examine the sea-level and isotopic records for this time period in some detail. This is especially true in view of the divergent interpretations of the various lines of evidence. Three general topics are relevant to the discussion: (1) the stratigraphic record of sea-level change, (2) the accuracy of dating methods involved and (3) the inter-

276

pretation of oxygen isotope data.

The abundance of U/Th dates within the range 120-140,000 BP recorded from many parts of the world, especially in stable or near-stable sites, has been pointed out repeatedly since the seminal study of Veeh (1966). Chappell and Veeh (1978a) examined the stratigraphic record at Atauro and the Huon Peninsula and concluded that there were two closely spaced high relative sea-level events in this period, separated by a minor phase of marine removal. They estimated that these high sea-level events culminated at ~ 135,000 and ~ 120,000 BP, both with relative palaeo sea levels slightly above present. Elsewhere in the Southern Hemisphere, stratigraphic evidence for this double event is recorded in Western Australia (Veeh *et al.*, 1979; Hewgill *et al.*, 1983) and tentatively identified at Evans Head (Chappell and Thom 1978), Aldabra Atoll (Thomson and Walton, 1972) and South Africa (Barwis and Tankard, 1983). In the Northern Hemisphere, supportive evidence comes from Jamaica (Moore and Somayajulu, 1974) and Hawaii (Ku *et al.*, 1974). Thus the double peak is identified from both emergent and stable coastlines (Figs. 9.1, 9.7). The stratigraphic evidence is therefore interpreted as representing a global (eustatic) change in sea level, which involved not simply a change in rate of sea-level movement but also a change in direction (sea-level rise/sea-level fall/sea-level rise).

The accuracy of the U/Th method is seldom sufficient adequately to distinguish samples taken from both stratigraphic units (Harmon *et al.*, 1979; Butzer, 1983; Kaufman, 1986) and the exact timing of the two high sea-level stands is uncertain. This is especially true when one considers that not all dated samples were taken from reef crest (culmination) sites; some of the dated corals clearly came from transgressive sedimentary facies which preceded the culminations (Marshall and Thom, 1976; Veeh and Chappell, 1978a; Drury and Roman, 1982). Some workers (e.g. Harmon *et al.*, 1979; CLIMAP Project Members, 1984) have suggested that terrace ages in the range 130-140,000 BP must be in error, probably as a result of contamination. However, this requires either post-depositional loss of U or accession of Th, for which there is no evidence in most of the samples concerned (Moore, 1982). In most instances, reworking of older deposits can also be rejected, as the samples are described as *in situ* specimens.

Curiously, evidence of the double event is almost entirely lacking in oxygen isotope records from deep-sea cores. This has previously been ascribed to blurring of the isotope record (Chappell and Veeh, 1978a), but even in closely sampled high sedimentation rate cores only one peak is normally present (CLIMAP Project Members, 1984). Correlation of this single peak (designated oxygen isotope stage 5e) is normally made with Barbados III coral terrace which has a mean U/Th age of 122,000 BP (Broecker *et al.*, 1968; Shackleton and Matthews, 1977). The principal support for such a correlation comes from a Pa/Th estimate of 127,000 BP

277

for the stage 5/6 boundary in Caribbean core V12-122 (Broecker and van Donk, 1970). Based upon an estimated duration of ~ 11,000 years for stage 5e in core V12-122 (Broecker and van Donk 1970), the age of Barbados III terrace lies close to the midpoint of the stage. Later estimates (Kominz *et al.*, 1979) report an older age for the 5/6 boundary in core V28-238; one of 145,000 BP, based on U-series dates, and another of 138,000 BP based on constant aluminium accumulation. Kominz *et al.* (1979) considered that the error limits of their age estimates are too small to account for the difference from the date of Broecker and van Donk (1970). They concluded that the age of the 5/6 boundary should be further investigated. Strongest support for the Broecker and van Donk chronology comes from proposed links between orbitally controlled insolation changes and global climate (Broecker *et al.*, 1968; Hays *et al.*, 1976; CLIMAP Project Members, 1984). Much weight has been placed on similar frequencies of insolation changes and $\delta^{18}0$ variation in deep-sea cores as well as constant phase relationships between the two (Hays *et al.*, 1976; Kominz *et al.*, 1979; Morley and Hays, 1981). Such studies all conclude that there are strong linking statistical relationships, although the arguments become somewhat circular in that considerable 'tuning' of $\delta^{18}0$ records is made in order to achieve a best fit between these and insolation curves.

Seasonal variations in the solar radiation budget of the earth for the last 150,000 years are illustrated in Figure 9.9. Four major peaks ($\delta^{18}0$ depleted) in the isotopic records from the Huon Peninsula and core V19-30 (Fig. 9.9) closely follow times of increased seasonal contrast in the Northern Hemisphere at ~ 127,000, 106,000, 82,000 and 11,000 BP. These are consistent with decreased continental ice volumes in the Northern Hemisphere, caused by increased summer ablation at these times (Chappell, 1978). A prominent peak in the Huon Peninsula isotopic record at ~ 135,000 BP has no corresponding peak in the record from core V19-30 (accepting the Broecker and van Donk (1970) age of 127,000 years for the 5/6 boundary), and also occurs at a time of reduced seasonal contrast in the Northern Hemisphere. If one accepts at face value: (1) the chronologic and stratigraphic evidence from on-land records which indicates successive high relative sea-level events at ~ 135,000 and ~ 120,000 BP, both with palaeo sea levels several metres higher than present, (2) the presence of only a single peak in oxygen isotope records from deep-sea cores and (3) strong links between insolation changes and global climate — then it must be concluded that the 135,000 BP high sea level was caused by factors other than changes in earth orbital parameters and that the event occurred in such a way as to leave little or no record in deep-sea cores (Moore, 1982).

In summary, whatever the precise sea-level history during the period 120-150,000 BP, the fact remains that it is universally identified as the last time prior to the Holocene that relative sea level reached to present or slightly higher levels. Furthermore, faunal evidence from many on-land

Figure 9.9: Upper: oxygen isotope records from benthic foraminifera *Uvigerina* in east equatorial Pacific core V19-30 (Shackleton *et al.*, 1983), and giant clam *Tridacna gigas* at the Huon Peninsula (Aharon, 1983), plotted on timescales (in ka) reported by the authors. Error bars for Huon data (heavy lines) are ±1σ. Oxygen isotope stages after Shackleton *et al.* (1983). Lower: Insolation budgets calculated as deviations from present in Langleys per day, after Berger (1978). See text for discussion

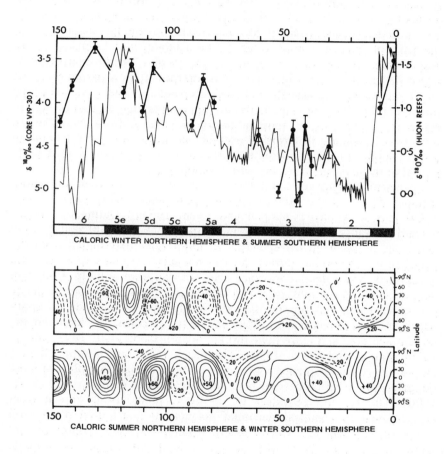

sequences indicates that this was the last time prior to the Holocene when water temperatures were as warm or slightly warmer than present (Valentine, 1965; Pickett, 1981). In the absence of good chronological control in some areas, faunal evidence has been used as an aid in distinguishing 120-140,000 age deposits from younger ones (e.g. Kennedy *et al.*, 1982).

BRUNHES NORMAL CHRON — THE LAST 730,000 YEARS

The Brunhes/Matuyama (B/M) palaeomagnetic boundary, which is dated radiometrically by the K/Ar method at ~ 730,000 BP (Mankinen and Dalrymple, 1979), is one of the most important chronostratigraphic markers in Quaternary sea-level studies. This arises because it enables correlation not only between deep-sea cores but also between cores and on-land sequences independent of any absolute dating technique (Kukla, 1977). In deep-sea cores the B/M boundary is consistently recognised near the base of oxygen isotope stage 19 (Shackleton and Opdyke, 1973). Above stage 19, stages 1, 5, 7, 9, 11, 13, 15 and 17 are interpreted as major deglacial events, while stage 3 is interpreted as a partial deglacial event (i.e. eight interglacials and one interstadial). The oxygen isotope record in core V28-238 (Fig. 9.10) and other cores suggests that successive major deglacial events, the terminations of Broecker and van Donk (1970), were broadly similar in terms of isotopic signature and hence ice–ocean water volume. In other words, the present interglacial (stage 1) is fairly typical of those occurring in the past 730,000 years.

Spectacular flights of coral terraces are inferred to span the entire Brunhes Chron at the Huon Peninsula (Chappell, 1974, 1983; Ghani, 1983) and at Atauro (Chappell and Veeh, 1978a). Dating of terraces beyond the range of U/Th dating methods (i.e. > 250,000 years) in the two areas is based on extrapolation of uplift rates determined from the height of the 120,000 BP terrace, for which a palaeo sea level of +5 to +8m is assumed. The calculated terrace ages are therefore entirely dependent on assumed uniform uplift rates and all high sea levels being approximately the same height as present-day sea level. By comparison, in South Australia, a set of beach sand-ridges can be traced some 250 km along the present coastline and more than 100 km inland (Cook et al., 1977; Sprigg, 1979; Idnurm and Cook, 1980). The beach ridges are interpreted as a prograding coastal sequence consequent upon slow regional uplift, with the ridges representing successive high sea-level stands (Idnurm and Cook, 1980). The relation of ridges to the B/M boundary is known from measurement of chemical remanent magnetisation acquired during soil formation (Idnurm and Cook, 1980). Eight major ridge complexes are known to post-date the B/M boundary on land and at least two more lie offshore (Sprigg, 1979; Idnurm and Cook, 1980). Approximate ages of ridges on land were determined by assuming a constant rate of progradation during the Brunhes Chron. These ages were then adjusted by Idnurm and Cook (1980) to match summer insolation maxima (at 65°N latitude) produced by orbital perturbations. Independent evidence of the age of one ridge system is provided by an amino acid age estimate of ~ 120,000 years calibrated to ^{14}C dated shells from the same region (von der Borch et al., 1980).

In South Taranaki, New Zealand, a broad flight of erosional terraces (cliff/platform type) is dated using amino acid racemisation dates calibrated to ~ 370,000-year-old fission-track dated tephra, coupled with reasoned land uplift models incorporating crustal tilting (Pillans, 1983). The chronology is also supported by recognition of two important bio-statigraphic datums within the sequence: the last appearance datum (LAD) of *Pseudoemiliania lacunosa* and first appearance datum (FAD) of *Emiliania huxlei* (Beu and Edwards, 1984) known to occur in deep-sea cores within oxygen isotope stages 12 and 8 respectively. The horizon of the B/M boundary has been located in emergent marine sediments in this area (Seward, 1974), but its relationship to the terrace sequence is not precisely known. The oldest dated terrace is estimated to be ~ 680,000 years old (Pillans, 1983).

These four on-land sequences described are plotted in Figure 9.10 according to ages calculated by the authors. Also shown in the figure is the $\delta^{18}O$ record from core V28-238 (Shackleton and Opdyke 1973), plotted on a linear timescale using depth of B/M boundary and constant sedimentation rate. Agreement between $\delta^{18}O$ maxima and dated high sea levels is

Figure 9.10: Chronology of sea-level changes during the last 750,000 years, as inferred from times of high sea level at South Taranaki (Pillans, 1983), Atauro (Chappell and Veeh, 1978a), the Huon Peninsula (Chappell, 1983; Ghani, 1983) and South Australia (Idnurm and Cook, 1980). High sea level at ~ 240,000 BP for South Taranaki included from Beu and Edwards (1984). Oxygen isotope data from equatorial Pacific core V28-238 (Shackleton and Opdyke, 1973) plotted on a uniform sedimentation timescale with the B/M boundary placed at 730,000 BP (Mankinen and Dalrymple, 1979)

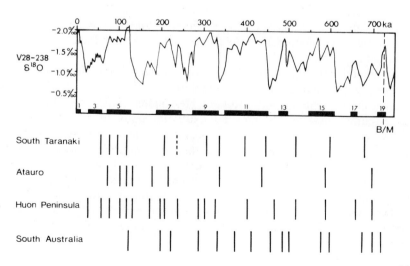

281

moderately good back to stage 11, but variable beyond that. Errors associated with terrace ages are not specifically stated by the authors, except for Atauro (Chappell and Veeh, 1978a), where they exceed ± 50,000 years for the oldest terrace (~ 700,000 BP). Errors of at least that magnitude are likely for South Taranaki, were 2σ error is ± 50,000 years for the fission track ages (Pillans, 1983). The correlation between these various records is analysed simply by comparing the timing of high sea-level events at each of the four localities with the timing of odd-numbered oxygen isotope stages. When plotted on a constant sedimentation timescale (Fig. 9.10), the odd-numbered oxygen isotope stages comprise some 56 per cent of the last 730,000 years. This means that if the ages of the terraces were randomly distributed, then approximately 56 per cent would coincide with odd numbered stages. In fact the percentage coincidence between odd-numbered stages and terraces ages are South Taranaki 77 per cent, Huon Peninsula 74 per cent, Atauro 70 per cent and South Australia 67 per cent, which are strongly suggestive of a non-random distribution, although a χ^2 test shows none of these is significant at the 95 per cent confidence level. (The percentage correspondence for the Huon Peninsula record would be significant at the 95 per cent confidence level with recognition of the prominent $\delta^{18}O$ maxima during stage 6 as an acceptable correlative of a terrace dated at ~ 175,000 BP.) Analysis of correlations between the four relative sea-level records themselves is complicated by the fact that the same events are not represented at each site. This is not unexpected since the potential for recording and preserving high sea-level events varies with uplift rate and environment of deposition. Wellman (1983) has compared the Huon Peninsula record of Ghani (1983) with the South Taranaki record of Pillans (1983) by assuming that all terraces present in South Taranaki must have equivalents in the Huon Peninsula record. Matching of terraces in this way, and based on correlation of terraces with most similar ages, led Wellman (1983) to conclude that either the reported ages for each sequence are right or that Pillans and Ghani consulted the same gypsy!

One could conclude from the above discussion that the evidence for synchronous relative sea-level maxima during the last 750,000 years in four widely scattered sites in the Southern Hemisphere is strong, albeit somewhat circumstantial, and that a similar situation holds for the correlation of these with $\delta^{18}O$ records in deep-sea cores. Certainly one must not overlook the fact that the errors associated with the various dating techniques clearly allow non-unique correlations (e.g. errors of ± 50,000 years in older terrace ages allow miscorrelation equivalent to at least one oxygen isotope stage). Furthermore, Johnson (1982) has recently suggested an age of ~ 790,000 years for the B/M boundary based on correlation of $\delta^{18}O$ records in cores V28-238 and V28-239 with the astronomical timescale of insolation variation.

THE LAST TWO MILLION YEARS

A major reference for environmental changes spanning the last two million years is the oxygen isotope record from equatorial Pacific core V28-239 (Shackleton and Opdyke, 1976). Primary chronology for this core comes from palaeomagnetic determinations which indicate the presence of the B/M boundary and Jaramillo and Olduvai subchrons, the ages of which are known from K/Ar dating and palaeomagnetic studies elsewhere (Mankinen and Dalrymple, 1979; Ness et al., 1980). With recent age estimates of the Plio/Pleistocene boundary converging at ~ 1.6 Ma (Tauxe et al., 1983; Backman et al., 1983), the isotopic record of core V28-239 clearly spans the entire Quaternary Period (Fig. 9.11). Several workers (Williams et al., 1981; Prell, 1982) have commented on the change in amplitude and frequency of isotopic fluctuations evident in V28-239 and other cores at around 900,000 BP. It appears that during the Early Quaternary less continental ice existed in glacial periods and more in interglacial periods than during the Late Quaternary. The lower amplitude fluctuations that occurred prior to ~ 900,000 BP exhibit a periodicity of ~ 40,000 years, while the higher amplitude fluctuations after 900,000 BP are characterised by a ~ 100,000 years periodicity. Prell (1982) has calculated that if these two modes of $\delta^{18}O$ variability represent solely changes in terrestrial ice volume, then the amplitudes of inferred sea-level change are ~ 100 m and ~ 180 m for the Early and Late Quaternary respectively. This suggests that the stratigraphic record of sea-level changes should be somewhat different in these two intervals.

Some of the few detailed on-land records of sea-level change which span the entire Quaternary Period are preserved within emergent, shallow marine basins of the North Island of New Zealand (Beu and Edwards, 1984). Probably the best-documented and most complete of these is the Wanganui Basin (Fig. 9.1), where progressive uplift accompanied by seaward tilting has resulted in simultaneous on-land preservation of marine terraces and their equivalent offshore marine strata (Fig. 9.11). Chronology is provided by a combination of fission-track dates on interbedded rhyolitic tephras (Seward, 1974; 1976, 1979; Boellstorff and Te Punga, 1977; Pillans and Kohn, 1981), biostratigraphy (Fleming, 1953; Beu and Edwards, 1984), magnetostratigraphy (Seward, 1974) and amino acid geochronology (Pillans, 1983). Correlations between Wanganui Basin sequence and $\delta^{18}O$ variations in core V28-239 are shown in Figure 9.11 (modified after Beu and Edwards, 1984).

Fleming (1953) was the first to try to estimate eustatic sea levels from the Wanganui Basin sequence. The fact that his thinking was constrained by the 'four glaciation myth' does not detract from his vision and advanced thinking for his time. Fleming recognised that periods of relative high sea level were represented by shore platform/marine cliff type terraces which

Figure 9.11: Correlation of Wanganui Basin strata with oxygen isotope stratigraphy in equatorial Pacific core V28-239 (Shackleton and Opdyke, 1976), partly modified after Beu and Edwards (1984). Naming and numbering of oxygen isotope stages follows Shackleton and Opdyke (1976) and Gardner (1982). Palaeomagnetic timescale after Ness *et al.* (1980). Lithostratigraphic names after Fleming (1953). Terrace names and stage correlations after Pillans (1983). Plio-Pleistocene boundary after Backman *et al.* (1983) and Tauxe *et al.* (1983). Note that Hautawa Shellbed in the Lower Okiwa Group marks the first incoming of cold-water species to the Wanganui Basin and has traditionally been regarded as the Plio-Pleistocene boundary in New Zealand.

⋀⋀⋀ Major unconformities; × × × (marine) and —.— (non-marine) tephric horizons labelled A–H. Fission track ages (in thousands of years) from Seward (1974, 1976, 1979), Boellstorff and Te Punga (1977) and Pillans and Kohn (1981): A. Rangitawa Pumice 370 ± 50 (zircon), 370 ± 70, 390 ± 10, 380 ± 40 (glass). B. Waiomio Shellbed 450 ± 90 (glass). C. Waitapu Shell Conglomerage 520 ± 80 (glass). D. Kaukatea Ash 570 ± 80 (glass). E. Kaimatira Pumice Sand (a) Potaka Pumice 640 ± 180 (zircon), 610 ± 60 (glass); (b) Rewa Pumice 740 ± 90 (glass). F. Makirikiri Tuff Formation (a) Mangapipi Ash 880 ± 130 (glass); (b) Ridge Ash 1040 ± 150 (glass). G. Mangahou Ash 1260 ± 170 (glass). H. Ohingaiti Ash 1780 ± 440 (zircon), 1500 ± 210 (glass). The Brunhes/Matuyama palaeomagnetic boundary is located between Rewa and Potaka Pumices (Seward, 1974: p. 245)

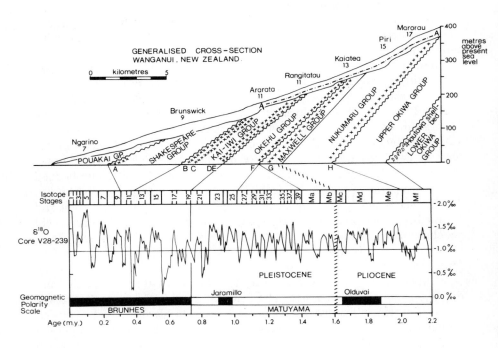

were uplifted around the margin of the basin depocentre. He was the first to point out (Fleming, 1953: p. 252) that the older marine terraces might be the shoreline equivalents of marine sediments deposited offshore but now exposed on-land. Fleming presented faunal evidence from mudstone and sandstone units within these offshore sediments which indicated deposition in 10 to 60 m of water. After allowance for subsidence and accumulated sediment thickness, amplitudes of inferred sea-level change are at least 100 m (Fleming, 1953: p. 301; Beu and Edwards, 1984). Fleming also recognised that emergence occurred during periods of relative low sea level and hence the sedimentary sequence, now exposed on land in the Wanganui Basin, was deposited almost entirely during high sea-level events (see also Beu and Edwards 1984). One obvious exception to this situation is the Maxwell Group (Fig. 9.11) which contains dominantly non-marine sediments, including lignites. Pollen evidence from these lignites indicates at least three phases of climate cooler and wetter than present (Fleming, 1953: pp. 160-8) and is consistent with deposition during eustatic low sea levels. Within the Wanganui sequence, widespread unconformities, with thin overlying conglomerates, were interpreted by Fleming to represent major phases of marine inundation which truncated older strata. These may account for the general absence of non-marine strata, and many were used by Fleming as boundaries for his major lithostratigraphic units (Fig. 9.11). According to Beu and Edwards (1984), one of these unconformities represents a time break of ~ 400,000 years in the depositional record, equivalent to oxygen isotope stages 27 to 39. Another unconformity, at the base of the Shakespeare Group represents at least 100 m of relative sea-level rise at the stage 11/12 boundary (after due allowance for subsequent seaward tilting) which culminated in the cutting of Rangitatau Terrace (Pillans, 1981).

In sediments older than Okehu Group (Fig. 9.11) inferred glacio-eustatic sea-level fluctuations are less obvious in Wanganui Basin (see Fleming 1953: p. 302). In other North Island basins, however, relative sea-level fluctuations of 30-50 m are indicated by faunal evidence from sediments of pre-Okehu age (Beu and Edwards, 1984). The stratigraphic record of sea-level fluctuations in North Island, New Zealand therefore provides possible support for the two different amplitudes of sea-level change inferred from $\delta^{18}O$ records by Prell (1982) (see also Beu and Edwards, 1984).

DISCUSSION

The Southern Hemisphere data described in preceding sections have clearly contributed much to our understanding of Quaternary sea-level changes. Regardless of a number of chronological difficulties, the data

285

confirm that major, periodic fluctuations in sea level have left their imprint in isotopic and stratigraphic records at many locations. Apart from more precise documentation of the magnitude and timing of sea-level changes, the ultimate challenge is to explain their occurrence. Rates of relative sea-level change in the Quaternary (Cronin, 1983) are high enough to exclude as primary causes changes in ocean basin volume and isostatic deformation. Such high rates, when coupled with the synchroneity of sea-level changes at widely scattered sites (Fig. 9.7) also argues against tectonism as a dominant cause of rapid, oscillating sea-level changes in the Quaternary. Mörner (1976) has pointed out that geoidal changes could account for sea-level changes of the order of $\pm$ 180 m; however, once again the apparent synchroneity of events (and their similar magnitudes) at widely scattered sites does not support this interpretation. On the other hand, the growth and decay of continental ice sheets is widely recognised as the dominant control of Quaternary sea-level changes. The glacio-eustatic control of sea level in turn invites explanations of Quaternary climatic history, principal among which is the astronomical theory of climatic change.

Periodic variations in the earth's orbit around the sun (tilt of earth's axis, precession and eccentricity of orbit) produce changes in the distribution of solar radiation received by the earth. Links between the geological record and these orbital perturbations have been sought since before the turn of the century (e.g. Gilbert, 1895); however, it is only within the last twenty years that convincing relationships have been established using, for example, Barbados coral reef data (Mesolella et al., 1969), the Huon Peninsula coral reefs (Chappell, 1973), Southern Indian Ocean cores RC11-120 and E49-18 (Hays et al., 1976) and South Atlantic core RC13-229 (Morley and Hays, 1981). This has resulted in more or less general (but by no means universal) acceptance of earth orbital perturbations as the dominant control on Quaternary climatic and sea-level history (Imbrie and Imbrie, 1979). Other potential controls of ice volume (and hence sea level) such as periodic explosive volcanism, changes in the earth's magnetic field and variation in solar output, for example, are seen as less important although their effects are by no means excluded (see this vol., Peltier, Ch. 3 and Andrews, Ch. 4 for further discussion).

In a recent review of global ice volumes during the last glaciation, Denton and Hughes (1981) point out deficiencies in our knowledge, especially with regard to the relative importance of marine-based versus land-based ice sheets. Such information is critical to understanding the $\delta^{18}O$ variations in deep-sea cores, because marine-based ice sheets will affect the $\delta^{18}O$ signal but have little effect on sea level (Broecker, 1975). Williams et al. (1981) have suggested that coincidence of marine and land-based ice sheets in the latter part of the Quaternary led to greater $\delta^{18}O$ fluctuations than in the Early Quaternary. If true, this certainly goes some way towards explaining the greater glacial/interglacial amplitudes in $\delta^{18}O$ records than can be accounted for in terms of estimated sea-level changes alone. Particularly relevant to a Southern Hemisphere perspective of Quaternary sea-level changes are the roles of the Antarctic ice sheets. Stuiver *et al.* (1981) have concluded that fluctuations in Northern Hemisphere ice sheets controlled the behaviour of the marine-based West Antarctic ice sheet through a eustatic (global) sea-level signal (see Andrews, this vol., Ch. 4). Interestingly, they report a U/Th date on mollusc shells from McMurdo Sound of 120,000 ± 6,000 BP (Stuiver *et al.*, 1981: p. 364) and suggest that McMurdo Sound may have been free of grounded ice at this time. Mercer (1981) has pointed out that since Northern Hemisphere deglaciation is almost complete at the present time, rising sea level could cause no further significant shrinkage in the West Antarctic Ice Sheet. Thus, if the West Antarctic Ice Sheet was absent at ~ 120,000 BP (compatible with a possible ice surge inferred by Aharon *et al.*, 1980, at this time), Mercer (1981) has argued that it was warmth rather than rising sea level which accomplished the final deglaciation. Similar eustatic sea-level control of the size of the East Antarctic Ice Sheet has been suggested by Budd (1981). However, snow budget observations in East Antarctica suggest net accumulation at the present time (Morgan and Jacka, 1981), while some stratigraphic and chronologic evidence (Hendy *et al.*, 1979) suggests maximum East Antarctic ice volume during interglacial (high sea-level) periods. These latter data indicate that the growth and decay of the East Antarctic Ice Sheet could be out of phase with fluctuations in Northern Hemisphere ice sheets and the West Antarctic Ice Sheet. Such a situation is not unlikely according to the astronomical theory of climatic change which predicts out-of-phase changes in solar radiation budgets for each hemisphere.

In summary, the isotopic and stratigraphic records of Quaternary sea-level change differ in detail because each is affected by factors other than eustacy. Because the effects of these other factors are not easily separated from eustatic effects, the isotopic and stratigraphic records have yet to be fully integrated. Limitations imposed by the accuracy of currently available radiometric dating techniques are an additional obstacle to integration and use of magneto-, bio- and tephrostratigraphy should be encouraged. More precise documentation of continental ice volumes for specific times in the Quaternary would enable non-eustatic factors affecting sea-level records to be investigated more thoroughly.

ACKNOWLEDGEMENTS

I particularly thank P. Vella, J. Chappell and P. Aharon for their comments on an early draft of this paper. V. Hibbert typed the manuscript. P. Hoverd and E. Hardy drew the diagrams.

287

REFERENCES

Aharon, P. (1983) '140,000 year isotope climate record from raised coral reefs in New Guinea', *Nature, 304*, 720-3.
—— Chappell, J. and Compston, W. (1980) 'Stable isotope and sea-level data from New Guinea supports Antarctic ice-surge theory of ice ages', *Nature, 283*, 649-51.
Backman, J. Shackleton, N.J. and Tauxe, L. (1983) 'Quantitative nanno-fossil correlation to open ocean deep-sea sections from Plio–Pleistocene boundary at Vrica, Italy', *Nature, 304*, 156-8.
Bada, J.L. and Deems, L. (1975) 'Accuracy of dates beyond the ^{14}C dating limit using the aspartic acid racemization reaction', *Nature, 255*, 218.
Barwis, J.H. and Tankard, A.J. (1983) 'Pleistocene shoreline deposition and sea-level history at Swartklip, South Africa', *J. Sed. Petrol., 53*, 1281-94.
Baulig, H. (1935) *The Changing Sea Level*, Inst. Br. Geogr. Publ. No. 3.
Berger, A.L. (1978) 'Long term variations of caloric insolation resulting from the earth's orbital elements', *Quat. Res., 9*, 139-67.
Beu, A.G. and Edwards, A.R. (1984) 'New Zealand Pleistocene and Late Pliocene glacio-eustatic cycles', *Palaeogeography, Palaeoclimatol., Palaeoecol., 46*, 119-42.
Bloom, A.L., Broecker, W.S., Chappell, J., Matthews, R.K. and Mesolella, K.J. (1974) 'Quaternary sea-level fluctuations on a tectonic coast: New ^{230}Th/^{234}U dates from the Huon Peninsula, New Guinea', *Quat. Res., 4*, 185-205.
Boellstorff, J.D. and Te Punga, M.T. (1977) 'Fission-track ages and correlation of Middle and Lower Pleistocene sequences from Nebraska and New Zealand', *NZ J. Geol. Geophys., 20*, 47-58.
Borch, C.C. von der, Bada, J.L. and Schwebel, D.L. (1980) 'Amino acid racemization dating of Late Quaternary strandline events of the coastal plain sequence near Robe, southeastern South Australia', *Trans. Roy. Soc. S. Australia, 104*, 167-70.
Bowden, A.R. and Colhoun, E.A. (1984) 'Quaternary emergent shorelines in Tasmania', in B.G. Thom (ed.) *Coastal Geomorphology in Australia*, Academic Press, North Ryde, Australia, pp. 313-42.
Broecker, W.S. (1975) 'Floating ice cap on the Arctic Ocean', *Science, 188*, 1116-18.
——, Thurber, D.L., Goddard, J., Ku, T.L. Matthews, R.K. and Mesolella, K.J. (1968) 'Milankovitch hypothesis supported by precise dating of coral reefs and deep-sea sediments', *Science, 159*, 297-300.
—— and van Donk, J. (1970) 'Insolation changes, ice volumes and the O^{18} record in deep-sea cores', *Rev. Geophys. Space Phys., 8*, 169-98.
Brothers, R.N. (1954) 'The relative Pleistocene chronology of the South Kaipara District, New Zealand', *Trans. Roy. Soc. New Zealand, 82*, 677-94.
Budd, W.F. (1981) 'The importance of ice sheets in long term changes of climate and sea level', in I. Allison (ed.), *Sea Level, Ice, and Climatic Change* (Proceedings of the Canberra Symposium, December 1979), Int. Ass. Hydrol. Sci., Publication No. 131, Washington, DC, pp. 441-71.
Bull, W.B. (1984) 'Tectonic geomorphology', *J. Geol. Educ., 32*, 310-24.
Butzer, K.W. (1983) 'Global sea-level stratigraphy: an appraisal', *Quat. Sci. Rev., 2*, 1-15.
—— and Helgren, D.M. (1972) 'Late Cenozoic evolution of the Cape Coast between Knysna and Cape St Francis, South Africa', *Quat. Res., 2*, 143-69.
Carter, R.M., Carter, L. and Johnson, D.P. (1986) 'Submergent shorelines in the SW

Pacific: evidence from an episodic post-glacial transgression', *Sedimentology, 33,* 629-50.

Chappell, J. (1973) 'Astronomical theory of climatic change: status and problem', *Quat. Res., 3,* 221-36.

—— (1974) 'Geology of coral terraces, Huon Peninsula, New Guinea: a study of Quaternary tectonic movements and sea level changes', *Bull. Geol. Soc. Am., 85,* 553-70.

—— (1975) 'Upper Quaternary warping and uplift rates in the Bay of Plenty and West Coast, North Island, New Zealand', *NZ. J. Geol. Geophys., 18,* 129-55.

—— (1978) 'Theories of Upper Quaternary ice ages', in A.B. Pittock, L.A. Frakes, D. Jenssen, J.A. Peterson and J.W. Zillman (eds), *Climatic Change and Variability: A Southern Perspective,* Cambridge University Press, Cambridge, pp. 211-25.

—— (1981). 'Relative and average sea level changes, and endo-, epi- and exogenic processes on the earth', in I. Allison, (ed.), *Sea Level, Ice, and Climatic Change* (Proceedings of the Canberra Symposium, December 1979), Int. Ass. Hydrol. Sci., Publication No. 131, Washington, DC, pp. 411-30.

—— (1983) 'A revised sea-level record for the last 300,000 years from Papua New Guinea', *Search, 4,* 99-101.

—— and Polach, H.A. (1976) 'Holocene sea-level change and coral-reef growth at Huon Peninsula, Papua New Guinea', *Bull. Geol. Soc. Am., 87,* 235-40.

—— and Shackleton, N.J. (1986) 'Oxygen isotopes and sea level', *Nature, 324,* 137-40.

—— and Thom, B.G. (1978) 'Termination of last interglacial episode and the Wilson Antarctic surge hypothesis', *Nature, 272,* 809-10.

—— and Veeh, H.H. (1978a) 'Late Quaternary tectonic movements and sea-level changes at Timor and Atauro Island', *Bull. Geol. Soc. Am., 89,* 356-68.

—— and Veeh, H.H. (1978b) '^{230}Th/^{234}U age support of an interstadial sea level of −40 m at 30,000 yr BP', *Nature, 276,* 602-4.

CLIMAP Project Members (1984) 'The last interglacial ocean', *Quat. Res., 21,* 123-224.

Cook, P.J., Colwell, J.B., Firman, J.B., Lindsay, J.M., Schwebel, D.A. and von der Borch, C.C. (1977) 'The Late Cainozoic sequence of southeast South Australia and Pleistocene sea-level changes', *BMR J. Austr. Geol. Geophys., 2,* 81-8.

Cooke, C.W. (1930) 'Correlation of coastal terraces', *J. Geol., 38,* 577-89.

Cronin, T.M. (1983) 'Rapid sea-level and climate change: evidence from continental and island margins', *Quat. Sci. Rev., 1,* 177-214.

Darwin, C.R. (1842) *The Structure and Distribution of Coral Reefs,* Smith, Elder, London.

Denton, G.H. and Hughes, T.J. (eds) (1981) *The Last Great Ice Sheets,* Wiley, Chichester and New York.

Drury, L.W. and Roman, D. (1982) 'Chronological correlation of interglacial sediments of the Richmond River Valley, New South Wales', in W. Ambrose and P. Duerden, (eds), *Archaeometry: An Australasian Perspective.* Australian National University Press, Canberra, pp. 290-6.

Dubois, J., Launay, J., Recy, J. and Marshall, J. (1977) 'New Hebrides trench: Subduction rate from associated lithospheric bulge', *Can. J. Earth Sci., 14,* 250-5.

Fairbanks, R.G. and Matthews, R.K. (1978) 'The marine oxygen isotope record in Pleistocene coral, Barbados, West Indies', *Quat. Res., 10,* 181-96.

Fleming, C.A. (1953) 'The geology of Wanganui subdivision', *NZ Geol. Surv. Bull., 52.*

Gardner, J.V. (1982) 'High-resolution carbonate and organic-carbon stratigraphies for the Late Neogene and Quaternary from the Western Caribbean and Eastern Equatorial Pacific', *Initial Reports, Deep Sea Drilling Project, 68,* 347-64.

Ghani, M.A. (1978) 'Late Cenozoic vertical crustal movements in the southern North Island, New Zealand', *NZ J. Geol. Geophys., 22,* 117-26.

—— (1983) 'Uplifted shorelines and geodynamics of the Cromwell Mountains, Papua New Guinea', in *XV Pacific Science Congress, Program, Abstracts and Congress Information* (Dunedin, February 1983), p. 80.

Gilbert, G.K. (1895) 'Sedimentary measurement of Cretaceous time', *J. Geol., 3,* 121-7.

Gill, E.D. and Amin, B.S. (1975) 'Interpretation of 7.5 and 4 metre Last Interglacial shore platforms in southeast Australia', *Search, 6,* 394-6.

Harmon, R.S., Ku, T.-L., Matthews, R.K. and Smart, P.L. (1979) 'Limits of U-series analysis: phase 1. Results of the Uranium-Series Intercomparison Project', *Geology, 7,* 405-9.

Hays, J.D., Imbrie, J. and Shackleton, N.J. (1976) 'Variations in the Earth's orbit: pacemaker of the ice ages', *Science, 194,* 1121-32.

Hendy, C.H., Healy, T.R., Rayner, E.M., Shaw, J. and Wilson, A.T. (1979) 'Late Pleistocene glacial chronology of the Taylor Valley, Antarctica, and the global climate', *Quat. Res., 11,* 172-84.

Hewgill, F.R., Kendrick, G.W., Webb, R.J. and Wyrwoll, K.-H. (1983) 'Routine ESR dating of emergent Pleistocene marine units in Western Australia', *Search, 14,* 215-17.

Hopley, D. (1982) *The Geomorphology of the Great Barrier Reef: Quaternary Development of Coral Reefs,* Wiley, Chichester and New York.

Idnurm, M. and Cook, P.J. (1980) 'Palaeomagnetism of beach ridges in South Australia and the Milankovitch theory of ice ages', *Nature, 286,* 699-702.

Imbrie, K. and Imbrie, K.P. (1979) *Ice Ages: Solving the Mystery,* Macmillan Press, New York.

Iso, N., Okada, A. Ota, Y. and Yoshikawa, T. (1982) 'Fission-track ages of Late Pleistocene tephra on the Bay of Plenty coast, North Island, New Zealand', *NZ J. Geol. Geophys., 25,* 295-303.

Johnson, R.G. (1982) 'Brunhes-Matuyama magnetic reversal dated at 790,000 yr BP by marine-astronomical correlations', *Quat. Res., 17,* 135-47.

Jongsma, D. (1970) 'Eustatic sea level changes in the Arafura Sea', *Nature, 228,* 150-1.

Jouannic, C. Taylor, F.W., Bloom, A.L. and Bernat, M. (1980) 'Late Quaternary uplift history from emerged reef terraces on Santo and Malekua Islands, Central New Hebrides Island Arc', *UNESCAP CCOP/SOPAC Technical Bulletin, 3,* 91-108.

Kaizuka, S., Matsuda, T., Nogami, M. and Yonekura, N. (1973) *Quaternary Tectonic and Recent Seismic Crustal Movements in the Arauco Peninsula and its Environs, Central Chile,* Geographical Reports of Tokyo Metropolitan University, 8.

Kaufman, A. (1986) 'The distribution of ^{230}Th/^{234}U ages in corals and the number of last interglacial high-sea stands', *Quat. Res., 25,* 55-62.

Kennedy, G.L., Lajoie, K. and Wehmiller, J.F. (1982) 'Aminostratigraphy and faunal correlations of Late Quaternary marine terraces, Pacific Coast, USA', *Nature, 299,* 545-7.

Kominz, M.A., Heath, G.R., Ku, T.-L. and Pisias, N.G. (1979) 'Brunhes time scales and the interpretation of climatic change', *Earth Planet. Sci. Lett., 45,* 394-410.

Ku, T.-L. (1976) 'The Uranium-series methods of age determination', *Ann. Rev. Earth Planet. Sci., 4,* 347-79.

——, Kimmel, M.A., Easton, W.H. and O'Neill, T.J. (1974) 'Eustatic sea level 120,000 years ago on Oahu, Hawaii', *Science, 183*, 959-61.

Kukla, G.J. (1977) 'Pleistocene land–sea correlations', *Earth Sci. Rev., 13*, 307-74.

McGlone, M.S., Neall, V.E. and Pillans, B.J. (1984) 'Inaha Terrace deposits: a Late Quaternary terrestrial record in South Taranaki, New Zealand', *NZ J. Geol. Geophys., 27*, 35-49.

Mankinen, E.A. and Dalrymple, G.B. (1979) 'Revised geomagnetic polarity time scale for the interval 0-5 my BP', *J. Geophys. Res., 84*, 615-26.

Marshall, J.F. (1983) 'Lithology and diagenesis of the carbonate foundations of modern reefs in the southern Great Barrier Reef', *BMR J. Austr. Geol. Geophys., 8*, 253-65.

—— and Launay, J. (1978) 'Uplift rates of the Loyalty Islands as determined by ^{230}Th/^{234}U dating of raised coral terraces', *Quat. Res., 9*, 186-92.

—— and Thom, B.G. (1976) 'The sea level in the last interglacial', *Nature, 263*, 120-1.

Matthews, R.K. (1972) 'Dynamics of the ocean-cryosphere system: Barbados data', *Quat. Res., 2*, 368-73.

Mercer, J.H. (1981) 'West Antarctic ice volume: the interplay of sea level and temperature, and a strandline test for absence of the ice sheet during the last interglacial', in I. Allison (ed.), *Sea Level, Ice and Climatic Change* (Proceedings of the Canberra Symposium, December 1979), Int. Ass. Hydrol. Sci., Publication No. 131, Washington, DC, pp. 323-33.

Mesolella, K.J., Matthews, R.K., Broecker, W.S. and Thurber, D.L. (1969) 'The astronomical theory of climatic change: Barbados data', *J. Geol., 77*, 250-74.

Moore, W.S. (1982) 'Late Pleistocene sea-level history', in M. Ivanovitch and R.S. Harmon (eds), *Uranium Series Disequilibrium: Applications to Environmental Problems*, Clarendon Press, Oxford, pp. 481-96.

—— and Somayajulu, B.L.K. (1974) 'Age determinations of fossil corals using ^{230}Th/^{234}Th and ^{230}Th/^{227}Th', *J. Geophys. Res., 89*, 5065-8.

Morgan, V.I. and Jacka, T.H. (1981) 'Mass balance studies in East Antarctica', in I. Allison (ed.), *Sea Level, Ice, and Climatic Change* (Proceedings of the Canberra Symposium, December 1979), Int. Ass. Hydrol. Sci., Publication No. 131, Washington, DC, pp. 253-60.

Morley, J.J. and Hays, J.D. (1981) 'Towards a high resolution, global, deep-sea chronology for the last 750,000 years', *Earth Planet. Sci. Lett., 53*, 279-95.

Mörner, N.-A. (1976) 'Eustasy and geoid changes', *J. Geol., 84*, 123-51.

Neef, G. and Veeh, H.H. (1977) 'Uranium-series ages and Late Quaternary uplift in the New Hebrides', *Nature, 269*, 682-3.

Ness, G., Levi, S. and Couch, R. (1980) 'Marine magnetic anomaly timescales for the Cenozoic and Late Cretaceous: A précis, critique and synthesis', *Rev. Geophys. Space Phys., 18*, 753-70.

Paskoff, R.P. (1977) 'Quaternary of Chile: the state of research', *Quat. Res., 8*, 2-31.

Pickett, J. (1981) 'A Late Pleistocene coral fauna from Evans Head, NSW', *Alcheringa, 5*, 71-83.

Pillans, B.J. (1981) 'Upper Quaternary landscape evolution in South Taranaki, New Zealand', unpublished PhD dissertation, Australian National University, Canberra.

—— (1983) 'Upper Quaternary marine terrace chronology and deformation, South Taranaki, New Zealand', *Geology, 11*, 292-7.

—— and Kohn, B. (1981) 'Rangitawa Pumice: A widespread (?) Quaternary marker bed in Taranaki-Wanganui', *Victoria University of Wellington Geology*

Department Publication No. 20, 94-104.

Pitman III, W.C. (1978) 'Relationship between eustasy and stratigraphic sequences of passive margins', *Bull. Geol. Soc. Am.*, *89*, 1389-1403.

Prell, W.L. (1982) 'Oxygen and carbon isotope stratigraphy for the Quaternary of Hole 502B: evidence for two modes of isotopic variability', *Initial Reports Deep Sea Drilling Project*, *68*, 455-64.

Schornick, J.C., Jr (1973) 'Th^{230}/U^{234} Geochronology of marine shells from near Sale, E. Victoria, Australia', *Proc. Roy. Soc. Vict.*, *86*, 35-7.

Seward, D. (1974) 'Age of New Zealand Pleistocene substages by fission-track dating of glass shards from tephra horizons', *Earth Planet. Sci. Lett.*, *24(2)*, 242-8.

—— (1976) 'Tephrostratigraphy of the marine sediments in the Wanganui Basin, New Zealand', *NZ J. Geol. Geophys.*, *19(1)*, 9-20.

—— (1979) 'Comparison of zircon and glass fission-track ages from tephra horizons', *Geology*, *7*, 479-82.

Shackleton, N.J., Imbrie, J. and Hall, M.A. (1983) 'Oxygen and carbon isotope record of East Pacific core V19-30: implications for the formation of deep water in the Late Pleistocene North Atlantic', *Earth Planet. Sci. Lett.*, *65*, 233-44.

—— and Matthews, R.K. (1977) 'Oxygen isotope stratigraphy of Late Pleistocene coral terraces in Barbados', *Nature*, *268*, 618-20.

—— and Opdyke, N.D. (1973) 'Oxygen isotope and paleomagnetic stratigraphy of equatorial Pacific core V28-238: oxygen isotope temperatures and ice volumes on a 10^5 and 10^6 year scale', *Quat. Res.*, *3*, 39-55.

—— and Opdyke, N.D. (1976) 'Oxygen-isotope and palaeomagnetic stratigraphy of Pacific core V28-239 Late Pliocene to Latest Pleistocene', *Geol. Soc. Am. Mem. 145*, 449-64.

Sprigg, R.C. (1979) 'Stranded and submerged sea-beach systems of southeast South Australia and the aeolian desert cycle', *Sed. Geol.*, *22*, 53-96.

Stuiver, M., Denton, G.H., Hughes, T.J. and Fastook, J.L. (1981) 'History of the marine ice sheet in West Antarctica during the last glaciation: a working hypothesis', in G.H. Denton and T.J. Hughes (eds), *The Last Great Ice Sheets*, Wiley, Chichester and New York, pp. 319-436.

Szabo, B.J. (1979) 'Uranium-series age of coral reef growth on Rottnest Island, Western Australia', *Mar. Geol.*, *29*, M11-M15.

Tauxe, L. Opdyke, N.D., Pasini, G. and Elmi, C. (1983) 'Age of the Plio–Pleistocene boundary in the Vrica section, southern Italy', *Nature*, *304*, 125-9.

Taylor, F.W. and Bloom, A.L. (1977) 'Coral reefs on tectonic blocks, Tonga Island Arc', in *Proceedings of the Third International Coral Reef Symposium, Miami*, pp. 275-81.

Thom, B.G. (1973) 'The dilemma of high interstadial sea levels during the last glaciation', *Prog. in Geogr.*, *5*, 170-246.

Thomson, J. and Walton, A. (1972) 'Redetermination of chronology of Aldabra Atoll by ^{230}Th/^{234}U dating', *Nature*, *240*, 145-6.

Thurber, D.L., Broecker, W.S., Blanchard, R.L. and Potraz, H.A. (1965) 'Uranium-series ages of Pacific atoll coral', *Science*, *149*, 55-8.

Trichet, J., Repellin, P. and Oustriere, P. (1984) 'Stratigraphy and subsidence of the Mururoa Atoll (French Polynesia)', *Mar. Geol.*, *56*, 241-57.

Vail, P.R. and Hardenbol, J. (1979) 'Sea-level changes during the Tertiary', *Oceanus*, *22*, 71-9.

Valentine, J.W. (1965) 'Quaternary mollusca from Port Fairy, Victoria, Australia, and their palaeoecological implications', *Proc. Roy. Soc. Victoria*, *78*, 15-73.

Veeh, H.H. (1966) 'Th^{230}/U^{234} and U^{234}/U^{238} ages of Pleistocene high sea-level stand', *J. Geophys. Res., 71*, 3379-86.

—— and Chappell, J. (1970) 'Astronomical theory of climatic change: support from New Guinea', *Science, 167*, 862-5.

—— Schwebel, D., van de Graaff, W.J.E. and Denham, P.D. (1979) 'Uranium-series ages of coralline terrace deposits in Western Australia', *J. Geol. Soc. Australia, 26*, 285-92.

—— and Veevers, J.J. (1970) 'Sea level at -175 m off the Great Barrier Reef 13,600 to 17,000 years ago', *Nature, 226*, 536-7.

Ward, W.T. (1965) 'Eustatic and climatic history of the Adelaide area, South Australia', *J. Geol., 73*, 592-602.

Wellman, H.W. (1983) 'Lucky numbers for the high sea levels of the last million years', *Geol. Soc. NZ Newsletter, 64*, 14.

Williams, D.F., Moore, W.S. and Fillon, R.H. (1981) 'Role of glacial Arctic Ocean ice sheets in Pleistocene oxygen isotope and sea level records', *Earth Planet. Sci. Lett., 56*, 157-66.

Williams, P. (1982) 'Speleotherm dates, Quaternary terraces and uplift rates in New Zealand', *Nature, 298*, 257-60.

Zeuner, F.E. (1959) *The Pleistocene Period*, Hutchinson, London.

10

Sea-level Changes During the Holocene: The North Atlantic and Arctic Oceans

R.J.N. Devoy

INTRODUCTION

The North Atlantic and Arctic oceans encompass a large (50.3M km²) and, in relation to Late Quaternary isostatic movements, a structurally dynamic zone of the earth's crust. Although difficult to define perhaps as a regional unit, the southern boundary of this zone may be placed arbitrarily at approximately Lat. 30°N (Fig. 10.1). Waters north of this line generally have a high turbidity and are characterised by mean annual surface temperatures of ~ 20°C, resulting in a lack of the coral reefs so typical of shallow water zones in the Pacific and Indian oceans. Coastlines north of 40°N are also dominated by meso- to macrotidal, high energy storm wave conditions, with lower energy swell wave mesotidal coasts occurring southward (Fig. 10.1) (Davies, 1980). Further distinctiveness is given to this region in two important respects. Firstly, in terms of sea-level research, the North Atlantic forms one of the most intensively studied of the earth's oceans. Most perhaps is known about conditions here during the Holocene and coastal–offshore areas retain some record, often a detailed one, of environmental changes from this time. Analyses of the data from this and earlier geological stages have given rise to many of the major concepts conditioning our current understanding of the processes and patterns of sea-surface changes. Secondly, the Quaternary history of the North Atlantic–Arctic oceans, together with their continental margins, has been distinctively influenced by the build-up and decay of the major Northern Hemisphere ice masses. This presence or absence of ice has been influential in determining the oceans' changing physical, chemical and biological characteristics (Imbrie, et al., 1983; Ruddiman and McIntyre, 1976, 1981; Kellogg, 1976; Mörner, 1974; Boulton, 1979a). Most importantly, the cyclic pattern of ice loading–unloading of the earth's crust, with subsequent return of water to the oceans, has resulted in a series of regional compensatory crustal movements conditioning constant changes in ocean basin shape. The results of these vertical movements are most clearly registered at the coastline in the

294

Figure 10.1: North Atlantic and Arctic oceans, showing coastal variations in tidal range

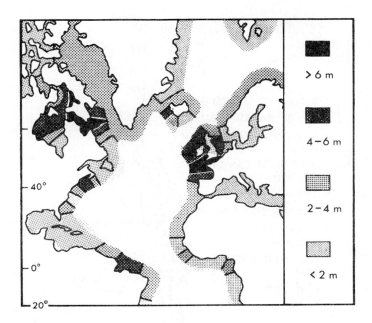

Source: Adapted from Davies (1980).

form of changing shoreline/sea-level positions. It is the record of these and associated relative sea-level changes in the North Atlantic during Holocene time that forms the focus for this chapter.

TECTONIC AND RELATIVE GEOID CHANGES

The tectonic setting of the North Atlantic–Arctic oceans was controlled from the beginning of the Mesozoic onward by sea-floor spreading and development of the mid-ocean ridge system (Le Pichon *et al.*, 1977; Laughton, 1975; Naylor and Shannon, 1982; Pitman and Talwani, 1972). Prior to ~ 200 Ma the North Atlantic did not exist as an ocean, the North American–Greenland and European continental plates being joined to form a single landmass. Early plate margin development between the North American and European plates began during the Middle Jurassic, initiating a series of faults in the basement rocks along the western Atlantic margin (Fig. 10.2). These faults subsequently formed the boundaries to a series of subsiding depositional basins paralleling the present continental

295

Figure 10.2: Major structural basins and fault zones of the North American Atlantic margin geosyncline

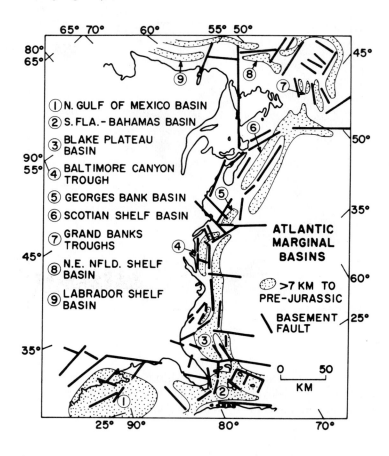

Source: Sheridan (1979).

slope and together forming the Atlantic Margin Geosyncline (Sheridan, 1979). The basins are separated by transverse arches and ocean fracture zones forming areas of relative tectonic stability, or marginal uplift. On the east of the developing ocean, tensional stresses associated with plate margin formation gave rise to the similar development of assymetric basins (Ziegler, 1982; Dobson, 1977a, b, 1979) and fault bounded troughs–grabens along the west coast of Britain, in the Irish–Celtic Seas and in the North Sea. These grabens formed a part of the northern North Atlantic–Norwegian–Greenland Sea–Eurasian Arctic mega-rift system, which gradually opened up during the Mesozoic (Fig. 10.3).

Figure 10.3: Schematic pattern of the Mesozoic Arctic–North Atlantic rift system. The continental fit shown is for approximately Early Cretaceous time

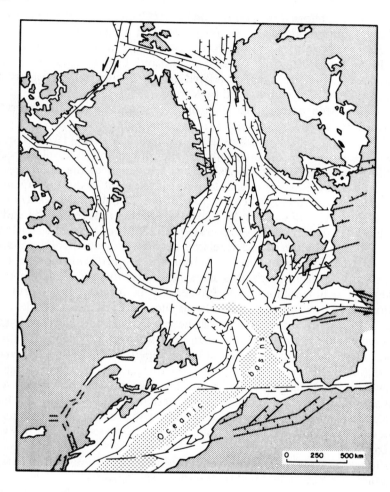

Source: Ziegler (1982).

From Mid-Jurassic to Mid-Cretaceous time, separation of the continental blocks of South Europe, North Africa and North America led to the generation of new ocean crust to the south. Northwards the opening of the North Atlantic between the Greenland and Northwest European plates did not get under way until the beginning of the Tertiary (~ 65 Ma ago). The axes of spreading (Fig. 10.3) here are aligned northeastwards between Ireland and Greenland, also extending northwards into the Norwegian Sea and northwestwards, initially as two arms, the Ireland–Labrador and Southern Labrador Sea arms, the latter becoming senescent ~ 40 million years ago. At this time (~ 38-40 Ma) the opening up of the Spitzbergen–Greenland area allowed the connection of Arctic and Atlantic ocean water. Similarly, later subsidence of the Faerö–Iceland ocean ridge in the Early Miocene (~ 22.5 Ma) admitted North Atlantic water into the Norwegian Sea (Pitman and Talwani, 1972; Talwani and Eldholm, 1977). Palaeoceanographic-climatic changes related to these later phases of crustal movement have been summarised by Mörner (1980a).

As can be seen from Figure 10.4, the N–S aligned axes of ocean ridge and sea-floor spreading transect the trend of Late Devensian–Holocene sea-level/crustal behaviour zones identified by Clark et al. (1978). Nevertheless, the longer-term mantle–lithospheric movements controlling ocean spreading do not appear to have been in opposition to the influences of glacial and hydro-isostatic crustal loading, at least not in the Late Quaternary. In the southern North Sea, western Britain and along the east coast of the United States and Canada, long-term subsidence induced by the underlying graben structures have exaggerated the effects of ice and water loading, as indicated by the recorded relative sea-level movements (Newman et al., 1980a, b). Continued tectonic subsidence coupled with the effects of collapsing ice margin forebulges have facilitated the accumulation of ~ 1,000 m of sediment during the Quaternary in the southern North Sea (Caston, 1979) and similar amounts off the northeast coast of the USA (Grant, 1980). In earlier phases of the opening of the North Atlantic, subsidence of the Faerö–Iceland ocean ridge was probably coupled with an eastward transfer of mantle material and compensatory upward movement by ~ 600 m of the Fennoscandian Shield (Mörner, 1979a, 1980a), smaller movements being recorded in the Spitzbergen and Barents Shelf areas. Later subsidence of these zones in the Quaternary (~ 0.9 Ma ago) may have led to uplift again of the Faerö–Iceland ridge, with possible repercussions for ocean circulation and the build-up of ice in the region (see Devoy, this vol., Ch. 16).

In long-term geological perspective, this tectonic framework has been important in conditioning the broad pattern of relative sea-level changes in the region (see Devoy, this vol., Ch. 16; Mörner, this vol., Ch. 8; Sheridan, 1979; Pitman, 1978). At shorter timescales tectonic activity may have continued to have been a component in influencing such regional changes in level, although factors of ice, water and sediment volume–load changes have been of greater importance here. For the Holocene analysis of relative sea-level data (Newman et al., 1980a, b, 1981) shows that, despite an unreliable data base and an uneven spatial–temporal spread of information (Tooley, 1985), there has been substantial regional deformation of sea-surface level between 0 to 12,000 BP (Marcus and Newman, 1983) (Fig. 10.5). The major zone of relative sea-surface uplift in the northern North

Figure 10.4: The six sea-level zones for the realistic ice melt model of Clark *et al.* (1978), showing also the general signature of Holocene relative sea-level recovery (RSL) predicted for zones I–III and VI discussed in the text

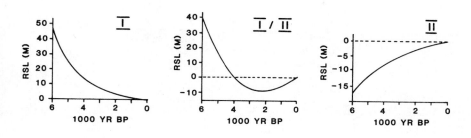

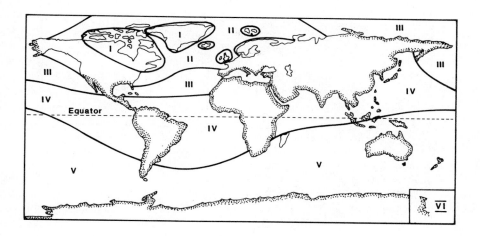

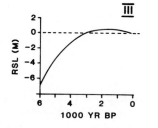

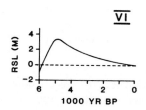

Source: Adapted from Clark and Lingle (1979).

Figure 10.5: Trend surface analysis of radiocarbon dated 'sea-level' index points between 1,000–6,000 BP for the North Atlantic–Arctic. Contours are drawn at 2 m intervals, with the + and − signs showing areas of relative sea-surface rise and depression respectively. Longer records of relative sea-surface variation are given in Marcus and Newman (1983)

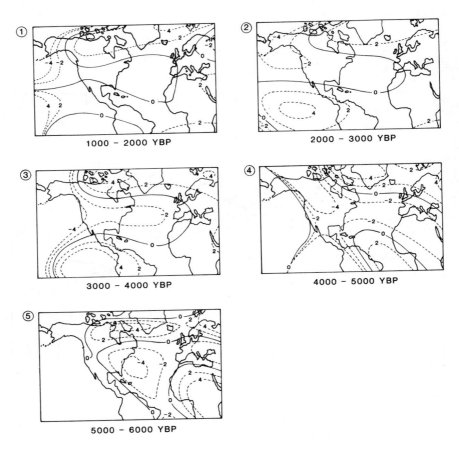

① 1000 – 2000 YBP

② 2000 – 3000 YBP

③ 3000 – 4000 YBP

④ 4000 – 5000 YBP

⑤ 5000 – 6000 YBP

Source: Adapted from Newman *et al.* (1980b).

Atlantic identified in these studies is explained in terms of isostatic rebound following ice retreat. In contrast, a zone of relative sea-surface depression, possibly migrating variably southward through time across the North Sea and British Isles, covers the bulk of the ocean area to the south. A number of causes for this have been proposed, namely, crustal downwarping along continental margins, collapse of former ice forebulge zones, factors of sediment loading and hydro-isostasy resulting from deglaciation and lastly geoid surface migration. The likelihood is that a combination of these processes was responsible. These broad patterns of relative sea-surface change support the expected subsidence trends and the postglacial zonal changes in sea level of Walcott (1972a) and Clark *et al.* (1978) (see Fig. 10.4). In the present day the geoid topography (the equipotential surface of the sea) of the North Atlantic–Arctic oceans varies spatially by $\geqslant 100$ m (see Fig. 1.4), resulting in a series of 'humps and depressions' in the sea surface (Mörner, 1976a, 1977, 1981; Gaposchkin, 1973; Geodynamics Programme Office, 1983). In addition the dynamic sea surface varies by 1.5 to 2.0 m, showing an east–west vertical variation of ~ 0.6 m (Defant, 1941; Lisitzin, 1965). Further discussion of the possible instability and migration of this geoid surface is given in Chapter 1.

SEA-LEVEL CHANGE PROBLEMS

As with any region containing a spectrum of earth processes, energy regimes and sediment inputs, many different types of coastal environment are to be found in this zone, ranging from beach-barrier–lagoon systems, sand dune, estuary-mudflat and delta, to open, rocky-cliffed coastline. This spatial pattern is further complicated by different trends in vertical crustal motion. These range from areas of subsidence, such as the southern North Sea or the Delaware-central east Coast of North America, to areas of land uplift as in Scandinavia, northeast Canada (Newfoundland), or the eastern Mediterranean. This diversity of present-day environments also reflects the changing coastal complex operating here throughout the Holocene. Consequently, the region has preserved within it a wide variety of sea-level indicators representing these diverse palaeoenvironments, which can be used as a data base for inferring former sea-level positions. Indicators occur in the form of shells or other carbonates in marine-brackish water sediments, marine inorganic sediments interleaved with freshwater-terrestrial biogenic material (peat, organic muds), beach structures, palaeovalleys, wave-cut rock platforms, marine notches, cliffs and other features of coastal erosion (Jardine, 1981). This range of palaeoenvironments and sea-level indicators, with each often conditioned by different tidal and energy regimes, has given rise to a number of major methodological problems in sea-surface reconstruction (see Chapter 1 for discussion; Kidson, 1982; Gray, 1983). Variations in the development of coastal environments, or in the types of data preserved, make comparison between even small areas of coastline difficult. Tooley (1978a, b, 1985) and Devoy (1982, 1983 and this vol., Ch. 1), amongst others have highlighted the problems of using a mixed data base from a non-homogeneous environment in regional sea-level representation. Debate has also centred upon the indicative meaning (van de Plassche, 1977) or water-level positions represented by particular sea-level data points/palaeoenvironments. A sea-level indicator's position may, in addition to wider regional influences, have been conditioned by local factors of coastal shape, sediment supply and tidal amplitude; parameters which over time and under the high energy conditions of the North Atlantic–Arctic oceans may have changed, altering the significance of the indicator for local–regional sea-level reconstruction. In eastern Ireland and western Scotland, for example, features formerly identified as lateglacial raised beaches and marine erosion platforms may have been the product of Holocene changes in tidal amplitude and wave climate. This has resulted here in the reassessment of the established regional sea-level models (Carter, 1983) (see Orford, this vol., Ch. 13 and Fig. 13.5 for further discussion).

Further, a now classic methodological argument has centred on the indicative meaning of *in situ* plant material preserved in marine sequences: namely, whether such materials can be related directly to a former sea level, or alternatively to a freshwater table, and this in turn to a contemporary sea level (Jelgersma, 1961: van de Plassche, 1982; Shennan, 1986; Tooley, 1978b). In a related context, researchers also commonly use different height datums in measuring sea level. These vary between use of local, arbitrary height points to use of national geodetic datums based on varying tidal positions. Such differences either make interregional comparison impossible or lead to significant height errors when converting between datums (Devoy, 1983). Difficulties in providing an objective chronology are also well displayed in this region (see Sutherland, this vol., Ch. 6).

These methodological problems are not all peculiar to the North Atlantic–Arctic oceans and have been referred to by other authors in this volume. However, as the birthplace of many of the early fundamental concepts in sea-level studies, the region has also seen the development of many of the approaches in research methodology. Of particular importance in this context, in considering how to examine the pattern of Holocene and relative sea-level changes, has been the problem of using the 'sea-level curve' or time-depth data plots in interregional correlations. The sea-level curve as a synthesis of data has been widely applied to many of the region's coastlines; the three main different postulated patterns of Holocene sea-level recovery having been examined elsewhere (see Devoy, this vol., Ch. 1). However, growing awareness from the mid-1970s onwards that no part of the earth's crust could be regarded as stable, the inadequacy of glacio-eustasy as the principle determinant of a zone's sea-level history and the possible influence of geoid changes on sea level (Mörner, 1976a; Kidson, 1982; Bloom, 1983, 1984) have made such curves of no more than local significance for purposes of correlation. In this North Atlantic–Arctic region the pattern of glacial change and related ocean basin history has not been areally uniform or synchronous. During the glacial maxima, and in the following phase of deglaciation which conditioned Early Holocene marine-coastal environment, ice thickness and extent varied enormously between the Northwest European and Laurentide ice sheets. This, coupled with the non-synchroneity in ice melt (Andrews and Barry, 1978; Andrews *et al.*, 1974; Boulton, 1979a), led to different scales of land uplift and timing of shoreline development between the eastern and western sides of the ocean. Attempts, therefore, to make correlations across the oceans on the basis of relative sea-level curves applied to local height and age data sets are probably erroneous. This point is reinforced by current knowledge of the changing patterns in relative sea level and possible geoid configuration (Newman *et al.*, 1980). Alternative methods for both the plotting and interregional comparison of data have been suggested, such as the use of time phase isobase-'palaeogeoid' maps (Newman *et al.*, 1980b, 1981; Marcus and Newman, 1983), or statistical analyses of height dimensionless sea-level events (Devoy, 1982; Shennan, 1983; Shennan *et al.*, 1983). Further, the production of mathematical-geophysical process-response models for crustal behaviour under changing ice and water loading have also allowed the definition of distinct zonal trends in sea-level recovery (Walcott, 1972a, b; Peltier and Andrews, 1976, 1983; Clark, 1980; Clark *et al.*, 1978; Clark and Lingle, 1977, 1979; Cathles, 1975) (see Peltier, this vol., Ch. 3 for further discussion). Within each zone, given the model constraints of variations in the global-hemispherical timing, rate and volume of ice melt, coastlines should display a characteristic relative sea-level response signature independent of specific local age and height influences. For this region five zonal types have been identified (see Fig. 10.4):

Zone I	Relative land emergence
Zone I/II	Early relative emergence followed by submergence
Zone II	Rapid and continuing land submergence
Zone III	Initial relative submergence followed by marginal emergence
Zone VI	Continental shorelines — Late Holocene emergence.

Some anomalies do occur in relation to characterising shoreline histories and the model is not perfect. In particular, the local glacial history of the British Isles has complicated the accurate fitting of coastal regions here into the zonal model, with local ice load factors superimposed on the wider North Sea and eastern North Atlantic trends.

Despite the possible inaccuracies these zones do provide an ideal framework for examining the enormous weight of Holocene sea-level and palaeogeographic evidence collected for this region. Treatment of data here will therefore be based on this zonation. Discussion of sea-level changes in the Middle–South Atlantic during this time can be found in Nunn (1984).

THE PATTERNS OF SEA-LEVEL CHANGE

Zone I

The common denominator linking coastlines within this zone is that of glacio-isostatic crustal uplift. Here the 'unloading of the ice through time causes the land to rise more rapidly than the ocean surface' (Clark *et al.*, 1978). The result is a persistent trend of falling relative sea level and of coastal emergence (Fig. 10.4), decaying in time toward the present as isostatic balance is achieved. Coastlines occurring within this zone are those of Highland Britain, the Baltic region — Fennoscandia, Spitzbergen, Iceland, Greenland and northeast Canada.

Despite the linking uniformity of process major differences in the pattern of relative sea-level change do occur within the zone. These result from local factors of tectonic uplift, differences in ice extent–volume-load covering the component regions (i.e. maximum extent of Scandinavian ice sheet = ~ 6.7 M km^2; Laurentide ice sheet = ~ 13.4 M km^2, Flint, 1971) and the timing of ice melt/coastal exposure. In northeast Canada, the New Quebec and Kewatin ice sheets existed until at least 7,000 BP (Peltier and Andrews, 1983; Bryson *et al.*, 1969), whilst deglaciation had been completed throughout Northwest Europe between 10,500 and 8,500 BP (Lundquist, 1965). In addition, as Gray *et al.* (1980), Stephens and McCabe (1977) and Boulton (1979b) have shown, the pattern of ice retreat, particularly across irregular, indented coastlines, was spatially irregular in form. This enabled the rising sea to invade the newly exposed land surface at different times within parts of the same area. This has resulted in the metachroneity of maximum shoreline heights, with development of spatially variable and complex shoreline histories, even within contiguous areas.

A further linking factor within this zone is the form in which the evidence for sea-level change occurs. This is the zone of raised shorelines, represented by exposed former beach deposits, rock-cut platforms, marine notches and washing limits. Morphological, height-levelling and mapping studies have formed common techniques in unravelling the sea-level story here. As discussed by Andrews (this vol., Ch. 4), the use of biostratigraphic techniques to examine raised coastal sediments, for example, from emergent and progressively isolated coastal basin environments, has not been so widespread, being restricted largely to work within Europe.

Highland Britain

In Britain, land uplift has centred over Scotland (Fig. 10.6). Despite the complexity here of the pattern of relative sea-level change (Jardine, 1975, 1977, 1982; Sissons, 1972; Sissons and Dawson, 1981), the basic picture is one of continued land emergence throughout the Holocene.

Figure 10.6: Present pattern of land uplift and subsidence in Northwest Europe based upon tide gauge data. The isobases show the rate of change (+ or −) in mm a^{-1}. Dashed isobases are less certain, whilst dotted isobases are based on interpolation (after West, 1968). Figure 10.6a shows pattern of relative sea-level change in northern and eastern Ireland (based on Taylor *et al.*, 1986)

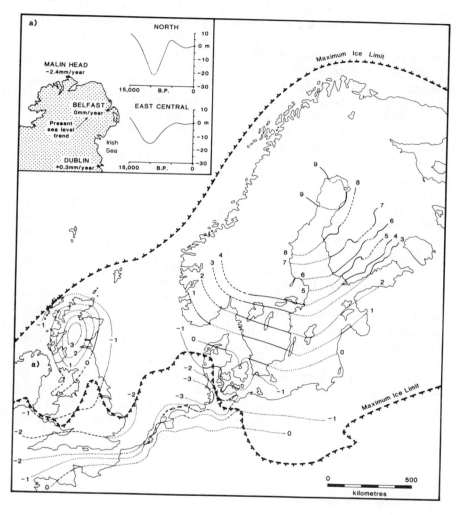

Interruption of the resultant relative sea-level fall through marine flooding evidences the interaction of glacio-isostatic/-eustatic recovery factors (Cullingford *et al.*, 1980; Sissons, 1983; Jardine, 1979; Dickson *et al.*, 1978; Donner, 1970; Haggart, 1986). The presence of extensive glacio-marine sediment sequences on both the continental shelf west of Scotland

(Davies *et al.*, 1984; Binns *et al.*, 1974a) and in the North Sea (Jansen, 1976; Jansen *et al.*, 1979; Jardine, 1982) shows the existence of a persistent cold water marine influence along the ice margin during Late Devensian times. At ~ 18,000 BP, sea level stood at ~ −110/130 m Ordnance Datum (OD) on the Scottish continental shelf edge and at −60 to −90 m OD in the northern North Sea (Jardine, 1982). Subsequent early marine flooding of these glacio-isostatically depressed surfaces by 'warmer' water is recorded at many sites (Peacock, 1983). In southwest Scotland the timing of this 'warm' marine influence is dated to between ~ 11,000 and 16,000 BP, at which time local relative sea-level is placed at ~ −30 m OD (Binns *et al.*, 1974a, b). By 13,000 BP land emergence had become dominant, developing the highest shorelines at levels of +50 m and +41 m OD in eastern and western Scotland respectively (Jardine, 1979, 1982).

In southeast Scotland the main phase of Holocene marine flooding began at ~ 8,500 BP and ended at ~ 6,500 BP as continued land emergence re-established the trend of falling relative sea level. On the periphery of ice loading, for example, in the Solway Firth, northwest England and the outer Firth of Forth, land crustal emergence has been gradual, leading here to prolonged marine inundation (Tooley, 1978a, b, 1982; Jardine, 1979, 1982). In northern Ireland sea-level data shows that this area also lies close to the margin of land uplift (Fig. 10.6), emergence continuing today at a maximum rate of 2.3 mm a^{-1} (Todd, 1981). Central Ireland is identified (Carter, 1982) as lying possibly within the ice forebulge zone, although research tying areas here into a transitional zone (I/II) is scanty (Devoy, 1983, 1985; see also McCabe, in press). A pattern of relative emergence prior to ~ 9,500 BP followed by submergence to ~ 6,000 BP (Fig. 10.6a) is recorded for northern and northeast coasts (Carter, 1982; Shaw, 1985). Along the east coast shorelines dated to ~ 9,600-13,500 BP (Synge, 1981) drop below present sea level within the ice limits of the last cold stage. These have been interpreted as evidence for ice marginal crustal downwarping (Synge, 1980). Questions as to the true age or reality of these as lateglacial shorelines have, however, been raised (Carter, 1982; Devoy, 1983). Further, whether this record and the sea-level pattern as a whole represents partly a northward migration of a collapsing forebulge is also open to question.

The Baltic–Fennoscandia

One of the oldest and perhaps best-documented regions within Zone I is that of the Baltic and associated Norwegian coastline (see Chapter 1 for discussion). Observations of sea-level change here date from the nineteenth century and earlier (Sars, 1865; Lindström, 1886; De Geer, 1890; Brøgger, 1900-1). The basic model of postglacial sea-level recovery in the region was established with the work of Sernander (1902) and Ramsay (1924). Subsequent studies (Lundquist, 1971; Eronen 1974; Eriksson, 1979; Donner, 1969; Hyvärinen, 1980) have added essential accuracy and definition to the pattern. The highest, although not the oldest, shorelines are located toward the centre of former ice loading at levels of ~ 290 m asl along the Bothnian coast in Sweden (Lundquist, 1965), falling eastwards to 220 m asl in Peräpohjola region, northern Finland (Eronen, 1983). Present rates of maximum land uplift of 9 mm a^{-1} (Fig. 10.6) occur to the north of this area in the Gulf of Bothnia. Ice retreat in the Late Weichselian took place from south to north along approximately a S–SE to N–NW axis, leaving the southeast part of Finland ice free by ~ 11,000 BP but submerged, together with the Baltic basin as a whole, beneath the freshwater Baltic Ice Lake (Fig. 10.7a). The interaction of the lake and conti-

Figure 10.7: Stages in the palaeoenvironmental and shoreline development of the Baltic–Fennoscandian region

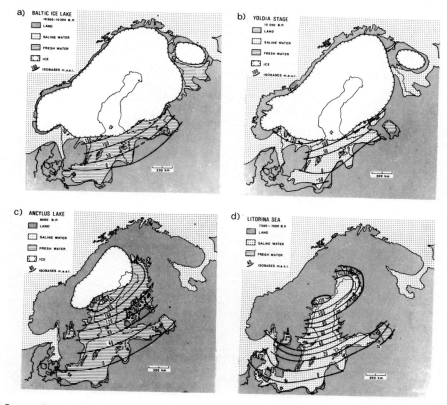

Source: Eronen (1983).

307

nental ice sheet is recorded in a complex shorelines series (Fig. 10.8), produced by variations in the water level and subsequent land tilting, best represented in the Salpausselka ice stillstand zone of southeast Finland (Donner, 1978). Formation upon ice retreat of a connection across central Sweden with the open ocean, consequent incursion of marine water and a drop of the Baltic Ice Lake level, dated to ~ 10,200 BP (Eronen, 1983), mark the onset of the Yoldia stage (10,200-9,500 BP) in the region's evolution (Fig. 10.7b). The marine effect appears to have declined eastward away from the zone of ocean water incursion and is represented by a brackish water diatom assemblage from sediments in southwest Finland. This time is essentially one of land emergence and falling relative sea-level, with shorelines developed as metachronous features.

Closure of the marine connection across central Sweden through continued glacio-isostatic uplift (Eronen, 1976) initiated the next, the Ancyclus Lake, stage (~ 9,500-9,000 BP) (Fig. 10.7c). This time is characterised, in the southern Baltic at least, by rising water levels and freshwater inundation of the newly emerged land surface. In the north rapid rates of land uplift continued the pattern of land emergence (Fig. 10.9) and shorelines formed at this time, as in the Yoldia stage, are strongly metachronous.

Figure 10.8: Shorelines developed during the Baltic Ice Lake stage at Kitee, eastern Finland. The surface at the top represents a glacio-fluvial delta plateau into which a later shoreline has been cut at ~ 7 m below this surface. (Photograph by M. Eronen.)

Figure 10.9: 'Staircase' type stony beach ridge sequence forms the Ancylus Lake stage at Puolakkavaara, northern Finland. (Photograph by M. Punkari.)

A marked drop (~ 30 m) in freshwater lake levels after ~ 9,000 BP is equated with the final disappearance of ice from the region and reconnection in the southwest with the open ocean, probably now through the straits of Denmark area (Eronen, 1983). The initial renewed marine incursion, the *Mastagloia* phase (beginning ~ 8,500 BP; Donner and Eronen, 1981), characterised by the appearance of brackish water diatoms including those of the genus *Mastagloia*, was at first gradual. Evidence from the south coast of Sweden indicates a saltwater influx here at 8,500-8,000 BP (Berglund, 1971), with a weak saline effect recorded at 8,000 BP from sediments in Finland (Eronen, 1974). The final, Litorina Sea, stage (Fig. 10.7d) begins in southern Baltic areas at 7,500-7,300 BP and is marked in sediment sequences by a clear stratigraphic boundary. Here, a rise in water levels of perhaps several metres led to a persistent marine incursion of lowland areas south of the 34 m isobase (Hyvärinen, 1980, 1982). North of this line, the marine effect reached the Gulf of Bothnia by ~ 7,000 BP, but inundation, if recorded at all, was only temporary before land emergence again overtook the rate of sea-level rise. The highest Litorina shoreline, marked by the upper limit of the common Litorina sea diatom *Campylodiscus clypeus* ('the Clypeus limit'), falls from ~ 100 m asl in Ostrobothnia to ~ 30 m asl in the Helsinki area. Shorelines are again metachronous throughout the region. By 6,500-6,000 BP continued glacio-isostatic land uplift had re-established the trend of falling relative sea level. A gradual slowing in the rate of shoreline retreat since 6,000 BP is equated with the decay in the rate of land uplift.

Whilst many would agree with these broad trends in coastal palaeo-geography, a number of important studies from this region recognise repeated phases of marine inundation since ~ 7,000 BP, separated by times of sea-level fall or stillstand. These 'alternations in sea level' are interpreted as evidence for an oscillatory behaviour for ocean water level recovery during the Holocene, with a ± 1 m amplitude for sea-level variations post ~ 5,000 BP. The most notable modern studies are those of Mörner (cf. 1969, 1971, 1976b, 1980b) and Berglund (1971). Mörner's work on the Swedish Kattegat coast is notable, if for no other reason than that it has been extensively quoted, widely used and worked on as a basis for defining regional eustatic (sea-level) curves for Northwest Europe (Mörner, 1980b). Biostratigraphic/morphological studies of enclosed coastal basins and raised shoreline deposits have led to the identification here of at least nine phases of marine incursion and inferred sea-level rise (PTM[2-10]). These have been correlated (Mörner, 1976b) with a similar shoreline sequence (Siretorp I-VI) (see Table 10.1) in the neighbouring area of Blekinge (Berglund, 1971) and with phases of marine inundation at Søborg Sø, north of Copenhagen, Denmark (Iversen, 1937).

Table 10.1: Tentative correlation of shoreline and sedimentary transgressive overlap data from southern Sweden and interbedded/lacustrine – marine/brackish sediments, Denmark

KATTEGAT		SØBORG SØ		SIRETORP		
Shoreline (1969)	Transgressive overlap (1976b)	Brackish – Marine phases 1937	1976b	^{14}C age of maxima (years BP)	Transgressive overlap (1971)	^{14}C ages stradling maxima (years BP)
PTM 2	2	I	I	7,000	I	6,950–6,650
PTM 3	3A	IIa	IIa	6,450 }	II	6,450–6,250
PTM 4A	3B	IIb	IIb	6,250 }		
PTM 4B	4	III	III	5,850	{ III IV	5,850–5,550 5,450–5,250
PTM 5A	5A	IV	IV	4,980		
PTM 5B	5B	IVb	V	4,600	V	4,650–4,450
PTM 6	6		VI	4,050	VI	4,050–3,850
PTM 7	7		VII	3,500		
PTM 8	8			2,300		
PTM 9	9			1,600		
PTM 10	10			1,000		

Sources: Based upon data from the Kattegat (Mörner, 1969, 1976b), Søborg Sø (Iversen, 1937; Mörner, 1976b) and Siretorp, Blekinge (Berglund, 1971).

Explanation of the sea-level evidence from southern Sweden as the product of discrete 'rises and falls' in water level is not accepted universally. The data have been interpreted variably as the product of localised variations in crustal uplift, dating errors, sedimentary changes and bio-stratigraphic misinterpretation (Eronen, 1974). Such criticisms have also been levelled at the recognition of separate phases of marine inundation on the open southwest coast of Norway. Formerly, two phases of raised relative sea level have been identified in some western and southwest areas (Faegri, 1944), reaching a height of 3 m asl between 6,000 and 5,500 BP and a later elevation of ~ 8 m asl between 4,500-4,000 BP. These changes are now re-interpreted as representing only a single phase of relative sea-level rise, the 'Tapes marine Transgression' (Kaland et al., 1984; Hafsten, 1979, 1983; Kaland, 1984). This major phase of sea-level rise is equated with the Litorina stage of the Baltic, reaching a maximum shoreline elevation of 11-12 m asl on the isostatically rising west coast. The long-term pattern of Holocene sea-level recovery for the south of Norway as a whole is, as in the Baltic, dominated by land emergence. The highest marine shorelines, of Late Weichselian age, lie to the east in areas of maximum ice loading, occurring at ~ 221 m asl in the Oslo fjord area (Fig. 10.10) and at ~ 175 m asl around Trondheims fjord. Glacio-isostatic uplift here exceeded the rates of sea-level rise throughout the Holocene. Further west, land uplift has not been as pronounced and the marine limit lies at levels of ~ 35-44 m asl in western Norway (Hafsten, 1979). In the southwest the highest shorelines at ~ 8 m asl in Lista province (Fig. 10.10) are of Holocene age. Lateglacial (12,000-10,500 BP) marine flooding of this west-southwestern area (Kaland et al., 1984; Kaland, 1984; Krzywinski and Stabell, 1984) was replaced by a time of rapid relative sea-level fall (~ 30 m), the minimum shoreline level of this phase reaching 4 m bsl in Jaeren province, but rising northwards and westwards to 4/5 m asl in Sotra and Bømlo provinces (Fig. 10.10). In these western areas marine deposits interleave terrestrial–freshwater peat/lacustrine muds within emergent coastal lake basins (Kaland, 1984) and these evidence the concluding stage (Tapes) in sea-level recovery (~ 8,500-7,000 BP). After ~ 6,000 BP relative sea level is again interpreted, as in the Baltic, as falling from ~ 12 m asl here to present levels.

Northern Norway, Spitzbergen, Greenland–Arctic Islands

Evidence from the Norwegian Sea (Kellogg, 1976; Jansen and Erlenkeuser, 1985) suggests that a broad belt of cold open water was maintained between Norway and Spitzbergen during the Late Weichselian. However, glaciomarine sediments from the Barents Sea and Norwegian continental shelves (Boulton, 1979b) indicate the widespread influence of the region's continental ice sheet. Foraminiferal and related studies of deep-sea cores from both these areas and the Greenland Sea (Kellogg, 1976; Hald and

Figure 10.10: Late Weichselian and Holocene shoreline displacement curves from southern Norway. The dotted line on the map shows the Younger Dryas glacial limit. On the graphs, periods of ice cover are shaded. ML = the marine limit.

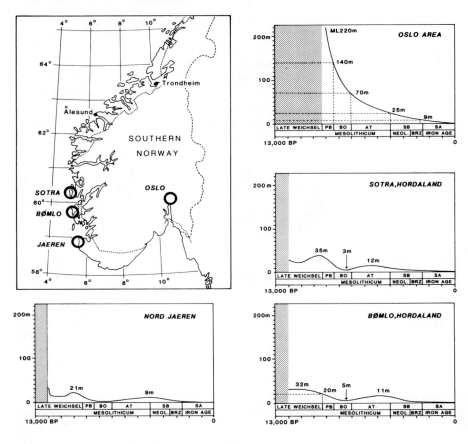

Sources: Derived from Hafsten (1983); Kaland (1984); Kaland *et al.* (1984).

Vorren, 1984; Peacock, 1983) show the subsequent early (pre-10,000 BP) influence of ocean warming. Cold glacier-influenced water from Norwegian shores is progressively replaced by warmer water of the Norwegian Current from the Atlantic. Shoreline sequences formed at this time in northern Norway and Spitzbergen are complex, representing a fine balance between localised regional glacier movements, glacio-isostatic uplift and sea-level rise. Marine limits of Late Weichselian age occur at ~ 87-96 m asl (Rose, 1978), in the Bugöyfjord and Varangerfjord areas (Fig. 10.11), although large areal differences in elevation exist and shorelines are

312

Figure 10.11: The Arctic. Inset map shows tentative shoreline emergence curves for sites referred to in the text — northern Norway (Synge, 1980), Spitzbergen (Feyling-Hanssen and Olsson, 1959; Boulton, 1979), West Greenland (Clark, 1976), Ellesmere Is. (Blake, 1975), Boothia peninsula, Igloolik Is. and Outer fjords, Baffin Is. (Walcott, 1972b).

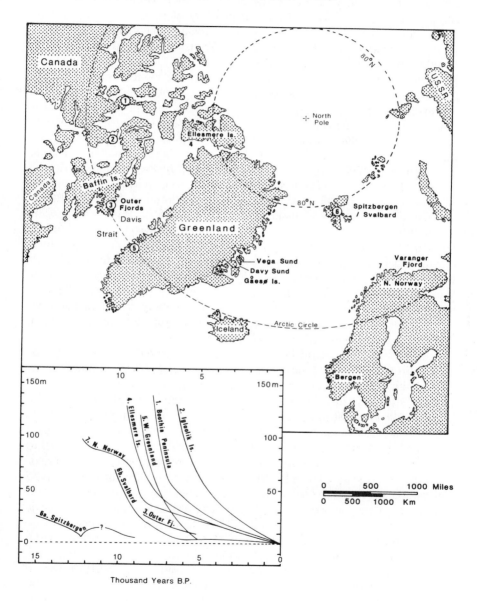

metachronous (Synge, 1980; Sollid *et al.*, 1973). Spectacular 'staircase' beach ridge–shore platform sequences record the subsequent pattern of predominant rapid land emergence, with Holocene isobases along this northern coast dipping steeply in a northerly direction (Sollid *et al.*, 1973). Phases of marine incursion interrupt this land uplift, occurring in the areas studied by Rose (1978) as the result possibly of reduced glacio-isostatic uplift rates consequent upon local ice advance/thickening. The main Holocene relative sea-level rise ('Tapes Transgression') in the region is timed between 6,000-7,500 BP (Donner *et al.*, 1977). It is represented at ~ 27-30 m asl by massive shoreline notches and the highest limit of drifted pumice in beach sediments (Rose, 1978; Synge 1980).

Further north in Spitzbergen (Fig. 10.11), marine limits are highest at ~ 130 m asl outside the Billefjorden stage moraine limits, central Spitzbergen (Linder *et al.*, 1984; Boulton, 1979b). A drop to levels of 80-90 m asl inside the moraine marks the point of local deglaciation, timed at ~ 10,000 BP (Boulton, 1979b). The subsequent shoreline history resembles that of northern Norway, in so far as showing a rapid but decelerating rate of land emergence from 10,000 BP to the present, with shorelines from ~ 4,000 BP at ~ 6 m asl (Feyling-Hanssen and Olsson, 1959). Phases of marine inundation are also recorded from Mid-Holocene time (6,000-4,000 BP) by Hyvärinen (1969) and Feyling-Hanssen (1964a, b). Controversy does, however, centre on the location of ice loading and deglacial history. Schytt *et al.* (1968) and Hughes *et al.* (1977) suggest the existence of a Late Weichselian Barents Shelf ice sheet to the east of Spitzbergen. Detailed analysis of Holocene shoreline data from the Spitzbergen archipelago does not support this view, but shows ice loading to be centred over Spitzbergen itself (Birkenmaier and Olsson, 1971; Boulton, 1979b).

Studies from Greenland indicate an early and rapid rate of postglacial land uplift (100 m a^{-1000}), between 9,000-7,000 BP (Washburn and Stuiver, 1962) with the ice sheet reaching its present position by ~ 5,500 BP. In eastern Greenland the highest marine limits exceed 65 m asl (Hjort, 1973a, b, 1979, 1981) whilst in the west shorelines indicate land uplift of ~ 135 m (Ten Brink, 1974; Clark, 1976) (Fig. 10.11). This rapid rate of land emergence (1.5 times that on Baffin Island, Canada) had been attributed to differences in mantle viscosity and crustal strength between the areas. Subsequent work by Clark (1976) concludes that ~ 54 m of the observed emergence in western Greenland results from the instantaneous uplift effects of collapse in ice–sea surface gravity attraction and crustal elastic rebound on ice wasting (Andrews, 1970), factors which should also be considered in other areas. Evidence of Holocene marine inundation from the coastline is limited. On Gåseø Island in the Vega Sund area (Fig. 10.11) marine sediments in a low cliff section (+4-5 m HWM) do record phases of relative sea-level change since deglaciation of the area ~ 7,000 BP. A transgressive onlap sediment sequence represents the main phase of

314

marine inundation, 'the Vega Transgression', beginning before ~ 5,640 BP and reaching a relative height of +6-8 m HWM before glacio-isostasy re-established a pattern of falling relative sea level.

Across the Davis Strait (Fig. 10.11) on the Arctic Islands of Canada work has been more extensive (Andrews, 1970, 1980; England and Andrews, 1973; Blake, 1964, 1970; Dyke, 1974; Løken, 1965), although many imponderables in the sea-level pattern remain (see Andrews, this vol. Ch. 4). Ice only reached the outer coast of Baffin Island in some places during the Late Wisconsin and was separated from the Laurentide ice sheet at an early stage (Andrews, 1980). The major phases of deglaciation here began at ~ 8,000 BP. In the Arctic Islands, maximum shoreline elevation at 8,000-9,000 BP lies in the range 120-130 m asl (Blake, 1964). In eastern Baffin Island the marine limit at the Late Wisconsin ice margin occurs commonly at 70-80 m asl. This dips seaward, with the zero isobase commonly passing offshore in the outer fjord zone (Dyke, 1974). Maximum heights on earlier shorelines (9,500-10,000 BP) here at the open coast lie at 46 m asl at Scott Inlet. The Holocene in the Baffin Island region probably opened with marine flooding of the isostatically depressed land surface (Andrews, 1980). This, the 'Cape Adair Transgression', extended to a height of 5-7 m asl on Broughton Island and is dated here to 9,850 ± 250 BP. Subsequent seaward removal of the marine effect, the 'Inugsuin Regression', took place between 8,000 and 9,000 BP, consequent upon glacio-isostatic land uplift. Overlap of shoreline elevations in the period 8,800-7,800/7,000 BP may result from initially slowed uplift rates follow-ing ice readvance stability at this time. Unlike the Vega Sund area, Green-land, no stratigraphic evidence has been found here for renewed phases of marine incursion during this period of relative sea-level fall (Andrews, 1980). Some evidence exists possibly, based on a closely ^{14}C controlled emergence curve from Ellesmere Island (Blake, 1975) for small-scale < ± 2 m oscillations of sea level over < 500-year periods between 8,000 and 3,500 BP. However, Andrews (1980) favours changes in storminess/ effective wave action and sediment influx changes as explanations of major shoreline features formed during this phase, rather than glacio-isostatic/-eustatic factors (Hillaire-Marcel and Ochietti, 1977; Blake, 1975). Since ~ 1,000 BP, submergence and a relative rise in sea level of ~ 1 m has taken place. Although the causes are unclear (Andrews, 1980), this may repre-sent the north and westward migration of the collapsing ice margin forebulge (Grant, 1980, Newman et al., 1980a, b).

Northeast Canada

It is estimated that the appearance and wasting of the Laurentide ice sheet that covered the region accounted for 60-80 per cent of Quaternary global ocean water volume change (Andrews, 1970; see also discussion in Andrews, this vol., Ch. 4). The task of modelling the changes in ice sheet dimension has been complex and has given rise to many problems in explanation of both regional palaeogeographic and regional-hemispherical patterns of Holocene sea-level recovery. Study of these problems has given rise to a number of fundamental papers in modelling relative land — sea-level behaviour (Daly, 1934; Andrews, 1970; Andrews and Peltier, 1976; Peltier and Andrews, 1976, 1983; Walcott, 1972a, b; Peltier et al., 1978; Clark, 1977, 1980). More extensive discussion of this work and the region overall has been given in Chapters 3 and 4. Key stratigraphic and shoreline studies have been summarised by Andrews (1970) and Peltier and Andrews (1983).

Based on these sources, reconstruction of the regional palaeogeography shows marine invasion of the St Lawrence valley ('Champlain Transgression') at ~ 11,500-12,000 BP, with approximately half of Canada east of the Rockies affected by marine submergence between 7,000 and 8,000 BP (Andrews, 1970). Shoreline uplift patterns (Peltier and Andrews, 1976) show the dissolution of the Laurentide ice sheet over Hudson Bay into two major ice centres between 8,000 and 10,000 BP (Fig. 10.12a), supporting earlier similar ice sheet models (Tyrrell, 1898; Lee, 1960). Rising post-glacial sea level led to marine incursion (Tyrrell Sea phase) between the two ice centres along the Hudson Strait at 8,100-8,300 BP. Ice readvance (Cockrane Ice advance) from a residual ice mass still over west-central Hudson Bay occurred between 8,000 and 8,500 BP. At ~ 8,000 BP the Tyrrell Sea broke southward into glacial Lake Obijway to create a deep (250-300 m) marine environment over an expanded Hudson Bay area. Clearance of the Foxe and Hudson Basins of ice took place rapidly and was complete by ~ 7,000 BP with final disappearance of the New Quebec (Labrador) ice by 6,500-6,800 BP (Hillaire-Marcel, 1980). Subsequent land emergence has been rapid. Analysis of raised 'staircase' shorelines (see Peltier, this vol., Fig. 3.3) from Hudson Bay — Newfoundland shows beaches at 120-140 m asl for ~ 7,000 BP. A decline in their height southeastwards across Ungava Bay to values of 25 m at outer Hamilton Inlet indicates a declining glacio-isostatic influence in this direction. Shore-lines of this age at 180 and > 200 m asl show maximum uplift for the region at Southampton Island and Richmond Gulf respectively (Fig. 10.12b). At Richmond Gulf the emergence rate declines from 6 cm a^{-1} at 8,000 BP to present values of 1.1 cm a^{-1} (Hillaire-Marcel, 1980). Inflection in the rate of emergence curve here at 6,000 BP is thought to reflect earth mantle responses to shifts in ice loading/deglaciation together with the factors of hydro-isostatic loading of the Tyrrell Sea and possibly local crustal anomalies (Peltier and Andrews, 1976). The 185 raised beaches recognised in the spectacular 'staircase' sequence from the area are now interpreted as indicative of a 45-year climatic cycle in beach development. This may result from fluctuating storm and high tide maxima superimposed on the glacio-isostatic uplift pattern (Hillaire-Marcel, 1980; Fairbridge,

Figure 10.12: (a) Deglaciation of the Hudson Bay region showing suggested ice margins and palaeogeography at ~ 8,000 BP. (b) Shoreline emergence curves since deglaciation from sites (A–H) around Hudson Bay. A, outer Hamilton Inlet; B, inner Hamilton Inlet; C, Cape Henrietta Maria; D, Ottawa Islands; E, Richmond Gulf; F, Ungava Bay; G, Southampton Is.; H, Churchill

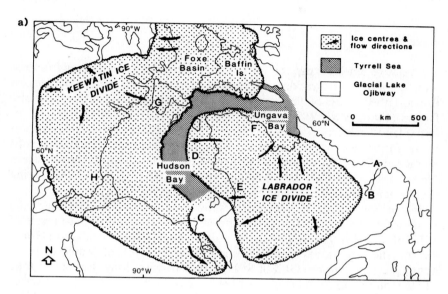

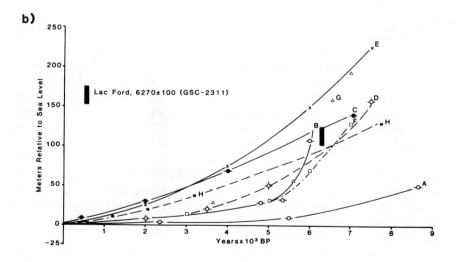

Source: Adapted from Peltier and Andrews (1983).

1983). Despite a decline in emergence rates toward the present, historical records and tide gauges show that relative sea level here is still falling. Estimates indicate that a further 160/180 m of recovery remain to be completed, although others place the value higher at 300 ± 120 m (Walcott, 1972b).

Zone I/II

The relative sea-level pattern for regions within this zone derives from the interaction of the migration and collapse of an ice marginal forebulge zone (toward the centre of former ice loading) with glacio-isostatic/-eustatic factors (Clark *et al.*, 1978). Here relative sea level describes a pattern of initial land emergence followed by later submergence. Those areas closest to the former ice margin experience the greatest emergence and least submergence. Two such regions occur here, situated (i) along the northeastern United States and Canadian Atlantic seaboards, and (ii) in areas bordering the former British–Fennoscandian ice sheet (see Fig. 10.6).

In the first region, shoreline data show a distinct NNW-SSE tilt of the land surface with Late Wisconsin shorelines rising northward to levels of 120-150 m asl in the St Lawrence Valley and SE Newfoundland (Grant, 1980). The zero isobase for this time crosses the Gulf of Maine to Nova Scotia, curving into the Gulf of St Lawrence to pass along the coastal zone of Newfoundland Island (Fig. 10.13).[1] Across this boundary a transgressive onlap sequence of beach deposits, interleaved freshwater and brackish-marine sediments thicken southward, marking the differential downwarp of the region to the south. Sea-level data indicate a recent rapid rise of relative sea level since ~ 5,500 BP of ~ 15 m with submergence at rates of 30 to 50 cm a^{-100} since ~ 7,000 BP in the Nova Scotia area (Grant 1970; Kranck, 1972). Biostratigraphic studies from the area of Prince Edward Island show continued modern submergence of this zone and maximum rate of 1.02 m a^{-1000} (Palmer, 1982). Similar, although reduced emergence–subsidence trends are noted along the Massachusetts and New Jersey coasts (Kaye and Barghoorn, 1964; Redfield, 1967; Stuiver and Daddario, 1963; Clark, 1981; Cinquemani *et al.*, 1982). Newman *et al.* (1980a) note that 'many areas north of Cape Hatteras which had been rising previously prior to 6,000 BP have subsequently appeared to be sinking'. Initial marine incursion of the New England area was overtaken by land uplift at ~ 12,500 BP (Stuiver and Borns, 1975). Subsequent gradual marine flooding and land submergence beginning again between 3,000 and 4,000 BP along the coasts of Maine (Bloom, 1984).

In Northwest Europe much of the eastern and northern margins of the forebulge zone lie beneath the waters of the Atlantic ocean–Norwegian Sea area and a detailed sea-level record does not exist. Off the coast of

318

Figure 10.13: St. Lawrence — Maritime provinces of Canada showing shoreline isobase lines for the Late Wisconsin marine limit

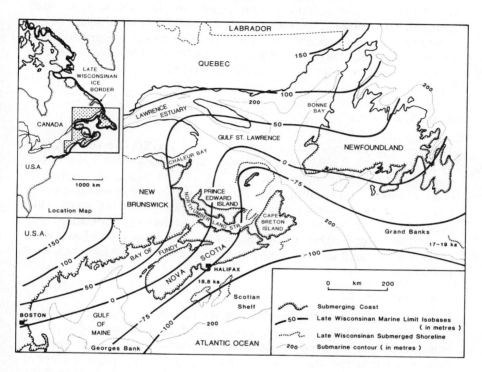

Source: Adapted from Grant (1980).

north Norway the predicted relative sea-level rise for sites close to the former ice margin (67°N, 12°E) (Clark *et al.*, 1978) shows an early rapid uplift of shorelines at ~ 18,000 BP to ~ 58 m asl, followed by submergence after ~ 12,500 BP of levels at ~ −13 m msl. In southeast Scotland a widespread planation surface, falling southeastwards from +6 m OD at Grangemouth through −9/10 m OD at Edinburgh to possibly correlatable levels of −18/27 m OD southward and dated to ~ 10,500 BP (Jardine, 1979), may mark broadly the line of subsidence to the collapsing forebulge zone of the central North Sea. Further north the Orkney and Shetland area may have been situated atop the forebulge itself (Flinn, 1964), undergoing uplift during glacial loading. Fragmentary evidence indicates subsequent submergence here from at least the Mid-Holocene following forebulge collapse (6,000-8,000 BP onwards; Walcott, 1972a). Peat at ~ −9 m OD beneath marine sediments at Whalsay, Shetland Island shows the position of relative sea level at ~ 6,670 BP (Hoppe *et al.*, 1965; Engstrand, 1967).

319

Submergence is similarly indicated for the Outer Hebrides, although the area lies within the maximum Devensian ice limits (see Fig. 10.6). Extensive beach deposits around present shores are thought to have been recruited with rising sea level between 8,400 and 5,500 BP (Ritchie, 1972). Submerged organic deposits from intertidal areas on the western coasts of Lewis and South Uist (von Weymarn, 1974; Ritchie, 1966) indicate gradual marine incursion from ~ 5,700 BP onwards.

Zone II

The record of sea-level change in this zone (see Fig. 10.4) is characterised by a pattern of land submergence throughout the Holocene. Crustal downwarping, consequent upon the collapse of ice marginal forebulges, combined with postglacial ocean volume recovery, result in maximising the effects of marine flooding phases on coastlines within the zone. This, together with its consequences for the large human population now concentrated upon these coastlines, makes the zone one of the most important for studying the processes and patterns of Holocene sea-level change.

The North Sea

Extensive sea-level research has been undertaken in this region (Oele *et al.*, 1979; Greensmith and Tooley, 1982) based on a variety of sea-level indicators, including those of marine erosion surfaces, palaeo-river valley–canyon systems and interleaved marine inorganic and freshwater-terrestrial organic sediment sequences. From these data sources it is clear that the North Sea has long formed a focus for crustal downwarping (Ziegler, 1982). In the short term, subsidence has been controlled by forebulge collapse. Although this process was largely complete ~ 3,000 years ago (Fairbridge, 1983) subsidence persists in the southern North Sea with continued sediment loading from inflowing rivers.[2] Based on geological-geophysical data and reconstructed shoreline tilt lines, Mörner (1979a, 1980b) estimates that the region has undergone ~ 170 m of 'absolute subsidence' (including calibration for ocean water-volume rise) since deglaciation. A minimum value of ~ 25 m for the isostatic subsidence component has been given for areas in the northern North Sea (Jansen, 1976). At the opening of the Holocene (~ 10,300 BP) relative sea level in the region lay at heights of −60 to −70 m OD (Jardine, 1979, 1982; Jelgersma, 1979), the contemporary coastline lying across present central-northern areas (Fig. 10.14). Behind the coast the bulk of the region formed a terrestrial zone cut by northward and southwestward (via the English Channel) flowing rivers. Areas of 'cover sand' and glacial sediments from the previous cold stage formed zones of relative upland, the more low-

Figure 10.14: Holocene shoreline changes in the North Sea Region, showing approximate (hypothetical) shoreline positions for (1) 10,000–10,300 BP, (2) 8,700 BP and (3) 7,800 BP with sea level at ~ 65, ~ 36 and ~ 20 m respectively below present mean sea level.

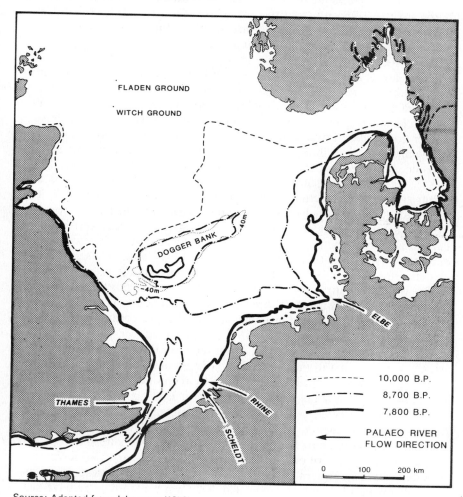

Source: Adapted from Jelgersma (1979).

lying, less well drained areas dominated by freshwater bogs/fen and lakes (Jelgersma, 1979; Jelgersma *et al.*, 1979). Further north in the present Witch and Fladen Grounds (Fig. 10.14) laminated marine clays and fine sands (Jansen, 1976; Jansen *et al.*, 1979), reaching a maximum thickness of ~ 15 m, represent a former cool/Arctic environment and evidence an earlier phase of inundation of the region. Erosion channels developed in the underlying Fladen deposits to depths of −110 m OD mark the known bases of relative sea level in the area (Jardine, 1979).

The subsequent pattern of marine flooding across the depressed land surface shows that incursion took place from two main directions: from the N toward the SWS and from the SW to NE. By ~ 8,700 BP ocean water recovery had displaced the northern shorelines further southward to the area of the Dogger Bank. Freshwater peats dated to between 8,900 and 9,900 BP record the rise of relative sea level to heights above ~ −46 m OD, reaching levels of ~ −36 m by 8,700 BP in the Leman and Ower Banks area (Behre *et al.*, 1979; Jelgersma *et al.*, 1979; Behre *et al.*, 1985). Offshore linear sand ridges developed parallel to the coastline, whilst erosion channels dated to this time from the Outer Silver Pit may represent former tidal gullies. Further south marine incursion had penetrated through the Straits of Dover into the area of the Southern Bight, with a marine influence recorded in the Thames estuary before 8,000 BP (Devoy, 1979, 1982). Rapid submergence of the region continued between 7,500 and 8,700 BP, the coast reaching a point close to its present position by 7,500-7,800 BP (Fig. 10.14). Linear sand-ridges again paralleled the coast in western areas marking the line of developing tidal stream and offshore current activity. Coastal beach-barrier systems were probably also in existence now (van Straaten, 1965; Hageman, 1969) although sedimentary evidence for these is lacking. Between this shore and the present Netherlands−West German coast developed an extensive depositional environment of mudflat and intertidal brackish water sediments. These overlie the basal peat of the region and give way landward to the continued growth of freshwater peat (Jelgersma, 1961, 1979; Hageman, 1969; van de Plassche, 1982) (see Shennan, this vol., Ch. 7). In the river estuaries and other low-lying zones marginal to the North Sea basin accumulation of similar inter-tidal-shallow water deposits begin now to record the effects of marine inundation in these areas (Behre *et al.*, 1979; Ludwig *et al.*, 1981; Krog, 1979; Petersen, 1985; Alderton, 1983; Shennan, 1982; Devoy, 1982; Greensmith and Tucker, 1973; Roeleveld, 1974; Denys, 1985). These sediment sequences show frequently a spatially variable pattern of repeated partial-full transgressive onlap alternating with regressive offlap sediments. The depositional surfaces developed, which are essentially time transgressive (Shennan, 1983 and this vol., Ch. 7), form a record of relative sea-level changes (Fig. 10.15). Further, the record forms a major focus for controversy on the nature of postglacial sea-level recovery, arguments centering on whether sequences represent the result of real sea-level 'oscillations', sedimentary influx changes, storminess-increased storm surge effects, climatically induced changes in river discharge, or a combination of these factors (Devoy, 1982).

The final phase in marine flooding took place more slowly after 6,000 BP (Fig. 10.15). Progressive onshore migration of the developing beach-

Figure 10.15: Time-depth plot of sea-level index points from southeast England, giving an estimate of past MHWST levels. The box around each point represents ± 1σ of the ^{14}C date, whilst the horizontal line gives the original height of the sample, the box here giving the possible vertical error in elevation. The plot serves to illustrate the trend of sea-level rise commonly found in the southern North Sea zone (see text for further references).

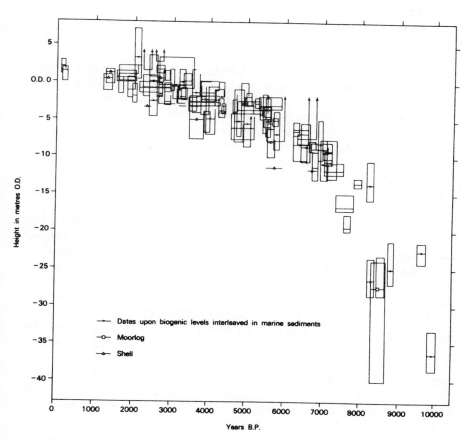

Source: After Devoy (1982).

barrier shoreline resulted in a broad landward progradation of the sediment sequence and partial to complete erosion of the exposed earlier brackish water sediments. On the Dutch coast sediments of the first preserved beach barriers formed at 4,500-5,000 BP. Stabilisation of these barriers and continued recruitment of sand to the shore resulted in development before ~ 4,100 BP of the 'Older Dune' series above these, a later 'Younger Dune' series forming by 800-900 BP (Jelgersma *et al.*, 1970). Elsewhere the slowed rate of sea-level rise, a complex of local depositional tidal–river discharge factors and progressively the influence of man, has resulted in a less clear and spatially varied pattern of coastal development.

Eastern coastline of the United States

Across the North Atlantic, the eastern seaboard of the USA from New England to Florida forms a comparable region of persistent Holocene submergence. The existence on the continental shelf here of marine erosion surfaces/shore notches at levels of -120 ± 30 m msl (Bloom, 1983), fresh-water peat, beach deposits and 'bedded' marine shells to depths $\geqslant -50$ m msl (Milliman and Emery, 1968; Emery, 1969; Emery and Garrison, 1967)[3] together mark the initial submergence of this coastline from 10,000-14,000 BP. Further, seismic profiling and stratigraphic work show, as in Northwest Europe, the existence of buried palaeovalley systems on the continental shelf. These evidence the seaward extension and migration of rivers such as the Hudson and Delaware to levels of -70 to -90 m msl at times of lower sea level (Bloom, 1984; Swift *et al.*, 1980). At $\sim 10,000$ BP relative sea level is estimated to have reached -25 to -32 m msl along the coastline (Bloom, 1984). An illustration of the broad pattern of Holocene shoreline positions has been given by Emery (1969) and others, although this cannot serve as a true palaeogeographic map (Fig. 10.16). For earlier, deeper shorelines the available radiocarbon data and accompanying shore-line indicators are often untrustworthy, due to problems in the analytical techniques used, data quality and retrieval (Bloom, 1983). Predictions based on ice–sea-level models (Clark, 1981; Cathles, 1975) show, however, a southward decline in the $\sim 12,000$ BP shoreline height from -50 m msl off New Jersey to -90 m msl off Florida, but rising to above present sea level northward into Maine and Zone I/II. Support for this predicted pattern comes from sea-level data in areas southeast of New England (Oldale and O'Hara, 1980) and from other parts of the conti-nental shelf (Cinquemani *et al.*, 1982). Together these indicate relative sea-level as lying at $-70/80$ m at $\sim 12,000$ BP, although information from other areas shows a much lower degree of submergence (Blackwelder, 1980). Since $\sim 6,000$ BP the trend of land submergence has continued, with relative sea-level rising variably by 3 to 17 m to present levels at sites between South Carolina to New York (Newman *et al.*, 1980a; Emery and Garrison, 1967; Bloom, 1984). At the interregional level, analysis of tide gauge levelling and sea-level data shows a consistent pattern of crustal downwarping along the coast, centred about a hinge line along the Cape Fear Arch (Fig. 10.16). Sites to the north of this line appear to be differen-tially subsiding relative to areas to the south at rates of ~ 1 m a^{-1000} since Early to Mid-Holocene time (Newman, *et al.*, 1980a, b; Bloom, 1984; Walcott, 1972b). Maximal downwarping since $\sim 12,000$ BP is centred on the area between New York to Delaware, with subsidence continuing in the

Figure 10.16: The United States Atlantic coast, illustrating broad changes in past shoreline position and its future location should all polar-glacial ice melt. Former exposure of the continental shelf is supported by finds of mammoth teeth (△), freshwater 'peats' (·) and oolites (○)

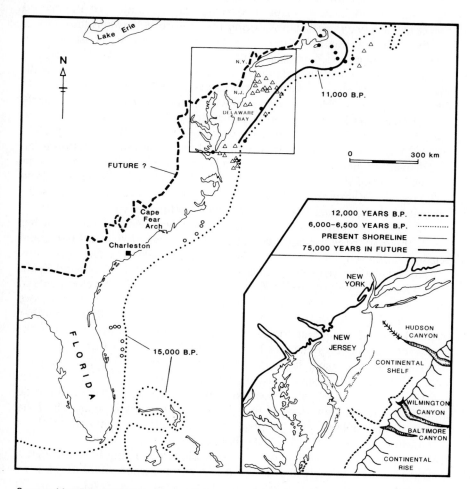

Source: After Emery (1969); inset map after Kraft (1979).

present at rates of 4-6 mm a^{-1} (Gable and Hatton, 1983; Walcott, 1972b). Sea-level data also record a further differential tilting toward the continental shelf edge with downwarping of at least 25 m since 12,000 BP, reaching possibly 40 m since 10,000 BP off Delaware (Belknap and Kraft, 1977; Newman *et al.*, 1980a). Explanation of these variations lies probably in the operation of a complex of factors, the chief ones being glacio-isostatic

325

deformation—forebulge collapse, geosynclinal subsidence (see chapters in Part Two), sediment loading and hydro-isostasy (Bloom, 1967).

Within this framework the pattern of relative sea-level change is recorded variably upon the coastline from a number of different coastal environments. A critical factor in determining the nature of the record preserved has been the balance during the Holocene between the rates of relative submergence and those of sediment influx to the coast (Bloom, 1984; Moslow and Colquhoun, 1981; Orford, this vol., Ch. 13). In New England coastal environments are dominated by mud and fine sized sediments, giving rise to the development of extensive mudflat and saltmarsh. The initial phase of Holocene water-level rise here is recorded commonly by the accumulation of fresh-brackish water peats upon the former land surface (Bloom, 1964; Bloom and Stuiver, 1963; Kaye and Barghoorn, 1964; Davis, 1910). This gives way upward upon marine inundation to an interleaved sequence of peat and estuarine—open water inorganic sediments. Reduced rates of submergence after ~ 3,000 BP to between 0.85 to 1.0 m a^{-1000} (Rampino, 1979), together with a relative increase in sedimentation rates, caused a general regrowth of saltmarsh vegetation over the mudflats, progressing to high tidal marsh and peat accumulation. Further south this pattern is repeated within confined areas of river estuaries and shallow embayed environments. From within Delaware Bay (see Fig. 10.16) detailed stratigraphic studies (Kraft 1971, 1979; Belknap and Kraft, 1977; Kayan and Kraft, 1979; Kraft et al., 1978) date the developments of peat growth over the former Pleistocene sand gravel surface to before 9,718 BP (Fig. 10.17a). Subsequent accumulation of mudflat and other intertidal marsh sediments took place from levels of −30 m msl to present sea level. Sedimentary changes here showing a sequence of transgressive—regressive—transgressive facies are attributed in part to changes in sedimentation rates (Belknap and Kraft, 1977).

Along the South Carolina coast sea-level studies again record estuarine intertidal sediment interleaved with fresh-brackish water peats. These lie above former coastal plain deposits in river valleys north and south of Charleston (see Fig. 10.16) (Colquhoun, 1981a; Colquhoun et al., 1980; Brooks et al., 1979). Relative sea level is shown as rising here through a height of ~ 9 m to the present high marsh surface since 9,000-10,000 BP. The identification of an oscillating pattern for relative sea level from 6,000 BP onward, based on a combination of archaeological, historical and stratigraphical data, is open to question on grounds of methodology. Support for this pattern through correlation with Northwest European records (Colquhoun et al., 1980; Brooks et al., 1979) is similarly dubious, as no attention has been given to the problems of interregional comparisons in levels and possible geoid changes (Newman et al., 1981; Kidson, 1982).

In contrast to these environments the open, mid-Atlantic coast is

Figure 10.17: A stratigraphic section-model of Holocene coastal sedimentary units from the Murderkill river valley and the open Atlantic coast, Delaware (^{14}C dates shown are calculated on a 5,730 a half life. The terms 'Transgression' and 'Regression' relate to phases of assumed relative sea-level rise and fall respectively)

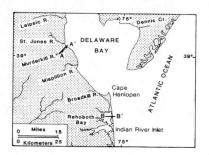

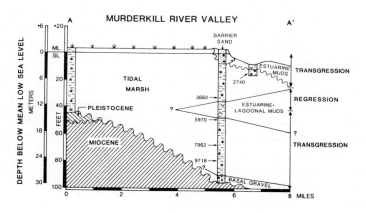

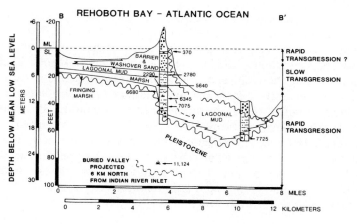

Source: After Belknap and Kraft (1977).

327

characterised by the dominance of sand-gravel sized sediments. Under the influence of Holocene sea-level recovery these have resulted in the formation of migrating beach–barrier coasts, backed by lower energy lagoon and tidal marsh environments (Swift, 1975). In Delaware, sea-level work (Kraft, 1979; Belknap and Kraft, 1977, 1985; Kraft and Chacko, 1979) shows westward advancing transgressive beach–spit–barrier systems (Fig. 10.17b). These form part of a much larger mid-Atlantic shelf transgressive sedimentary unit developed over the past 12,000-14,000 years (Kraft and Chacko, 1979). The Delaware system is bounded on the east by the subsiding Baltimore Canyon trough geosyncline. A reduced rate of fine sediment supply coupled with increased subsidence toward the geosyncline axis control the onshore stratigraphy and sea-level record (Belknap and Kraft, 1977). Further, more detailed discussion of this coastline and offshore zone is given by Orford in Chapter 13 of this volume.

In the recent geological past relative sea-level rise rates have fallen on this central eastern USA seaboard to 15 cm a^{-100} (Kraft and Chacko, 1979). In the past 60 years, however, tide-gauge records show an acceleration to 33 cm a^{-100} (Hicks *et al.*, 1983; Nummedal, 1983). This may be coupled with an average coastal erosion rate of 1.5 m a^{-1} (Dolan *et al.*, 1979) on Atlantic coasts, reaching maximum values of 6 m a^{-1} on the Delaware estuary coast (Kraft, 1984). This apparent increase in the rate of sea-level rise and erosion is thought to result from a combination of factors, including long-term sea-level rise itself, storminess, tidal-coastal shape changes and particularly the influence of man. Whatever the causes, the changes have caused great concern amongst those involved in coastal management and this is reflected in surveys by the US Environmental Protection Agency attempting to predict short-term sea-level changes (Hoffman *et al.*, 1983; Titus *et al.*, 1984; Titus *et al.*, 1985) (see Titus, this vol., Ch. 15).

South of this region greater tectonic stability, steady sediment supply and a reduced rate of relative sea-level rise after 3,000-4,000 BP have been important in developing a regressive barrier island sediment sequence (Moslow and Colquhoun, 1981). In the Kiawah and Seabrook island area of South Carolina the Holocene stratigraphy shows initial landward migration of a transgressive beach–barrier sediment sequence between 4,000 and 5,800 BP, similar to that shown in the Delaware model. Protected back barrier marsh sediments formed, coupled with an eroding shore face environment. Between 3,500 and 4,000 BP shore face accretion began against the primary barrier as a series of beach ridges, leading to seaward progradation of the barrier system and expansion of the tidal flat–saltmarsh. After 3,500 BP progradation continued periodically forming a beach ridge complex 2-3 km wide.

Zones III and VI

Coastlines within Zone III occur in the southern part of the North Atlantic region (see Fig. 10.4). Prediction of relative sea-level behaviour here (Clark *et al.*, 1978) shows a pattern of initial relative land submergence on deglaciation followed by slight land emergence, with shorelines formed after 11,000 BP reaching maximum elevations of ≤ 0.5 m above present sea level. The pattern is again conditioned by the time-dependent shape of the migrating ice marginal forebulge areas together with the redistribution of water load on the earth's surface following ice melt (Clark and Lingle, 1977, 1979). The timing of the change in direction and scale of relative land movements depends on location of the coastline. Final emergence is progressively delayed for sites closer to the former ice margins, although these should also record the maximum degree of eventual uplift. In reality, the geological record of shoreline evolution does not support exactly the predicted pattern. Many areas show relative land submergence throughout the Holocene (Coleman and Smith, 1964; Curray, 1960; Emery and Garrison, 1967) although at greatly reduced rates toward the present day (Scholl and Stuiver, 1967). Apart from inaccuracies within the model itself, explanation of this disparity may lie in the masking of the net low amplitude vertical movements by the effects of local and regional subsidence factors.

At the northwest margin of the zone inundation of the Florida continental shelf, from below levels of −21 m msl at 12,000 BP (Clausen *et al.*, 1979) across an emerged Karstic landscape, developed during the previous cold stage (Bloom, 1984). Marine flooding of the limestone depressions and cavern systems gave rise to the now famous 'blue-holes' of the Florida 'reef'. At the shelf edge initial submergence of this carbonate dominated environment allowed development of a series of narrow coral reefs before 9,000 BP. Subsequent increase in turbidity, through erosion and reworking of shelf sediments as sea-level rose, resulted probably in their death between 7,000 and 9,000 BP. By 7,000 BP relative sea level is indicated as lying at ~ 4 m below the regional mean sea level (msl) and thereafter rising slowly to present levels (Fig. 10.18) (Scholl and Stuiver, 1967; Scholl *et al.*, 1970). Further west, into the Gulf of Mexico, this picture of relative shoreline stability changes under the influence of tectonic downwarping and sediment loading from inflowing rivers. The location of the Mississippi and its accompanying delta which dominates the centre of this region, is controlled by an underlying zone of thinned and fractured earth crust (Sleep and Snell, 1976). Thermal cooling of this crust over the last 20 Ma coupled with surface fluvial and deltaic loading has resulted in the relative subsidence of this area throughout the Holocene. The tectono-isostatic components of this subsidence has been estimated at ~ 150 m since ~ 10,000 BP (Fairbridge, 1983), based on the 200 m thick sequence of Holo-

Figure 10.18: Gulf of Mexico, showing sea-level date plotted for sites within the region: (a) ¹⁴C ages on shells of nearshore environments from the continental shelf west of 93°W. Vertical and horizontal bars give possible depth and age ranges (Curray, 1960); (b) Louisiana (submerged peats) (Coleman and Smith, 1964); (c) Sabine — High Island area (Nelson and Bray, 1970); (d) South Florida (Scholl *et al.*, 1970). Lines, where shown on the time-depth plots (c) and (d), should be regarded only as authors' estimated trends on relative land/sea-level movements

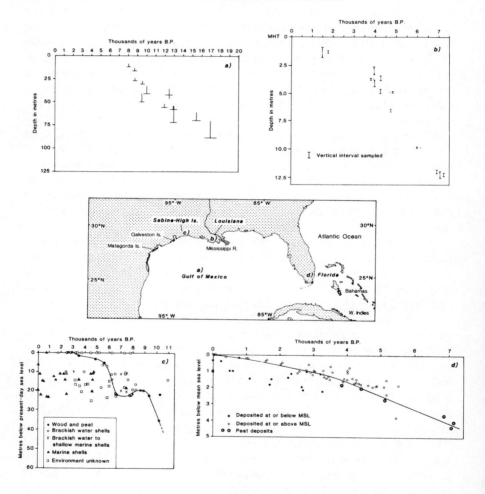

cene sediments found in the centre of the delta. Analysis of sea-level indicators from the delta and neighbouring Louisiana coast shows relative sea level lying here at ~ −40 m at ~ 7,000 BP (Coleman and Smith, 1964; Gould and McFarlan, 1959; McFarlan, 1961). The wide scatter of time-depth points from the delta itself is attributed to the effects of subsidence

330

and sediment compaction (McFarlan, 1961). Study of the delta's depositional record shows the development since the Mid-Holocene ($\sim$ 5,000 BP onward) of six overlapping sedimentary lobes (Kolb and van Lopik, 1966; Morgan, 1970). Authorities differ as to the significance of these for the sea-level record, sedimentologists seeing the lobes as a function of random changes in river load and the point of principal sediment discharge. Alternatively, Fairbridge (1968, 1983) interprets the lobes as a response to an 'oscillating' pattern of Holocene sea-level recovery. Peat and soil cover develop over the surface of each lobe during a phase of stillstand, or falling relative sea level. Subsequent subsidence of each lobe coupled with renewed relative sea-level rise allows progradation of the delta and initiation of a new lobe. Continued deposition had led, by the Late Holocene (after 550 BP), to progradation of the delta across the continental shelf, allowing sediment discharge now directly onto the continental slope (Morgan, 1970).

During the maximum lowering of relative sea level in the Wisconsin cold stage, shorelines were displaced across the western shelf areas of the Gulf of Mexico to levels of $\leqslant$ -80 m msl, although lower elevations may have been reached (Bloom, 1984; Curray, 1960; Emery and Garrison, 1967). Remnant buried channels evidence the extension of the Mississippi and other rivers across the exposed Coastal Plain (Fisk and McFarlan, 1955; Wilkinson and Basse, 1978). Away from the Mississippi area the influences of sediments and tectonic subsidence decline. Study of *in situ* marine shells and fresh-brackish water peats interleaved in the marine sediments from the Texas shelf, show relative sea-level rising from a minimum of $\sim$ -70 m at 17,000 BP, to reach a position close to present levels only late in the Holocene (Fig. 10.18) (Curray, 1960; Nelson and Bray, 1970). Progressive submergence led, as on the open Atlantic coast, to recruitment of large quantities of sand and fine sediment. This moved primarily westward across the continental shelf under the processes of littoral drift to form migrating transgressive barrier-beach systems (Curray, 1960; Curray *et al.*, 1969; Otvos, 1981). The present barrier islands on the Texas coast represent the final stage in this process. Studies from the Matagorda (Wilkinson, 1975) and Galveston Island areas (Bernard *et al.*, 1962; Wilkinson and Basse, 1978; Bernard and Le Blanc, 1965; Bernard *et al.*, 1970) show transgressive beach ridges and overlying deltaic back-barrier lagoon sediments, formed behind an earlier stage in barrier development. By $\sim$ 4,000 BP, onshore movement of the back ridges had reached its most landward position. After this time, seaward progradation of the beach ridge system has formed a ridge complex 1.6-4.5 km wide over the last 3,500 years. Re-interpretation of the radiocarbon dating sequence at Galveston Island by Kraft (1978) suggests that progradation started earlier here at $\sim$ 5,300 BP.

One of the principal effects of ice meltwater return to the world's oceans has been an increase in water load on the sea floor and depression of the ocean basins (Clark *et al.*, 1978; Cathles, 1975; Bloom, 1967). This should result in progressive development of crustal tilt across the continental margins toward the ocean basins and the relative emergence of these areas as deglaciation is completed. Areas undergoing such tilt are recognised as forming a separate sea-level signature zone (Zone VI). On many continental edges the effect on shoreline altitude of this process is exaggerated, dampened or offset by overriding tectonic, sediment subsidence or glacio-isostatic effects. On the Atlantic swell wave coasts of west Africa within Zone VI (Fig. 10.19), a widespread pattern of Mid-Holocene emergence is observed and may evidence the operation of this mechanism (Walcott, 1972a; Faure *et al.*, 1980). Sea-level data here is derived commonly from radiocarbon dates on shell and carbonate material of raised shell banks, shoreline terraces, other beach structures and lagoonal deposits. Along the coastlines of Morocco, the western Sahara and Mauritania shoreline studies show that relative sea level had reached heights of -1 to > 2 m above the regional msl between 3,000 and 6,000 BP (Colquhoun, 1981b; Faure and Hébrard, 1977; Einsele *et al.*, 1974; Delibrias, 1974; Elouard *et al.*, 1969, 1977; Elouard, 1968). At Agadir relative sea level is recorded at a height of $\sim +2$ m msl between 4,800 and 6,200 BP (Rhodenburg, 1977). Further south in Mauritania, msl positions are similarly found at elevations of $>$ $+2$ m msl at $\sim$ 5,500 BP (Einsele *et al.*, 1974). From these 'maximum' heights relative sea level seems to have remained variably above present levels during the Late Holocene, although falling progressively toward its current position (Fig. 10.19). Apparent variations in water level evidenced by the data may result from a complex of local-regional scale sedimentary, climatic and isostatic changes (Ortlieb, 1975; Ortlieb and Petit-Maire, 1976; Hébrard, 1972; Petit-Maire *et al.*, 1977). Removal of the marine influence and dessication of coastal/lake environments recorded from sites in Mauritania at $\sim$ 3,500 BP may have been due to changes in wave height and coastal geometry, building blocking dune–beach structures (Hébrard, 1972).

Toward the southeast margin of Zone III and across into Zone IV the picture of Mid- to Late Holocene emergence is repeated (Faure and Elouard, 1967; Faure and Hébrard, 1973; Faure *et al.*, 1980; Tastet, 1975; Pomel, 1979). Detailed sea-level levelling work from the Senegal river and delta areas indicate that relative sea level has remained close to its present position (within 1-2 m) since 6,000 BP (Faure *et al.*, 1980). Fresh-brackish water peats show the sea surface here at $\sim -22 \pm 2$ m (IGN) between $\sim$ 8,450 and 8,870 BP (Monteillet, 1977; Faure *et al.*, 1980) with the palaeo-shoreline some 10 km seaward on the continental shelf. From this position marine flooding had penetrated > 120 km inland along the Senegal river after $\sim$ 7,000 BP, forming shallow bays and lagoons in former Pleistocene sand dunes. Between 1,800-4,000 BP progradation of the coast took place

Figure 10.19: Sea-level signature zonation in the southeast Atlantic–West Africa region. Altitudinal distribution of ^{14}C dated sea-level index points (bars giving age and height errors) are shown for four locations, each showing relative sea-level rising above its present position after ~ 6,000 BP. The lines drawn through the data points are a suggested trend for relative mean sea level in each locality

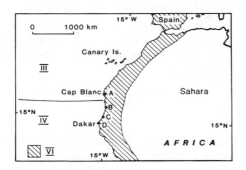

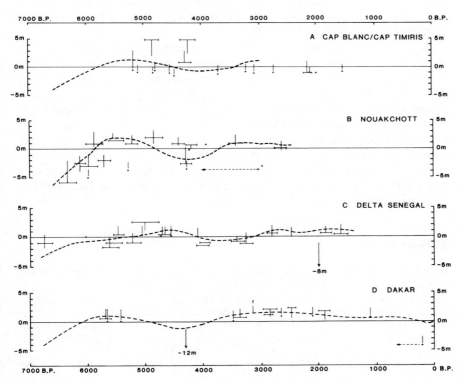

Source: After Faure (1980).

333

under the influence of changing coastal deposition, isolating former brackish-marine environments. After ~ 2,000 BP marine inundation again affected the river estuary. Shoreline transects taken here across the continental edge have been used to test the models of crustal tilt for the region. These predict a total marginal tilt of ~ 5.5 m since 5,500 BP with the land rising by ~ 4 m. The shoreline transects, representing palaeogeoid surfaces, indicate a tilt of $< \pm 1$ m since 6,500 BP although limitations of the size and inequalities of the data base must be borne in mind (Faure *et al.*, 1980). Remarkably, the shoreline planes remain relatively parallel for this period, suggesting that the lithosphere may behave here in a more rigid fashion than the crustal models assume.

CONCLUSION

Within a limited survey it is neither possible nor fair to attempt a characterisation of all coastlines contained in the region. Indeed, discussion may have gone too far as it is! Introduction to information and literature sources on coastal environments that have not been mentioned can be gained from Bloom (1977) and Colquhoun (1981b).

In broad terms the survey does show that relative sea-level behaviour in the region during the Holocene can be explained by the interaction of ocean water volume recovery, tectonic, isostatic and sedimentary factors. Areas formerly covered by ice in the last cold stage have been dominated by the process of glacio-isostatic recovery, causing a pattern of predominant coastal emergence with falling relative sea levels. Away from former ice areas, collapse of the ice marginal forebulge coupled with ocean water-volume rise has led to continuous coastal submergence and a long-term trend of rising relative sea level. Further away still, the effects of forebulge migration and hydro-isostasy result again in a pattern of relative land emergence from the Mid-Holocene onward.

REFERENCES

Alderton, A. (1983) 'Flandrian vegetational history and sea-level change of the Waveney Valley', PhD dissertation, University of Cambridge.
Andrews, J.T. (1970), *A Geomorphological Study of Postglacial Uplift, with Particular Reference to Arctic Canada*, Inst. Br. Geogr. Spec. Publ. No. 2.
——— (1980) 'Progress in relative sea level and ice sheet reconstructions, Baffin Island NWT for the last 125,000 years', in N-A Mörner (ed.) *Earth Rheology, Isostasy and Eustasy*, Wiley, Chichester and New York, pp. 175-200.
——— and Barry, R.G. (1978) 'Glacial inception and disintegration during the last glaciation', *Ann. Rev. Earth Planet. Sci.*, 6, 205-28.
——— Funder, S., Hjort, C. and Imbrie, J. (1974) 'Comparison of the glacial chronology of eastern Baffin Island, East Greenland and the Camp Century accumulation record', *Geology*, 2, 355-8.

—— and Peltier, W.R. (1976) 'Collapse of the Hudson Bay ice centre and glacio-isostatic rebound', *Geology, 4,* 73-5.

Behre, K.-E., Dorjes, I. and Irion, G. (1985) 'A dated Holocene core from the bottom of the southern North Sea', *Eiszeit, Gegenw., 35,* 9-13.

—— Menke, B. and Streif, H. (1979) 'The Quaternary geological development of the German part of the North Sea', in E. Oele, R.T.E. Schüttenhelm and A.J. Wiggers (eds) *The Quaternary History of the North Sea,* Acta. Univ. Ups. Symp. Univ. Ups. Annum Quingentesimum Celebrantis: 2, Uppsala, pp. 85-113.

Belknap, D.F. and Kraft, J.C. (1977) 'Holocene relative sea-level changes and coastal stratigraphic units on the northwest flank of the Baltimore Canyon trough geosyncline', *J. of Sed. Petrol, 47,* 610-29.

—— and Kraft, J.C. (1985) 'Influence of antecedent geology on stratigraphic preservation potential and evolution of Delaware's barrier system', *Mar. Geol., 63,* 235-62.

Berglund, B.E. (1971) 'Litorina transgressions in Blekinge, south Sweden: a preliminary survey', *Geol. Fören. Stockh. Förh., 93,* 625-52.

Bernard, H.A. and Le Blanc, R.J. (1965) 'Résumé of the Quaternary geology of the northwestern Gulf of Mexico Province', in H.E. Wright, Jr and D.G. Frey (eds), *The Quaternary of the United States,* Princeton University Press, Princeton, NJ, pp. 137-85.

Bernard, H., Le Blanc, R.J. and Major, C.F. (1962) 'Recent and Pleistocene geology of southeast Texas', in E.H. Rainwater, and R. Zingula, (eds), *Geology of the Gulf Coast and Central Texas, Guidebook Excursions,* Houston Geological Society, Houston, pp. 175-224.

—— Major, C.F., Jr, Parrott, B.S. and Le Blanc, R.J. (1970) *Recent Sediments of Southeast Texas: A Fieldguide to the Brazos Alluvial and Deltaic Plains and the Galveston Barrier Island Complex. Guidebook II,* University of Texas, Austin.

Binns, P.E., Harland, R. and Hughes, M.J. (1974a) 'Glacial and postglacial sedimentation in the Sea of the Hebrides', *Nature, 248,* 751-4.

—— McQuillin, R. and Kenolty, N. (1974b) 'The geology of the Sea of the Hebrides', *Rept. Inst. Geol. Sci., No. 73/14.*

Birkenmaier, K. and Olsson, I.U. (1971) 'Radiocarbon dating of raised marine terraces at Hornsund, Spitsbergen and the problem of land uplift', *Norsk Polarinstitutt Årbok 1969.*

Blackwelder, B.W. (1980) 'Late Wisconsin and Holocene tectonic stability of the United States mid Atlantic coastal region', *Geology, 8,* 534-7.

Blake, W., Jr (1964) 'Preliminary account of the glacial history of Bathurst Island, Arctic Archipelago', *Geol. Surv. Can. Pap., 64,* 1-8.

—— (1970) 'Studies of glacial history in Arctic Canada, 1, Marine radiocarbon dates and differential postglacial uplift in eastern Queen Elizabeth Island', *Can. J. Earth Sci., 7,* 634-64.

—— (1975) 'Radiocarbon age determinations and postglacial emergence at Cape Storm, southern Ellesmere Island, Arctic Canada', *Geogr. Ann., 57A,* 1-71.

Bloom, A.L. (1964) 'Peat accumulation and compaction in a Connecticut coastal marsh', *J. Sed. Petrol, 34,* 599-603.

—— (1967) 'Pleistocene shorelines: a new test of isostasy', *Bull. Geol. Soc. Am., 78,* 1477-94.

—— (1977) *Atlas of Sea-level Curves,* IGCP Project 61, Cornell University, Ithaca, NY.

—— (1983) 'Sea-level and coastal morphology through the Late Wisconsin glacial maximum', in S.C. Porter (ed.), *Late Quaternary Environments of the United States, Vol. 1. The Late Pleistocene,* Longman, London, pp. 215-29.

—— 'Sea-level and coastal changes', in H.E. Wright (ed.), *Late Quaternary*

Environments of the United States, Vol. 2, The Holocene, Longman, London, pp. 42-51.

—— and Stuiver, M. (1963) 'Submergence of the Connecticut coast', *Science, 139*, 332-4.

Boulton, G.S. (1979a) 'A model of Weichselian glacier variation in the North Atlantic region', *Boreas, 8,* 373-95.

—— (1979b) 'Glacial history of the Spitzbergen archipelago and the problem of a Barents Shelf ice sheet', *Boreas, 8,* 31-57.

Brøgger, W.C. (1900-1) 'Om de senglaciale og postglaciale nivåförändringer i Kristianiafeltet (Molluskfaunaen)', *Nor. Geol. Unders., 31,* 1-731.

Brooks, M.J., Colquhoun, D.J., Pardi, R.R., Newman, W.S. and Albott, W.H. (1979) 'Preliminary archaeological and geological evidence for Holocene sea-level fluctuations in the lower Cooper river valley, S.C.', *The Florida Anthropologist, 32,* 85-103.

Bryson, R.A., Wendland, W.M., Ives, J.D. and Andrews, J.T. (1969 'Radiocarbon isochrones on the disintegration of the Laurentide ice sheet', *Arctic and Alpine Res., 1,* 1-13.

Carter, R.W.G. (1982) 'Sea-level changes in Northern Ireland', *Proc. Geol. Ass., 93,* 7-23.

—— (1983) 'Raised coastal landforms as products of modern process variations and their relevance in eustatic sea-level studies: examples from eastern Ireland', *Boreas, 12,* 167-82.

Caston, V.N.D. (1979) 'A new isopachyte map of the Quaternary of the North Sea', in E. Oele, R.T.E. Schüttenhelm, and A.J. Wiggers, (eds) *The Quaternary History of the North Sea,* Acta Univ. Ups. Symp. Univ. Ups. Annum Quingentesimum Celebrantis: 2, Uppsala, pp. 23-8.

Cathles III, L.M. (1975) *The Viscosity of the Earth's Mantle,* Princeton University Press, Princeton, NJ.

Cinquemani, L.J., Newman, W.S., Sperling, J.A., Marcus, L.F. and Pardi, R.R. (1982) 'Holocene sea-level changes and vertical movements along the east coast of the United States: a preliminary report', in D.J. Colquhoun (ed.), *Holocene Sea-level Fluctuations, Magnitude and Causes,* IGCP Project 61, Department of Geology, University of South Carolina, Columbia, SC, pp. 13-33.

Clark, J.A. (1976) 'Greenland's rapid postglacial emergence: a result of ice — water gravitational attraction', *Geology, 4,* 310-12.

—— (1977) 'Global sea-level changes since the last glacial maximum and sea-level constraints on the ice sheet disintegration history', PhD dissertation, University of Colorado, Boulder, Col.

—— (1980) 'A numerical model of worldwide sea-level changes on a viscoelastic Earth', in N.-A. Mörner (ed.), *Earth Rheology, Isostasy and Eustasy,* Wiley, Chichester and New York, pp. 525-34.

—— (1981) 'Predicted relative sea-level changes, glacial history and neotectonics of eastern North America', Abstract from IGCP Project 61, *Variations in Sea-level in last 15,000 years,* Conference, Department of Geology, University of South Carolina, Columbia, SC.

—— Farell, W.E. and Peltier, W.R. (1978) 'Global changes in postglacial sea level: a numerical calculation', *Quat. Res., 9,* 265-87.

—— and Lingle, C.S. (1977) 'Future sea-level changes due to West Antarctic ice sheet fluctuations', *Nature, 269,* 206-9.

—— and Lingle, C.S. (1979) 'Predicted relative sea-level changes (18,000 years BP to present) caused by lateglacial retreat of the Antarctic ice sheet', *Quat. Res., 11,* 279-98.

Clausen, C.J., Cohen, A.D., Emiliani, C., Holman, J. and Stipp. J.J. (1979) 'Little

Salt Spring, Florida: a unique underwater site', *Science*, *203*, 609-14.

Coleman, J.M. and Smith, W.G. (1964) 'Late recent rise of sea level', *Bull. Geol. Soc. Am.*, *75*, 833-40.

Colquhoun, D.J. (1981a) 'Variation in sea level on the south Carolina Coastal Plain', in D.J. Colquhoun (ed.), *Variation in Sea Level on the South Carolina Coastal Plain*, Department of Geology, University of South Carolina, Columbia, SC, pp. 1-44.

Colquhoun, D.J. (ed.) (1981b) *World Shorelines Map, Euro-African Sector*, INQUA Quaternary Shorelines Commission, Department of Geology, University of South Carolina, Columbia, SC.

Colquhoun, D.J., Brooks, M.J., Abbott, W.H., Stapor, F.W., Newman, W.S. and Pardi, R.R. (1980) 'Principles and problems in establishing a Holocene sea-level curve for South Carolina', in J.D. Howard, C.B. Depratter and R.W. Fey (eds), *Excursions in Southeastern Geology. The Archaeology-Geology of the Georgia Coast*, Guide Book 20, Geol. Soc. Am., pp. 143-59.

Cullingford, R.A., Caseldine, C.J. and Gotts, P.E. (1980) 'Early Flandrian land and sea-level changes in Lower Strathearn', *Nature*, *284*, 159-61.

Curray, J.R. (1960) 'Sediments and history of the Holocene transgression, continental shelf, northwest Gulf of Mexico', in F.P. Shepard, F.B. Phleger and Tj.H. van Andel (eds), *Recent Sediments, Northwest Gulf of Mexico*, Am. Assoc. Petrol Geol., Tulsa, Okla., pp. 221-66.

Curray, J.R., Emmel, F.J. and Crampton, D.J.S. (1969) 'Holocene history of a strand plain, lagoon coast, Nayarit, Mexico', in A.A. Castanares and F.B. Phleger (eds), *Coastal Lagoons — A Symposium*, Universidad Nacional Autónoma, Mexico, pp. 63-100.

Daly, R.A. (1934) *The Changing World of the Ice Age*, Yale University Press, New Haven, Conn.

Davies, H.C., Dobson, M.R. and Whittington, R.J. (1984) 'A revised seismic stratigraphy for Quaternary deposits on the inner continental shelf west of Scotland between 55° 30'N and 57° 30'N, *Boreas*, *13*, 49-66.

Davies, J.L. (1980) *Geographical Variation in Coastal Development*, 2nd edn, Longman, London.

Davis, C.A. (1910) 'Saltmarsh formation near Boston and its geological significance', *Econ. Geol.*, *5*, 623-39.

Defant, A. (1941) 'Untersuchungen zue Statik und Dynamik des Atlantischen Ozeans, 5', *Wissensch. Ergebn. Dtsch. Atlant. Exped. 'Meteor'*, *6*, 191-260.

De Geer, G. (1890) 'Om Skandinaviens nivåförändringar under Quartärperioden', *Geol. Fören. Stockh. Förh.*, *31*, 511-56.

Delibrias, G. (1974) 'Variation du niveau de la mer, sur la côte ouest Africaine, depuis 26,000 ans', *Colloques Int. du CNRS, No. 219*, 127-34.

Denys, L. (1985) 'Diatom analysis of an Atlantic — Sub Boreal core from Slijpe (Western Belgian Coastal Plain)', *Rev. Palaeobot. Palynol.*, *46*, 33-53.

Devoy, R.J. (1979) 'Flandrian sea-level changes and vegetational history of the lower Thames estuary', *Phil. Trans. Roy. Soc. Lond., B.*, *285*, 355-410.

—— (1982) 'Analysis of the geological evidence for Holocene sea-level movements in southeast England', *Proc. Geol. Ass.*, *93*, 65-90.

—— (1983) 'Late Quaternary shorelines in Ireland: an assessment of their implications for isostatic land movement and relative sea-level changes', in D.E. Smith, and A.G. Dawson (eds), *Shorelines and Isostasy*, Academic Press, London and New York, pp. 227-54.

—— (1985) 'Holocene sea-level changes and coastal processes on the south coast of Ireland: corals and the problems of sea-level methodology in temperate waters', *Proc. Fifth Int. Coral Reef Congr. Tahiti, 1985*, *3*, 173-8.

Dickson, J.H., Stewart, D.A., Baxter, M.S., Drndarsky, N.D., Thompson, R., Turner, G. and Rose, J. (1978) 'Palynology, palaeomagnetism and radiometric dating of Flandrian marine and freshwater sediments of Loch Lomond', *Nature*, 274, 548-53.

Dobson, M.R. (1977a) 'The geological structure of the Irish Sea', in C. Kidson and M.J. Tooley (eds), *The Quaternary History of the Irish Sea*, Seal House Press, Liverpool, pp. 13-26.

—— (1977b) 'The history of the Irish Sea basins', in C. Kidson and M.J. Tooley (eds), *The Quaternary History of the Irish Sea*, Seal House Press, Liverpool, pp. 93-8.

—— (1979) 'Aspects of the Post-Permian history of the aseismic continental shelf to the west of the British Isles', in F.T. Banner, M.B. Collins and K.S. Massie (eds), *The Northwest European Shelf Seas: The Seabed and the Sea in Motion, I Geology and Sedimentology*, Elsevier, Amsterdam, pp. 25-41.

Dolan, R., Hayden, B.P., Rea, C. and Heywood, J.E. (1979) 'Shoreline erosion rates along the middle Atlantic coast of the United States', *Geology*, 7, 602-6.

Donner, J.J. (1969) 'Land/sea-level changes in southern Finland during the formation of the Salpausselka end-moraines', *Bull. Geol. Soc. Finland*, 41, 135-50.

—— (1970) 'Land/sea-level changes in Scotland', in D. Walker and R.G. West (eds), *Studies in the Vegetational History of the British Isles*, Cambridge University Press, Cambridge, pp. 23-29.

—— (1978) 'The dating of the levels of the Baltic Ice Lake and the Salpausselka moraines in south Finland', *Soc. Sci. Fennica. Comment. Phys.-Math.*, 48, 11-38.

—— and Eronen, M. (1981) *Excursion Guide*, Stencil 5, Dept. of Geology, University of Helsinki.

—— Eronen, M. and Junger, H. (1977) 'The dating of the Holocene relative sea-level changes in Finnmark in northern Norway', *Norsk Geogr. Tidsskr.*, 20, 1-70.

Dyke, A.S. (1974) 'Deglacial chronology and uplift history: northeastern sector, Laurentide ice sheet', Occ. Pap. 12, Inst. Arctic and Alpine Res., University of Colorado, Boulder, Col.

Einsele, G., Herm, D.F. and Schwarz, H.U. (1974) 'Holocene eustatic sea-level fluctuation at the coast of Mauritania', *Quat. Res.*, 4, 282-9.

Elouard, P.C. (1968) 'Le Nouakchottien, étage du Quaternaire de Mauritanie', *Ann. Fac. Sci. Univ. Dakar, Ser.*, 2, 121-37.

—— Faure, H. and Hébrard, L. (1969) 'Quaternaire du littoral Mauritanien entre Nuakchott et Port-Etienne', *Bull. ASEQUA*, 23, 15-24.

—— Faure, H. and Hébrard, L. (1977) 'Variations du niveau de la mer au cours des 15,000 dernières années de la presqui'île de Cap Vert', *Bull. ASEQUA*, 50, 29-49.

Emery, K.O. (1969) 'The continental shelves', *Sci. Am.*, 221, 107-26.

—— and Aubrey, D.G. (1985) 'Glacial rebound and relative sea levels in Europe from tide gauge records', *Tectonophys.*, 57.

—— and Garrison, L.E. (1967) 'Sea-levels 7,000 to 20,000 years ago', *Science*, 157, 684-7.

England, J.H. and Andrews, J.T. (1973) 'Broughton Island — a reference area for Wisconsin and Holocene chronology and sea-level changes on eastern Baffin Island', *Boreas*, 2, 17-32.

Engstrand, L.G. (1967) 'Stockholm natural radiocarbon measurements VII', *Radiocarbon*, 9, 387-438.

Erikssen, K.G. (1979) 'Late Pleistocene and Holocene shorelines on the Swedish west coast', in E. Oele, R.T.E. Schüttenhelm and A.J. Wiggers (eds), *The*

Quaternary History of the North Sea, Acta Univ. Ups. Symp. Univ. Ups. Annum Quingentesimum Celebrantis: 2, Uppsala, pp. 61-74.

Eronen, M. (1974) 'The history of the Litorina Sea and associated Holocene events', *Comment. Phys.-Math.*, *44*, 79-195.

—— (1976) 'A radiocarbon dated transgression site in southeastern Finland', *Boreas*, *5*, 65-76.

—— (1982) 'The course of shore displacement in Finland', in D.J. Colquhoun (ed.), *Holocene Sea-level Fluctuations, Magnitude and Causes*, IGCP Project 61, Dept. of Geology, University of South Carolina, Columbia, SC, pp. 43-60.

—— (1983) 'Late Weichselian and Holocene shore displacement in Finland', in D.E. Smith and A.G. Dawson (eds), *Shorelines and Isostasy*, Academic Press, London and New York, pp. 183-207.

Faegri, K. (1944) 'Studies on the Pleistocene of western Norway III. Bømlo', *Bergens Mus. Arb. 1943 Naturvitensk. R.*, *8*, 1-100.

Fairbridge, R.W. (ed.) (1968) *The Encyclopedia of Geomorphology*, Reinhold, New York.

—— (1983) 'Isostasy and eustasy', in D.E. Smith and A.G. Dawson (eds), *Shorelines and Isostasy*, Academic Press, London and New York, pp. 3-25.

Faure, H. (1980) 'Late Cenozoic vertical movements in Africa', in N.-A. Mörner (ed.), *Earth Rheology, Isostasy and Eustasy*, Wiley, Chichester and New York, pp. 465-9.

—— and Elouard, P. (1967) Paléo-océanographie. Schéma des variations du niveau de l'océan Atlantique sur la côte de l'ouest de l'Afrique depuis 40,000 ans', *Acad. Sci. Paris. Comptes Rendus*, *265*, 784-7.

—— and Hébrard, L. (1977) 'Variations des lignes de rivages au Sénégal et en Mauritanie au cours de l'Holocène', *Studia Geologica Polonica, LII*, Warszawa, 144-57.

—— Hébrard, L., Monteillet, J. and Pirazolli, P.A. (1980) 'Geoidal change and shore-level tilt along Holocene estuaries: Senegal river area, West Africa', *Science*, *210*, 421-3.

Feyling-Hanssen, R.W. (1964a) 'Shoreline displacement in central Spitsbergen', in J. Büdel and A. Wirtman (eds), *Vorträge des Fridtjof-Nansens- Gedächtnis-Symposions über Spitsbergen*, Frans Steiner, Weisbaden, pp. 24-38.

—— R.W. (1964b) 'A marine section from the Holocene of Talavera on Barent-soya in Spitsbergen', in J. Büdel and A. Wirtmann (eds), *Vorträge des Fridtjof-Nansens- Gedächtnis-Symposions über Spitsbergen*, Franz Steiner Verlag, Weisbaden, pp. 30-58.

—— and Olsson, I. (1959) 'Five radiocarbon datings of postglacial shorelines in central Spitzbergen', *Norsk Geogr. Tidsskr.*, *17*, 121-31.

Fisk, H.N. and McFarlan, E., Jr (1955) 'Late Quaternary deltaic deposits of the Mississippi river', in A. Poldervaart (ed.), *Crust of the Earth*, Geol. Soc. Am., Spec. Pap. 62, pp. 279-302.

Flinn, D. (1964) 'Coastal and submarine features around the Shetland Islands', *Proc. Geol. Ass.*, *75*, 321-39.

Flint, R.F. (1971) *Glacial and Quaternary Geology*, Wiley, New York.

Gable, D.J. and Hatton, T. (1983) Maps of Vertical Crustal Movements in the Coterminous United States over the last 10 million years. *Map 1-1315, Miscellaneous Investigation Series. Restan, Department of the Interior, United States Geological Survey.*

Gaposchkin, E.M. (1973) 'Satellite dynamics', in E.M. Gaposchkin, *Standard Earth III*, Smithsonian Astron. Obs. Spec. Rept. 353, pp. 85-192.

Geodynamics Program Office (1983) *The NASA Geodynamics Program: An Overview*, NASA Technical Paper, No. 2147.

339

Gould, H.R. and McFarlan, E., Jr (1959) 'Geologic history of the chenier plain southwestern Louisiana', *Gulf Coast Ass. Geol. Soc. Trans.*, *9*, 201-70.

Grant, D.R. (1970) 'Recent coastal submergence of the Maritime Provinces', *Can. J. Earth Sci.*, *7*, 679-89.

—— (1980) 'Quaternary sea-level change in Atlantic Canada as an indication of crustal delevelling', in N.-A. Mörner (ed.), *Earth Rheology, Isostasy and Eustasy*, Wiley, Chichester and New York, pp. 201-14.

Gray, J.M. (1983) 'The measurement of shoreline altitudes in areas affected by glacio-isostasy, with particular reference to Scotland', in D.E. Smith and A.G. Dawson (eds), *Shorelines and Isostasy*, Academic Press, London and New York, pp. 97-127.

—— Boutray, B. de, Hillaire-Marcel, C. and Lauriol. B. (1980) 'Postglacial emergence of the west coast of Ungava Bay, Quebec', *Arctic and Alpine Res.*, *12*, 19-30.

Greensmith, J.T. and Tooley, M.J. (eds) (1982) 'IGCP Project 61, Sea-level movements during the last deglacial hemicycle (about 15,000 years): final report of the UK working group', *Proc. Geol. Ass.*, *93*, 3-125.

—— and Tucker, E.V. (1973) 'Holocene transgressions and regressions on the Essex coast outer Thames estuary', *Geol. Mijnb.*, 52, 193-202.

Hafsten, V. (1979) 'Late and Post Weichselian shore-level changes in south Norway', in E. Oele, R.T.E. Schüttenhelm and A.J. Wiggers (eds), *The Quaternary History of the North Sea*, Acta Univ. Ups. Symp. Univ. Ups. Annum Quingentesimum Celebrantis: 2, Uppsala, pp. 45-59.

—— (1983) 'Biostratigraphical evidence for Late Weichselian and Holocene sea-level changes in southern Norway', in D.E. Smith and A.G. Dawson (eds), *Shorelines and Isostasy*, Academic Press, London and New York, pp. 161-81.

Hageman, B.P. (1969) 'Development of the western part of the Netherlands during the Holocene', *Geol. Mijnb.*, *48*, 373-88.

Haggart, B.A. (1986, in press), 'Flandrian relative sea-level change in the Beauly Firth: evidence from Barnyards near Beauly, Inverness, *Boreas.*

Hald, M. and Vorren, T.O. (1984) 'Modern and Holocene foraminifera and sediments on the continental shelf off Troms, north Norway', *Boreas, 13*, 133-54.

Hébrard, L. (1972) 'Fichier des âges absolus du Quaternaire d'Afrique au nord de l'Equateur', *Bull. ASEQUA, 31-2*, 45-68.

Hicks, S.D. Debaugh, H.A., Jr and Hickman, L.E., Jr (1983) *Sea-level Variations for the United States 1855-1980*, Tides and Water Levels Branch, US Department of Commerce, National Ocean Service, NOAA.

Hillaire-Marcel, C. (1980) 'Multiple component postglacial emergence eastern Hudson Bay, Canada', in N.-A. Mörner (ed.), *Earth Rheology, Isostasy and Eustasy*, Wiley, Chichester and New York, pp. 215-30.

—— and Ochietti, S. (1977) 'Fréquence des datations au ^{14}C de faunes marines postglacières de l'Est de Canada et variations paléoclimatiques', *Palaeogeography, Palaeoclimatol., Palaeoecol.*, *21*, 17-54.

Hjort, C. (1973a) 'The Vega transgression: a hypsithermal event in central East Greenland', *Bull. Geol. Soc. Denmark*, *22*, 25-38.

—— (1973b) 'A sea correction for East Greenland', *Geol. Fören. Stockh. Förh.*, *95*, 132-4.

—— (1979) 'Glaciation in northern East Greenland during the Late Weichselian and Early Flandrian', *Boreas, 8*, 281-96.

—— (1981) 'A glacial chronology for northern East Greenland', *Boreas, 10*, 259-74.

Hoffman, J.S., Keyes, D. and Titus, J.G. (1983) *Projecting Future Sea-level Rise,*

US Environmental Protection Agency, Washington, DC.

Hoppe, G., Fries, M. and Quennerstedt, N. (1965) 'Submarine peat in the Shetland Islands', *Geogr. Ann.*, *45A*, 195-203.

Hughes, T., Denton, G.H. and Grosswald, H.G. (1977) 'Was there a Late Wurm Arctic ice sheet?', *Nature*, *266*, 596-602.

Hyvärinen, H. (1969) 'Trullvatnet: a Flandrian stratigraphical site near Murchinson-fjorden, Nordaustlandet, Spitzbergen', *Geogr. Ann.*, *51A*, 42-5.

—— (1980) 'Relative sea-level changes near Helsinki, southern Finland, during early Litorina times', *Bull. Geol. Soc. Finland*, *52*, 207-19.

—— (1982) *Ann. Acad. Sci. Fennicae*, *AIII*, *134*, 139-49.

Imbrie, J., McIntyre, A. and Moore, T.C., Jr (1983) 'The ocean around North America at the last glacial maximum', in S.C. Porter (ed.), *Late Quaternary Environments of the United States. Vol. 1, The Late Pleistocene*, Longman, London, pp. 230-6.

Iversen, J. (1937) 'Undersögelser over Litorinatransgressioner i Danmark', *Medd. Dan. Geol. Fören.*, *9*, 223-32.

Jansen, E. and Bjorklund, K.R. (1985) 'Surface ocean circulation in the Norwegian Sea 15,000 BP to present', *Boreas*, *14*, 243-57.

—— and Erhenkeuser, H. (1985) 'Ocean circulation in the Norwegian sea during the last deglaciation: isotopic evidence', *Palaeogeography, Palaeoclimatol. Palaeoecol.*, *49*, 189-206.

Jansen, J.H.F. (1976) 'Late Pleistocene and Holocene history off the northern North Sea, based on acoustic reflection records', *Neth. J. Sea Res.*, *10*, 1-43.

—— Weering, van T.C.E. and Eisma, D. (1979) 'Late Quaternary sedimentation in the North Sea', in E. Oele, R.T.E. Schüttenhelm, and A.J. Wiggers (eds), *The Quaternary History of the North Sea*, Acta. Univ. Ups. Symp. Univ. Ups. Annum Quingentesimum Celebrantis: 2, Uppsala, pp. 175-87.

Jardine, W.G. (1975) 'Chronology of Holocene marine transgression and regression in southwestern Scotland', *Boreas*, *4*, 173-96.

—— (1977) 'The Quaternary marine record in southwest Scotland and the Scottish Hebrides', in C. Kidson and M.J. Tooley (eds), *The Quaternary History of the Irish Sea*, Seal House Press, Liverpool, pp. 99-118.

—— (1979) 'The western (United Kingdom) shore of the North Sea in Late Pleistocene and Holocene times', in E. Oele, R.T.E. Schüttenhelm and A.J. Wiggers (eds), *The Quaternary History of the North Sea*, Acta. Univ. Ups. Symp. Univ. Ups. Annum Quingentesimum Celebrantis: 2, Uppsala, pp. 159-74.

—— (1981) 'Holocene shorelines in Britain: recent studies', *Geol. Mijnb.*, *60*, 297-304.

—— (1982) 'Sea-level changes in Scotland during the last 18,000 years', *Proc. Geol. Ass.*, *93*, 25-41.

Jelgersma, S. (1961) 'Holocene sea-level changes in the Netherlands', *Med. Geol. Sticht.*, *Series C-VI*, 1-101.

—— (1979), 'Sea-level change in the North Sea basin. In E. Oele, R.T.E. Schüttenhelm, and A.J. Wiggers (eds), *The Quaternary History of the North Sea*, Acta Univ. Ups. Symp. Univ. Ups. Annum Quingentesimum Celebrantis: 2, Uppsala, pp. 233-48.

——, de Jonge, J., Zagwijn, W.H. and van Regteren Altena, J.F. (1970) 'The coastal dunes of the western Netherlands; geology, vegetational history and archaeology', *Meded. Rijks Geol. Dienst, N.S.*, *21*, 93-167.

——, Oele, E. and Wiggers, A.J. (1979) 'Depositional history and coastal development in the Netherlands and the adjacent North Sea since the Eemian', in E. Oele, R.T.E. Schüttenhelm, and A.J. Wiggers (eds), *The Quaternary History of the North Sea*, Acta Univ. Ups. Symp. Univ. Ups. Annum Quingentesimum

341

Celebrantis: 2, Uppsala, pp. 115-42.

Kaland, P.E. (1984) 'Holocene shore displacement and shorelines in Hordaland, western Norway', *Boreas, 13*, 203-42.

——, Krzywinski, K. and Stabell, B. (1984) 'Radiocarbon dating of transitions between marine and lacustrine sediments and their relation to the development of lakes', *Boreas, 13*, 243-58.

Kayan, I. and Kraft, J.C. (1979) 'Holocene geomorphic evolution of a barrier salt-marsh system, southwest Delaware Bay', *Southeastern Geology, 20*, 79-100.

Kaye, C.A. and Barghoorn, E.S. (1964) 'Late Quaternary sea-level change and crustal rise at Boston, Massachusetts, with notes on the autocompaction of peat', *Bull. Geol. Soc. Am., 75*, 63-80.

Kellogg, T.B. (1976) 'Late Quaternary climatic changes: evidence from deep-sea cores of Norwegian and Greenland Seas', in R.M. Cline and J.D. Hays (eds), *Investigations of Late Quaternary Palaeo-oceanography and Palaeoclimatology*, Geol. Soc. Am. Mem., 145, pp. 77-110.

Kidson, C. (1982) 'Sea level changes in the Holocene', *Quat. Sci. Rev.*, 1, 121-51.

Kolb, C.R. and Lopik, J.R. van (1966), in M.L. Shirley and J.A. Ragsdale (eds), *Deltas*, Houston Geological Society, Houston, pp. 17-61.

Kraft, J.C. (1971) 'Sedimentary facies patterns and geological history of a Holocene marine transgression', *Bull. Geol. Soc. Am., 82*, 2131-58.

—— (1978) 'Coastal stratigraphic sequences', in R.A. Davis, Jr. (ed.), *Coastal Sedimentary Environments*, Springer Verlag, New York, pp. 361-83.

—— (1979) 'The coastal environment', in J.C. Kraft and W. Carey (eds), *Selected Papers on the Geology of Delaware, Transactions of the Delaware Academy of Science, Vol. 7*, Delaware Academy of Science, Newark, Del., pp. 31-66.

—— (1984) 'Sea-level rise and prediction of coastal erosion and landform change: Delaware coastal zone', unpublished report, Dept. of Geology, University of Delaware, Newark, Del.

——, Allen, E.A. and Maurmeyer, E.M. (1978) 'The geological and palaeogeo-morphological evolution of a spit system and its associated coastal environments: Cape Henlopen spit, Delaware', *J. Sed. Petrol, 48*, 211-26.

—— and Chacko, J.J. (1979) 'Lateral and vertical facies relations of transgressive barrier', *Am. Ass. Petrol Geol. Bull., 63*, 2145-63.

Kranck, K. (1972) 'Geomorphological development and post Pleistocene sea-level changes, Northumberland Strait, Maritime Provinces', *Can. J. Earth Sci., 9*, 835-44.

Krog, H. (1979) 'Late Pleistocene and Holocene shorelines in western Denmark', in E. Oels, R.T.E. Schüttenhelm and A.J. Wiggers (eds), *The Quaternary History of the North Sea*, Acta Univ. Ups. Symp. Univ. Ups. Annum Quingentesimum Celebrantis: 2, Uppsala, pp. 75-84.

Krzywinski, K. and Stabell, B. (1984) 'Late Weichselian sea-level changes at Sotra, Hordaland, western Norway', *Boreas, 13*, 159-202.

Laughton, A.S. (1975) 'Tectonic evolution of the North-eastern Atlantic ocean, a review', *Nor. Geol. Unders. 316, Bull., 29*, 169-94.

Lee, H.A. (1960) 'Lateglacial and postglacial Hudson Bay sea episode', *Science, 131*, 1609-11.

Le Pichon, X., Sibuet, J-C. and Francheteau, J. (1977) 'The fit of the continents around the North Atlantic ocean', *Tectonophys., 38*, 169-209.

Linder, L., Marks, L. and Pekala, K. (1984) 'Late Quaternary glacial episodes in the Hornsund region of Spitzbergen', *Boreas, 13*, 35-47.

Lindström, G. (1886), 'Om postglaciala sankningar af Gotland', *Geol. Fören. Stockh. Förh.*, 8, 251-81.

Lisitzin, E. (1965) 'The mean sea level of the world oceans', *Comment. Phys.-Math.*

Helsingf., 30, 1-35.

Løken, O. (1965) 'Postglacial emergence at the south end of Inugsuin Fiord, Baffin Island', *Geogr. Bull., 7,* 242-58.

Ludwig, G., Muller, H. and Streif, H. (1981) 'New dates on Holocene sea-level changes in the German Bight', *Spec. Publs. Int. Ass. Sedim., 5,* 211-19.

Lundquist, J. (1965) 'The Quaternary of Sweden', in A. Rankama (ed.), *The Quaternary, 1,* Wiley, London, pp. 139-98.

—— (1971) 'Kvartargeologisk forskring Sverige 1946-1970', *Geol. Fören. Stockh. Förh., 93,* 303-34.

McCabe, A.M., Haynes, J.R. and MacMillan, N.F. (in press) 'Late Pleistocene glacio-isostatic sequences, sea-level data and glacio-isotatic provences in north Co. Mayo, Republic of Ireland', *J. Quat. Sci.*

McFarlan, E., Jr (1961) 'Radiocarbon dating of the Late Quaternary deposits, South Louisiana', *Bull. Geol. Soc. Am., 72,* 129-58.

Marcus, L.F. and Newman, W.S. (1983) 'Hominid migrations and the eustatic sea-level paradigm: a critique', in P.M. Masters and N.C. Flemming (eds), *Quaternary Coastlines and Marine Archaeology,* Academic Press, London and New York, pp. 63-85.

Milliman, J.D. and Emery, K.O. (1968) 'Sea levels during the past 35,000 years', *Science, 162,* 1121-3.

Monteillet, J. (1977), *Bull. Assoc. Sénégal. Etude Quat., 50,* 23-36.

Morgan, J.P. (1970) 'Deltas: a résumé', *J. Geol. Educ., 18,* 107-17.

Mörner, N.-A. (1969) 'The Late Quaternary history of the Kattegat Sea and the Swedish west coast: deglaciation, shore level displacement, chronology, isostasy and eustasy', *Sveriges Geol. Unders. Series C., 640,* 1-487.

—— (1971) 'Eustatic changes during the last 20,000 years and a method of separating the isostatic and eustatic factors in an uplifted area', *Palaeogeography, Palaeoclimatol., Palaeoecol., 19,* 63-5.

—— (1974) 'Ocean palaeotemperature and continental glaciations', *Colloques Intern. CNRS. No. 219,* 43-9.

—— (1976a) 'Eustasy and geoid changes', *J. Geol., 84,* 123-51.

—— (1976b) 'Eustatic changes during the last 8,000 years in view of radiocarbon calibration and new information from the Kattegatt region and other North-western European coastal areas', *Palaeogeography, Palaeoclimatol., Palaeoecol., 19,* 63-85.

—— (1977) 'Eustasy and instability of the geoid configuration', *Geol. Fören. Stockh. Förh., 99,* 369-76.

—— (1979a) 'The Fennoscandian uplift and Late Cenozaic geodynamics: geological evidence', *Geo-Journal, 3,* 287-318.

—— (1980a) 'Earth's movements, plaeoceanography, palaeoclimatology and eustasy: major events in the Cenozoic of the North Atlantic', *Geol. Fören. Stockh. Förh., 102,* 261-8.

—— (1980b) 'The Northwest European Sea-level Laboratory and regional Holocene eustasy', *Palaeogeography, Palaeoclimat. Palaeoecol., 29,* 281-300.

—— (1981) 'Space geodesy, paleogeodesy and paleogeophysics', *Ann. de Géophys., 37,* 69-76.

Moslow, T.F. and Colquhoun, D.J. (1981) 'Influence of sea-level change on barrier island evolution', in D.J. Colquhoun (ed.), *Variation in Sea Level on the South Carolina Coastal Plain,* Dept. of Geology, University of South Carolina, Columbia, SC, pp. 104-20.

Naylor, D. and Shannon, P. (1982) *The Geology of Offshore Ireland and West Britain,* Graham and Trotman, London.

Nelson, H.F. and Bray, E.E. (1970) 'Stratigraphy and history of the Holocene sediments in the Sabine-High Island area, Gulf of Mexico', in J.P. Morgan (ed.), *Deltaic Sedimentation, Modern and Ancient*, Soc. Econ. *Pal. Min.*, Spec. Publ. No. 15, pp. 48-77.

Newman, W.S., Cinquemani, L.J., Pardi, R.R. and Marcus, L.F. (1980a) 'Holocene delevelling of the United States' east coast', in N.-A. Mörner (ed.), *Earth Rheology, Isostasy and Eustasy*, Wiley, Chichester and New York, pp. 449-63.

—— Marcus, L.F. and Pardi, R.R. (1981) 'Palaeogeodesy: Late Quaternary geoidal configurations as determined by ancient sea levels', in I. Allison (ed.), *Sea Level, Ice and Climatic Change*, Proceedings of the Canberra Symposium, December, 1979, Int. Ass. Hydrol. Sci., Publication No. 131, Washington, DC, pp. 263-75.

—— Marcus, L.F., Pardi, R.R., Paccione, J.A. and Tomecek, S.M. (1980b) 'Eustasy and deformation of the geoid: 1000-6000 radiocarbon years, BP' in N.-A. Mörner (ed.), *Earth Rheology, Isostasy and Eustasy*, Wiley, Chichester and New York, pp. 555-67.

Nummedal, D. (1983) 'Rates and frequency of sea-level changes: review with an application to predict future sea levels in Louisiana', in *Trans. 33rd Annual Meeting of the Gulf Coast Ass. Geol. Socs.*, Jackson, Miss., pp. 361-5.

Nunn, P.D. (1984) 'Occurrence and ages of low-level platforms and associated deposits on South Atlantic coasts: appraisal of evidence for regional Holocene high sea level', *Prog. Phys. Geogr.*, 8, 32-60.

Oele, E. and Schüttenhelm, R.T.E. (1979) 'Development of the North Sea after the Saalian glaciation', in Oele, E., R.T.E. Schüttenhelm, and A.J. Wiggers (eds), *The Quaternary History of the North Sea*, Acta Univ. Ups. Symp. Univ. Ups. Annum Quingentesimum Celebrantis: 2, Uppsala, pp. 191-215.

Oldale, R.N. and O'Hara, C.J. (1980) 'New radiocarbon dates from the inner shelf off southeastern Massachusetts and a local sea-level rise curve for the past 12,000 years', *Geology*, 8, 102-6.

Ortlieb, L. and Petit-Maire, N. (1976) 'The Atlantic border of the Sahara, in Holocene time', in E.M. van Zindern Bakker (ed.) *Palaeoecology of Africa, the Surrounding Islands and Antarctica*, Balkema, Cape Town, pp. 4-5.

Otvos, E.G. (1981) 'Barrier island formation through nearshore aggradation — stratigraphic and field evidence', *Mar. Geol.*, 43, 195-243.

Palmer, A.J.M. (1982) 'Modern coastal submergence in Atlantic Canada recorded by diatom microfossils', in D.J. Colquhoun (ed.), *Holocene Sea-level Fluctuations, Magnitude and Causes*, pp. 135-42. Dept. of Geology, University of South Carolina, Columbia, SC, pp. 135-42.

Peacock, J.D. (1983) 'A model for Scottish interstadial marine palaeotemperature 13,000 to 11,000 BP', *Boreas*, 12, 73-83.

Peltier, W.R. and Andrews, J.T. (1976) 'Glacial isostatic adjustment, 1. The forward problem', *Geophys. J. Roy. Astion. Soc.*, 46, 669-705.

—— and Andrews, J.T. (1983) 'Glacial geology and glacial isostasy of the Hudson Bay region' in D.E. Smith and A.G. Dawson (eds), *Shorelines and Isostasy*, Academic Press, London and New York, pp. 285-319.

—— , Farrell, W.E. and Clark, J.A. (1978) 'Glacial isostasy and relative sea level: a global finite element model', *Tectonophys.*, 50, 81-110.

Petersen, K.S. (1985) 'Late Weichselian and Holocene marine transgressions in northern Jutland, Denmark', *Eiszeit. Gegenw.*, 35, 71-8.

Petit-Maire, N., Delibrias, G. and Ortlieb, L. (1977) 'New radiometric data for the Atlantic Sahara (Holocene, 19° to 28°N): tentative interpretations', *X INQUA Congress, Birmingham, Abstracts Volume*, Geo-Abstracts, Norwich, p. 353.

Pitman, W.C., III (1978) 'Relationship between eustasy and stratigraphic sequences of passsive margins', *Bull. Geol. Soc. Am.*, 89, 1389-403.

—— and Talwani, M. (1972) 'Sea floor spreading in the North Atlantic', *Bull. Geol. Soc. Am.*, *83*, 619-46.

Plassche, O. van de (1977) *A Manual for Sample Collection and Evaluation of Sea-level Data*, Institute for Earth Sciences, Free University, Amsterdam.

—— (1982) 'Sea-level change and water-level movements in the Netherlands during the Holocene', *Meded. Rijks. Geol. Dienst*, *36*, 1-93.

—— (1986), *Sea-level Research: a Manual for the Collection and Evaluation of Data*, Geo Books, Norwich.

Pomel, R. (1979) 'Géographie physique de la Basse côte d'Ivoire', dissertation, University of Caen.

Rampino, M.R. (1979) 'Holocene submergence of southern Long Island, New York', *Nature*, *280*, 132-4.

Ramsey, W. (1924) 'On relations between crustal movements and variations of sea-level during the Late Quaternary time, especially in Fennoscandia', *Bull. Comm. Geol. Finlande 66, Fenia*, *44*, 1-39.

Redfield, A.C. (1967) 'Postglacial change in sea level in the western North Atlantic ocean', *Science*, *157*, 687-91.

Ritchie, W. (1966) 'The postglacial rise in sea-level and coastal changes in the Uists', *Trans. Inst. Br. Geogr.*, *39*, 79-86.

—— (1972) 'The evolution of coastal sand dunes', *Scott. Geogr. Mag.*, *88*, 19-35.

Roeleveld, W. (1974) 'The Groningen coastal area: a study in Holocene geology and lowland physical geography', *Berichten van de Rijks. voor bet Oudheid-kundig Bodemonderzoek*, *20-21*, 7-25 and *24*, 7-132.

Rohdenburg, H. (1977) 'Neue ^{14}C-daten aus Marokko und Spanien'. *Catena*, *4*, 215-28.

Rose, J. (1978) 'Glaciation and sea-level change at Bugöjfjord, South Varanger-fjord, north Norway', *Norsk Geogr. Tidsskr.*, *32*, 121-35.

Ruddiman, W.F. and McIntyre, A. (1976) 'Northeast Atlantic palaeoclimatic changes over the past 600,000 years', in R.M. Cline and J.D. Hays (eds), *Investigations of Late Quaternary Palaeoceanography and Palaeoclimatology*, Geol. Soc. Am. Mem., 145, pp. 111-46.

—— and McIntyre, A. (1981) 'The North Atlantic during the last deglaciation', *Palaeogeography, Palaeoclimatol., Palaeoecol.*, 35, 145-214.

Sars, M. (1865), *Om de i Norge Forekommende Fossile Dyrelevninger fra Qvartaer-perioden*, Christiana (Oslo).

Scholl, D.W., Craighead, F.C. and Stuiver, M. (1970) 'Florida curve revised: its relation to coastal sedimentation rates', *Science*, *163*, 562-4.

—— and Stuiver, M. (1967) 'Recent submergence of southern Florida: a comparison with adjacent coasts and other eustatic data', *Bull. Geol. Soc. Am.*, *78*, 437-54.

Schytt, V., Hoppe, G., Blake, W., Jr and Grosswald, M.G. (1968), 'The extent of the Wurm glaciation in the European Arctic', *Int. Ass. Hydrol. Sci., General Assembly of Berne, 1967, Publ. 79*, 207-16.

Sernander, R. (1902) 'Bidrag till den vastskandinaviska vegetationens historia i relation till nivåförändring-anna', *Geol. Fören. Stockh. Förh.*, 24, 125-44 and 415-66.

Shaw, J. (1985) 'Aspects of Holocene coastal evolution, Co. Donegal, Ireland', unpublished D. Phil. dissertation, University of Ulster.

Shennan, I. (1982) 'Interpretation of Flandrian sea-level data from the Fenland, England', *Proc. Geol. Ass.*, *93*, 53-63.

—— 'Flandrian and Late Devensian sea-level changes and crustal movements in England and Wales', in D.E. Smith and A.G. Dawson (eds), *Shorelines and Isostasy*, Academic Press, London and New York, pp. 255-83.

—— (1986) 'Flandrian sea-level changes in the Fenland', *J. Quat. Sci.*, *1*, 119-79.

—— Tooley, M.J., Davis, M.J. and Haggart, B.A. (1983) 'Analysis and interpretation of Holocene sea-level data', *Nature*, *302*, 404-6.

Sheridan, R.E. (1979) 'Geology of the Atlantic continental margin of Delaware', in J.C. Kraft and W. Carey (eds), *Selected Papers on the Geology of Delaware, Transactions of the Delaware Academy of Science, Vol. 7*, Delaware Academy of Science, Newark, Del., pp. 67-86.

Sissons, J.B. (1972) 'Dislocation and non-uniform uplift of raised shorelines in the western part of the Forth valley', *Trans. Inst. Br. Geogr.*, *55*, 145-59.

—— (1983) 'Shorelines and isostasy in Scotland', in D.E. Smith and A.G. Dawson (eds), *Shorelines and Isostasy*, Academic Press, London and New York, pp. 209-25.

—— and Dawson, A.G. (1981) 'Former sea-levels and ice limits in part of Wester Ross, northwest Scotland', *Proc. Geol. Ass.*, *92*, 115-24.

Sleep, N.H. and Snell, N.S. (1976) 'Thermal contraction and flexure of mid-continent and Atlantic marginal basins', *Geophys. J. Roy. Astron. Soc.*, *45*, 125-54.

Sollid, J.L., Andersen, S., Hamre, N., Kjeldsen, O., Salvigsen, O., Sturod, S., Tveita, T. and Wilhelmsen, A. (1973) 'Deglaciation of Finnmark, north Norway', *Norsk Geogr. Tidsskr.*, *27*, 233-325.

Stephens, N. and McCabe, A.M. (1977) 'Late Pleistocene ice movements and patterns of late and postglacial shorelines on the coast of Ulster (Ireland)', in C. Kidson and M.J. Tooley (eds), *The Quaternary History of the Irish Sea*, Seal House Press, Liverpool, pp. 179-98.

Straaten, L.M. van (1965) 'Coastal barrier deposits in south and north Holland, in particular in areas around Scheveningen and Ijmuiden', *Med. Geol. Sticht, N.S. 17*, 41-75.

Stuiver, M. and Borns, H.W., Jr (1975) 'Late Quaternary marine invasion in Maine: its chronology and associated crustal movement', *Bull. Geol. Soc. Am.*, *86*, 99-104.

—— and Daddario, J.J. (1963) 'Submergence of the New Jersey coast', *Science*, *142*, 951.

Swift, D.J.P. (1975) 'Barrier island genesis: evidence from the central Atlantic shelf, eastern USA', *Sed. Geol.*, *14*, 1-43.

—— Moir, R. and Freeland, G.L. (1980) 'Quaternary rivers on the New Jersey shelf: relation of seafloor to buried valleys', *Geology*, *8*, 276-80.

Synge, F. (1980) 'A morphometric comparison of raised shorelines in Fennoscandia, Scotland and Ireland', *Geol. Fören. Stockh. Förh.*, *102*, 235-49.

—— (1981) 'Quaternary glaciation and changes of sea level in the south of Ireland', *Geol. Mijnb.*, *60*, 305-15.

Talwani, M. and Eldholm, O. (1977) 'Evolution of the Norwegian–Greenland Sea', *Bull. Geol. Soc. Am.*, *88*, 969-99.

Tastet, J.P. (1975) 'Variations Holocènes du niveau marin en Côte d'Ivoire', *IX Congrès Int. de Sedimentologie, Nice.*

Taylor, R.B., Carter, R.W.G., Forbes, D.L. and Orford, J.D. (1986) 'Beach sedimentation in Ireland: contrasts and similarities with Atlantic Canada', *Current Research, Geological Survey of Canada, 1986A*, 55-64.

Ten Brink, N.W. (1974) 'Glacio-isostasy: new data from West Greenland and geophysical implications', *Bull. Geol. Soc. Am.*, *85*, 219-28.

Titus, J.G., Henderson, T.R. and Teal, J.M. (1984) 'Sea-level rise and wetlands loss in the United States', *National Wetlands Newsletter*, *6*, 3-6.

—— Leatherman, S.P., Everts, C.H. and Kriebel, D.L. (1985) *Potential Impacts of Sea-level Rise on the Beach at Ocean City, Maryland.* US Environmental Protection Agency, Washington, DC.

Todd, N.W. (1981) 'Recent sea-level changes in north and east Ireland', unpublished MSc dissertation, University of Ulster, Coleraine.

Tooley, M.J. (1978a) 'Interpretation of Holocene sea-level changes', *Geol. Fören. Stock. Förh.*, 100, 203-12.

—— (1978b) *Sea-level Changes in Northwest England During the Flandrian Stage*, Clarendon Press, Oxford.

—— (1982) 'Sea-level changes in northern England', *Proc. Geol. Ass.*, 93, 43-51.

—— (1985) 'Sea levels', *Prog. Phys. Geogr.*, 9, 113-20.

Tyrrell, J.B. (1898) 'Report on the Dubawnt, Dazan and Ferguson rivers and the northwest coast of Hudson Bay and on the two overland routes from Hudson Bay to Lake Winnipeg', *A. Rep. Geol. Surv. Can.*, 9, IF-218F.

Walcott, R.I. (1972a) 'Past sea levels, eustasy and deformation of the earth', *Quat. Res.*, 2, 1-14.

—— (1972b) 'Late Quaternary vertical movement in eastern North America: quantitative evidence of glacio-isostatic rebound', *Rev. Geophys. Space Phys.*, 10, 849-84.

Washburn, A.L. and Stuiver, M. (1962) 'Radiocarbon dated postglacial delevelling in northeast Greenland and its implications', *Arctic*, 15, 66-73.

Weymarn, J. von (1974) 'Coastline development in Lewis and Harris, Outer Hebrides, with particular reference to the effects of glaciation', unpublished PhD dissertation, University of Aberdeen.

Wilkinson, B.H. (1975) 'Matagorda Island, Texas: the evolution of a Gulf Coast barrier complex', *Bull. Geol. Soc. Am.*, 86, 959-67.

—— and Basse, R.A. (1978) 'Late Holocene history of the central Texas coast from Galveston Island to Pass Cavallo', *Bull. Geol. Soc. Am.*, 89, 1592-1600.

Woude, J.D. van der (1981) *Holocene Palaeoenvironmental Evolution of a Peri-marine Fluviatile Area*, Free University, Amsterdam.

Ziegler, P.A. (1982) 'Faulting and graben formation in Western and Central Europe', *Phil. Trans. Roy. Soc. Lond., A.*, 305, 113-43.

NOTES

1. New data by Grant (1986) is now available for this region, in Fulton, R.J. (1986), *Quaternary Geology of Canada and Greenland.* Geological Survey of Canada, Ottawa.

2. Reinterpretation of sea-level data from the Thames (Devoy, 1979) might suggest the northeastward migration of a collapsing forebulge zone in this area of the southern North Sea prior to ~ 3,000 BP (Tooley, M.J., 1986, Dept. of Geography, University of Durham, personal communication).

3. More recent evidence may be found cited in Emery, K.O. and Uchupi (1984), *Geology of the Atlantic Ocean.*

11

Sea-level Changes during the Holocene: The Northwest Pacific

Yoko Ota

INTRODUCTION

In this chapter Holocene sea-level changes in the region of the north-western Pacific rim are discussed. This includes consideration of the coastal areas of China and Korea as part of a relatively stable intraplate continent, of Taiwan in a tectonically active collision zone, as well as discussion of the Japanese Islands representing active island arcs in a crustal subduction zone (Fig. 11.1). Due to the major differences in the tectonic setting of these areas the Holocene sea-level changes are recorded in either a variety of coastal landforms or in marine deposits, the form of occurrence depending on the rate of vertical displacement of each area and the effects of other environmental factors. For example, flights of Holocene marine terraces form a characteristic feature of coasts of uplift. In contrast, coastal plains accompanied by a series of beach ridges are developed in more stable areas, whilst small lowland embayments underlain by marine or estuarine deposits burying drowned valleys are typical landforms in the subsiding areas (Yoshikawa et al., 1981). In addition, to the south of the northern limit of coral reef growth shown later in Figure 11.10, coral reefs are developed. These form one of the best indicators of palaeo sea level with abundant datable material.

Much work on Holocene sea-level changes has been carried out in these areas, especially in the Japanese Islands where despite the relatively small land area the length of coastline reaches nearly 27,000 km. However because most of the work undertaken in these areas has usually been published in the indigenous languages of the region little has found its way into the wider international literature. Few publications from the region were included in the *Atlas of Sea-Level Curves* (Bloom, 1977), only two coming from Korea and its vicinity, four from Japan and none from China. In these circumstances, the Japanese Working Group of IGCP Project 61 compiled all the Holocene sea-level curves for the region, together with data used in

Figure 11.1: Location map of areas referred to in the Northwest Pacific

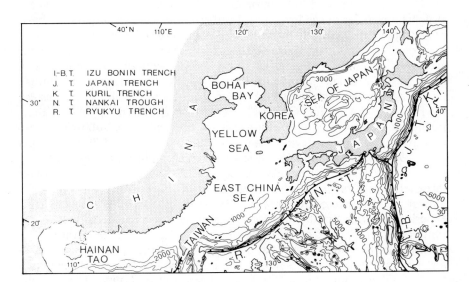

their construction, in English (Ota *et al.*, 1981) in order to introduce the results internationally. In addition, a compilation of the Quaternary shoreline data in these areas has been started as a part of the programme of IGCP Project 200 and the Indian–Pacific Quaternary Shorelines Subcommission (INQUA).

The chapter begins with a comparison and review of the patterns of Holocene sea-level curves in the designated areas, based on recent studies in the northwestern Pacific rim. This is followed by discussion of the significance and problems of the regional differences in the sea-level curves presented.

PATTERNS OF HOLOCENE SEA-LEVEL CHANGES IN THE NORTHWESTERN PACIFIC RIM

The Japanese Islands

The Holocene sea-level changes in Japan have been summarised and reviewed several times (e.g. Ota, 1982; Ota *et al.*, 1982; Naruse and Ota, 1984; Ota and Machida, in press) and readers are referred to these for more detailed discussion of data.

Japan has a great variety of Holocene coastal landforms despite the

349

smallness of the land area. In large coastal plains of tectonic origin, such as the Kanto Plain in central Japan or the Osaka Plain in southwestern Japan, Holocene changes in sea level have been reconstructed mainly on the basis of borehole data, or partly by direct observation of sediments at the time of construction works (e.g. Kaizuka *et al.*, 1977; Maeda, 1976). These plains areas are generally underlain by thick marine sediment formations as a result of a continuous subsidence throughout the Quaternary. In contrast, Holocene marine terraces fringe the coast of central and southwestern Japan along the Pacific Ocean. Those at the southern tip of the Boso Peninsula, Kanto, are especially well defined. These are subdivided into four groups of terraces formed at (1) 6,200-6,900 BP, (2) 4,300-5,500 BP, (3) 2,900-4,000 BP, and (4) the lowest terrace, uplifted co-seismically at the time of the 1703 earthquake. A probable co-seismic origin for the older three terraces is suggested also (Matsuda *et al.*, 1978; Nakata *et al.*, 1980). Similar subdivision of terraces can be seen in the other coastal areas experiencing high uplift rates, for example near the northern limit of the Philippine Sea Plate. The maximum height of the Holocene terraces here reaches nearly +30 m msl (Ota, 1982). Much farther to the southwest a series of raised coral reefs is developed in the Ryukyu Islands. Raised coral reefs of Kikai Island (Fig. 11.2), being closest to the Ryukyu Trench (Fig. 11.1), are divided here into a fourfold sequence, dated at (1) 6,100-6,600 BP, (2) 4,000-5,200 BP, (3) 2,800-3,500 BP, (4) 1,500-2,500 BP, in descending order. This implies that intermittent drops in relative sea-level have occurred since the postglacial Hypsithermal (~ 5,100-7,100 BP; see later discussion, pp. 364-71) (Ota *et al.*, 1978; Nakata *et al.*, 1978). Relative sea-level curves from this island have been constructed using more than 50 radiocarbon dates (Fig. 11.3). Holocene marine terraces are also well developed on the western coast of Hokkaido and northern Honshu. This area lies close to a newly proposed plate boundary occurring offshore (Nakamura, 1983). Here, the Holocene marine terrace attains heights of up to +10 m msl and co-seismic uplift associated with historical earthquakes and their repeated activity is known to have affected the area (e.g. an emerged bench was formed by the 1802 earthquake on Sado Island; Ota *et al.*, 1976).

The Japanese Islands forming active island arcs are also active volcanic zones from which many tephra layers have been produced. Among these tephras, Kikai-Akahoya ash (K-Ah), dated at 6,300 BP from the Kikai Caldera of south Kyushu, is a very useful widespread time marker in Holocene sea-level studies (Machida and Arai, 1978), as will be discussed later.

Based on geomorphological and geological studies of the coastal areas, nearly 50 Holocene sea-level curves have been produced in Japan since the early 1960s. However, as some curves prepared in the early stages were drawn using data from various areas which have different tectonic histories,

Figure 11.2: Airphoto mosaic of Kikai Island, southwest Japan. This island is typical of areas of land uplift in Japan, close to the Ryukyu Trench. The highest part of the island is composed of the coral reef terrace of the last Interglacial maximum, at ~ 200 m high. It is dislocated by a series of faults. Successive lower shorelines from ~ 100 ka to ~ 40 ka are also preserved in the form of lower coral reef terraces. The Holocene raised coral reef occurs at a height of +13 m msl at maximum and is divided into four steps from ~ 6 ka to 1.5 ka. The Holocene terrace surrounds the whole coast and is separated by a steep cliff from Late Pleistocene reefs. Source: Geographical Survey Institute, Japan.

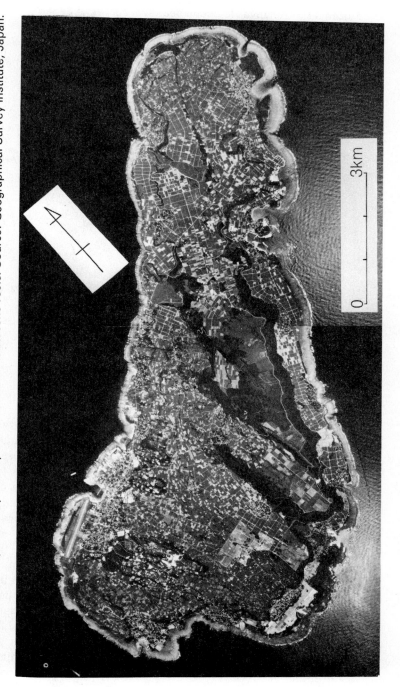

Figure 11.3: Relative Holocene sea-level curves in Japan
1: A, Tohoku Nosei-Kyoku, 1979; B, Omoto, 1979; C, Chida *et al.*, 1984.
2: A, Fuji and Fuji, 1981; B, Toyoshima, 1978.
3: A, Maeda, 1976; B, Kaizuka *et al.*, 1977.
4: A, Furukawa, 1972; B, Umitsu, 1979; C, Iseki and Moriyama, 1981; D,
   Maeda *et al.*, 1983.
5: A, Nakata *et al.*, 1980; B, Yokota, 1978; C, Moriwaki, 1979
6: A, Ota *et al.*, 1978; B, Nakata *et al.*, 1978.

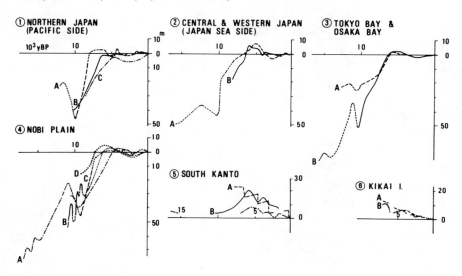

Source: Modified from Ota *et al.* (1982).

and some were based on only a few dated samples, these curves have been excluded from further examination here. Figure 11.3 summarises selected sea-level data presented in the form of time-depth curves. These are based on intensive areal studies of Holocene sequences/deposits, in terms of age, height and other environmental controls. They usually cover different time spans. Those from areas of uplift mainly show sea-level changes since the postglacial Hypsithermal event; in such areas the Holocene marine sequences are often exposed as terrace structures and are easily observable. In contrast, sea-level curves from subsiding zones usually include older material pre-dating the Hypsithermal event.

A considerable local difference exists in the curves shown in Figure 11.3 within the Japanese Islands. This implies that each curve must only represent a 'relative sea-level curve', combining an eustatic change of sea level and local vertical movements of various types. It is still difficult to distinguish here tectonic, isostatic and eustatic effects from a given relative sea-level curve quantitatively. At present, therefore, it is important that as precise a relative sea-level curve (data plot) as possible is constructed for each defined area (see Shennan, this vol., Ch. 7). These should be based on sufficient radiocarbon dates, in which careful consideration is given to sample height, stratigraphic position and associated palaeontological evidence.

On examining the sixteen curves shown in Figure 11.3, the following common features (tendencies) can be recognised:

(1) Relative sea level rose very rapidly from ~ 15,000 BP to 6,000 BP. This sea-level rise, correlated with the early to main postglacial rise in sea level, is called the 'Jomon Transgression' in Japan, after the presence of early Jomon middens in the vicinity of the estimated shoreline of this period of marine inundation. During this general rise in sea level, a time of marine removal or negative tendency at ~ 10,000 BP is observed commonly in many areas. Evidence for this is based on the presence of a coarse sand layer occurring within fine silts on a transgressive sediment facies.

(2) All the relative sea-level curves, except for IB from Sendai Plain (Omoto, 1979), reached to a slightly higher level than those of the present at about 6,000 BP, even in relatively stable or slightly subsiding areas. In the following discussion this high sea-level position is described as the 'culmination of the Jomon (or postglacial) Transgression' or simply 'culmination'. Most of the Japanese relative sea-level curves, therefore, do not fit the pattern in which the present sea-level position is regarded as being the highest one since the low sea-level stand of the last glaciation (see also Hopley, this vol., Ch. 12).

(3) The age and height of the culmination of the 'Jomon Transgression', corresponding to the postglacial Hypsithermal, shows regional and local differences reflecting the different tectonic history of each area. This event ranges in age from ~ 5,000 to 7,000 BP and attains maximum heights of ~ 30 m msl.

(4) There are at least two minor negative tendencies in sea level after the 'culmination'. A relative fall at ~ 2,000-3,000 BP is common and an earlier fall at ~ 4,000-5,000 BP is also recognised in several locations. Oscillations of relative sea level probably occurred, therefore, to some extent after the 'culmination'.

China

Coastal areas of China extend from about 41°N to 18°N (Fig. 11.1) and cover a variety of environmental regions. As summarised by Lin (1983), many works on Holocene sea-level changes here have been carried out. The following are regarded as the main palaeo sea-level indicators in Chinese studies (Zaho and Zhang, 1982):

(1) Marine layers along the coast and non-marine layers on the sea floor
(2) Coastal-submarine peat and mud beds
(3) Subaerial and submerged beach ridges and chenier ridges
(4) Ostracean reef and drowned estuaries/river valleys
(5) Raised and submerged beaches and beach rock
(6) Littoral and submarine marine terraces and coral reefs.

Among these features chenier ridges, composed of many fossil shells, and developed on coastal plains in northern and eastern China, are one of the most distinguishable landforms related to former sea level. As they are products of breaking waves at or near the high tide position, the base of these features is considered to represent nearly the level of contemporary high tides at the time of chenier formation. Main chenier ridges are mapped in Figure 11.4, where former shorelines at several stages are also shown. A series of chenier ridges in the coastal area along Bohai Bay forms the most spectacular feature on this coastline. It is composed of four rows from the innermost IV to the outermost I, becoming younger outwards. Chenier ridge II is the largest, reaching 300 m in width and 8 m in height. Based on ^{14}C dates of shells from these ridges, the age of the chenier formation is estimated to be between 4,000 and 4,700 BP for IV, 3,000 and 3,800 BP for III, and 1,100 and 2,500 BP for II (Zaho et al., 1980). These ages are consistent with most of the ancient cultural sites discovered on the chenier ridges.

Figure 11.5 shows a synthesised sea-level curve from the coastline of eastern China covering the past 20,000 years. This is based on more than 60 radiocarbon dates from various parts of China (Zaho et al., 1979). Sea level was lowest here at ~ 15,000 BP, reaching levels of −150 to −160 m msl. Since then, relative sea level is shown as rising very rapidly at a rate of 16.7 mm a^{-1} to the highest level, positioned some 2 to 4 m above the present sea level at 5,000-6,000 BP. On the basis of evidence from submarine terraces, submerged chenier ridges and buried peat layers, relative sea level is indicated as having been stable at ~ 14,000 BP, forming shorelines at −100 m to −120 m msl, and also subsequently at ~ 12,000 BP at levels between −65 to −77 m msl. After the peak of the Holocene relative sea-level rise between 5,000 and 6,000 BP, it is suggested that sea level fell to its present position through a series of oscillatory movements with maxima of +1 to +2 m higher than at present, as represented by the chenier ridges mentioned earlier, timed at 4,000-4,800 BP, 3,000-3,800 BP and 1,000-2,500 BP. A similar, though not identical sea-level curve has been constructed by Zaho and Zhang (1982) using data from beach rock, coral reefs and other information in addition to data from chenier ridges, as shown by the thin dotted line in Figure 11.5. Movements in the position of Middle to Late Holocene relative sea level are recorded in a beach rock series located on the South China Sea Islands and the northern coast of the

Figure 11.4: Map showing distribution of main chenier ridges and former shorelines in northern and eastern China. 1. Post-lastglacial shorelines with age ($\times$ 10 ka). 2. Post last-interglacial shorelines with age ($\times$ 10 ka). 3. Chenier ridges. 4. ^{14}C dates

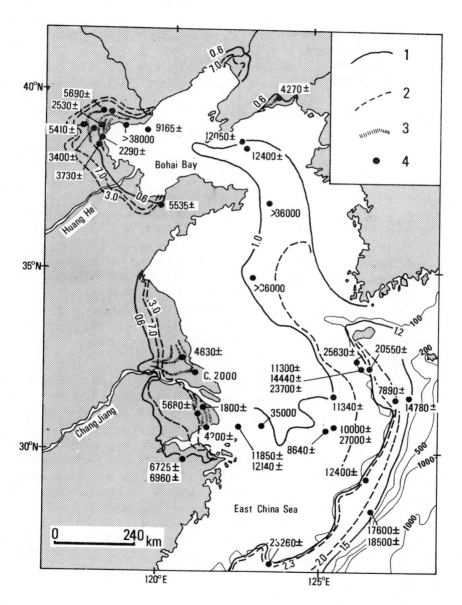

Source: Compiled by Yanagida and Kaizuka (1982), based on Wang and Wang (1980) and others.

355

Figure 11.5: Sea-level changes in China

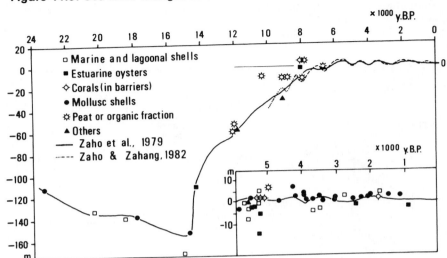

South China Sea. On the Fujian coast, beach rock may be divided into five groups of the following ages: 4,700-5,400 BP, 4,000-4,200 BP, 3,200-3,600 BP, 2,300-2,800 BP and 1,500-2,000 BP. Among these, the period of 3,100-3,600 BP is considered to be a maximal period for the formation of beach rock, judging from its occurrence (Xie, 1983). Problems remain, however, as to why on the Fujian coast no beach rock dated to 5,000-6,000 BP and corresponding to the Hypsithermal is known; also, as to why the peaks of the relative 'oscillations' in sea level recorded on this coast are slightly different from those obtained by the study of chenier ridges and associated deposits. A series of minor 'oscillations' in the Holocene sea-level curves has similarly been observed in Taiwan. Here phases of marine flooding, timed to 5,000-6,500 BP, 3,500-4,000 BP, 2,000-2,700 BP and 1,200-1,700 BP, together with four phases of marine removal in between them, were recognised on the basis of both ^{14}C dates from Holocene marine deposits, particularly from raised coral reef data, and from archaeo-logical — 'cultural' sites (Lin, 1969). In addition to these movements in relative sea level an older phase of marine inundation has also been recorded in Taiwan as occurring between ~ 7,000 to 8,500 BP (Lin, 1969).

In southern China coral reefs have formed an important focus for study-ing Holocene sea-level changes. In the South China Sea Islands, atolls underlain by more than 1,000 m of thick reef limestone have been developed since the end of the Tertiary. Holocene fringing reefs and barrier reefs can be seen commonly in the northern coastal province of the South China Sea, as well as on the coast of Hainan Island (Zaho *et al.*, 1983). In contrast, a flight of Holocene raised coral reefs characterises the coast of Taiwan, except its northern part (Lin, 1969; Peng *et al.*, 1977; Hashimoto *et al.*, 1970, 1972; Konishi *et al.*, 1968).

The formation age of coral reefs can be divided into the following groups based on about 100 radiocarbon dates, two-thirds of them coming from Taiwan: 7,800-8,500 BP (I), 6,900-7,300 BP (II), 4,800-6,300 BP (III), 4,000-4,500 BP (IV), 3,100-3,800 BP (V), 1,300-2,800 BP (VI) and younger than 1,000 BP (VII) (Figure 11.6). Geological observation of coral reefs suggests that stages I and II represent the initial period in coral growth occurring at the time of temperature rise to the Hypsithermal and stage III represents the period of maximum growth–flourishing. After stage IV coral reef formation enters a final phase of senescence and decline (Zaho, 1983). In Taiwan, formed as it is in the tectonically active collision zone between the Eurasian Plate and Philippine Sea Plate, every stage in coral reef growth occurs at a different height as shown in Figure 11.6, where the relation between ^{14}C dates and altitude is plotted. In contrast, coral reefs in the other two areas mentioned are very low in height. Despite the difference in recent vertical deformation in these areas, a synchroneity in stages of coral growth can be recognised between these areas, suggesting that the 'oscillations' mentioned earlier should be of eustatic origin (Zaho, 1983).

Korean Peninsula

In the relatively stable crustal area of the Korean Peninsula, Holocene marine terraces are rather limited in distribution and are low in height, compared to the Japanese Islands or Taiwan. A study of Holocene sea-level changes started in Korea in the late 1960s (Park, 1969). These changes are recorded on the east coast in the landforms and sediments of small coastal plains which lie less than several metres above sea level and are characterised by the presence of two or three rows of beach ridges and inter-ridge lowlands, underlain by peaty deposits. On the western Yellow Sea coast, however, estuarine deposits containing transgressive sedimentary facies, deposited in small drowned valleys, record the Holocene sea-level changes here.

Two sea-level curves are shown in Figure 11.7, based on the recent studies along these eastern and western coasts. The sea-level curve for the east coast is supported by nineteen ^{14}C dates obtained from borehole and surface section material from the coastal plains of Jumoonjin, Kangareung and Ulsan (Jo, 1980). Results of particle size analyses were used for supplemental correction of the curve before ~ 6,000 BP and those of 'pollen analysis' for the curve after this date. Judging from the curve (time-depth data plot) shown in Figure 11.7, sea level at ~ 10,000 BP was at a

Figure 11.6: Relation between radiometric ages and height of Holocene coral reefs in China. 1. Coast of Guandong and Guangxi. 2. Coast of Hainan I. 3. South China Sea Islands. 4. North east of Taiwan. 5. Diaoyu I. 6. Hengchun Peninsula, Taiwan. 7. Southern Taiwan. 8. East coast of Taiwan. 9. Position of palaeo sea level

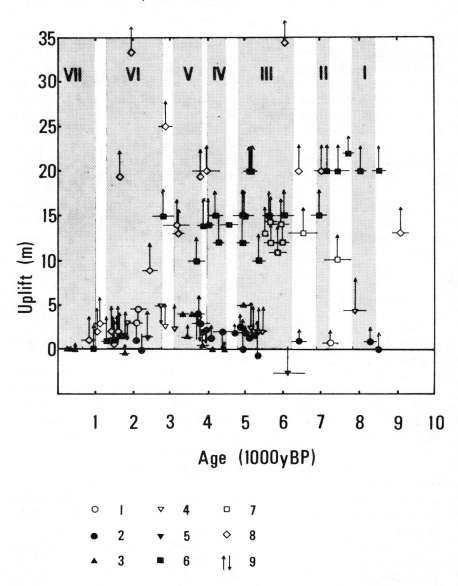

Source: Simplified from Zaho (1983).

Figure 11.7: Sea-level changes in the Korean Peninsula eastern coast (Jo, 1980): 1. Summarised sea-level curve. 2. Minor fluctuations deduced from grain size analysis for Holocene deposits. 3. Minor fluctuation deduced from palynological examination of back marsh deposits. Western Coast (Bloom and Park, 1985)

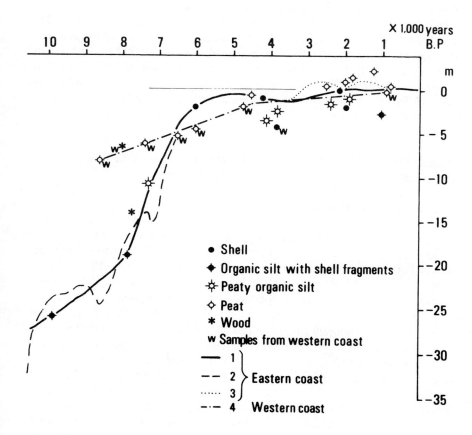

level of −25 m msl rising to −10 m msl at ~7,000 BP, at an average rate of ~ 5 mm a⁻¹. At ~ 6,000 BP it reached to a height close to its present position. Since then, minor sea-level fluctuations appear to have occurred resulting in the development of a series of beach ridges and inter-ridge lowlands. It is estimated that minor phases of marine inundation occurred at ~ 3,000 BP and also at ~ 1,800 BP, with minor negative tendencies occurring at ~ 4,000 BP and ~ 2,000 BP. On the west coast, the trend for sea-level change is based upon eight ¹⁴C dates from the Kimje, Kunsan and Baenaru areas. These show sea-level rising at a rate of 1.6 mm a⁻¹ from ~ 8,500 BP to 4,800 BP and at the reduced rate of 0.4 mm a⁻¹ since ~ 4,800

BP. No data have been obtained to show any minor 'oscillations' of sea level (Bloom and Park, 1985). Bloom and Park (1985) have suggested that because the pattern of sea-level change on this Yellow Sea coast is very similar to that of aseismic areas in the mid latitudes, such as the southern coast of the USA or of Northwestern Europe and is also similar to the Clark *et al.* (1978) theoretically predicted curve for sea-level Zone IV areas, where the Korean Peninsula is located (Clark *et al.*, 1978), then data for this coastline should represent the regional pattern of eustatic sea-level change. However, the relative sea-level position at ~ 9,000 BP on the west coast is about 20 m higher than that on the east coast, despite the south-westward crustal downtilting that is assumed to have occurred in the Korean Peninsula during the Late Quaternary (Oh, 1981). This is but one of the problems to be resolved in this area.

DISCUSSION OF THE PROBLEMS OF HOLOCENE SEA-LEVEL CHANGES IN THE NORTHWESTERN PACIFIC RIM

The recognition of palaeo sea level

An examination of palaeo sea-level indicators is essential in the discussion of any sea-level studies (readers are referred to Part Two and other chapters in Part Three of this volume for consideration of sea-level indicators in other regions). An accurate determination of palaeo sea-level position is especially important for the Holocene, where the various stages in sea-level recovery are recorded both over short intervals of time and with low amplitudes of vertical (height) change in position. In this region, on rocky coasts, emergent coastal landforms such as emerged sea caves, marine notches — shoreline angles, or the inner edge of marine terraces, are used as indicators of palaeo sea levels higher than those of the present. However, a problem remains as to which part of these emergent landforms really represents the palaeo 'mean sea level' (Sakaguchi, 1983a). Further consideration, in particular an examination of modern nearshore-submarine topography, paying special attention to the affects of wave action corresponding to the various tidal levels, is required to solve this problem. Another difficulty is a lack of datable material from these landforms, except for the occurrence in some instances of shells which live in a rocky intertidal environment and raised coral reefs. Fossil barnacles *Balanus* spp. and worm tubes, derived from *Pomatoleios krausii* or *Pomatoleios cressladi*, are often found from rocky coasts. Usually, however, they are unsuitable for dating purposes due to the small amount of material provided and possible contamination consequent upon the very thin crust of these fossils. Similar sorts of problems exist in the case of coastal plains composed of beach ridge series. Although the tops of the ridges are known to have formed above mean sea level it is not known by exactly how much, or which part of beach ridge corresponds to the mean sea level.

Former indicators of sea-level change associated with the postglacial rise in sea level are often preserved as coastal lowlands of varying dimension underlain by a sequence of marine deposits. The surfaces of these lowlands are usually overlain by terrestrial deposits after their emergence. Accordingly, it is necessary to undertake extensive stratigraphic studies to recognise the real marine limit in such areas.

Intensive excavation works have been carried out in several places in Japan in order to collect systematic samples for the location of the marine limit as well as to identify the pattern here of Holocene sea-level recovery. In the southern Kanto area, central Japan, a dendritic pattern of shorelines associated with drowned river valleys, formed at the time of the 'culmination' of the 'Jomon Transgression', has been reconstructed on the basis of detailed stratigraphic surveys and molluscan assemblage analysis (Matsushima, 1984). Examination of palaeoenvironmental changes occurring in association with the sea-level rise has also been undertaken.

To the southeast of the Plain excavations at twelve localities were carried out in the small Takagami lowland at +9 to 10 m msl (Fig. 11.8). This forms a depositional surface underlain by recent sediments, representing a former drowned valley which dissects the Late Pleistocene marine terrace of the Choshi Peninsula. The excavations show that the top of the Takagami lowland is underlain by a terrestrial peat layer 4 to 6 m thick, the basal part of which is dated at between ~ 4,500 and 5,000 BP. It is suggested, therefore, that the emergence of the Takagami lowland took place between ~ 5,500 and 6,000 BP, immediately prior to the deposition of the peat layer. Figure 11.9 represents the results of sediment and palaeontological analyses, such as those of molluscs, diatoms, pollen and ostracods, which were obtained from the longest core taken from the central part of the lowland. A brackish water to shallow-marine molluscan assemblage representing intertidal conditions is found in the lower part of the core. In the middle part, a faunal assemblage consisting of species common to a muddy bottom, greater than several metres in depth, predominates. The upper part of the core shows a return to the shallow-marine assemblage characteristic of the base of the core, as well as the occurrence of other assemblages indicative of shallow, sandy-bottom intertidal environments. These changes in water depth and environment, deduced from the molluscan assemblages, are concordant with those estimated from the examination of the diatom, ostracod and pollen material. The heights of the upper limit of marine deposition obtained by the different indices are also in good agreement. This level occurs at +3.3 m msl on the basis of sediment and diatom analyses, +2.4 m from the molluscs, more than +1.8 m by ostracod analysis and +3.2 m from the pollen data (Ota et al., 1985). By comparison, however, the top of the marine limit near the valley wall in the

Figure 11.8: Map showing landforms of the Choshi Peninsula, Kanto district. The area of Late Pleistocene marine terrace is shaded. Solid circles with figures are excavation sites and height of the upper limit of Holocene marine deposits in metres. The solid square is the regional location of Figure 11.9

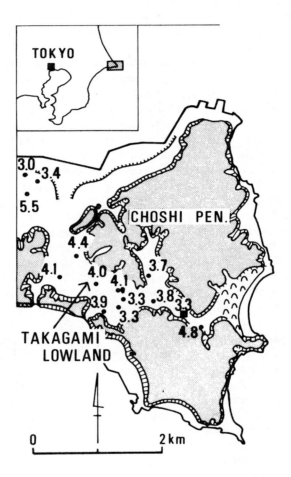

Takagami lowland is at nearly +5 m msl. Thus the marine limit in the central part of the lowland, at +3.3 m msl, is considerably lower than this level. It is pointed out, therefore, that the 'marine limit' as such sometimes does not represent the real former sea-level position. This observation is especially important when considering the magnitude of sea-level fluctuation accurately in a depositional lowland.

362

Figure 11.9: Geological section, ^{14}C dates and biological assemblages of the core at the Takagami lowland, Choshi Peninsula

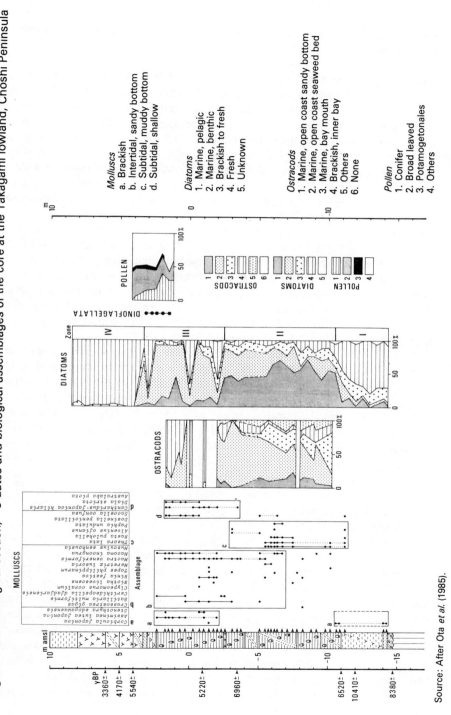

Source: After Ota *et al.* (1985).

363

Former sea-level age and height at the 'culmination' of postglacial sea-level recovery

According to the predictions by Clark *et al.* (1978) all the areas discussed in this chapter belong to sea-level Zone IV, where relative sea-level heights at ~ 5,000-6,000 BP should be lower than the present position. However, most of the relative sea-level data in these areas show that sea level was higher than today at that period (cf. Figs 11.3 and 11.4). In particular, a significantly higher relative sea-level position at ~ 6,000 BP is recorded by abundant coastal landforms and marine deposits in many parts of Japan and Taiwan. In addition, even in relatively stable areas, or those characterised by subsidence in a longer time span, sea level of this period is often preserved at a slightly higher position than the present. This evidence is taken as support for the now generally accepted view in Japan and China (e.g. Ota *et al.*, 1982; Zaho and Zhang, 1982; Wang and Wang, 1980) that sea level was actually higher than the present at that time. An exact height for the palaeosea-level is not yet confirmed, because of the difficulty in completely eliminating any tectonic effect from a given relative sea-level data set. However, the idea that the present sea-surface position is the highest sea level since the last glacial period is still supported by some researchers (e.g. Xia, 1981; Bloom and Park, 1985).

Figure 11.10 summarises available data on the height of terraces and deposits dated to between 4,500 and 7,000 BP. These are regarded as corresponding approximately with the 'culmination' of the postglacial rise in sea level. In areas where a number of ^{14}C dates and height data were obtained only representative ones are plotted in the figure. It is obvious from this that a considerable regional–local difference can be recognised in both data height and age. In the tectonically active zone of Japan the following notable local differences particularly are observed. The Pacific coasts of central and southwestern Japan such as south Kanto, south Shikoku or Kikai Island are characterised by a great amount of uplift, ranging in height from 8 to 25 m asl. In these areas the uplift rate since the last Interglacial has also been high, reaching values of 1 to 1.5 mm ka^{-1} although this is complicated here by a landward crustal tilt; a peculiar deformation pattern occurring so close to the northern limit of the Philippine Sea Plate (Ota and Yoshikawa, 1978). The western coast of northern Honshu as well as that of Hokkaido also show high elevations for shorelines formed during the postglacial 'culmination', reaching to a maximum height of +11 m msl. A new plate boundary off these coasts, proposed by Nakamura (1983) and others, may be responsible for such a high uplift rate. By comparison, in most of Hokkaido and some other places the former shoreline height for this period is placed at +5 m msl or slightly lower. In addition, an apparent subsidence can be seen on the Seto Island Sea coast and in north Kyushu, where a submergence of early Jomon

Figure 11.10: Map showing the distribution of the present height of dated marine terraces or deposits between 4,500 BP and 7,000 BP (compiled from various sources). HK, Hokkaido; Hs, Honshu; S, Shikoku; K, Kyushu; RI, Ryukyu Islands.

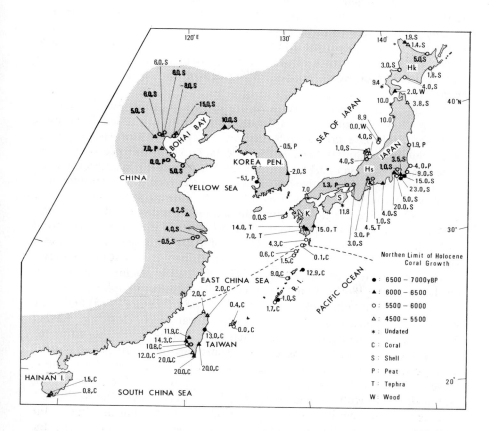

archaeological sites is known (Tachibana and Sakaguchi, 1971). In the areas of land uplift the 'culmination' of the postglacial rise in sea level is usually ^{14}C dated to between ~ 6,000 and 7,000 BP. In contrast, this event is younger than ~ 6,000 BP in Hokkaido and in other places recording a low shoreline height. Accordingly, this approximate positive correlation between the height and age of the 'culmination' event indicates a differential pattern of emergence, with shorelines of this age forming earliest in areas of land uplift.

Akahoya ash (K-Ah), dated radiometrically to ~ 6,300 BP, forms an excellent time marker horizon for recognition of the 'culmination' in areas where no datable material is available (Machida and Arai, 1978). In addition, it forms an important dating plane which helps in the understanding

365

of different environments associated with the 'Jomon Transgression' and affected by differences in tectonic movement. K-Ah is found in the fluvial deposits overlying marine Holocene sequences in the areas of strong crustal uplift (> 3 m ka^{-1}), whilst it rests on the top of Holocene marine sediments in the areas of less extreme emergence. In the stable or subsiding zones, it is usually found within marine deposits at elevations of -2 to -40 m msl. Overall it is hoped that the use of the K-Ah horizon may provide more data for the identification of stable coastal areas and that, accordingly, it can be used to elucidate regional differences in Holocene crustal movements.

The distribution of dated surfaces or deposits in Japan is still not sufficient to allow presentation of a detailed local picture of relative sea-level data and Holocene crustal deformation, except in a few specific areas where intensive studies have been undertaken. However, data are even more limited in China and Korea, as shown in Figure 11.10. An outline of the age and height of relative sea level here at the time of the 'culmination' event is tentatively summarised as follows. In Taiwan, the position of the former sea level at the time of the 'culmination' is recorded in raised coral reefs at high elevations of up to $+20$ m msl in many places. A raised coral reef dated to $2,850 \pm 120$ BP at $+25$ m msl in the vicinity of Hualien, east Taiwan (Konishi *et al.*, 1968), places the former 'culmination' sea-level event here at an even higher level. The Holocene marine terrace itself reaches to nearly $+40$ m in height (Lin, 1969). Changes in uplift rate in this region during the Holocene have been discussed by Taira (1975) and Peng *et al.* (1977), based on the dated raised coral reefs. However, further discussion of results should be avoided at this stage, since the estimated land uplift rates depend on different 'eustatic' sea-level curves and there is no agreement on the values (or existence) of a eustatic sea-level curve. Nevertheless, it can be said that Taiwan, except in its northern part, forms one of the areas with the highest relative land uplift rates in the world.

There is a notable difference in the height of sea level at the time of the 'culmination' event between the Japanese Islands or Taiwan and the Korean Peninsula, where relative sea-level at $\sim 6,000$ BP was slightly lower than the present. In China Holocene sea-level data are known from the northern and eastern parts. No information, however, is available from southern China except from Hainan Island. This may be caused partly by an incomplete collection of publications. As far as can be seen (Fig. 11.10), the former relative sea-level height at the 'culmination' is very low in China, usually less than $+7$ m msl, and no obvious differential vertical movement is recognised over an extensive area, except a relatively high altitude of $\sim +10$ m msl near Liaotung Peninsula, northern China. It seems that such a pattern reflects the 'stable' tectonic character of this continental region.

Minor fluctuations of holocene sea-level curves and their significance

Figures 11.11 and 11.12 are presented in order to examine the relationship between the sea-level changes, especially fluctuations since the Hypsithermal, and climatic changes in eastern Asia. The following discussion is based on these figures and other data already shown.

A negative sea-level tendency at 10,000 BP

A minor phase of marine removal recorded during the period of postglacial rapid sea-level rise to the Hypsithermal is recognised at ~ 10,000 BP in most of the Japanese sea-level data (cf. Figure 11.3). A relative fall or stagnation of sea-level rise of a similar age is also known in eastern Korea (Jo, 1980). This negative sea-level tendency may have resulted from the onset of cold climatic conditions with a peak at about that period, as revealed by

Figure 11.11: Sea-level fluctuations and climatic changes in Japan. Shaded zones represent the 'middle Jomon Regression' and the 'Yayoi Regression'

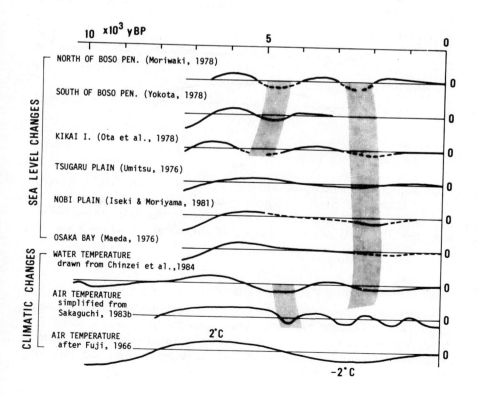

367

Figure 11.12: Sea-level changes, formation age of coral reefs/beachrock and climatic changes in Korea and China

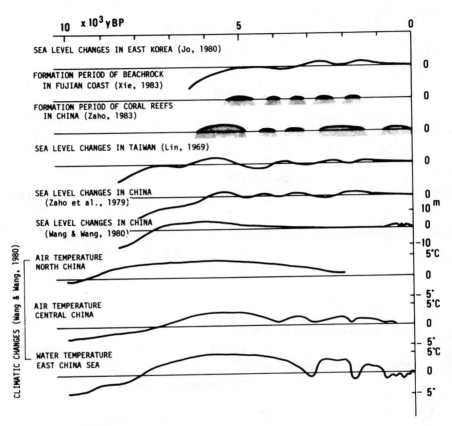

palynological and microfossil analyses of deep-sea cores taken from the Pacific Ocean off Kanto (Chinzei *et al.*, 1984). In China evidence to support such a tendency is hard to find, however. Reassessment of the sea-level record here is required if this cold climate effect was real and if it covered an extensive area including China.

High relative sea level at the Hypsithermal

Sea level during the period 5,000-7,000 BP is commonly regarded as being slightly higher than the present in many cases in Japan and China (cf. Fig. 11.3). This corresponds to the Hypsithermal period during which optimal-peak environmental conditions existed, as supported by the following phenomena in the areas discussed:

368

(a) Palynological data from the Japan Sea side of Honshu, Japan, show that the air temperature in the region was 2-3°C higher than the present with a 200-300 m rise of the timber line (Fuji, 1966).

(b) Littoral molluscan assemblages on the Pacific coast of Japan suggest that the water temperature was ~ 5°C higher than the present (Matsushima and Oshima, 1974).

(c) A warm climate is indicated by detailed palynological data from a peat core taken from Ozegahara, central Japan (Sakaguchi, 1983b) and from microbiological assemblages in deep-sea cores (Chinzei et al., 1984), as shown in Figure 11.11.

(d) Palaeo air temperature, reconstructed by bioclimatic, floral and archaeological data, was 0.5°C higher than the present in north and central China (Wang and Wang, 1980).

(e) Palaeo water temperature in the Yellow Sea and East China Sea zone, estimated from marine faunal analyses, was also ~ 5°C higher than the present (Wang and Wang, 1980).

Sea-level fluctuations since the Hypsithermal

As already stated, minor changes in relative sea level after the 'culmination' of the postglacial sea-level recovery are recognised in most of the Japanese sea-level curves (Figs. 11.3 and 11.11). The younger marine withdrawal (negative tendency), at ~ 2,000-3,000 BP has been reported from 20 areas of Japan, based on the presence of shallow buried channels beneath the Holocene marine deposits, submerged trees and falling water levels in coastal lagoons. These features are observed regardless of the nature of tectonic setting, implying a probable eustatic origin for this negative sea-level tendency, named the 'Yayoi Regression' (cf. Yoshikawa et al., 1981). Subsequent studies show another earlier relative fall in sea level prior to the 'Yayoi Regression', timed at ~ 4,000-5,000 BP, followed by a later rise in relative sea level.

In south Boso and Kikai Island, the northern Ryukyus, the Holocene marine terraces and coral reefs here are composed of four well-defined steps, the highest of which records the 'culmination' of the postglacial sea-level recovery (Fig. 11.2). The second higher terrace of the Boso Peninsula is underlain by relatively thick marine deposits dated to ~ 3,600-4,700 BP, which in turn overlie unconformably the older Holocene sequences (Yokota, 1978). The second higher raised coral reef of Kikai, dated at ~ 3,800-5,000 BP, is the widest among the four Holocene coral reefs and consisted of a very flat lagoonal surface fringed by higher reef edges (Ota et al., 1978). No fossil corals of between 5,000 and 5,800 BP in age have been found in Kikai Island. These data suggest that the second higher terraces formed were accompanied by a minor rise in relative sea level, subsequent to a minor removal of the marine effect, which followed the culmination of the 'Jomon (Postglacial) Transgression'. This relative fall is called the 'middle Jomon Regression', and is probably the result of an eustatic sea-level lowering. The presumed 'eustatic' origin for these two phases of negative sea-level tendency seems to be supported by the following facts, which indicate the presence of cold climatic conditions corresponding to the age of these two events. For the 'middle Jomon Regression':

(a) Warm water species of a shallow water molluscan assemblage began to decrease in number at ~ 5,000 BP and disappeared completely around 4,500 BP in the embayment in south Kanto (Matsushima, 1979).

(b) The $\delta^{18}O$ values of *Meretrix lamarcki* shells which occur at prehistoric shell middens in south Kanto indicate that the sea water temperature had fallen consistently since ~ 6,000 BP, reaching its minimum at about 4,500 BP (Chinzei *et al.*, 1980).

(c) Microfossil analysis and $\delta^{18}O$ values in deep-sea cores taken from eastern and southern Kanto indicate the presence of a weak but persistent cold episode between 5,000 and 4,000 BP (Chinzei *et al.*, 1984).

(d) A palynological analysis of the Ozegahara peat shows the presence of a remarkable cold stage of 'Mid/Late Jomon' age (Sakaguchi, 1983b).

For the 'Yayoi Regression':

(a) A cold stage at ~ 2,000 BP is evidenced from microfossil analyses of deep-sea cores (Fig. 11.11) (Chinzei *et al.*, 1984), and

(b) A recent cold stage at 865 BC to 398 BC is recorded by the palynological work on the Ozegahara peat; this cold stage is registered on a global basis (Sakaguchi, 1983b).

Thus, the two periods of negative sea-level tendency coincide approximately with the phases of cold climate, as shown in the dotted zone of Figure 11.11, although each cold period is slightly different depending on the number and distance of sampling horizons taken. The magnitude of these relative sea-level fluctuations seems to be small, probably less than several metres judging from the thickness of the marine deposits. However, the exact vertical amplitude of these changes has yet to be determined. Figure 11.10 indicates that two cycles of relative sea-level rise and fall are observed in areas with high uplift rates and only one in other areas. Accordingly, these areas of land emergence can be regarded as being suitable locations for studying minor sea-level fluctuations, since the process of continuing land uplift will separate each successive sea-level trace into clearly definable positions.

Minor sea-level fluctuations are also recognised in east Korea (Jo, 1980), Taiwan (Lin, 1969) and in parts of China (e.g. Zaho *et al.*, 1979). It seems that the relative fall in sea level at ~ 2,500-3,000 BP corresponds to a cold climatic period shown in temperature curves by Wang and Wang (1980) and that this phase is correlated with the 'Yayoi Regression' in Japan. Although older cycles of fluctuations in sea level are indicated here (Figure 11.11) corresponding climatic change is not obvious. A problem with the fluctuation of relative sea level in the record from China occurs if all the changes shown in Figure 11.11 represent eustatic sea-level changes. The presence of several relative sea-level fluctuations in Taiwan seems to be reasonable as this forms an area of strong land uplift. However, similar changes are also shown for northern and central China (Zaho *et al.*, 1979) where the present height of the palaeo sea level at the 'culmination' of postglacial sea recovery is $<$ +6 to +7 m msl. It may be possible that a series of chenier ridges (which in part form the data base here) can be formed by an intermittent coastal progradation, due to discontinuous supply of deposits by large rivers, without significant oscillation in sea level taking place. Further study is necessary here on the correlation of minor fluctuations over an extensive area and their causes.

CONCLUSIONS

(1) It is probable that sea level during the time of the postglacial Hypsithermal was slightly higher than the present one. The age of the 'culmination' in postglacial sea-level recovery, however, is different from place to place, ranging from ~ 7,000 BP in areas of land uplift to ~ 5,000 BP in relatively stable or slightly subsiding zones.

(2) Present heights of sea level/shorelines formed during the 'culmination' phase also vary regionally–locally, attaining heights of +20 to +30 m in areas along subducting plate boundaries, such as the Pacific coast of central and southwestern Japan or in the collision zone of Taiwan.

(3) At least two phases of negative sea-level tendency are recognised in Japan after the 'culmination': the earlier 'middle Jomon Regression' and the later 'Yayoi Regression'. It is likely that these events are eustatic in origin, caused by deterioration to cold climate and consequent ice sheet expansion. It is uncertain, however, as to the correlation of these fluctuations throughout the areas discussed, or the vertical magnitude of the changes recorded.

REFERENCES

Bloom, A.L. (ed.) (1977) *Atlas of Sea Level Curves*, ICGP Project 61, Cornell University, Ithaca, NY.

——— and Park, Y.A. (1985) 'Holocene sea-level history and tectonic movements, Republic of Korea', *Quat. Res. Japan*, 24, 77-83.

Chida, N., Matsumoto, H. and Obara, S. (1984) 'Recent alluvial deposits and Holocene sea-level change on Rikuzentakada coastal plain, northern Japan', *Ann. Tohoku Geogr. Ass.*, *37*, 232-9.

Chinzei, K., Oba, T., Koike, Y., Matsushima, Y. and Kitazato, H. (1980), 'Change in oxygen isostage of the shells from shell mounds and palaeoenvironments in the prehistoric age', in *Researches on Archeological Sites, Cultural Properties and so on by Natural Scientific Methods*, 103-17. Unpublished report, Tokyo, pp. 103-17.

—— Okada, H., Oda, M., Oba, T., Kitazato, H., Koizumi, J., Sakai, T., Tanimura, Y., Fujikoka, K. and Matsushima, Y. (1984) 'Paleooceanography since the last glacial in the Pacific along the east coast of Honshu', in *Comprehensive Report of Scientific Researches on Preservation of Cultural Properties*, unpublished report, Tokyo, pp. 441-57.

Clark, J.A., Farrel, W.E. and Peltier, W.R. (1978) 'Global changes in postglacial sea level: a numerical calculation', *Quat. Res.*, *9*, 265-87.

Fuji, N. (1966) 'Climatic changes of postglacial age in Japan', *Quat. Res. Japan*, *5*, 149-56.

Fuji, S. and Fuji, N. (1981) 'Sea-level curve in Hokuriku District during the last 20,000 years', in Y. Ota, Y. Matsushima and H. Moriwaki (eds) *Atlas of Holocene Sea-level Records in Japan*, ICGP Project 61, Cornell University, Ithaca, NY, pp. 43-4.

Furukawa, H. (1971) 'Alluvial deposits of the Nobi Plain, central Japan', *Mem. Geol. Soc. Japan*, *2*, 39-59.

Hashimoto, W., Taira, K., Kurihara, K., Imai, T. and Makino, K. (1970) 'Studies on the Younger Cenozoic deposits in Taiwan (Formosa): Part I, the Younger Cenozoic deposits of the middle part of west Taiwan', *Geol. Paleont. Southeast Asia*, *8*, 237-52.

—— Taira, K., Kurihara, K., Imai, T. and Makino, Y. (1972) 'Studies on the Younger Cenozoic deposits in Taiwan (Formosa): Part II, The Younger Cenozoic deposits in south and east Taiwan', *Geol. Palaeont. Southeast Asia*, *10*, 215-303.

Iseki, H. and Moriyama, A. (1981) 'Sea-level changes in the Nobi Plain and adjacent area', in Y. Ota, Y. Matsushima and H. Moriwaki (eds), *Atlas of Holocene Sea-level Records in Japan*, IGCP Project 61, Cornell University, Ithaca, NY, pp. 67-8.

Jo, W. (1980) 'Holocene sea-level changes on the east coast of the Korea Peninsula', *Geogr. Rev. Japan*, *53*, 317-28.

Kaizuka, S., Naruse, Y. and Matsuda, I. (1977) 'Recent formations and their basal topography in and around Tokyo Bay, central Japan', *Quat. Res.*, *8*, 32-50.

Konishi, K., Omura, A. and Kimura, T. (1968) '$^{234}U/^{230}Th$ dating of some Late Quaternary coral limestone from southern Taiwan (Formosa)', *Geol. Paleont. Southeast Asia*, *5*, 211-24.

Lin, C.C. (1969) 'Holocene geology of Taiwan', *Acta Geol. Taiwanica. Sci. Rep. National Taiwan Univ.*, *13*, 83-126.

Lin, D.K. (1983) 'Progress of sea-level variations research in China during the Late Quaternary', *Abstracts of Papers; International Symposium on Coastal Evolution in the Holocene, Tokyo*, pp. 69-77.

Machida, H. and Arai, F. (1978) 'Akahoya Ash: A Holocene widespread tephra erupted from the Kikai caldera, south Kyushu, Japan', *Quat. Res. Japan*, *17*, 143-64.

Maeda, Y. (1976) 'The sea-level changes of Osaka Bay from 12,000 BP to 6,000 BP. Environmental changes in the Osaka Bay area during the Holocene. Part I,' *J.*

Geosci., Osaka City Univ., 20, Art. 3, 43-59.

—— Yamashita, K., Matsushima, Y. and Watanabe, M. (1983) 'Marine transgression over Mazukari Shell Mound on the Chita Peninsula, Aichi Prefecture, central Japan', *Quat. Res. Japan, 22,* 213-22.

Matsuda, Y., Ota, Y., Ando, M. and Yonekura, N. (1978) 'Fault mechanism and recurrence time of major earthquakes in southern Kanto District, Japan, as deduced from coastal terrace data', *Bull. Geol. Soc. Am., 89,* 1610-18.

Matsushima, Y. (1979) 'Littoral molluscan assemblages during the postglacial Jomon Transgression in the southern Kanto, Japan', *Quat. Res. Japan, 17,* 243-65.

—— (1984) 'Shallow marine molluscan assemblages of postglacial period in the Japanese Islands: its historical and geographical changes induced by the environmental changes', *Bull. Kanagawa Pref. Mus., 15,* 37-109.

—— and Oshima, K. (1974) 'Littoral molluscan fauna of the Holocene climatic optimum (5,000-6,000 a.BP) in Japan', *Quat. Res. Japan, 13,* 135-59.

Moriwaki, H. (1979) 'The landform evolution of the Kujukuri coastal plain, central Japan', *Quat. Res. Japan, 18,* 1-16.

Nakamura, K. (1983) 'Possible nascent trench along the eastern Japan Sea as the convergent boundary between Eurasian and North American plates', *Bull. Earthquake, Res. Inst., Univ. Tokyo, 58,* 711-22.

Nakata, T., Takahashi, T. and Koba, M. (1978) 'Holocene emergent coral reefs and sea-level changes in the Ryukyu Islands', *Geogr. Rev. Japan, 51,* 87-108.

—— Koba, M., Imaizumi, T., Jo, W.R., Matsumoto, H. and Suganuma, T. (1980) 'Holocene marine terraces and seismic crustal movements in the southern part of Boso Peninsula, Kanto, Japan', *Geogr. Rev. Japan, 53,* 29-44.

Naruse, Y. and Ota, Y. (1984) 'Sea-level changes in the Quaternary in Japan', S. Horie (ed.), *Lake Biwa,* Dr W. Junk Publishers, Dordrecht, pp. 461-73.

Oh, G.H. (1981) 'Marine terraces and their tectonic deformation in the coast of the southern part of the Korean Peninsula', *Bull. Dept. Geogr. Univ. Tokyo, 13,* 11-61.

Omoto, K. (1979) 'Holocene sea-level changes: a critical review', *Sci. Rep. Tohoku Univ.,* 7th ser., *29,* 205-22.

Ota, Y. (1982) 'Holocene marine terraces of uplifting areas in Japan', in D.J. Colquhoun (ed.), *Holocene Sea Level Fluctuations, Magnitude and Causes,* Dept. of Geology, University of South Carolina, Columbia, SC, pp. 118-34.

—— and Machida, H. (in press) *Quaternary Sea-Level Changes in Japan.*

—— Machida, H., Hori, N., Konishi, K. and Omura, A. (1978) 'Holocene raised coral reef of Kikai-jima (Ryukyu Islands): an approach to Holocene sea-level study', *Geogr. Rev. Japan, 51,* 109-30.

—— Matsuda, T. and Naganuma, K. (1976) 'Tilted marine terraces of the Ogi Peninsula, Sado Island, central Japan, related to the Ogi earthquake of 1802', *Zishin,* 2nd ser., *29,* 55-70.

—— Matsushima, Y., Miyoshi, M., Kashima, K., Maeda, Y. and Moriwaki, H. (1985) 'Holocene environmental changes in the Choshi Peninsula and the surroundings, easternmost Kanto, central Japan', *Quat. Res. Japan, 24,* 13-30.

—— Matsushima, Y. and Moriwaki, H. (eds) (1981) *Atlas of Holocene Sea-Level Records in Japan,* ICGP Project 61, Cornell University, Ithaca, NY.

—— Matsushima, Y. and Moriwaki, H. (1982) 'Notes on the Holocene sea-level study in Japan: on the basis of the "Atlas of Holocene Sea-Level Records in Japan"', *Quat. Res. Japan, 21,* 133-43.

—— and Yoshikawa, T. (1978) 'Regional characteristics and their geodynamic implications for Late Quaternary tectonic movement deduced from former shorelines in Japan', *J. Phys. Earth*, 26, Supp., S379-S389.

Park, A.Y. (1969) 'Submergence of the Yellow Sea coast of Korea and stratigraphy of the Sinpyenngheon, Kimje, Korea', *J. Geol. Soc. Korea*, 5, 57-66.

Peng, T., Li, Y. and Wu, F.T. (1977) 'Tectonic uplift of the Taiwan Island since the Early Holocene', *Mem. Geol. Soc. China*, 2, 57-69.

Sakaguchi, Y. (1983a), 'On the postglacial sea-level changes in Japan', *J. Geogr.*, 92, 448-54.

—— (1983b) 'Warm and cold stages in the past 7,600 years in Japan and their global correlation, especially in climatic impacts to the global sea-level changes and the ancient Japanese history', *Bull. Dept. Geogr. Univ. Tokyo*, 15, 1-31.

Tachibana, K. and Sakaguchi, K. (1971) 'Age of beach rock containing the Jomon pottery in the Goto Islands', *Quat. Res. Japan*, 10, 54-9.

Taira, K. (1975) 'Holocene crustal movements in Taiwan as indicated by radio-carbon dating of marine fossils and driftwood', *Tectonophys.*, 28, T1-T5.

Tohoku Nosei-Kyoku (1979) 'A report on the earth surface submergence at Haranomachi District, Fukushima Prefecture, north Japan', unpublished report, Tokyo.

Toyoshima, Y. (1978) 'Postglacial sea-level change along San'in District, Japan', *Geogr. Rev. Japan*, 51, 147-57.

Umitsu, M. (1976) 'Geomorphologic development of the Tsugaru Plain in the Holocene period', *Geogr. Rev. Japan*, 49, 714-35.

—— (1979) 'Geomorphic development of the Nobi Plain in the Late Quaternary', *Geogr. Rev. Japan*, 52, 199-208.

Wang, J. and Wang, P. (1980) 'Relationship between sea-level changes and climatic fluctuation in east China since the Late Pleistocene', *Acta Geogr. Sinica*, 35, 299-312.

Xia, D. (1981) 'Whence comes the high sea level during the Holocene?' *Acta Oceanol. Sinica*, 3, 601-9.

Xie, Z. (1983), 'The beach rock stages in the Holocene and sea-level changes along the Fujian coast', *Abstracts of Papers: International Symposium on Coastal Evolution in the Holocene, Tokyo*, pp. 149-50.

Yanagida, M. and Kaizuka, S. (1982) 'Recent Chinese studies on the sea-level changes since the last interglacial in the Sea of Bohai, Yellow Sea, and east China Sea', *Quat. Res. Japan*, 21, 115-22.

Yokota, K. (1978) 'Holocene coastal terraces of the southeast coast of the Boso Peninsula', *Geogr.Rev. Japan*, 51, 349-64.

Yoshikawa, T., Kaizuka, S. and Ota, Y. (1981) *The Landforms of Japan*, University of Tokyo Press, Tokyo.

Zaho, X. (1983) 'Development of Holocene coral reefs in China and their relations on sea-level changes and tectonic movement', *Scientia Sinica, ser. B.*, 26, 413-23.

—— and Zhang, J. (1982) 'Basic characteristics of the Holocene sea-level changes along the coastal areas in China', in T. Liu, *et al.* (eds), *Quaternary Geology and Environment of China. China Ocean Press*, pp. 155-60.

—— Geng, X. and Zhang, J. (1979) 'Sea-level changes of eastern China during the past 20,000 years', *Acta Oceanol. Sinica*, 1, 269-81. (Translated into English in *Sea Level*, 8, 4-15, 1982).

—— Zhang, J., Jlao, W. and Li, G. (1980) 'Chenier ridges on the west coast of the Bohai Bay', *Xexue Tongbao*, 25, 243-7.

—— Zhang, J. and Li, G. (1983) 'Development of the Holocene coral reefs along the southern coast of Hainan Island', *Scientia Geol. Sinica*, 2, 150-9.

12

Holocene Sea-level Changes in Australasia and the Southern Pacific

David Hopley

INTRODUCTION

The vast oceanic area covered by this chapter encompasses over 25 per cent of the earth's surface (Fig. 12.1). It stretches east to west across 19,000 km or 170° of longitude, and over 14,000 km in a north to south direction from just north of the equator to the coastline of Antarctica. It is bounded by continental land masses; Antarctica lies to the south, Australia and archipelagic Southeast Asia to the west and South and Central America to the east. This, the world's largest ocean, also contains up to 25,000 islands (Douglas, 1969) including the Melanesian, Micronesian and Polynesian groups. The amount of land upon which sea-level signatures may prevail is minute and many of the islands are coral atolls whose very existence has depended on the stabilisation of sea level in the last 6,000 years. A large proportion of the sea-level evidence comes from tropical or sub-tropical shores with coral reefs and associated features providing much of this evidence. However, a wide range of coastal types are found, though with the exception of the north–south oriented coastline of South America, there is a large expanse of almost unbroken ocean in temperate and sub-polar latitudes, the temperate coastlines of southern Australia and New Zealand not extending further than 48° from the equator.

In the light of current knowledge of earth behaviour and global sea-level change, as discussed earlier in this volume, uniformity in evidence for Holocene sea levels cannot be expected over such an enormous area. Centred on the southern portion of the Pacific plate, it also includes parts of the South American, Nasca, Cocos, Antarctic, Indo–Australian, Asian and Philippine plates (Fig. 12.2) with a variety of plate marginal conditions. Included are some of the most active tectonic areas in the world but also the Australian island continent, regarded as one of the most stable areas of the earth's crust. Although distant from the major areas of glacial loading and unloading in the Northern Hemisphere, isostatic rebound from the deglaciation of Antarctica is now considered significant (see, for

Figure 12.1: The Australasian-southern Pacific region, with major cyclone belts

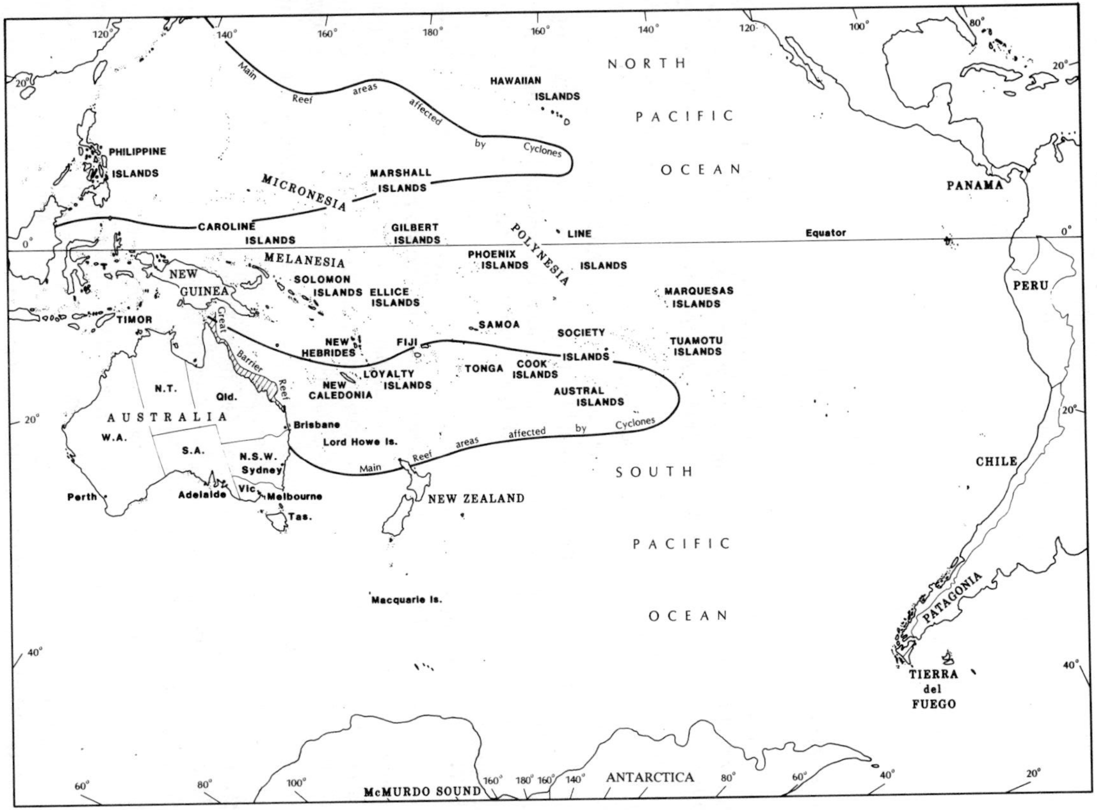

Source: After Stoddart (1971).

Figure 12.2: The tectonic setting of the Pacific, with tsunami generating earthquakes

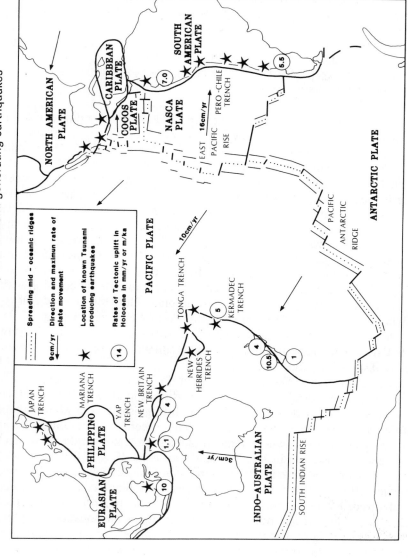

Sources: Davies (1972); Macdonald and Abbott (1977).

377

example, Clark and Lingle, 1979) whilst more local glacio-isostatic effects may be found in Patagonia, New Zealand and the few high-latitude islands such as Macquarie. The region is also affected by numerous other factors which may influence the apparent sea-level record (such as vertical movements associated with mid-plate hot spots or melting anomalies) or misrepresent the record with misleading evidence (as produced by tsunami or cyclonic storm surges).

THE TECTONIC AND ISOSTATIC SETTING: ITS EFFECT ON THE HOLOCENE SEA-LEVEL RECORD

The accepted model of plate tectonics and sea-floor spreading envisages the Pacific plate moving northwestwards at rates up to ~ 10 cm a^{-1} (McDougall, 1971; Gromme and Vine, 1972; Winterer, 1973; Hammond et al., 1974; Duncan and McDougall, 1974). Its northwestern boundary is formed by the subduction zone of the Japan, Mariana and Yap trenches where it collides with the eastward-moving Eurasian plate. Further south the collision with the northward-moving Indo–Australian plate (3 cm a^{-1}) is less direct but is still marked by significant subduction zone trenches (New Britain, New Hebrides, Tonga and Kermadec trenches) and associated active island arcs. The large islands of New Guinea and New Zealand sit astride the plate margin. In the east the Nasca and Cocos plates, moving eastward at up to 16 cm a^{-1}, come into direct collision with the continental crust of the South American plate. Here the oceanic crust is subducted beneath the sialic crust in the Peru–Chile trench (Fig. 12.2). Only the two major continental areas of Australia and Antarctica are located in a mid-plate position and have therefore escaped the major tectonic dislocation found at the margins.

The tectonic factor becomes important for Holocene sea levels only when the vertical crustal movements are sufficient in amplitude to produce displaced sea-level evidence which can be distinguished from what may be regarded as the regional eustatic pattern. There seems little doubt of this at the active plate margins, where some of the highest tectonic rates of vertical movement are recorded in Pacific island arcs (Fig. 12.2). Maximum rates of up to ~ 10 mm a^{-1} are indicated here, with Holocene shoreline features exceeding 30 m in height above modern sea level. Significantly these rates are of an order of magnitude lower than those recorded in areas of maximum glacio-isostatic rebound. However, there are locations at plate margins where insignificant vertical movements appear to have taken place in the Holocene, for instance at Tongatapu and Eua in the Tonga group (Taylor and Bloom, 1977). Very large variations in rates are also noted over horizontal distances of only a few kilometres, for example at Santo and Malakula Islands in Vanuatu, 0.3-5.0 mm a^{-1} (Taylor et al., 1980), the

378

Huon Peninsula, New Guinea, 1.9-4.0 mm a^{-1} (Chappell, 1974a), or along the Alpine fault of New Zealand, negligible to 10.5 mm a^{-1} (various sources in Soons and Selby, 1982).

Plate margin-island arc movements are also very discontinuous in a time span as short as the Holocene, although over a longer period such as 125,000 years of a glacial–interglacial cycle, the irregularity may become so small as to allow the reconstruction of glacial period eustatic sea levels (e.g. Bloom et al., 1974). The discontinuity may be illustrated by vertical displacements associated with single tectonic events. Emergence of up to 1.5 m was associated with an earthquake on 1 August 1961 at Marau Sound, Guadalcanal (Grover, 1965). Taylor et al. (1980, 1982) report up to 1.2 m of emergence associated with earthquakes in 1965 and 1971 in Vanuatu. In New Zealand an 1855 earthquake uplifted shore platforms 2.4 m near Wellington; the Murchison earthquake of 1928 involved 4.5 m of vertical movement and the 1968 Inangahua event involved vertical movements of similar magnitude (various sources in Soons and Selby, 1982). Clearly in these highly tectonic regions of the south Pacific, Holocene shoreline history is more likely to illuminate the nature of earth movements than provide data on glacio-eustatic or macro-regional hydro-isostatic patterns (see Berryman, this vol., Ch. 5 for further discussion).

As might be expected vertical movements decrease with distance from the subduction zones. This is illustrated by New Caledonia and the Loyalty Islands (Bird and Iltis, 1985), situated up to 300 km west of the Australia–Pacific plate margin (Fig. 12.1). The Loyalties show a significant degree of Quaternary tilting in a southeast to northwest direction. The southernmost islands (Walpole and Mare Islands) are atolls uplifted by as much as 138 m asl. To the north the elevated atolls decline in height to 104 m asl at Lifu, 46 m asl at Ouvea and 4 m asl at Beautemps–Beaupré. Further north the Astrolabe Reefs are non-emergent. Similar tilting is suggested by reef development on the main islands of New Caledonia (Chevalier, 1973a) where emerged reefs are found only on the southeast coast and Isle of Pines. The amount of Holocene movement is unknown but some variations of 1.0 m or so may be suggested by the work of Baltzer (1970) and Coudray and Delibrias (1972).

Further east, Australia in the centre of the Indo-Australian plate displays a scale of tectonism of an order of magnitude lower than that at the plate boundaries (see Hopley, 1983a, for discussion). However, at least some tectonic movement is suspected, sufficient to warp Holocene shore-lines, although this may be due in part to hydro-isostatic responses (see later discussion). Hopley (1974a, 1978) suggested that variations in the maximum height of the Holocene relative sea-level recovery along the Queensland coastline were closely associated with regional structure. In support Cook and Mayo (1977) have indicated up to 6 m of displacement in the last 6,000 years on the eastern side of Broad Sound, central Queensland. Further, local seismo-tectonism is quoted as the most likely cause of evidence of sea levels 2.5-3.8 m above present at the head of Spencer Gulf in South Australia (Belperio *et al.*, 1983; Hails *et al.*, 1983). In Western Australia tectonism is considered a likely factor in the sea-level history of Rottnest Island (Playford, 1977) and the adjacent Perth coastline (Brown, 1983). Significant seismic events have been recorded in Australia (Denham, 1976), one of the most recent involving over 1 m of uplift at Meckering in Western Australia in 1968 (Conacher and Murray, 1969). Even on the most stable of continents it is hazardous to interpret glacio-eustatic events from evidence of former land—sea contacts alone!

Lithospheric movements associated with oceanic hot spots

Much of the evidence for sea-level change in the Pacific comes from volcanic islands or from coral atolls with proven or presumed volcanic foundations. Many of these islands and atolls have a linear pattern in the direction indicated by Pacific plate movement and they have been presumed to form from volcanic activity over a mid-oceanic plate hot spot, or melting anomaly (McDougall, 1971; McDougall and Duncan, 1980). As the volcanic platforms move away from the hot spot thermal contraction of the lithosphere with age produces subsidence (Sclater *et al.*, 1971) and simultaneously, the opportunity for development of a fringing reef-barrier reef-atoll sequence similar to that suggested by Darwin (1842). However, the pattern of vertical movement is not uniform in direction or rate. Lithospheric loading from a new volcanic mass produces isostatic subsidence, which in turn produces a 'moat-arch' development analagous to the isostatic responses of glacio-isostasy. Peripheral to the volcanic mass there develops a crustal moat. Beyond the outer edge of the moat, flexuring develops an arch which experiences uplift measured in tens of metres (McNutt and Menard, 1978; Scott and Rotondo, 1983). At further distance from the hot spot subsidence resumes at the general oceanic plate rate, although further uplift may take place if the volcanic foundations pass over an asthenospheric bump (Scott and Rotondo, 1983; Crough, 1984).

These vertical movements will have direct implications for sea-level records, although for the effects to show up in the Holocene the rates need to be significant (Fig. 12.3). Current knowledge suggests that the rates may just fall in the levels detectable from sea-level evidence. Standard sea floor subsides at a rate of 0.11 mm a^{-1} in the first 10 Ma of cooling (Parsons and Sclater, 1977). This is sufficient to produce two-thirds of a metre of movement in the 6,000 years of Holocene sea-level stillstand (see later sections for definition of this pattern), although mid-plate rates are about a quarter of this. Along the Hawaiian chain, on the island of Hawaii within the maximum loading area, Scott and Rotondo (1983) suggest that the subsi-

Figure 12.3: Scott and Rotondo's (1983) model for island-atoll development on the Pacific lithospheric plate. Vertical arrows indicate relative rates of vertical movement; rates given from literature quoted in this chapter

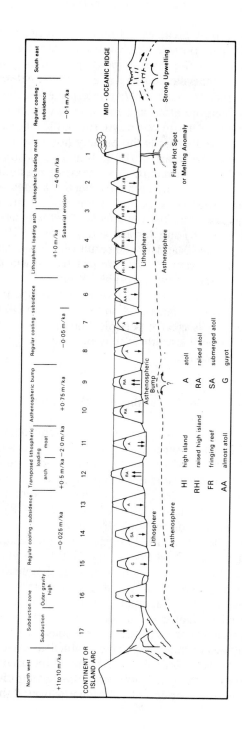

dence rates are as great as 4.4 mm a^{-1}, sufficient to have caused 0.3 m drowning of the spot on which Captain James Cook was killed in 1779. Maui, still within the crustal moat, is subsiding at 1.7 mm a^{-1}. Oahu, beyond the moat, is stable and Kauai is rising.

McNutt and Menard (1978) interpreted small elevation differences among the Tuamotu atolls as being due to volcanic loading within the past 1.5 Ma. However, more recently Crough (1984) has suggested that as the atolls have moved towards the Society hot spot elevation of up to 115 m has occurred at the mean rate of 0.76 mm a^{-1}, again sufficient to be noticeable in the Holocene record. Variations in maximum Holocene sea-level variations in French Polynesia have been recorded by Pirazzoli and Fontes (1982), the order of variation being up 1.7 m, well within the range suggested as possible from the effects of hot spot activity. Likewise McNutt and Menard (1978) and Jarrard and Turner (1979) consider that variable uplift in the Cook Islands is the result of lithospheric loading from nearby volcanoes. The recent HIPAC project has shown that the Holocene sea-level record in the Cooks is variable, Yonekura *et al.* (1984) suggesting that at Mangaia sea levels reached at least 2 m above present in the last 5,000 years, but on Rarotonga and Aitutaki there is no evidence of a level higher than present. It is clearly possible that small but distinctive variations in Holocene sea-level signatures may be seen along volcanic hot spot islands and associated atolls.

Glacio—isostasy in the southern Pacific region

The major ice mass of the Southern Hemisphere is the 3,000 m-thick Antarctic ice cap. Some ice has existed on this southern continent for at least 27 Ma but until recently its waxing and waning in Late Quaternary times has been enigmatic and probably underestimated (e.g. Kaula, 1980: p. 584). For example, Antarctic changes were not incorporated in the world hydro-isostatic model of Clark *et al.* (1978) (see also Chapters 3, 4 and 9 for further discussion). However, lateglacial ice sheet reconstructions by Hughes *et al.* (1981) suggest that the Antarctic ice sheet expanded to approximately 37 M km^3 at the 18,000 BP glacial maximum, approximately 9.8 M km^3 greater than present dimensions and enough to contribute to as much as 25 m of glacio-eustatic sea-level change (Clark and Lingle, 1979). Thus major glacio-isostatic rebound may be expected for Antarctica. However, unlike the Northern Hemisphere regions of rebound, most of the Antarctic land mass remains covered by ice even today and deglaciated areas upon which a sea-level signature could be recorded are very limited. Even these latter areas have become exposed possibly only in the last 7,000 years, thus giving the opportunity to record only part of the rebound story. Hence elevated shorelines reported from

Antarctica are younger and lower than for the Laurentide and Fenno-scandian ice sheet areas. None the less, the inferred uplift rates are impressive; for example, a 23 m emerged shoreline occurs at Wilkes Station, dated to 6,040 BP (Cameron and Goldthwaite, 1961). Further, although outside the area of this chapter, the report by Sugden and John (1973) of shorelines up to 7 m asl and dated to 9,670 BP in the South Shetlands (which lay within the extended 18,000 BP Antarctic ice sheet) is indicative of significant crustal rebound.

In the revised global model of Clark and Lingle (1979) a collapsing proglacial forebulge area is indicated around the Antarctic similar to that in the Northern Hemisphere (Fig. 12.4). Unfortunately, apart from the Antarctica Peninsula, there is little or no land within this zone. Macquarie Island (54° 36′S) (Fig. 12.1) may lie marginally within the zone but had its own extensive glaciers (Löffler and Sullivan, 1980) and has clearly elevated postglacial beaches (see, for example, front cover of *Search*, vol. 11, no. 7-8, 1980). Other areas of discrete glaciation, possibly large enough to cause isostatic response during deglaciation, occur in South Island, New Zealand and in Patagonia–Tierra del Fuego in southernmost Chile and Argentina. Both areas display evidence of land emergence (for New Zealand see Soons and Selby, 1982; for South America see Auer, 1970). However, as both also have plate-marginal locations (New Zealand's maximum glaciation area straddles the Great Alpine Fault) the possibility of recognising purely glacio-isostatic movements in the Holocene seems unlikely, although this factor should be considered in interpreting the sea-level record.

Predicted hydro-isostatic response for the southern Pacific

Response of the oceans and continental shelves to load variations produced by fluctuating sea levels, initially suggested by Daly (1925), more recently has become an acceptable explanation for worldwide variation in Holocene sea-level patterns (Bloom, 1967; Walcott, 1972; Mörner, 1972; Chappell, 1974b). A sophisticated spherical viscoelastic earth model was developed by Clark *et al.* (1978) which recognised six sea-level response zones based largely on distance away from the areas of major glacio-isostatic rebound in the Northern Hemisphere (see Peltier, this vol., Ch. 3). However, their results also reinforce the earlier work, particularly that of Chappell (1974b), in indicating a critical hinge zone between oceans (average depression about 8 m in the last 7,000 years) and continents (mean upward movement about 16 m). Continental margins as little as 100 km apart can have contrasting relative sea-level curves. Clark and Lingle (1979) refined this earlier model by taking into account the deglaciation of the Antarctic continent which has an effect particularly on the Southern Hemisphere, a

383

critical factor being whether or not melting has continued through the last 5,000 years. If it has, then the previously predicted broad zone of emergence becomes restricted to a relatively narrow band of the southern oceans.

The pattern predicted by Clark and Lingle (1979) for the Australasian–southern Pacific region based on no melting over the last 5,000 years is seen in Figure 12.4. The relevant zones are:

Zone I Antarctica, a zone of emergence as the result of glacio-isostatic rebound.

Zone II A narrow halo around Antarctica, largely landless, characterised by submergence as the result of proglacial forebulge collapse.

Zone IV Covering much of the northern Pacific, but extending into the Southern Hemisphere near New Guinea and more extensively off the South American coastline, this zone should show continuous submergence of 1 to 2 m in the last 5,000 years.

Zone V Most of the southern Pacific where initial submergence was followed by slight emergence of up to 2 m once water was no longer being added to the ocean.

Zone VI Continental margins where Mid-Holocene emergence of the continental mass is predicted as the result of ocean loading and is superimposed on the regional pattern.

What is particularly striking in the original model of Clark *et al.* (1978) is the pattern of decline in the maximum relative sea-level height in the Holocene (+2 m at 52°S to 0 m at 11°N) and a parallel progression towards younger ages for the first achievement of modern sea level (6,000 BP in the southern Pacific to modern times at 11°N) as seen in Figure 12.5. Predictions for particular stations and comparison with field data from Clark and Lingle (1979) are shown in Figure 12.6, reinforcing the previous pattern but adding the new dimension of Antarctic rebound. Also on Figure 12.6 are selected examples from Nakiboglu *et al.* (1983) who attempted to reassess postglacial sea levels in the Pacific largely on the basis of a shift in the time of deglaciation. The envelopes shown are based on deglaciation between 17,000 and 6,000 BP, the upper curve relying on Northern Hemisphere melting alone, the lower curve corresponding to Arctic plus Antarctic melting. These data also reinforce the original pattern predicted by Clark *et al.* (1978).

FACTORS INVOLVED IN THE INTERPRETATION OF PACIFIC SEA-LEVEL HISTORY

Numerous factors other than those already referred to influence either sea levels or the evidence used to interpret past sea-level histories. Many have

Figure 12.4: Distribution of sea-level change zones based on Antarctic and Northern Hemisphere ice retreat histories

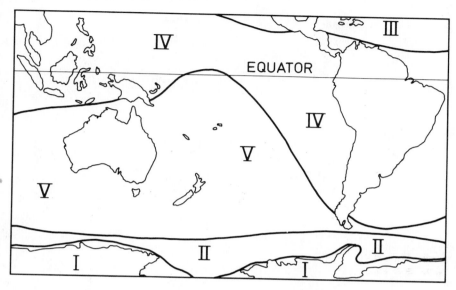

Source: Clark and Lingle (1979).

Figure 12.5: Predicted relative sea-level curves from 52°S to 11°N in the Pacific along meridian 165°W

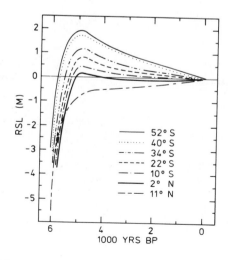

Source: Clark *et al.* (1978).

385

Figure 12.6: Some predicted Australasian, southern Pacific and Antarctic relative sea-level curves, or envelopes compared to site evidence (dots)

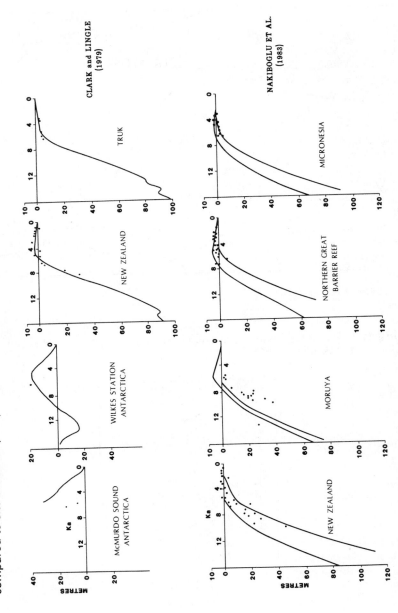

Sources: Clark and Lingle (1979); Nakiboglu *et al.* (1983).

influences which are so minor or act so slowly that it is highly unlikely that they will be identified in the Holocene record, particularly in view of the longer variations considered earlier (see for example the discussions in Mörner, 1980; Pirazzoli, 1976; Lisitzin, 1974). Others have yet to be proved significant, for example Schofield's (1967) suggestion that changes in sea level in the equatorial Pacific are likely to be later than those at higher latitudes, due to probable magnified differences in oceanic salinities during postglacial melting and subsequent delayed oceanic mixing. Of greater importance are the short-term changes in sea level which may leave depositional evidence above the levels of highest predicted astronomical tides. The Pacific is particularly prone to two such agencies, tsunamis and cyclonic storm surges. Seismic sea waves are related to large dip-slip motion along zones of plate convergence, particularly beneath deep oceanic trenches, thus explaining their overwhelmingly Pacific distribution (Fig. 12.2) (Isacks *et al.*, 1968; Cheng, 1965). As Davies (1972) has noted, coasts fronted by trenches which are important generators of tsunamis also tend to be important receivers because they have comparatively simple outlines and are fronted by deep water. Thus the long waves arrive with comparatively little loss of energy due to refraction and diffraction. Japan, the Kuriles, Kamchatka, the Aleutians and Chile are particularly prone, as are some of the larger islands of the central Pacific such as Hawaii which lie between the Asian and South American sources. For example, 32 tsunamis have been recorded in Hawaii between 1819 and 1975 (Macdonald and Abbot, 1977) and 61 on the South American coast between 1562 and 1960, 31 in Chile alone (Cheng, 1965). Close to source, water levels may rise up to 30 m above normal sea level. Hawaii's 1946 tsunami, originating in the Aleutians, reached 17 m asl. Smaller islands, including most atolls, appear to be transparent to tsunamis and there is fortunately little amplification of their small deep water height. New Zealand may be occasionally affected but heights do not appear to exceed 2 m whilst Australia appears to be well protected by off-lying oceanic rises and islands.

By coincidence the pattern for tropical cyclone activity, with accompanying storm surges which may raise water levels by up to 6 m asl, indicates an area of maximum incidence in the southwest Pacific extending eastwards to Polynesia but not to South America, almost the reverse of the tsunami pattern (Fig. 12.1). Close to the equator insufficient Coriolis force produces a low cyclone incidence although very occasional storms may occur even here. The northern Australian coastline is particularly prone to cyclones and surges (Hopley and Harvey, 1978); the effects on continental shorelines are illustrated by Hopley (1974b). Effects on reefs and reef islands may be particularly drastic, as reviewed by Stoddart (1971) and illustrated on Funafuti by Baines *et al.* (1974).

The result of these short-term variations is that throughout much of the region covered by this chapter depositional materials may be thrown many

387

metres above levels attained even during strong trade wind conditions. Certainly in the past this has led to confusion and to much discussion on the validity of many types of evidence produced for the Pacific (e.g. Shepard et al., 1967; Newell and Bloom, 1970). Resulting from these and other controversies concerning Holocene sea levels, it is now more customary to examine the validity and accuracy range of evidence used in sea-level reconstruction (see van de Plassche, 1986, for detailed discussion, and Hopley and Thom, 1983 for discussion in relation to Australia). Whilst almost every type of evidence used may be found in this enormous region a large proportion discussed below comes from tropical or subtropical latitudes, and it may be useful for readers from higher latitudes to outline some of the problems of interpretation of the major sea-level indicators.

Corals and coral reefs

In the southern Pacific coral reefs are found as far south as Lord Howe Island (31°35′S). Most Indo-Pacific corals have a wide water-level range over which they can grow and use of particular species as in the Caribbean (Lighty et al., 1982) is not possible, though some zonation may be recognisable in colonial morphology due to combinations of environmental stresses (Chappell, 1980). In general terms the uppermost limit of coral growth approximates to Mean Low Water-Mark of Spring Tides (MWLST) in open water situations, but variations of up to 25 cm may occur with exposure and species variation. However, at low tide many reef flats become moated behind shingle ramparts or algal rims to levels well above MLWST, sometimes as high as Mean Sea Level, which on parts of the Great Barrier Reef is some 2 m higher. None the less Chappell et al. (1983), for example, have been able to reconstruct palaeo sea levels in north Queensland using micro atolls with an accuracy of ±0.2 m. Such corals have a distinctive micro atoll form but still reduce the range of accuracy for past sea-level determination (Hopley, 1982a, 1986a). Problems also arise in proving in situ location for corals as it appears that a large proportion are removed during storms and deposited in rubble banks (Fig. 12.7). Exposed sites may be interpreted relatively easily but cores from reefs suggest that much of the Holocene veneer of reefs growing over older Pleistocene foundations (reef rock) is detrital in origin (Davies and Hopley, 1983). A further complication also arises from the fact that coral reef growth apparently lagged behind sea-level rise in the Holocene, giving rise to a period of higher wave energy in areas subsequently protected by reef growth. The deposits of this 'high energy window' may be higher than those of the more recent period of protection (Hopley, 1984).

Figure 12.7: Eroded blocks of coral reef deposited by storm/cyclone activity above ancient reef flat on the northern 'high energy' shore of Tupai atoll, Leeward Islands, Society Archipelago. Blocks are cemented into the reef flat and are thought to have been deposited when the reef was still live. A coral block included in the emerged reef flat and standing ~ 0.8 m above its surface has yielded a date of 1,950 ± 60 BP. (Pirazzoli *et al.*, 1985). (Photograph by R.J.N. Devoy.)

Beach rock and cemented rubble

A wide range of cementation processes occur in tropical and subtropical coastal environments. Submarine cementation is possible (Bricker, 1971) and supratidal cementation by carbonate or phosphatic cements is common. Beach rock is supposedly intertidal but much discussion (Hopley, 1982a, 1986b) still exists about the uppermost level of cementation (Mean High Water Springs, Highest Astronomical Tide, or higher?). This is particularly accentuated by a large tidal range, as along the Great Barrier Reef, but fortunately the micro tidal range found on most Pacific islands minimises the problem. Storm rubble and shingle ramparts are also cemented as breccias, conglomerates or rampart rocks (e.g. Newell and Bloom, 1970; Schofield, 1977a; Scoffin and McLean, 1978). Again the upper level of cementation is subject to discussion but would appear to be at least the MHWST, although present levels of 'back reef breccias' have been related to low tide levels in reconstruction (Schofield, 1977a). The problem has been addressed recently by Montaggioni and Pirazzoli (1984) who also recognise the difficulties of attributing the upper level of cementation to a particular level of the sea. However, their petrographic studies indicated distinctive cementation structures related to the vadose and phreatic zones. As the uppermost level of phreatic structures commonly lies above the modern phreatic zone (in some cases above the highest Spring tide level) and has more recent vadose structures superimposed upon it, they suggest that sea-level lowering by as little as 0.6 m may be distinguishable. Dating of bioclastic rubble, however, also causes problems for sea-level change interpretation, as constituent materials in the beach rocks and rampart rocks may be considerably older than the cements.

Mangrove peats

Although mangroves occupy an intertidal position and preservation of *in situ* stumps commonly occurs in older mangrove deposits, they do not always provide accurate sea-level data. Their generally fine-grained deposits with high water content make the mangrove peats subsequently formed particularly vulnerable to compaction after burial (Gill and Lang, 1977). This may be overcome by sampling basal sequences (Bloom, 1970). Migration of tidal creeks within the mangroves also causes considerable horizontal reworking of organic deposits, emphasising the preference for *in situ* stumps. However, even these may be difficult to identify due to the common problem of slumping of creek margin communities. Finally, the degree of height precision suggested by the intertidal zonation of mangrove species has been disproved in north Queensland. Here Spenceley (1982) has shown that in areas only 20 km apart, with similar tides and mangroves, the heights of the specific zones in relation to tidal datum vary by up to 0.5 m.

Limestone notches

Intertidal limestone erosional features, particular notch and visor structures, are particularly common in the Pacific due to the widespread occurrence of emerged Pleistocene coral reefs. These are often referred to as solutional in origin (e.g. Stoddart, 1969a, b). However, tropical surface sea waters are supersaturated with respect to calcium carbonate (Revelle and Emery, 1957). Nocturnal lowering in pH due to variations in photosynthesis may allow minor solution to take place in enclosed pools and selective dissolution of magnesium may produce disintegration of the limestone, but this too is small (Trudgill, 1976). Intense bioerosion appears to be the major process as indicated by Trudgill (1976) on Aldabra, where notch cutting rates were as high as 4-7 mm a^{-1} on exposed coasts. The deepest part of the notch coincides with maximum concentration of bio-erosion, dependent on concentration of bioeroders and erosion rates of individuals, both criteria being ones which may vary regionally and/or through time at particular locations.

HOLOCENE SEA-LEVEL HISTORY IN THE SOUTHERN PACIFIC AND AUSTRALASIA

The following regional review attempts to assess the literature objectively in the light of comments made about the nature of the evidence used. The division is based on a combination of crustal plate location and the zonation from isostatic deformation of Clark and Lingle (1979), as these factors in combination would appear to have the greatest potential for producing meso-scale regional variation in postglacial sea-level behaviour.

Australia, stable mid-plate continent

Considerable detail is available from the 'island continent' and Hopley (1983a) summarises all the recent work on Australian Holocene sea levels. From a minimal level of about −160 m at ~ 18,000 BP sea level rose to within 25 m of its present position by ~ 10,000 BP. From that time the envelope produced by Thom and Chappell (1975), revised by Thom and Roy (1983), appears to be generally applicable to most of the continent, with a 'smooth' rise to the modern sea-level position being achieved by ~ 6,000 BP at the latest (Fig. 12.8). However, discrepancies between different sites suggest that relative levels may have varied by up to 5 m at different locations. The most advanced of marine inundation rates suggest levels within 5 m of present at or shortly after 7,500 BP, within 3 m by ~ 7,000 BP and achieving modern sea level by 6,500 BP.

Greatest variations take place between those areas with and without evidence for higher Holocene sea levels. Many areas show evidence for at least a 1 m higher relative sea level, in places up to 5 m higher (Fig. 12.9). Tectonic uplift, particularly near Perth, Spencer Gulf and central Queensland is considered to be responsible for the highest levels, but Chappell *et al.* (1982) and Hopley (1982b, 1983a, b) have demonstrated cross shelf and regional variation almost certainly related to hydro-isostatic deformation in north Queensland. However, parts of all states show evidence for sea levels no higher than or only negligibly above present (i.e. ±1 m of present sea level). These areas include most of New South Wales, the Kimberly region and southwest corner of Western Australia and the Gippsland region of Victoria. The large sea-level oscillations of Fairbridge

Figure 12.8: The eastern Australian relative sea-level envelope, from Thom and Chappell (1975) modified after Thom and Roy (1983) and relative sea-level curves for New Caledonia (after Baltzer, 1970 and Coudray and Delibrias, 1972) and for New Zealand (after Gibb, 1983). Note: direct correlation of the curves is misleading due to the different ways in which ^{14}C dates have been correlated

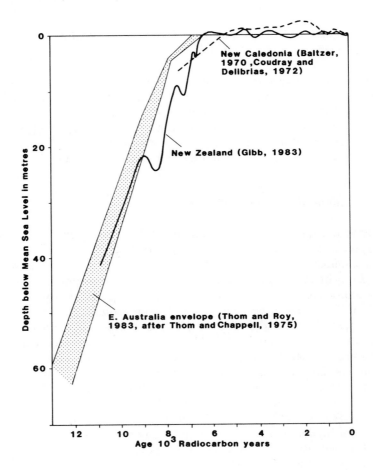

(1961), based in part on Western Australian evidence, are generally rejected by recent work. This shows that within a one metre envelope, the final part of the Holocene marine inundation and stillstand period, sea-level changes were smooth, not oscillatory.

Figure 12.9: A. Earliest purported attainment within one metre of modern sea level (ka) and B. Highest reported Holocene sea levels (m) in the Australasian–Pacific region

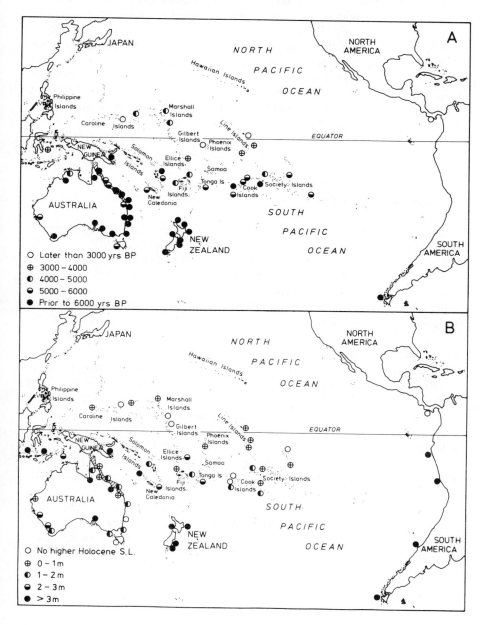

Antarctica, glacio-isostatically rising continent

Only sparse evidence is available from Antarctica, only 1.5 per cent of which is unglaciated, with many higher shorelines being identified but undated. Zivago and Esteev (1970) summarised the data available in 1969, 94 per cent of 137 observations on raised shorelines being not more than 60 m high, probably indicative of the late retreat of ice as indicated earlier. Those higher than this level were considered to be of doubtful origin or certainly older than the Holocene. Pirazzoli (1976) and Clark and Lingle (1979) discuss some dated evidence which included higher levels between 2 and 13 m above present sea level, dated between 3,000 and 6,500 BP. However, it is evident that the pattern of isostatic rebound in Antarctica is known only sketchily at present.

Western South America, plate-margin continent

Numerous reports of higher marine terraces come from the tectonically active west coast of South America (e.g. Tricart *et al.*, 1969; Pirazzoli, 1976; Paskoff, 1980). Generally the movements have been upward, but on the coast of central Chile some subsidence is considered to have taken place. Subsidence relative to the rest of the South Pacific is also suggested from the Pacific coast of Panama, where Glynn and Macintyre (1977) have dated a series of coral reef cores. No material from within 2 m of the present surface here is older than 2,500 BP. An age of 5,555 BP from 6.2 m depth may relate to a sea-level position.

Tectonic dislocation, possibly with some glacio-isostatic uplift in the far south, is clearly indicated along the entire coastline to the southernmost tip of the continent. Recent movements have been quite significant, in Chile, Pirazzoli (1976) reporting an uplift of 3 m during the 1835 earthquake at Isla Santa Marcu (see Berryman, this vol., Ch. 5) and 5.7 m at Isla de Guamblin during the 1960 disturbance. At Ica in Peru elevated beaches between 6 and 8.5 m asl produced ages of between 500 and 600 BP. Levels of 10 to 14 m at Isla Mocha, Chile were dated between 1,990 and 2,590 BP, and at Talava in northern Peru an age of 3,000 BP was obtained for a 4.5 m level. South of Ushuaia in Tierra del Fuego, Auer (1970) reported a radiocarbon age of 7,660 BP for a peat lying between marine deposits indicative of sea levels of +6 m and +10 m psl. The record of sea-level change along the Andean coastline is one of discontinuity in time and space, tectonic events dominating over the eustatic indicators.

Western Pacific, plate-margin island arcs

From the uplift data provided in Figure 12.2 it is clear that the island arcs of the western Pacific plate margin have undergone numerous episodic distortions affecting discrete blocks of varying sizes. Most surveys from the region report numerous emerged or submerged marine levels (e.g. in the Solomons, Stoddart, 1969b), the majority of which are undated. Even where dates are available, the numerous emerged reefs and associated deposits indicate that the seismicity of the region conceals any eustatic or general isostatic influence. For example, in eastern Indonesia Tjia *et al.* (1974, 1975) report a series of radiocarbon ages for materials of Holocene age ranging up to 33 m above present sea level (with an age of only 3,320 BP). Ages less than 500 BP are associated with levels of up to +3 m psl. On the Huon Peninsula in New Guinea, where the Late Pleistocene record is so clear (see Chapters 5 and 9), the Holocene emerged reefs, which range in elevation from +2.5 to +15 m psl, provide a series of problems of inter-pretation related particularly to radiocarbon and uranium series ages (Chappell and Polach, 1972; Chappell, 1974a; Bloom *et al.*, 1974). An age of about 6,800 BP for the crest of the Holocene reefs appears to be the most reasonable compromise. High rates of deformation also apply to Vanuatu where, on Malekula and Santo, there are up to nine Holocene reef terraces. Within the relatively small time range of 5,470 to 5,940 BP maximum levels vary from 8 to 28 m above present reef growth levels (Taylor *et al.*, 1980).

Within the apparent chaos of sea-level/land relationships at the Pacific plate margin there may be some areas which have not been influenced by recent tectonism. For example, on Atairo Island and Timor, Chappell and Veeh (1978) report no emerged Holocene reefs, even though extensive suites of emerged Pleistocene reefs exist. Even in Vanuatu, Guilcher (1974) reported that on Reef Island *in situ* corals and clams, which provided radiocarbon ages between 4,150 and 6,640 BP, indicate a maximum emergence of 1.5 m, a pattern not very different from that for areas of the southwest Pacific considered stable. In New Ireland, which is violently seismic, vertical uplift of even Mid-Holocene reefs is slight. Here the trend of Holocene marine inundation (from peat dates), passing through −8 to −5 m psl between 7,400 and 7,800 BP, is similar to that of Australia (Bloom, 1980). This may merely indicate the long periodicity of seismic events, the Holocene being too brief to have experienced major uplift. Within the complex structure of the plate margin, however, it has been recognised that 'fossil', essentially inactive island arc sectors exist, for instance, the inner Melanesian arc of Avias (1973) which contains New Caledonia and much of New Zealand. The sialic Fijian mass also seems most stable in spite of its proximity to the subduction zone. Thus the slight emergence of up to 1.6 m asl reported on Viti Levu by Sugimura *et al.*

(1983) and Matsushima *et al.* (1984), dated at between 2,000 and 3,000 BP, together with the small emergence indicated by McLean (1979) on Lakeba at ~4,500 BP, may represent the regional eustatic- hydro-isostatic record rather than local uplift. A similar conclusion was reached by Taylor and Bloom (1977) for an emergence of 2.2 m dated between 5,700 and 5,900 BP on both Tongatapu and Eua in Tonga (Fig. 12.9). In contrast Upolu in western Samoa appears to have subsided relative to these 'stable' sites. Bloom (1980) indicates sea level here at about −6 m psl at 4,800 BP and −3 m psl at 1,600 BP which is consistent with drowned archaeological sites on the island and is probably the result of effusive basalt volcanism.

A more complete record comes from New Caledonia. Baltzer (1970), using mangrove peats as evidence, showed that sea level was ~ 5 m below present at 7,300 BP, and only 0.15 m below at 5,600 BP. Subsequently Coudray and Delibrias (1972) suggested that modern sea level was exceeded by ~ 5,400 BP and was more than 1 m higher between 4,400 and 3,000 BP (Fig. 12.8) throughout the New Caledonia region, including the Isle of Pines and Loyalty Islands where Late Pleistocene warping is recorded (Dubois *et al.*, 1973).

The New Zealand record of sea-level change is certainly affected by differential movements of the land which have been the cause of some dispute and more recent revision. A general envelope of rise from levels of ~ −33 m 10,000 years ago was constructed by Pickrill (1976) from the evidence of Cullen (1967) and Suggate (1968). This evidence suggested that modern sea level was first achieved about 5,000 BP. Schofield, working in the Firth of Thames area near Auckland and in northern North Island, areas he claimed were tectonically stable, suggested post 5,000 BP for the attainment of modern sea level and that subsequently sea level reached a maximum of 2.1 m at ~ 3,900 BP, there being up to six 'oscillations' of less than 1 m amplitude subsequently (Schofield, 1960, 1973). More recently the evidence from the Firth of Thames has been challenged and Woodroffe *et al.* (1983) indicate that the maximum level was in the range +0.7 to 0.9 m at ~ 3,600 BP and fell gradually to its present position 1,200 years ago. Gibb (1983) has reassessed much of the New Zealand evidence and has produced a composite sea-level curve (Fig. 12.8) which he considers applicable to tectonically stable areas of New Zealand. It shows that from a −30 m level at ~ 10,000 BP sea level rose via a series of −22 m, −9 m and −4 m stillstands to reach its present position by 6,500 BP. Subsequently there are small 'oscillations', the maximum level being ~ 0.7 m at 4,600 BP. In areas of deformation the pattern diverges from this generalisation. For example, on the Mahia Peninsula of North Island, Berryman (1983) reports an age of ~ 8,380 BP for a level ~ +3.5 m psl. Gibb (1983) also reports data indicating ages earlier than ~ 6,500 BP for elevated sea-level evidence, extreme examples being ~ 7,590 BP for a +3.6 m level on the east coast of North Island and ~6,740 BP for a +22.5 m level at Pakarae River near

Gisborne. In contrast, only 175 km to the south, subsidence has apparently resulted in relative sea level being 11 m below present at ~ 6,580 BP (Gibb, 1983). It is evident that a proportion of New Zealand has sufficient seismicity for tectonic instability to influence apparent sea levels, although many areas are apparently unaffected and the sea-level pattern is similar to that of southeastern Australia.

Mid-Pacific volcanic islands and atolls

Only a very small sample of the myriad of Pacific volcanic islands and atolls have been researched in sufficient depth to produce convincing evidence of Holocene sea-level change. Much of this research has been carried out in the context of proving or otherwise the hypothetical 'Daly two metre terrace' (Daly, 1920a, b; Stearns, 1961; Newell and Bloom, 1970). Radiometric dating techniques have shown that many examples of emerged coral reef are Pleistocene in age (Veeh, 1966; Chappell et al., 1974). None the less there remains a variation in sea-level evidence over the last 6,000 years even from these islands considered by Bloom (1967) as the ideal 'dip sticks' of eustatic sea level.

Using basal peat sequences Bloom (1970) suggested that in the eastern Caroline Islands in the central western Pacific (Fig. 12.9) the sea-level pattern was one of continued submergence from ~ −6.2 m at 6,500 BP, the rate declining from 1.9 m ka^{-1} to 4,100 BP to 0.4 m ka^{-1} since. However, Shepard et al. (1967) and Newell and Bloom (1970) also report ^{14}C ages of up to 4,475 BP from cemented reef flat material of detrital origin. Although this may have been thrown up from greater depths during storms, elsewhere the age of conglomerates and in situ reef has been shown to be similar and some doubt may exist on the widespread validity of the submergence curve on the Carolines. Also on Ifaluk in the western Carolines, Tracey (1968) has described Holocene corals up to 1 m above present reef flat levels. Only 500 km to the north, Buddemeier et al. (1975) have shown that on Enewetak atoll modern sea level was achieved by or earlier than ~4,000 BP and was as much as a metre higher between ~ 3,500 and 2,000 BP. Tracey and Ladd (1974) support this pattern, extending the higher level back to ~ 4,125 BP and indicating sea level within at least 2 m of its present position at ~6,220 BP. To the east, at Bikini atoll a similar pattern was demonstrated with a +0.6 m peak at ~ 4,050 BP following a phase of marine inundation which may have been at or close to modern sea level at ~ 5,750 BP. Further south the pattern for the Gilbert and Ellice Islands has been confused by Schofield's (1977a) attempt to relate a series of ^{14}C dates from a variety of detrital and in situ materials to his New Zealand sea-level curve. There is undoubted emergence indicated in these atolls, but the in situ provenance of some materials and rejection of coral dates is dubious

397

and the question of moating of corals is not addressed by Schofield. His pattern of fluctuating levels from ~ 4,000 BP reaching a maximum of ~ +2.4 m psl at ~ 2,760 BP may need revision. Marshall and Jacobson (1985) suggest that at Tarawa modern sea level was reached prior to ~ 4,500 BP probably as early as 6,000 BP and are dubious about higher than present levels.

Data for the equatorial central Pacific is sparse, but what little there is (Tracey, 1972) suggests a pattern similar to that of the Gilbert and Ellice Islands. Emerged corals of up to +1 m psl have been dated from Enderbury Island in the Phoenix Group (~ 2,150 to 2,650 BP), Jarvis (up to ~ 2,800 BP), Starbuck (~ 3,950 BP) and Malden (~ 3,550 BP) in the Line Islands. Moating of corals at these elevations is possible but a low tidal range may minimise the degree of misinterpretation. The ubiquitous '1 m emergence' has also been described in the Cook Islands of the southwest Pacific (e.g. Wood, 1967; Schofield, 1970; Wood and Hay, 1970) although much of the material described and dated is coral conglomerate (see Stoddart, 1975 for critical discussion). On Rarotonga, Stoddart (1972) reported a 1 m emerged reef dated at ~ 2,030 BP, but this was not recognised by Yonekura et al. (1983, 1984). These latter authors drilled into the reef flat and obtained an age of ~ 6,010 BP for coral from only ~ 1.4 m depth, indicating the attainment of modern sea level at or shortly after 6,000 BP. They obtained similar results from Aitutaki and only on Mangaia in the south did Yonekura et al. (1983, 1984) accept evidence of a higher sea level in the form of emerged micro atolls and notches up to ~ 1.8 m higher than present and dated between ~ 3,410 and 5,020 BP. The various discussions on the Cook Islands epitomise the problems of using coral conglomerates and micro atolls for evidence of relatively small sea-level changes.

The only other area of the central Pacific for which there is much literature on sea-level changes is French Polynesia. Again, problems arise in the interpretation of some data involving conglomerates and micro atolls presented in earlier literature (see Guilcher, 1969 for discussion). An early attainment for modern sea level appears to be indicated by the dates from the deeper cores from Muroroa atoll (Lalou et al., 1966; Labeyrie et al., 1969). A basal Holocene date of ~ 8,200 BP from only 6 m depth would suggest a pattern of Early Holocene marine inundation similar to that of Australia. Presuming that this and other materials dated from the cores were in situ then the Muroroa evidence suggests modern sea level was achieved shortly after ~ 6,000 BP. Lalou et al. (1966) and Labeyrie et al. (1969) also claim that emergence of up to 4 m occurred at about 3,600 BP but this has since been disproved. However, there are numerous reports of higher levels, most within 1 m of present sea level and depending on conglomerate or micro atoll evidence. Such reports come from the Austral Islands (Chevalier, 1973b; Pirazzoli and Fontes, 1982), the Society Islands (Guilcher et al., 1969; Chevalier, 1973b; 1977; Pirazzoli, 1983; Montaggioni and Pirazzoli, 1984), and the Tuamotus (Montaggioni and Pirazzoli, 1984; Pirazzoli and Montaggioni, 1984, 1986), but not from the Marquesas. The age of the emergence at most sites is between 3,500 and 3,000 BP. Pirazzoli and Fontes (1982) suggested that there was indication of tilting from the Australs (+1.7 m) to the Tuamotus (<1 m). He has also suggested (1983) that there is a progressive decrease in the indicated emergence along the Society Islands from Maupiti (1.0 m) to Tahiti (0.15 m) which he ascribed to an active flexuring of the lithosphere resulting from volcanic loading. Tilting of Moorea of up to 1.4 m in the last 1,500 years was also suggested by Chevalier (1977). Recent movements have also been advocated for Taiaro in the last 1,100 years (Salvat et al., 1977) and in the last 640 years on Rurutu in the Australs (Pirazzoli and Fontes, 1982).

DISCUSSION AND CONCLUSIONS

A considerable amount of literature has been quoted in this review, but because of the size of the area covered, only broad generalisations can be made about the relationships between land and sea in the Holocene. The patterns recognised suggest that the major controls on deviations from a glacio-eustatic sea-level curve recognised in the early part of the chapter are correct. Superimposed upon a pattern of early attainment of modern relative sea level in the Southern Hemisphere, produced by a global scale glacio- and hydro-isostatic effect, are major departures produced by tectonic activity at plate margins and by independent glacio-isostasy in Antarctica. Minor departures may be affected by local lithospheric loading and associated moat effects at oceanic volcanic hot spots.

Away from these areas of clear earth movements the present evidence (Fig. 12.9) suggests that the southern Pacific generally may have seen the attainment of relative modern sea level earlier than anywhere else. Broadly it appears to have been achieved prior to ~ 6,000 BP in the Australasian region, between ~ 5,000 and 6,000 BP further east from the Cook Islands to French Polynesia and possibly as late as ~ 4,000 BP as one moves into the Northern Hemisphere (Marshall Islands). Evidence for higher Holocene sea levels, although not ubiquitous, is very widespread. Levels achieved show no broad regional pattern away from tectonically active areas, but in general the time of the peak of relative sea-level rise appears to follow a pattern similar to that shown for the first attainment of modern sea level. The high levels in Australasia generally date from between ~ 6,000 and 5,000 BP, being later in the equatorial Pacific (e.g. ~ 4,000 in the Marshalls) and apparently becoming progressively younger eastwards across the Pacific (~ 4,000 BP, in the Gilbert and Ellice Islands, ~ 5,000 to 3,400 BP in the Cooks, ~ 3,900 to 3,500 BP in the Line Islands and

399

~ 3,500 to 3,000 BP in French Polynesia).

Knowledge of the nature of relative and absolute sea-level change is of value to a wide range of research including studies of glaciation, palaeo-climates, tectonic and isostatic movements and the properties of the upper lithosphere. In the Pacific, an understanding of sea-level change and particularly the response in terms of coral reef growth have further import-ance as this is the only major part of the earth where initial colonisation by Man has taken place since the Mid-Holocene. Although some of the earliest humanoid remains are found in southeast Asia, and Australia, New Guinea and many of the islands of western Micronesia were settled before 30,000 BP, much of Polynesia was settled only in the last 2,000 years. Bellwood (1979) suggests that initially during the postglacial submergence of the wide Sunda shelf, the shallow seas supported a high biomass and a rapid expansion of a clinal population. However, as population grew so the land disappeared beneath the sea and Bellwood believes that the loss of some 3 million km^2 of land between ~ 14,000 and 2,000 BP was instru-mental in triggering migrations into new lands, including the western Pacific.

Further expansion into Melanesia and Micronesia was possible as a majority of the islands are volcanic or emerged reefal limestone, existing well before the Holocene. Settlement of the Marianas, Tonga and Samoa probably took place prior to ~ 3,500 BP. However, further extension by the skilled early Polynesian navigators to the similar volcanic and emerged reefal limestone islands of the Austral Pacific possibly relied on the development of low islands on the atolls of the intervening central Pacific. As many of these atolls have Pleistocene foundations, as much as 20 m below present sea level, and reef growth is known to lag behind sea-level rise (see e.g. Hopley, 1982a), reef flat and reef island development may have been as much as 2,000 years behind the first attainment of modern sea level in these areas. As Schofield (1977b) has previously suggested it may be more than coincidence that the rapid migrations of Polynesian peoples took place after 3,000 BP when atoll island building may have been aided by a small drop in relative sea level.

ACKNOWLEDGEMENT

The author is grateful to his many colleagues who have provided data and drawn his attention to papers used in this review, and to Professor B.G. Thom for helpful comments on the manuscript.

REFERENCES

Auer, V. (1970) 'The Pleistocene of Patagonia: V Quaternary problems of southern South America', *Ann. Acad. Sci. Fennicae, 4, III Geol.-Geograph., 100.*

Avias, J. (1973) 'Major Features of the New Guinea–Louisiade–New Caledonia–Norfolk Arc System', in P.J. Coleman (ed), *The Western Pacific: Island Arcs, Marginal Seas, Geochemistry,* pp. 113-26.

Baines, G.B.K., Beveridge, P.J. and Maragos, J.E. (1974) 'Storms and island building at Funafuti atoll, Ellice Islands', *Proc. Second Int. Coral Reef Symp.,* vol. 2, pp. 485-96.

Baltzer, F. (1970), Datation absolue de la transgression Holocène sur la Côte de Nouvelle-Calédonie sur des échantillons de tourbes à paletuviers. Interprétation Néotectonique. *Comptes Rendus Acad. Sci. Paris, 271, D,* pp. 2251-4.

Bellwood, P. (1979) *Man's Conquest of the Pacific,* Oxford University Press, New York.

Belperio, A.P., Hails, J.R. and Gostin, V.A. (1983) 'A review of Holocene sea levels in South Australia', in D. Hopley (ed.), *Australian Sea Levels in the Last 15,000 Years: A Review,* Monogr. Ser., Occ. Paper No. 3, Dept. of Geography, James Cook University of North Queensland, pp. 37-47.

Berryman, K. (1983) 'Tectonic implications of the Mid-Late Holocene geology of Mahia Peninsula, east coast, North Island, New Zealand', in *Abstracts of Papers, International Symposium on Coastal Evolution in the Holocene, Tokyo, Japan,* pp. 1-3.

Bird, E.C.F. and Iltis, J. (1985) 'New Caledonia and the Loyalty Islands', in E.C.F. Bird and M.L. Schwartz (eds), *The World's Coastline,* Van Nostrand Reinhold, New York, pp. 995-1002.

Bloom, A.L. (1967) 'Pleistocene shorelines: a new test of Isostasy', *Bull. Geol. Soc. Am., 78,* pp. 1477-94.

—— (1970), 'Paludal stratigraphy of Truk, Ponape and Kusaie, Eastern Caroline Islands', *Bull. Geol. Soc. Am., 81,* pp. 1895-904.

—— (1971) 'Glacial eustatic and isostatic controls of sea level since the last glaciation', in K.K. Turekian (ed), *The Late Cenozoic Glacial Ages,* Yale University Press, New Haven, Conn., pp. 355-79.

—— (1980) 'Late Quaternary sea-level change on south Pacific coasts: a study in tectonic diversity', in N.-A. Mörner (ed), *Earth Rheology, Isostasy and Eustasy,* Wiley, Chichester and New York, pp. 505-16.

—— Broecker, W.S., Chappell, J., Matthews, R.K. and Mesollella, K.J. (1974) 'Quaternary sea level fluctuations on a tectonic coast: new Th[230] dates from New Guinea', *Quat. Res., 4,* pp. 185-205.

Bricker, O.P. (ed) (1971), *Carbonate Cements,* Johns Hopkins University Studies in Geology 19, Johns Hopkins University Press, Baltimore, Md.

Brown, R.G. (1983) 'Sea level history over the past 15,000 years along the western Australian coastline', in D. Hopley (ed), *Australian Sea Levels in the Last 15,000 Years: A Review,* Dept. of Geography, James Cook University of North Queensland, Monogr. Ser., Occ. Paper No. 3, pp. 28-36.

Buddemeier, R.W., Smith, S.V. and Kinzie, R.A. (1975) 'Holocene windward reef-flat history, Enewetak atoll', *Bull. Geol. Soc. Am., 86,* pp. 1581-4.

Cameron, R.L. and Goldthwaite, R.P. (1961) 'The US–IGY contribution to Antarctic glaciology', *Union Géod. et Géophys. Ass. Int. d'Hydrol. Sci., Publ.* 55, pp. 7-13.

Chappell, J. (1974a) 'Geology of coral terraces, Huon Peninsula, New Guinea: a study of Quaternary tectonic movements and sea-level changes', *Bull. Geol. Soc.*

Am., *85*, pp. 553-70.

—— (1974b) 'Late Quaternary glacio- and hydro-isostasy on a layered earth', *Quat. Res.*, 4, pp. 429-40.

—— (1980) 'Coral morphology, diversity and reef growth', *Nature, 286*, pp. 249-52.

—— Broecker, W.S., Polach, H.A. and Thom, B.G. (1974), 'Problem of dating upper Pleistocene sea levels from coral reef areas', *Proc. Second Int. Coral Reef Symp.*, vol. 2, pp. 563-71.

—— Chivas, A., Wallensky, E., Polach, H.A. and Aharon, P. (1983) 'Holocene palaeoenvironmental changes, central to north Great Barrier Reef inner zone', *BMR J. Austr. Geol. Geophys.*, *8*, pp. 223-35.

—— and Polach, H.A. (1976) 'Holocene sea-level change and coral-reef growth at Huon Peninsula, Papua New Guinea', *Bull. Geol. Soc. Am.*, *87*, pp. 235-40.

—— Rhodes, E.G., Thom, B.G. and Wallensky, E. (1982) 'Hydro-isostasy and the sea level base of 5,500 BP in north Queensland, Australia', *Mar. Geol.*, *49*, pp. 81-90.

—— and Veeh, H.H. (1978) 'Late Quaternary tectonic movements and sea-level changes at Timor and Atauro Island', *Bull. Geol. Soc. Am.*, *89*, pp. 356-68.

Cheng, T.T. (1965) *Tsunamis*, Royal Observ., Hong Kong, TN 7.

Chevalier, J.P. (1973a) 'Coral reefs of New Caledonia', in O.A. Jones and R. Endean (eds), *Biology and Geology of Coral Reefs, I, Geology, I*, Academic Press, London and New York, pp. 143-67.

—— (1973b) 'Geomorphology and geology of coral reefs in French Polynesia', in O.A. Jones and R. Endean (eds), *Biology and Geology of Coral Reefs, I, Geology I*, Academic Press, London and New York, pp. 113-41.

—— (1977) 'Origin of the reef formations of Moorea island (Archipelago of La Société)', in *Proc. Third Int. Coral Reef Symp.*, vol. 2, pp. 283-7.

Clark, J.A., Farrell, W.E. and Peltier, W.R. (1978) 'Global changes in postglacial sea level: a numerical calculation', *Quat. Res.*, *9*, pp. 265-87.

—— and Lingle, C.S. (1979) 'Predicted relative sea level changes (18,000 years BP to present) caused by late glacial retreat of the Antarctic ice sheet', *Quat. Res.*, *11*, pp. 279-98.

Conacher, A.J. and Murray, I.D. (1969) 'The Meckering earthquake, western Australia, 14 October 1968', *Austr. Geogr.*, *11*, pp. 179-84.

Cook, P.J. and Mayo, W. (1977) 'Sedimentology and Holocene history of a tropical estuary (Broad Sound, Queensland)', *Bur. Min. Res. Geol. Geophys. Bull.*, *170*.

Coudray, J. and Delibrias, G. (1972) 'Variations du niveau marin au-dessus de l'actuel en Nouvelle-Calédonie depuis 6000 ans', *Comptes Rendus Acad. Sci. Paris, 275, D*, pp. 2623-6.

Crough, S.T. (1984) 'Seamounts as recorders of hot-spot epeirogeny', *Bull. Geol. Soc. Am.*, *95*, pp. 3-8.

Cullen, D.J. (1967) 'Submarine evidence from New Zealand for a rapid rise in sea level about 11,000 years ago', *Palaeogeography, Palaeoclimatol., Palaeoecol.*, *3*, pp. 289-98.

Daly, R.A. (1920a) 'A recent worldwide sinking of ocean level', *Geol. Mag.*, *57*, pp. 246-61.

—— (1920b) 'A general sinking of sea level in recent time', *Proc. Nat. Acad. Sci.*, *6*, pp. 246-50.

—— (1925) 'Pleistocene changes of level', *Am. J. Sci.*, *5th Ser.*, *10*, pp. 281-313.

Darwin, C.R. (1842) *The Structure and Distribution of Coral Reefs*, Smith, Elder, London.

Davies, J.L. (1972) *Geographical Variation in Coastal Development*, Oliver &

Boyd, Edinburgh.

Davies, P.J. and Hopley, D. (1983) 'Growth facies and growth rates of Holocene reefs in the Great Barrier Reef', *BMR J. Austr. Geol. Geophys.*, *8*, pp. 237-51.

Denham, D. (1976) 'Earthquake Hazard in Australia', *Bur. Min. Res. Geol. Geophys. Rec.*, 1976/31.

Douglas, G. (1969) 'Checklist of Pacific Oceanic Islands', *Micronesica, 5*, pp. 327-464.

Dubois, J., Launay, J. and Recy, J. (1973) 'Emersion de traces de niveaux marins quaternaires dans la région de Nouvelle-Calédonie—Iles Loyauté et tentative d'explication du Phénomène par un Bombement de la Lithosphère', *Le Quaternaire, géodynamique, stratigraphie et environnement, traveaux français récents'*, pp. 163-7.

Duncan, R.A. and McDougall, I. (1974) 'Migration of volcanism with time in the Marquesas Islands, French Polynesia', *Earth Planet. Sci. Lett., 21*, pp. 410-14.

Fairbridge, R.W. (1961) 'Eustatic changes in sea level', in L.H. Ahrens, F. Press, K. Rankama and S.K. Runcorn (eds), *Physics and Chemistry of the Earth*, Pergamon Press, London, vol. 4, pp. 99-185.

Gibb, J.G. (1983) 'Sea levels during the past 10,000 years BP from the New Zealand Region, South Pacific Ocean', in *Abstracts of Papers, International Symposium on Coastal Evolution in the Holocene, Tokyo, Japan*, pp. 28-31.

Gill, E.D. and Lang, J.G. (1977) 'Estimation of compaction in marine geological formations from engineering data commonly available', *Mar. Geol., 25*, M1-M4.

Glynn, P. and Macintyre, I.G. (1977) 'Growth rate and age of coral reefs on the Pacific coast of Panama', in *Proc. Third Int. Coral Reef Symp. Miami*, vol. 2, pp. 251-9.

Gromme, S. and Vine, F.J. (1972) 'Palaeomagnetism of Midway Atoll lavas and northward movement of the Pacific plate', *Earth Planet Sci. Lett., 17*, pp. 159-68.

Grover, J.C. (1965) 'Seismological and volcanological studies in the British Solomon Islands to 1961', *Br. Solomon Is. Geol. Rec., 2*, pp. 183-8.

Guilcher, A. (1969) 'Les Etudes françaises sur le Quaternaire dans le Pacifique. Etudes Franc. sur le Quaternaire', *Supp. Bull. de l'AFEQ, VIII Congr. INQUA*, pp. 247-51.

—— (1974) 'Coral reefs of the New Hebrides, Melanesia, with particular reference to open-sea, not fringing reefs', in *Proc. Second Int. Coral Reef Symp.*, vol. 2, pp. 523-35.

—— Berthois, L., Doumenge, F., Michel, A., Saint-Requier, A. and Arnold, R. (1969) 'Les Recifs et Lagons coralliens de Mopelia et de Bora-Bora (Iles de la Société)', *Mem. ORSTOM, 38*.

Hails, J.R., Belperio, A.P. and Gostin, V.A. (1983) 'Holocene sea levels of Upper Spencer Gulf, South Australia', in D. Hopley (ed), *Australian Sea Levels in the Last 15,000 Years: A Review*, Dept. of Geography, James Cook University of North Queensland, Monogr. Ser., Occ. Paper 3, pp. 48-53.

Hammond, S.R., Theyer, F. and Sutton, G.H. (1974) 'Palaeomagnetic evidence of northward movement of the Pacific plate in deep-sea cores from the central Pacific basin', *Earth Planet. Sci. Lett., 22*, pp. 22-8.

Hopley, D. (1974a) 'Investigations of sea level changes along the coast of the Great Barrier Reef', *Proc. Second Int. Coral Reef Symp.*, vol. 2, pp. 551-62.

—— (1974b) 'Coastal changes produced by tropical cyclone Althea in Queensland, December, 1971', *Austr. Geogr., 12*, pp. 445-56.

—— (1978) 'Sea level change on the Great Barrier Reef: an introduction', *Phil. Trans. Roy. Soc. London, A, 291*, pp. 159-66.

—— (1982a) *Geomorphology of the Great Barrier Reef: Quaternary Development of Coral Reefs*, Wiley-Interscience, New York.

—— (1982b) 'Holocene sea levels of the central Great Barrier Reef, Australia', in D.J. Colquhoun (ed), *Holocene Sea Level Fluctuations, Magnitude and Causes*, IGCP Project 61, Dept. of Geology, University of South Carolina, Columbia, SC, pp. 81-95.

—— (ed) (1983a) *Australian Sea Levels in the Last 15,000 Years: A Review*, Dept. of Geography, James Cook University of North Queensland, Monogr. Ser., Occ. Paper 3. (Australian Contribution to IGCP 61 and 200.)

—— (1983b) 'Deformation of the North Queensland continental shelf in the Late Quaternary', in D.E. Smith and A.G. Dawson (eds), *Shorelines and Isostasy*, Academic Press, London and New York, pp. 347-66.

—— (1984) 'The Holocene "high energy window" on the central Great Barrier Reef', in B.G. Thom (ed), *Coastal Geomorphology in Australia*, Academic Press, London, pp. 135-50.

—— (1986a) 'Corals and coral reefs as indicators of palaeo-sea levels with special reference to the Great Barrier Reef', in O. van de Plassche (ed), *Sea-level Research: Manual for the Collection and Evaluation of Data*, Geo-Abstracts, Norwich, pp. 195-228.

—— (1986b) 'Beach rock as a sea level indicator', in O. van de Plassche (ed), *Sea-level Research: Manual for the Collection and Evaluation of Data*, Geo-Abstracts, Norwich, pp. 157-73.

—— and Harvey, N. (1979) 'Regional variations in storm surge characteristics around the Australian coast: a preliminary investigation', in R.L. Heathcote and B.G. Thom (eds), *Natural Hazards in Australia*, Australian Academy of Science, Canberra, pp. 164-85.

—— and Thom, B.G. (1983) 'Australian sea levels in the last 15,000 years: an introductory review', in D. Hopley (ed), *Australian Sea Levels in the Last 15,000 Years: A Review*, Dept. of Geography, James Cook University of North Queensland, Monogr. Ser., Occ. Paper, 3, pp. 3-26.

Hughes, T.J., Denton, G.H., Anderson, B.G., Schilling, D.H., Fastook, J.L. and Lingle, C.S. (1981) 'The last great ice-sheets: a global view', in G.H. Denton and T.J. Hughes (eds), *The Last Great Ice Sheets*, Wiley, Chichester and New York, pp. 275-317.

Isacks, B.L., Oliver, J., and Sykes, L.R. (1968) 'Seismology and the new global tectonics', *J. Geophys. Res.*, *73*, pp. 5855-99.

Jarrard, R.D. and Turner, D.L. (1979) 'Comments on "Lithospheric Flexure and Uplifted Atolls" by M. McNutt and H.W. Menard', *J. Geophys. Res.*, *84*, pp. 5691-4.

Kaula, W.M. (1980) 'Problems in understanding vertical movements and earth rheology', in N.-A. Mörner (ed), *Earth Rheology, Isostasy and Eustasy*, Wiley, Chichester and New York, pp. 577-89.

Labeyrie, J., Lalou, C. and Delibrias, G. (1969) 'Etudes des transgressions marines sur l'atoll de Mururoa par la datation des différents niveaux de corail', *Cahiers du Pacifique*, *13*, pp. 59-68.

Lalou, C., Labeyrie, J. and Delibrias, G. (1966) 'Datations des calcaires coralliens de l'atoll de Muroroa (Archipel des Tuamotu) de l'epoque actuelle jusqu'à −500,000 ans', *Comptes Rendus Acad. Sci. Paris, 263-D*, pp. 1946-9.

Lighty, R.G., Macintyre, I.G. and Stuckenrath, R. (1982) '*Acropora palmata* reef framework: a reliable indicator of sea level in the western Atlantic for the past 10,000 years', *Coral Reefs*, *1*, pp. 125-30.

Lisitzin, E. (1974), *Sea Level Changes*, Oceanography Series, Elsevier, Amsterdam.

Löffler, E. and Sullivan, M.S. (1980), 'The extent of former glaciation on Macquarie Island', *Search*, *11*, pp. 246-7.

Macdonald, G.A. and Abbott, A.T. (1977) *Volcanoes in the Sea: The Geology of Hawaii*, Hawaii University Press, Honolulu.

McDougall, I. (1971) 'Volcanic chains and sea floor spreading', *Nature*, *231*, pp. 141-4.

—— and Duncan, R. (1980) 'Linear volcanic chains-recording plate motions?', *Tectonophys.*, *63*, pp. 275-95.

McLean, R.F. (1979) 'The coast of Lakeba: a geomorphological reconnaissance', in *Lakeba: Environmental Change, Population Dynamics and Resource Use*, UNESCO–UNFPA Population and Environment Project in the Eastern Islands of Fiji, Island Reports, 5, pp. 65-81.

McNutt, M. and Menard, H.W. (1978) 'Lithospheric flexure and uplifted atolls', *J. Geophys. Res.*, *83*, pp. 1206-12.

Marshall, J.F. and Jacobson, G. (1985), 'Holocene growth of a mid-Pacific atoll: Tarawa Atoll, Kiribati', *Coral Reefs*, *4*, pp. 11-17.

Matsushima, Y., Sugimura, A., Berryman, K.I., Ishii, T., Maeda, Y., Matsumoto, E. and Yonekura, N. (1984) 'Research report of B Party: Holocene sea-level changes in Fiji and Western Samoa, in *Sea Level Changes and Tectonics in the Middle Pacific. Report of the HIPAC Project 1981-3*, pp. 137-85.

Montaggioni, L.F. and Pirazzoli, P.A. (1984) 'The significance of exposed coral conglomerates from French Polynesia (Pacific Ocean) as indicators of recent relative sea level changes', *Coral Reefs*, *3*, pp. 29-42.

Mörner, N.-A. (1972) 'Isostasy, eustasy and crustal sensitivity', *Tellus*, *24*, pp. 586-92.

—— (ed) (1980) *Earth Rheology, Isostasy and Eustasy*, Wiley Interscience, Chichester and New York.

Nakiboglu, S.M., Lambeck, K. and Aharon, P. (1983) 'Postglacial sea levels in the Pacific: implications with respect to deglaciation regime and local tectonics', *Tectonophys.*, *91*, pp. 335-58.

Newell, N.D. and Bloom, A.L. (1970) 'The reef flat and "Two-meter eustatic terrace" of some Pacific atolls', *Bull. Geol. Soc. Am.*, *81*, pp. 1881-94.

Parsons, B. and Sclater, J.G. (1977) 'An analysis of the variation of ocean floor bathymetry and heat flow with age', *J. Geophys. Res.*, *82*, pp. 803-27.

Paskoff, R.P. (1980) 'Late Cenozoic crustal movements and sea level variations in the coastal area of northern Chile', in N.-A. Mörner (ed), *Earth Rheology, Isostasy and Eustasy*, Wiley, Chichester and New York, pp. 487-95.

Pickrill, R.A. (1971) 'Evolution of coastal landforms of the Wairau Valley', *New Zealand Geogr.*, *32*, pp. 17-29.

Pirazzoli, P.A. (1976) 'Les Variations du niveau marin depuis 2,000 ans', *Mem. Lab. Geomorph. EPHE, 30*.

—— (1983) 'Mise en évidence d'une flexure active de la lithosphère dans l'Archipel de la Société (Polynésie Française) d'après la position des rivages de la fin de l'Holocène, *Comptes Rendus Acad. Sci. Paris*, *296*, pp. 695-8.

——, Brousse, R., Delibrias, G., Montaggioni, L.F., Sachet, M.H. Salvat, B. and Sinoto, Y.H. (1985) 'Leeward islands (Maupiti, Tupai, Bora Bora, Huahine), Society archipelago', in B. Delesalle, R. Galzin and B. Salvat (eds), *French Polynesian Coral Reefs, Reef Knowledge and Field Guides*, Fifth Int. Coral Reef Congr., Tahiti, vol. 1, pp. 17-72.

—— and Fontes, J.C. (1982) 'Late Holocene sea level changes in the central Pacific: a 1000 km-long north–south transect of French Polynesia', *Abstract, XI INQUA Congress, Moscow*.

—— and Montaggioni, L.F. (1984) 'Variations Récentes du niveau de l'ocean et du bilan hydrologique dans l'atoll de Takapoto (Polynésie Française)', *Comptes Rendus Acad. Sci. Paris*, *299*, pp. 321-6.

—— and Montaggioni, L.F. (1986) 'Late Holocene sea-level changes in the northwestern Tuamotu Islands, French Polynesia', *Quat. Res.*, *25*, pp. 350-68.

Plassche, O. van de (ed) (1986), *Sea-level research: a manual for the collection and evaluation of data*, Geo-Abstracts, Norwich.

Playford, P.E. (1977) 'Geology and groundwater potential', in P.E. Playford and R.E.J. Leech (eds), *Geology and Hydrology of Rottnest Island*, Geol. Surv. W. Austr. Rept. 6.

Revelle, R. and Emery, K.O. (1957) 'Chemical erosion of beach rock and exposed reef rock', *US Geol. Surv. Prof. Rept.*, *260-T*, pp. 699-709.

Salvat, B., Richard, G., Poli, G., Chevalier, J.P. and Bagnis, R. (1977) 'Geomorphology and biology of Taiaro atoll, Tuamotu Archipelago', in *Proc. Third Int. Coral Reef Symp.*, vol. 2, pp. 289-95.

Schofield, J.A. (1960) 'Sea level fluctuations during the last 4000 years as recorded by a chenier plain, Firth of Thames, New Zealand', *NZ J. Geol. Geophys.*, *3*, pp. 467-85.

—— (1967) 'Post-glacial sea-level maxima a function of salinity?' *J. Geosci. Osaka Univ.*, *10*, pp. 115-20.

—— (1970) 'Notes on Late Quaternary sea levels, Fiji and Rarotonga', *NZ J. Geol. Geophys.*, *13*, pp. 199-206.

—— (1973) 'Post-glacial sea levels in Northland and Auckland', *NZ J. Geol. Geophys.*, *16*, pp. 359-66.

—— (1977a) 'Late Holocene Sea level, Gilbert and Ellice Islands, west central Pacific Ocean', *NZ J. Geol. Geophys.*, *20*, pp. 503-29.

—— (1977b) 'Effect of Late Holocene sea-level fall on atoll development', *NZ J. Geol. Geophys.*, *20*, pp. 531-6.

Sclater, J.G., Anderson, R.N., and Bell, M.L. (1971) 'The elevation of ridges and evolution of the central eastern Pacific'. *J. Geophys. Res.*, *76*, pp. 7888-915.

Scoffin, T.P. and McLean, R.F. (1978) 'Exposed limestones of the Northern Province of the Great Barrier Reef', *Phil. Trans. Roy. Soc. London.*, *A, 291*, pp. 119-38.

Scott, A.J. and Rotondo, G.M. (1983) 'A model to explain the differences between Pacific plate island-atoll types', *Coral Reefs*, *1*, pp. 139-50.

Shepard, F.P., Curray, J.R., Newman, W.A., Bloom, A.L., Newell, N.D., Tracey, J.I. and Veeh, H.H. (1967) 'Holocene changes in sea level: evidence in Micronesia', *Science*, *157*, pp. 542-4.

Soons, J.M. and Selby, M.J. (eds) (1982) *Landforms of New Zealand*, Longman Paul, Auckland.

Spenceley, A.P. (1982) *The Geomorphological and Zonational Development of Mangrove Swamps in the Townsville Area, North Queensland*, Dept. of Geography, James Cook University of North Queensland Monogr. Ser. 11.

Stearns, H.T. (1961) 'Eustatic shorelines on Pacific islands', in R.J. Russell (ed), *Pacific Island Terraces: Eustatic?*, *Zeits. für Geomorph. Suppl. Bd.*, *3*, pp. 3-16.

Stoddart, D.R. (1969a) 'Geomorphology of the Solomon Islands Coral Reefs', *Phil. Trans Roy. Soc. Lond*, *B, 255*, pp. 355-82.

—— (1969b) 'Geomorphology of the Marovo elevated barrier reef, New Georgia', *Phil. Trans. Roy. Soc. Lond.*, *B, 255*, pp. 383-402.

—— (1971) 'Coral reefs and islands and catastrophic storms', in J.A. Steers (ed), *Applied Coastal Geomorphology*, pp. 155-97.

—— (1972c) 'Reef islands of Rarotonga', *Atoll Res. Bull.*, *160*, pp. 1-7.

—— (1975) 'Almost Atoll of Aitutaki: Geomorphology of reefs and islands', *Atoll Res. Bull.*, *190*, pp. 31-57.

Sugden, D.T. and John, B.S. (1973) 'The ages of glacier fluctuations in the South Shetland Islands, Antarctica', *Palaeoecology of Africa*, *8*, pp. 139-59.

Suggate, R.P. (1968) 'Postglacial sea level rise in the Christchurch Metropolitan Area, New Zealand', *Geol. Mijnb.*, *47*, pp. 291-7.

Sugimura, A., Matsushima, Y., Matsumoto, E., Maeda, Y., Berryman, K., Ishii, T., Yonekura, M., Ida, Y. and Miyata, T. (1983) 'Holocene marine deposits in south coast of Viti Levu, Fiji', in *Abstracts of Papers, International Symposium on Coastal Evolution in the Holocene, Tokyo, Japan*, pp. 123-6.

Taylor, F.W. and Bloom, A.L. (1977) 'Coral reefs on tectonic blocks, Tonga island arc', in *Proc. Third Int. Coral Reef Symp.*, vol. 2, pp. 275-81.

——, Isacks, B.L., Jouannic, C., Bloom, A.L. and Dubois, J. (1980), 'Coseismic and Quaternary vertical tectonic movements, Santo and Malekula Islands, New Hebrides island arc', *J. Geophys. Res.*, *85*, pp. 5367-81.

——, Jouannic, C., Gilpin, L. and Bloom, A.L. (1982) 'Coral colonies as monitors of change of relative level of the land and sea: applications to vertical tectonism', in *Proc. Fourth Int. Coral Reef Symp.*, vol. 1, pp. 485-92.

Thom, B.G. and Chappell, J. (1975) 'Holocene sea levels relative to Australia', *Search*, *6*, pp. 90-3.

—— and Roy, P.S. (1983) 'Sea level change in New South Wales over the past 15,000 Years', in D. Hopley (ed), *Australian Sea Levels in the Last 15,000 Years: A Review*, Dept. of Geography, James Cook University of North Queensland, Monogr. Ser., Occ. Paper 3, pp. 64-84.

Tjia, H.D., Fujii, S., Kigoshi, K. and Sugimura, A. (1974) 'Late Quaternary uplift in eastern Indonesia', *Tectonophys.*, *23*, pp. 427-33.

——, Fujii, S., Kogoshi, K. and Sugimura, A. (1975) 'Additional dates on raised shorelines in Malaysia and Indonesia', *Sains Malaysiana*, *4*, pp. 69-84.

Tracey, J.I. (1968) 'Reef features of Caroline and Marshall Islands. *US Geol. Surv. Prof. Paper.*, *600-A*.

—— (1972) 'Holocene emergent reefs in the central Pacific', in *Abstracts Second Conf. Am. Quat. Ass.*, pp. 51-2.

—— and Ladd, H.S. (1974) 'Quaternary history of Eniwetok and Bikini atolls, Marshall Islands', in *Proc. Second Int. Coral Reef Symp.*, vol. 2, pp. 537-50.

Tricart, J., Dollfuss, O. and Clooths-Hirsch, A.R. (1969) 'Les Etudes françaises sur le Quaternaire sud-américain. Etudes franc. sur le Quaternaire', *Supp. Bull. de l'AFEQ, VIII Congr. INQUA*, pp. 215-34.

Trudgill, S.T. (1976) 'The marine erosion of limestones on Aldabra atoll, Indian Ocean', *Zeits. für Geomorph.*, *Suppl. Bd.*, *26*, pp. 164-200.

Veeh, H.H. (1966) 'Th^{230}/U^{238} and U^{234}/U^{238} ages of Pleistocene high sea level stand', *J. Geophys. Res.*, *71*, pp. 3379-86.

Walcott, R.I. (1972) 'Past sea levels, eustasy and deformation of the Earth', *Quat. Res.*, *2*, pp. 1-14.

Winterer, E.L. (1973) 'Sedimentary facies and plate tectonics of equatorial Pacific', *Am. Ass. Petrol. Geol. Bull.*, *57*, pp. 265-82.

Wood, B.L. (1967) 'Geology of the Cook Islands', *NZ J. Geol. Geophys.*, *10*, pp. 1429-45.

—— and Hay, R.F. (1970) 'Geology of the Cook Islands', *Bull. NZ Geol. Surv.*, *N.S.*, *82*, pp. 1-103.

Woodroffe, C.D., Curtis, R.J. and McLean, R.F. (1983) 'Development of a chenier plain, Firth of Thames, New Zealand', *Mar. Geol.*, *53*, pp. 1-22.

Yonekura, N., Matsushima, Y., Maeda, Y., Matsumoto, E., Togashi, S., Sugimura,

A., Ida, Y., and Ishii, T. (1983) 'Holocene sea-level changes in the southern Cook islands', in *Abstracts of Papers, International Symposium on Coastal Evolution in the Holocene Tokyo, Japan*, pp. 151-4.

——, Matsushima, Y., Maeda, Y. and Kayanne, H. (1984) 'Holocene sea-level changes in the southern Cook Islands', in *Sea-level Changes and Tectonics in the Middle Pacific. Report of the HIPAC Project 1981-83*, pp. 113-36.

Zivago, A. and Esteev, S. (1970) 'Shelf and marine terraces of Antarctica', *Quaternaria*, 12, pp. 89-105.

An Introduction
to Parts Four and Five

The Impact and Application of
Sea-surface Changes

R.J.N. Devoy

Part Four of this book, 'The Coastline: Processes, Planning and Management', is concerned with the interrelationships between sea-level change, coastal processes–dynamics and the impact these have upon mankind. Man who, both historically and increasingly today, lives in the coastal zone (Tooley, 1971, 1979; Bird, 1985; Bird and Koike, 1985; Walker 1984) is potentially at high risk from changes in sea-surface level, such as those of storm activity or long-term flooding and coastal submergence (Hoffman, 1985; Barth and Titus, 1984; Devoy, 1980). It is important, therefore, that consideration is given in this book to the factors responsible for coastal change and to the problems facing Man's habitation of this dynamic environment.

Variations in sea level first begin to have their impact upon the 'land' over the continental shelf. Here a complex of factors, including sea-floor topography, sediment composition, shelf geology/structural history, shelf gradient and configuration and factors of water movement (wave, tidal currents), interact with changes in sea level to produce alterations in the sediment pattern and development of the offshore zone over time. Discussion of these interrelationships is given in Kennett (1982), Greenwood and David (1984) and others (Hayes, 1967; Stanley and Swift, 1976; Roy, 1985), although limitation of space prevents their further treatment here. Part Four thus begins by focusing on the interaction of these and related hydrodynamic factors, as they affect development of the nearshore zone, in a discussion of sea-level change and coastal processes by Orford. The succeeding chapters pick up the theme of Man and the coastline. Chapter 14 by Carter takes a wide perspective of Man's responses to coastal change, set at differing timescales and in differing environmental situations, commenting upon both the engineering and alternative 'passive' approaches to coastal management. The final chapter by Titus forms a link to the theme(s) of Part Five. An examination is made by Titus of the consequences, in terms of sea-level change, of the connection between atmosphere and ocean. Past changes in climate, for example, in the form of

409

hemispherical–global glaciation, have led to large fluctuations in relative sea level and coastal position. Future shoreline changes may result both from Man's past and current industrial effect upon climate, and this impact may be the primary cause of the widely observed trend today of global coastal submergence (Bird and Koike, 1985; Gornitz *et al.*, 1982). Titus follows this theme by developing further some of the elements of coastal management, discussed earlier, which might follow on from such future climatologically induced changes in sea level.

As indicated in Part Four and elsewhere in the book, sea-level changes occurring at differing time, vertical and spatial scales can have profound effects upon both the ocean environment and also upon mankind. In Part Five, in a perhaps more positive vein, consideration is given to the repercussions and applications of changes in sea-level, showing how these can lead to potential benefits for mankind. For example, long-term changes in relative sea-level may have been instrumental in the formation of hydro-carbon resources, determining the location and timing of their formation. Discussion of this theme forms the focus of Chapter 16 by Devoy, together with a consideration of how the study of global relative sea-level move-ments has potential value in the discovery of hydrocarbons. Treatment here also includes an overview of biostratigraphic techniques, as these have a bearing upon the interpretation of sea-level change records as preserved in sedimentary environments. In Chapter 17 Sutherland goes on to show how sea-level movements over time have had consequences for the formation and distribution of a variety of economic mineral deposits. Again, the study of sea-level patterns is seen as an aid to their discovery and exploitation. The final chapter of Part Five, by Lewis, focuses upon an as yet poorly developed application of sea-level change — the production of power from wave and tidal action. Following initial definition of the form of short duration sea-surface movements, Lewis considers the types and genesis of engineering devices that have been constructed to harness the potential energy locked up in the moving water column of the oceans.

REFERENCES

Barth, M.C. and Titus, J.G. (1984) *Greenhouse Effect and Sea-level Rise*, Van Nostrand Reinhold, New York.

Bird, E.C.F. (1985) *Coastline Changes: A Global Review*, Wiley-Interscience, Chichester and New York.

—— and Koike, K. (1985) 'Man's impact on sea-level change: a review', unpublished paper. Abstract, *Proc. Fifth Int. Coral Reef Congr.*, Tahiti, vol. 3, pp. 91-2.

Devoy, R.J.N. (1980) 'Postglacial environmental change and Man in the Thames estuary: a synopsis', in F.H. Thompson (ed), *Archaeology and Coastal Change*, Society of Antiquaries, London, pp. 134-48.

Gornitz, V., Lebedeff, S. and Hansen, J. (1982) 'Global sea-level trend in the past century', *Science, 215*, 1611-14.

Greenwood, B. and Davis, R.A., Jr (1984) 'Hydrodynamics and sedimentation in wave dominated coastal environments', *Mar. Geol., 60*, 1-473.

Hayes, M.O. (1967) 'Relationship between coastal climate and bottom sediment type on the inner continental shelf', *Mar. Geol., 5*, 111-32.

Hoffman, D. (1985) 'The Holocene marine transgression in the region of the North Fresian Islands', *Eiszeit. Gegenw., 35*, 61-9.

Kennett, J.P. (1982) *Marine Geology*, Prentice-Hall, Englewood Cliffs. NJ.

Roy, P.S. (1985), *Marine Sand Bodies on the South Sydney Shelf, Southeastern Australia*, Coastal Studies Unit Technical Rept. No. 85/1, Dept. of Geography, University of Sydney.

Stanley, D.J. and Swift, D.J.P. (1976) *Marine Sediment Transport and Environmental Management*, Wiley, New York.

Tooley, M.J. (1971) 'Changes in sea-level and the implications for coastal development', in *Ass. River Auth. Year Book and Directory*, pp. 220-5.

—— (1979) 'Sea-level changes during the Flandrian stage and the implications for coastal development', in *Proc. 1978 Int. Symp. on Coastal Evolution in the Quaternary, São Paulo, Brazil (IGCP Project 61)*, pp. 502-33.

Walker, H.J. (1984) 'Man's impact on shorelines and nearshore environments; a geomorphological perspective', *Geoforum, 15*, 395-417.

411

Part Four

The Coastline: Processes, Planning and Management

13

Coastal Processes: The Coastal Response to Sea-level Variation

J. Orford

INTRODUCTION

Investigations into the coastal response to sea-level changes have long been a salient feature of Pleistocene geomorphological/geological research. However, of late a renewed interest has been generated by contemporary coastal problems, particularly of flooding and erosion related to sea-level changes on a worldwide scale.

Over the last hundred years, sea-level changes tended to be of interest only to geologists concerned with the reconstruction of past environments, particularly those related to changes since the last major ice extension. As shown in earlier chapters (Part Three), the primary means of obtaining information concerning past sea levels is by examination of the morpho-sedimentary deposition directly engendered by past sea-level positions. These fragmentary deposits of past shorelines, representing varying shoreline positions in space and time, often prove a handicap to major chronological reconstructions of palaeoenvironments. The complexity of coastal development, particularly in the vertical plane, was not always appreciated by a generation of Pleistocene investigators who attempted to establish stratigraphic chronologies based on limited shoreline morpho-sedimentary evidence; witness Synge's (1981) debate on the Irish Pleistocene succession (see Devoy, this vol., Ch. 10). As a consequence of an often inadequate understanding of the capacity of shoreline processes to establish a wide range of apparent morpho-sedimentary responses, erroneous conclusions were made. A classic example of this type of problem was the lack of appreciation concerning the height to which low-frequency storm wave activity can leave a permanent morpho-sedimentary response (Carter, 1983).

In recent years there has been a growing awareness of increasing shoreline erosion on a world scale (Bird, 1976). Such widespread erosion has been coupled with rising sea level, although the causes of the latter remain equivocal. The probability, however, of coastal damage to natural and man-made structures as a function of sea-level rise, has served to concentrate attention on to what actually happens to the shoreline when sea level varies. Both interpretation of past shorelines and extrapolation of future shorelines require a clear understanding of coastal processes *per se*, in that sea level is only a passive factor in coastal development. Variation in sea level only spatially displaces the point of process application, and as a consequence it is possible to advocate that there is no coastal process variation as a direct function of sea-level change. However, from our geomorphological interpretation of depositional evidence, it is clear that some coastal responses do alter as coastal processes vary spatially with sea-level change. Such changes in the degree, or even kind, of coastal response alteration appear related to the rate at which sea level varies and the effect that has on the major factors controlling coastal morphology.

This chapter examines such controls on coastal morphology which are influenced by sea-level change. This implies that coastal processes as such will *not* be examined, in that they are governed by physical laws which are controlled by factors independent of sea-level position. Discussion of such processes may be found in Komar (1976, 1983), Davies (1980), Greenwood and Davis (1984) and King (1972). The purpose of this chapter is (1) to outline the controls on coastal morphology which are directly affected by sea-level change and (2) to consider how these controls may interact with varying rates of sea-level change to effect coastal morphology.

CONTROLS ON COASTAL MORPHOLOGY

There are a number of factors which control the macro-scale development of coastal morphology: (1) the rate of sea-level variation, (2) the morphology and angle of the slope over which the sea level rises or falls, (3) the textural nature and volume of sediment available in the littoral zone, (4) the wave climate which indirectly measures the degree of storminess experienced at the shoreline and, finally, (5) the interaction between these factors leading to the preservation of the morpho-sedimentary facies upon which the coastal morphology is based. Each factor needs elaboration, although a combination of factors is required to provide a framework of reference for coastal response over the last 20 ka. Variation in each factor induced by sea-level change (SLC) will be emphasised where appropriate.

Rates of sea-level variation

At the core of any analysis of sea-level variation must be some assessment of the rate of sea-level rise or fall. Sea-level change rate (Δsl) is important as an independent control on the reaction of any coastal environment to process alteration. For example, the possibility of drowning or overstepping a beach face unit by a surf/swash zone carried landward on a rising sea level, rather than displacing the units landward in step with the rising sea level, is a key issue in Holocene shoreline survival (Swift, 1975; Rampino and Sanders, 1980; Staubblefield *et al.*, 1984). Curray (1969) argues that the presence of major barrier islands off of the eastern USA seaboard is a function of a major decrease in Δsl (8-9 ka BP). A reduction in Δsl enabled shoreline processes to persist spatially beyond the reaction time of the barrier morphology, allowing the bulk of the shoreline to keep step with sea-level rise.

Obtaining information on contemporary rates of sea-level change, let alone those from the Holocene or Late Pleistocene, highlights the inherent difficulties of defining and measuring such a concept. Part of this difficulty lies in defining sea-level change. The separation of extreme events and tidally induced periodic Δsl from the time-filtered (> 20 a) long-term secular Δsl, which concerns this chapter, is possible with sea-level data recorded over the last 100 years (Pugh and Faull, 1983). But such separation is not possible when based on fragmentary evidence of past morpho-sedimentary environments. Recent studies by Emery (1980a), Gornitz *et al.* (1982) and Barnett (1983, 1984) of Δsl over the last century (estimates vary between 1 and 3 mm a^{-1}) show that worldwide variation in regional-scale trends offer little hope that any unified global estimate can be found. This is despite global scale identification of factors and processes likely to cause variations in Δsl, such as increasing atmospheric temperature due to rising levels of atmospheric CO_2 (Hansen *et al.*, 1983), or thermal expansion of the upper oceanic layer (see Titus, this vol., Ch. 15). This view of variable regional rates is also considered appropriate for the Holocene (Kidson, 1982) due to variable non-eustatic controls on sea-level position. This non-eustatic variation can only serve to confuse the specification of coastal process, given the usual situation of fragmentary coastal response data without adequate time control.

Yet at the scale of geological time, there is a filtered simplification of Δsl which shows that a global smoothing may occur, despite variation in other independent factors such as tectonism and sediment supply. Vail *et al.* (1977) offer a unifying framework of oscillating eustatic variation with sea-level shifts leading to sequential phases of relative sea-level rise/fall which encompass the geological column (see Devoy, this vol., Ch. 16). Worsley *et al.* (1984) indicate that 'Vail type' sea-level change is of the order of 1-2 mm Ma^{-1} during cycles of continental movement of 25-50 Ma. Clearly the operation of such long-term rates of Δsl are well beyond the reaction time of all of today's coastal morpho-sedimentary environments, regardless of the level of rock/sediment strength involved. At this long-term level of Δsl, resultant coastal response must incorporate a substantial continental derived sediment supply, giving rise to a major control on transgressive or regressive sedimentary structures (Cant, 1984). Such a major control is missing in recent Δsl related sedimentation, due to the much greater rate of Δsl of the immediate past (Pleistocene) compared to the geological past (Guidish *et al.*, 1984). Thus wave climate, marine generated sediment supply and bathymetric controls are more likely to dominate present coastal responses. Further discussion of the sedimentary implications and problems of the Vail *et al.* model is given in Chapters 2 and 16.

Cronin (1983) has attempted to produce a generalised chart of Δsl for the last 150 ka (Fig. 13.1) despite Kidson's (1982) reservations about this type of approach. A peak Δsl of 20-30 cm a^{-1} (Hollin, 1977) is shown, based on an ice surge model leading to short-term ($<$100a) high level shorelines. This is a dramatic and speculative low frequency event with related evidence of its activity in the last Interglacial being too fragmentary for unequivocal signs of coastal response variation. More prolonged peak Δsl rates of 2.5-3.5 cm a^{-1} are given by Tooley (1974) and Ludwig *et al.* (1981), but still only applied for short periods $<$1 ka. Such peak rates are only recorded as a response, as investigators have no certain knowledge concerning their causes.

Following the last deglaciation, and regardless of site, the second derivative of Δsl is usually regarded as being negative exponential in form. For example, Kraft (1971) working on the Delaware coast suggests a Δsl of $>$ 10 cm a^{-1} before 8,000 BP, 0.32 cm a^{-1} between 8,000 and 3,700 BP and 0.15 cm a^{-1} from 3,700 BP to the present. Kraft advocates that the present coastal response is inherited from previous coastal environments dictated by this variation in Δsl.

As most of our contemporary coastal morpho-sedimentary environments have been both directly effected and affected by events of the last 20 ka, then they must reflect the change in coastal response associated with Δsl's of $<$10 cm a^{-1} to 0.1 cm a^{-1}. To what extent transitive thresholds of response, that is where a change of geomorphic state is caused by a persistent change of external boundary conditions (Chappell, 1983), can be recognised between these limits is unknown. Work on the USA barrier islands shows, however, that there is a high probability that such coastal response thresholds have occurred (see later discussion). However, estimates of Δsl associated with such thresholds have not, as yet, been specified adequately.

Coastal slopes

Coastal slopes affect coastal morphology on the basis of three factors: shoreline wave energy budget, tidal range at the shoreline and rate of shoreline migration. Note that in this chapter the coastal slope is regarded as being virtually synonymous with the continental shelf, though steeper

Figure 13.1: Eustatic sea-level elevation and rates of sea-level change over the last 150 ka

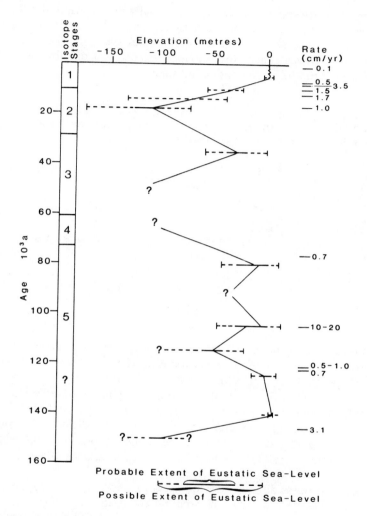

Source: After Cronin (1983).

slopes in the littoral zone may reflect either local variation in the pre-existing terrestrial relief or surface morphology of nearshore-beach sedimentation, though the latter will still be some function of the former.

The rate of change in water depths in a shorewards direction, particularly in the nearshore where depth is less than half the incident wave's wavelength, controls the rate at which wave deformation takes place as

419

waves move into shallow water. Figure 13.2 illustrates an example of how different offshore gradients induce breaker height variation at the shore-line, for storm waves with identical deep-water characteristics. Although short period storm waves are affected principally by nearshore depths, such that beach–nearshore sedimentation slopes may influence shoaling rates rather than the overall slope of the continental shelf, the same does not hold for swell waves. These feel shoaling effects well offshore where the coastal slope is in effect the continental shelf.

Figure 13.2: The effect of varying bathymetric gradients on the shoaling transformation of a storm wave. Note wave height, and hence wave energy, attenuation on low angle slopes

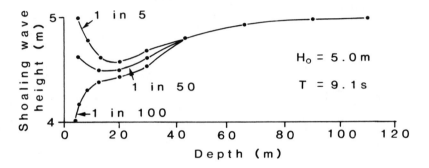

Longshore bathymetric variation generates through wave refraction (Komar, 1976) major longshore variation in wave direction and energy at the break-point. These longshore gradients of wave energy are responsible for the longshore movement of sediment. The more undulating, hence embayed, the inherited relief of the coastal slope is, partially due to prior terrestrial erosion, the greater the likelihood of disturbed longshore sediment transport patterns due to the development of littoral process cells (May and Tanner, 1973). Figure 13.3 shows a schematic diagram of this point.

A second process directly affected by slope and indirectly by the degree of surface dissection is that of tidal range at the shoreline. An early assumption that shorelines of a similar age would be at similar elevations on tectonically stable and undeformed coasts (cf. Zeuner, 1946) led to inconsistencies in postglacial chronologies based on palaeo sea levels. Carter (1983) indicates how the inclination of a suite of potential post-glacial shorelines, identified by Synge (1980) along the eastern coast of Ireland, shows correlation with present-day longshore tidal ranges (Fig. 13.4). Because in general the most reliable indicators of former sea levels are those based on deposits that form in the upper register of the tidal

Figure 13.3: The development of coastal cells as a function of slope variability of a pre-transgressive surface. P_1 is the longshore component of wave power (= ability to transport sediment). Positions a, c and e are key points in sediment cells as recognised by May and Tanner (1973)

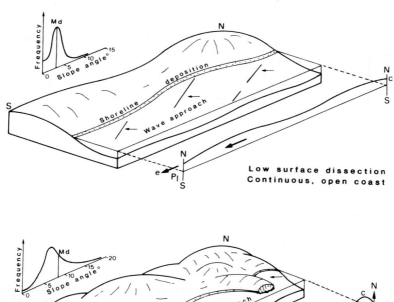

Low surface dissection
Continuous, open coast

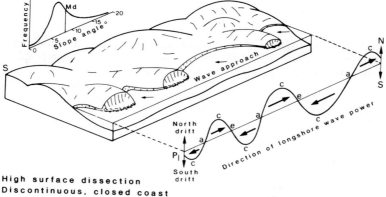

High surface dissection
Discontinuous, closed coast

range (Scott and Medioli, 1978), it is clear that longshore variation in tidal amplification could cause problems for sea-level correlation. From modern observation, it is clear that tidal range varies alongshore as a function of tidal wave amplification induced by nearshore bathymetry. Straight or open coasts are more likely to receive similar longshore tidal ranges within short distances than crenellated or closed coasts. The width of the continental shelf can contribute to major tidal range differences (Fig. 13.5) in that the range is amplified as a low-angle shelf is drowned under a rising sea (Cram, 1979). Enclosed seas and indented coasts can induce longshore tidal range

421

Figure 13.4: Tidal ranges can be amplified within small sea basins (e.g. Irish Sea). Carter (1983) shows that such tidal amplification may contribute to the apparent gradient of a suite of postglacial shorelines identified along the eastern seaboard of Ireland by Synge (1980)

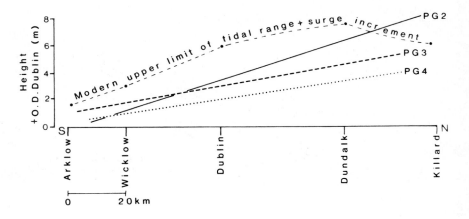

Source: After Carter (1983).

Figure 13.5: The gradient of the continental shelf can control the tidal range at the shoreline, such that palaeo-tidal ranges may have been less when shorelines were closer to the shelf-break

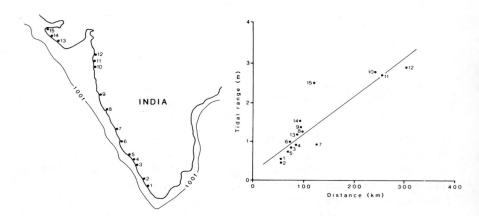

Source: After Cram (1979).

422

variation (Jardine, 1975). Morpho-sedimentary remnants of old shorelines are likely to be fragmentary along coasts which are crenellated and show such variable tidal ranges, therefore synchronous shorelines can show highly variable longshore heights under such conditions.

Sea-level change will also cause fluctuations in the amount of tidal amplification given an embayed coastal system. Such a case is identified by Scott and Greenberg (1983) for the Gulf of Maine and the Bay of Fundy. They show that change in tidal amplitude at the shoreline was linearly related to change in relative sea level over George's Bank in the entrance to the Gulf.

The third aspect, of horizontal shoreline migration rate, is initially determined by the coastal slope over which the SLC occurs. Obviously lower slopes will see faster rates of shoreline migration than steep slopes for a given SLC, constant regolith thickness and resistance. Figure 13.6 shows how recession rate (R) and eroded sediment volume (V) depend on coastal slope angle (Eq.13.1 and Eq.13.2 respectively). On a world scale it is interesting to note whether coastal slopes can be spatially separated. King

Figure 13.6: Shoreline recession (R) and the volume of sediment available for shoreline reworking (V) can be seen as a function of coastal slope angle (α). Steeper slopes show lower recession rates and smaller reworked sediment volumes

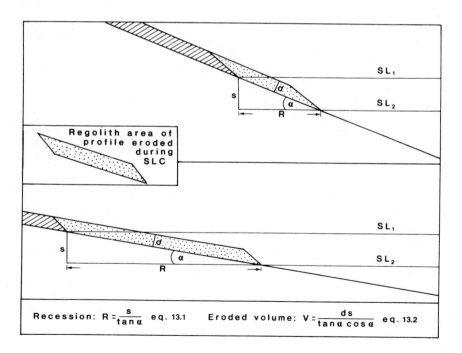

Recession: $R = \dfrac{s}{\tan \alpha}$ eq. 13.1   Eroded volume: $V = \dfrac{ds}{\tan \alpha \cos \alpha}$ eq. 13.2

423

(1972) noted that low-gradient coasts were mainly a low/middle latitude phenomenon. Bruun (1962) observed that the western edge of the North American continent exhibited a steeper and narrower continental slope than that observed along the eastern continental edge. This follows from the position of the west and east coasts as leading edge and trailing edge coasts respectively in plate tectonics terms. Emery (1980b) recognises a similar division based on the modification of the original tectonic continental displacement by continental shelf sedimentation. He classifies the east and west USA coasts' shelves as mature (cloaked by sediments) and young (exposed rifting) respectively. Inman and Nordstrom (1971) show that coastal slopes can be classified on the basis of plate tectonics theory. The USA west coast forms an example of a coast dominated by transform faults related to a subduction zone, while the east coast represents a stable plate zone. Denudation processes on the long-term tectonically stable eastern USA coastal plain are reflected in the overall very low angle coastal slope over which a rising Holocene sea level has allowed shoreline migration, at rates of up to 8 m a^{-1} in New Jersey, and up to 13 m a^{-1} in Louisiana. On this coast as a whole the Wisconsin glacial shoreline was up to 320 km from the present USA coastline (Nummedal et al., 1984). The Indian and Australian coasts are further examples of stable plate coasts.

High rates of shoreline migration have implications for the ability of littoral processes to achieve equilibrium of sediment morphology and texture. Full reworking of sediments over which the shoreline passes, and which the littoral zone receives from longshore and terrestrial sources, may not be feasible. Thus the likelihood of a spatially, well differentiated nearshore/offshore sedimentation pattern (textural maturity), related to the interaction of fairweather and storm conditions, decreases with declining overall continental shelf angle. The low gradient continental shelf of eastern USA shows a range of textural and morphological zones that may relate to earlier shoreline positions. Such relict patterns have not yet been totally reworked by modern shelf processes (Swift et al., 1984). Shoreline horizontal migration on leading edge coastlines is low, such that textural differentiation and shelf maturity is more likely to be found (Curray, 1964).

Though hard to generalise, the greater the median gradient of the coastal slope, the greater the variability of the slope distribution making up the coastal slope (Fig. 13.3). This depends on the degree of lithological and structural variation, as well as the inherited landscape dissection exhibited by the continental shelf. Vanney and Stanley (1983) show that shelf break morphology (the spatial link between continental shelf and slope) as a function of plate induced structure plus geomorphic environment encountered during lower sea levels, has a coherent world scale zonation. By spatial contiguity, the continental shelf may also exhibit world scale zonation reflecting global variation in weathering and transport regimes.

This coastal slope differentiation has implications for sediment yield as well as for the morphology to be encountered by a rising sea level. Morphological variability can be identified with varying plan-view scales of shoreline embayment or at an extreme form, crenellation, as the shoreline migrates over the shelf. Bloom (1978) uses the degree of geological resistance to identify bold or low coasts which mirrors this morphological variability theme. Areas of subdued morphology are more commonly associated with stable cratons like the eastern USA, though stable old basement complexes with major igneous and metamorphic elements can show low amplitude surfaces but also expose crenellation of the shoreline due to differential rock resistance, for example the Guinea and Ghana coasts.

The degree of coastal crenellation will partially dictate the nature of subsequent depositional morphology and its longshore and on/offshore variability. In particular, the lack of any coherent longshore drift is a probable result of crenellate coastal morphology. Numerous littoral cells, defined by longshore sediment movement, with reversals of drift direction are likely as wave attack will be very uneven along indented coasts. This is likely to lead to a range of beach types from rocky platforms (sediment free), to pocket beaches (with longshore sediment size differentiation: fine to coarse in lee to exposed beach sites), to wide sandy beaches or gravel structures. King (1972: Chs. 19 and 20) provides a comprehensive review of accumulation forms that occur on crenellate and open coasts.

An exception to this steepness–variability rule relates to low-lying coastal plains formed from, or veneered by, glacigenic material and found in high and middle latitudes, for example the eastern New Brunswick coast of Canada (Forbes and Taylor, in press). Small-scale morphological differentiation (e.g. formed by drumlins) and textural facies variation is often maximised in such situations. Assuming low-medium wave energy, variation of this type can lead to a highly interdigitated shoreline, if not a drowned one. Complex sequences of beach sedimentation occur as spatially variable sources of sediment are exposed upon shoreline retreat. Examples of this can be seen at Clew Bay and Galway Bay on the western Irish seaboard (Guilcher, 1962).

Sediment type and volume

Coastal depositional morphology is determined by the availability of sediment as well as by the textural composition of such material. By geological standards the volumes of sediment currently available in the littoral zone, per unit time, are considerably greater than that available within most former depositional cycles of the geological column. This increased beach volume is a partial function of a rapidly rising postglacial sea level sweeping up the continental shelf debris (Swift, 1970). On low-angle shelves the

425

volume swept up for a given sea-level increment will be more than that obtained on a steeper shelf (Fig. 13.6). However, the coastal sediment volume is not related solely to shelf slope, as the sediment texture will determine the proportion of the volume residing in the littoral zone. Silt and clays will disperse over the shelf leaving sand and gravel in position, whilst any contemporary terrestrial sediment input may also influence coastal depositional patterns.

Differences in textural variation are initially resolved into the relative proportions of gravel and sand in the littoral system. Given the importance of rising sea level during the Holocene, the relative proportions of sand and gravel will be determined principally by the environmental regime operating on the shelf prior to inundation. Hayes (1967) shows the relative proportions of sediment type found on the present continental shelf as a function of latitude (Fig. 13.7). From his results we can make some inferences about previous environmental regimes. The predominance of gravel in high latitudes reflects the deposition of extensive till plains during Pleistocene sea level minima and, in turn, is now reflected by the major presence of fringing gravel beaches around high-latitude coastlines. Sand size sediment volume peaks in the low–middle temperate and high tropical latitudes. This high arenaceous element is considered by Tricart (1972) as

Figure 13.7: Composition of continental shelf sediment by size and type. Note the large rock/gravel element in high latitudes as part of the 'Pleistocene Inheritance'

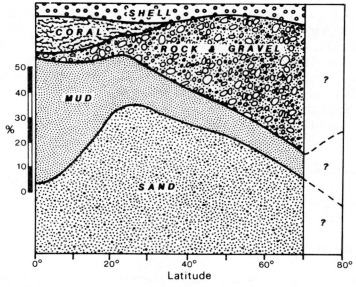

Source: After Hayes (1967).

426

the remnant of deep *in situ* weathering that occurs under humid/tropical conditions. Basement complexes exposed in low latitudes are usually quartz-rich and upon weathering provide large sand volumes (Faniran and Jeje, 1983). The rapidly decomposing clays from this weathering process are quickly washed offshore, whilst the arenaceous element is incorporated in the shoreline. So much sediment was produced in this way that many low-latitude coasts show major barrier beach construction. The likelihood of the barriers being principally spit based, with material derived from recent fluvial sediment loads, is discounted by Galliers (1968) and Usoroh (1977) on the grounds that barrier sediment particle morphology reflects an offshore rather than a fluvial origin.

In equatorial regions mud assumes an important sediment constituent, principally through the high silt and clay loads carried by the 'big' rivers of northern South America, West Africa and monsoon Asia. This leads to the second point concerning the relative balance between marine and terrestrial loads. In general it is only in the low latitudes, within areas locally influenced by fluvial deposition, that terrestrial sediment volume is sufficient today to produce apparent regressive sedimentary coastal sequences, despite regional transgressive shorelines. It is rare to find such areas in middle/high latitudes now, as fluvial sediment loads are usually insufficient to generate the sediment volume required for this type of deposition. However, it should be remembered that during deglaciation, high fluvial discharges of water and sediment in high and middle latitudes would have been available for such regressive deposition, as shown by the Dutch sand barriers around the Rhine and Scheldt estuaries (van Straaten, 1965) (see Devoy, this vol., Ch. 10).

Rapid phases of marine inundation can promote the development of littoral sedimentation at all points along the shoreline, albeit at variable sedimentation rates due to variable longshore wave energy. It is only when Δsl is very low or stationary that this feeding from offshore sources breaks down, due to either the attainment of profile equilibrium or the exhaustion of offshore sediment, and that longshore distribution–sorting of sediment may predominate (Orford, 1986). At such times point source entry of sediment due to fluvial means can set up regions of sediment poverty/enrichment (cf. Penland *et al.*, 1981; Chapman, 1981). The reworking of existing sediment into distinctive sediment cells by longshore variation in wave power, caused by wave refraction, will have similar effects.

Wave climates

Although peripheral to the problem of sea-level influence on coastal morphology, it is clear that the type and distribution of waves moving onshore will condition the nature of the coastal morphology. Davies (1972)

427

indicates that the world's wave climates fall into a series of zonal types (Fig. 13.8) that are determined by the ocean's atmospheric climate. During a glacial maximum, atmospheric circulation would not induce the same wave climate recognised today. Consequent upon a rising sea level, relatively enclosed sea basins, as exemplified particularly in areas of the Northern Hemisphere such as the North Sea, would have expanded their physical dimensions and generally shown an increase in the wave climate characteristics. This would be reflected in terms of longer wave periods (a generally constructive shoreline influence) and higher waves (generally a destructive influence). This, in conjunction with the establishment of cyclogenesis congruent with that observed today, would have seen a shift in the high/middle latitudes towards more stormy conditions. In the low latitudes the increasing presence of swells emanating from the expanding ice-free and increasing wind-dominated southern oceans is probable. Examples of morphological response to such wave climate changes are as yet only tentative and await more precise specification of atmospheric circulatory changes as a consequence of global warming. Thom (1978) gives a possible example of how spatially variable episodes in beach ridge

Figure 13.8: Coastal wave-climate zones as suggested by Davies (1972)

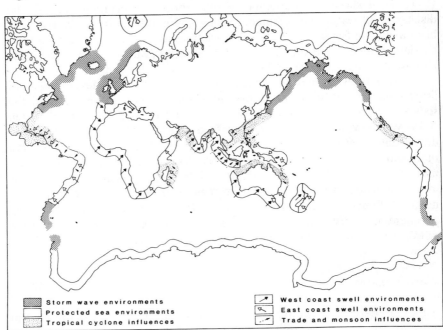

Storm wave environments
Protected sea environments
Tropical cyclone influences
West coast swell environments
East coast swell environments
Trade and monsoon influences

Source: After Davies (1972).

428

building along the southeast Australian coast during the Holocene are difficult to correlate directly with sea-level change. Instead, he suggests that the increased wave activity required for beach ridge building was a function of enhanced Tasman Sea cyclogenesis that impinged preferentially onto parts of the Australian coast. Increased cyclogenesis was caused by the presence of a blocking high pressure system, related to abnormal sea surface temperature gradients generated when the Antarctic convergence zone was at a lower latitude than it currently occupies.

COASTAL MORPHOLOGICAL RESPONSES TO SEA-LEVEL CHANGE

Changes in coastal morphology are best classified under a twofold heading: macro-scale and micro-scale. On the macro-scale, the relative response of coastal units to sea level needs to be considered in terms of the balance between sea-level change, sediment budget, wave characteristics and coastal slope. Although coastal morphology is often discussed at this generalised scale, it is important to recognise that coastal morphologies are dynamic entities and at the micro-scale show differential responses to SLC. The principal reference unit at this second level is the cross-beach profile. Variations here, when integrated over time, show the nature and direction of macro-coastal change. An understanding of beach profile variation underpins an appreciation of these macro-coastal changes.

Beach profile response to sea-level change

In order to simplify the nature of the process-response model of coastal variation as a function of sea-level change, investigators use the two-dimensional cross-beach (on–offshore) profile as the basic frame of reference. This means that two important assumptions have to be made: (1) that net longshore sediment movement through the plane of the cross-beach profile is zero and (2) that the profile is typical of the population of profiles that define the coastal unit in the longshore direction. On relatively straight or open coasts this latter point can be assumed, but on crenellate or closed coasts the effects of wave refraction will not be constant in a longshore direction in time; hence these two assumptions, (1) and (2), may be invalid. Therefore, models proposed for profile response to SLC are in general developed for relatively straight beaches. A third assumption concerns beach equilibrium, whereby the model's beach profile form is the net residual response to processes that operate on the profile over unit time.

Despite the importance of onshore–offshore sediment transport models in explaining processes of coastal development, there is a distinctive gap between these models and studies examining the relationship between SLC

429

and coastal morpho-sedimentary structure. Most of the latter studies are concerned only with unconsolidated material (beach sediments), and explanatory approaches to profile development on indurate materials (rock coasts) have been virtually neglected. Exceptions to this have been discussed by Dawson (1980) and Trenhaile (1983). These two approaches reflect either, (1) short-term SLC as a function of atmospheric disturbances of water level ($<10^2$h), referred to as *autogenetic* change, or, (2) long-term secular SLC leading to shoreline recession due to landward displacement of equilibrium profiles, referred to as *allogenetic* change. The first results from change inherent in the dynamics of the wave–beach system and the second through change induced by secular shifts in variables external to the coastal system. Our present capacity to understand autogenetic changes is superior to our knowledge of allogenetic change, in that the former is easier to study empirically, while often the latter has to be deductive in the light of the fragmentary and unequivocal stratigraphic evidence available.

Autogenetic beach profile adjustment: models of exchange

This type of profile adjustment has been analysed on the basis of distinctive profile type morphology as a function of wave/sediment interaction. The recognition of destructive, bar or storm generated profiles and constructive, step or fairweather (swell) generated profiles (Fig. 13.9) is a common theme of such studies (Johnson, 1949; Sunamura and Horikawa, 1974; Takeda and Sunamura, 1982) which seek to discriminate between profile type as a function of sediment size and/or wave type expressed in

Figure 13.9: The two main beach profile types associated with autogenetic profile change: step type (swell wave) and bar type (storm wave)

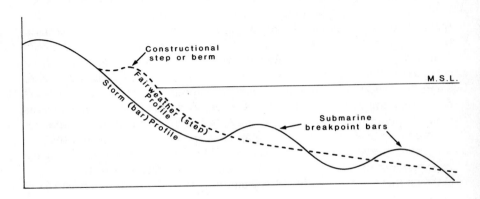

dimensionless form. The effects of short-term, storm-generated increases in sea level are marginal when compared to the significant differences in near-shore wave dynamics associated with the change in profile type.

Shoreline recession of a more extreme form is a principal factor in the analysis of beach profiles which result from severe storms superimposed on surges in water level, induced by the presence of the storm itself (Edelman, 1970; Dean, 1982; Vellinga, 1983). Cross-beach profiles are normally set in the context of dune-beach environments, though the vertical and land-ward extension of coarse clastic beach sediments by extreme wave run-up have also been studied (Orford, 1977, 1979). The former type of profile analysis is useful, in that dune front recession rates can be modelled, as the analysis explicitly expects shoreline displacement as a function of tempor-ary SLC. Figure 13.10 illustrates one approach to modelling such reces-sion, as advocated by Dean (1982) (Eq.13.3). It should be noted that this model type is solely a function of wave generated erosion, where waves are carried forward on the incremented water level. Eroded sediment is usually displaced seawards either into a nearshore bar or spread evenly over the bed of the nearshore zone. There is no overall long-term rise in nearshore

Figure 13.10: An example model of autogenetic profile change in a dune front related to erosion caused by a temporary rise in sea-level due to a storm. R is dune recession distance, B is the height of the dune above normal sea-level, d_{bs} is the depth of water in which the characteristic storm wave breaks, h(X) is the depth of water at position X on the profile and A is a scaling factor

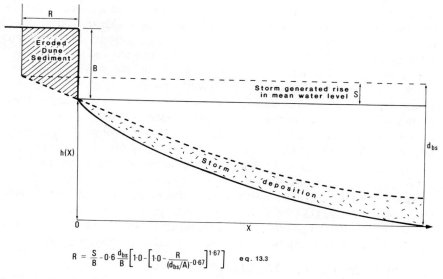

$$R = \frac{S}{B} - 0.6 \frac{d_{bs}}{B} \left[1.0 - \left[1.0 - \frac{R}{(d_{bs}/A)^{-0.67}} \right]^{1.67} \right] \quad \text{eq. 13.3}$$

Source: After Dean (1982).

bed elevation. For with the onset of post-storm constructional conditions (i.e. swell waves), this sediment is usually moved back onshore to heal over the recession scar.

Allogenetic beach profile adjustment: models of translation

This section includes two specific models which have some degree of empirical observation in their construction and testing, as well as a number of approaches which consider macro-sediment volume displacement in the littoral zone as a function of sea level. Due to the diachronous nature of the morpho-sedimentary evidence, the latter models are based principally on stratigraphic relationships. Most models relate to rising SLC given the trends of the last 20 ka. It should not be assumed that the same models can be used to explain both rising and falling SLC.

Shoreline displacement or encroachment model

This does not consider the change of profile *per se* with rising SLC, rather it considers shoreline recession (R) as a direct function of onshore shoreline displacement, caused by a rising sea-level that carries the shoreline forward and upward. Given an overall coastal slope angle (α), then recession is some function of tangent α ($R = s/\tan \alpha$, see Fig. 13.6), indicating that shallow slopes have the greatest shoreline displacement. Although this process is intuitively obvious, only Galvin (1983) has considered it as a major explanation of shoreline retreat. He regards much of the worldwide recession of sandy shorelines indicated by Bird (1976) as a function of this process. Galvin indicates that on the US coastal plain with an overall slope of 0.94 m km^{-1}, there will be +120 m recession per century given current USA east coast Δsl. However, he notes that the east coast shows a distinctive scarp on the landward edge of the barrier bay which would mitigate much of the effects of this SLC. If Galvin's theory is valid, barriers with wide back-barrier lagoons should show the greatest erosion rates, as they represent the flattest coastal slopes. Field tests in New Jersey, on the Delmarva peninsula, and along the Outer Banks of Carolina do not verify this theory. Further field tests by Dubois (1975) and Hands (1983) in the Great Lakes show that <25 per cent of shoreline recession can be attributed to shoreline encroachment. Galvin argues that the lack of field test success is due to shoreline displacement effects, in terms of cross-beach sediment exchange (Q_x) being of orders of magnitude less important than the process of longshore sediment dispersal (Q_y). This imbalance ($Q_x < 0.05Q_y$) would seriously mask the encroachment effects of SLC.

Discrete sea-level change: The Bruun Rule

Although the encroachment rule must define a base line for recession of sandy shores, other factors may account for variation in R. Most obvious are the erosion and deposition rates of wave processes associated with the SLC, as well as the control on recession imposed by the assertion of a new equilibrium profile when SLC is completed.

The Bruun Rule (BR), nominated as such by Schwartz (1967) (Schwartz and Milicic, 1980), is a basic concept first expounded by Bruun (1962) to account for variation in shoreline recession observed along the Florida coast (see Titus, this vol., Ch. 15) The basic concept (Fig. 13.11) is that

Figure 13.11: The elements of the Bruun Rule. Stages b and c are assumed to be synchronous. Parameters are defined in the text. Shoreline recession is related to the product of the profile length and sea-level change, and inversely related to profile height.

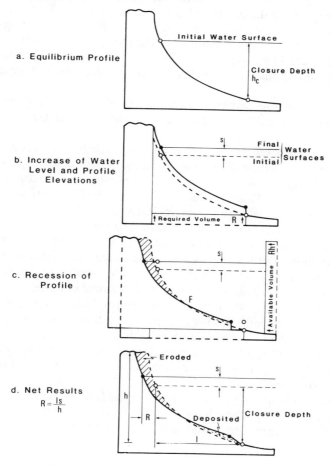

Source: After Hands, 1983.

433

after a discrete rise in sea level, the equilibrium beach profile occurring at SL_1 (initial) is re-established in the same form at SL_2. In the process the profile must now extend further onshore as a consequence of encroachment. In addition, however, shoreline recession in perimarine sediments will continue, until the nearshore sediment bed level is elevated by an offshore moving sediment volume, sufficient to re-establish a new nearshore depth equal to that existing under SL_1. Thus the rise in nearshore bed elevation is equal to the rise in sea-level. The mechanisms of this sediment transfer were not elaborated upon by Bruun, as the concept was descriptive of the responses to processes that were occasioned by Δsl. The principal implication of BR is that rates of shoreline recession are governed by the volume of sediment required to raise the nearshore bed to the required height (i.e. the amount of SLC). Obviously erosion of high coastal bluffs as opposed to low sand dunes would lead to recession rate differences, assuming that all other factors controlling sediment transport were constant.

A number of boundary conditions to the model were initially stipulated and later extended by Bruun (1980, 1983) in the light of field testing:

(1) that the initial and re-established profiles are net equilibrium profiles.

(2) that longshore sediment transport is absent or, if present, that sediment entering and leaving the beach profile in the longshore direction balances.

(3) that eroded sediment carried into the nearshore is not allowed to leak seaward of a certain depth, known as the closure depth (h_c). Variation in h_c will account for widely differing recession rates. Bruun (1962) suggested that $h_c = 18$ m, on the basis of the Florida data, while Hands (1980) from Lake Michigan data indicates $h_c = 11$ m. This difference relates to the varying wave energy for the two sites, in that Hands (1983) identifies h_c as equal to twice the breaker height of a five-year return period wave. The position of h_c is highly debatable, as the interaction of nearshore and shelf sediment is spatially and temporally variable to an extent that definitions of h_c, as a function of wave base, become impossible to quantify with certainty (Swift, 1980). In the more controllable environment of the Great Lakes, Dubois (1980a) identifies h_c as a function of nearshore bar position, rather than in terms of water depth *per se.*

(4) that the sediment volume eroded from the back beach, which is too fine in size for residence in the nearshore, must be accounted for in the recession estimate.

(5) that the shape of the equilibrium profile follows the form of $h(X) = AX^m$ (see Fig. 13.11), where h is the profile height measured at x— distance in a seaward direction and A and m are scale constants.

$$R = ls/h \qquad \text{(Eq. 13.4)}$$

where h = height of equilibrium profile, l = length of equilibrium profile, hence the importance of h_c, and s = the value of SLC (Fig. 13.11). Both Dubois (1977) and Rosen (1978) argue that l should be replaced by c. This represents the cross-beach length of the deposition zone of the new profile, that is, the position where the new profile intersects with the old one (i.e. the fulcrum position, F). If F is crucial to recession it implies that Bruun's Rule must be dependent on the overall shape of the profile. Allison (1980) shows by a mathematical first principles approach that this modification is not required and that profile shape is irrelevant to the application of the rule. He argues that BR is invariant as long as the ratio $s/l < \ll 1$ holds true. This means that for small incremental rises in sea level the recession rate may be monitored by means of the Bruun Rule.

A number of studies have been undertaken to measure the predictive power of the Bruun Rule. Fisher (1980a) gives a review of papers which have attempted to evaluate BR and indicates that a range of approaches, including model testing (Schwartz, 1965, 1967) and prototype testing at a range of timescales, have offered partial support for Bruun's contention (Dubois, 1975, 1977, 1980a; Rosen, 1980). As Bruun (1983) points out, the real difficulty in testing the rule is finding situations where the boundary conditions are matched.

Probably the most exhaustive field study of BR is that reported by Hands (1983), concerning his seven-year study of 25 profiles over 50 km of sandy beaches also along the Lake Michigan shoreline. It is this kind of tideless environment, having a secular mean water-level variation sufficient to engender shoreline recession, that provides the best testing ground of BR. Hands suggests that the generalised Bruun Rule should be

$$R = (sl \ (Ra)^m)/h - Qt/Yh \qquad \text{(Eq. 13.5)}$$

where s,l and h are defined as in Eq. 13.4 $(Ra)^m$ is the overfill ratio of sediment that may be missing from nearshore deposition, due to both its fine size and leakage beyond some notional closure point, while m = 1 if s > 0 and m = −1 if s < 0. This latter weighting means that with a falling sea-level the overfill ratio becomes inverse and represents erosion of the nearshore bed. Qt/Yh is the longshore (Y) sediment discharge rate. In the sites that Hands used, the silt content is < 1 per cent and longshore sediment discharge is zero. Hence Eq. 13.5 reduces to the basic BR of Eq. 13.4.

The changes in beach profile relative to mean water-level variation measured by Hands (1983) in Lake Michigan are shown in Figure 13.12. Note the complication that 1976 saw a fall in mean water level. Between 1969 and 1976 mean recession was 17.9 m while recession over 1969-75 was 13.6 m. In the same period encroachment predicted 3-4 m of recession, underlining the inefficiency of slope alone to predict recession. On the

Figure 13.12: Shoreline, beach profile and lake level changes in Lake Michigan (USA) between 1967 and 1976 (Hands, 1983). These data constitute the basis of a thorough test of the Bruun Rule

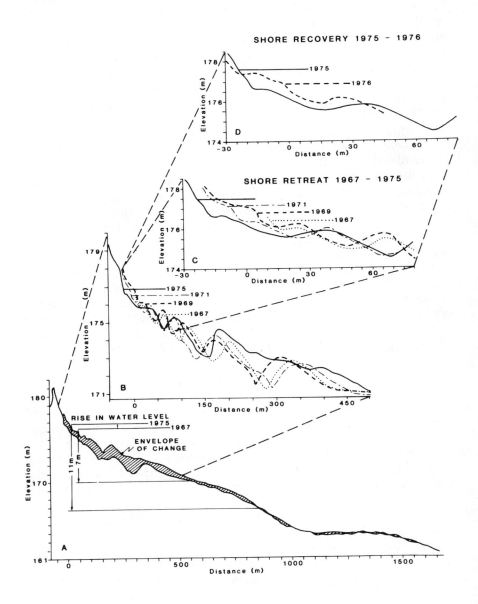

Source: After Hands (1983).

436

basis of the profile measurements, Hands indicates a predicted recession of 13.6 m, given the observed net mean water level change between 1969 and 1976, which appears to support Bruun's Rule. However, taking a shorter period within 1969-76, estimated R is too large; only by 1976 with a fall in mean water level do observed values match predicted rates. This leads Hands to recognise that profile adjustment lags by several years behind the forcing event of water-level rise. Hands recognises that short-term studies will be of little value in BR analysis as in the long term the major control of closure depth on nearshore deposition, and hence shoreline erosion, will show extreme variation as a function of low frequency, high magnitude wave height. The possibility of a long return period storm occurring which will have a major effect on near/offshore sediment dispersal, therefore, needs to be included in any analysis. Hands suggests that $h_c = 4H_{5a}$ would be required for long-term SLC analysis, where h_{5a} is the wave height associated with the 5-year return period wave (twice that used in his study).

Hands's study underlines a limitation of Bruun's Rule: it indicates a potential ultimate recession with discrete changes in sea level, but not at the time of occurrence. That will depend on the site's storminess which controls the rate of profile adjustment. Even if storminess can be predicted, there is still a time lag. How these changes can occur when sea-level rise is continuous is far from clear.

Evidence of considerable violation of the boundary conditions to Bruun Rule operation is provided by the last decade of process-response research undertaken on coastal morphodynamics of barrier islands along the USA eastern seaboard (Leatherman, 1979). It is evident that the Bruun Rule needs major modification when used on a coastline which is dominated by longshore sediment transport and high lateral variability of depositional environments. Barrier islands remain a specific problem in this context, as the onshore movement of storm-related sediment (overwash sedimentation) is completely contrary to the Bruun Rule expectation (Dubois, 1980b), as is the major lateral movement of sediment to depositional sinks (tidal deltas) unrelated to the beach profile (Walton, 1978).

Fisher (1980b) has also examined the Bruun Rule in the context of barrier island retreat under continuous SLC. The rate of barrier beach retreat along Rhode Island between 1939 and 1975 is stated to be around 0.2 m a^{-1}. Given the current sl of 3 mm a^{-1}, he shows that 85 per cent of the shoreline retreat is due to direct wave-related erosion. From analysis of air-photography coverage Fisher calculates that storm washover processes account for the removal of 26 per cent of the eroded beach volume, while tidal delta accretion accounts for a further 36 per cent of the eroded beach volume. This leaves about 38 per cent of the eroded sediment available for deposition offshore according to the Bruun Rule. Without adequate bathymetric chart control, Fisher can only see if a potential offshore sink exists which could absorb that volume. This potential depends on the position of the fulcrum and closure position on the equilibrium profile and their relationship to identifiable thresholds in the transformation of shoaling waves. Fisher calculates that this volume could be absorbed in a zone seaward from the breaker point (F) up to the regional wavebase at ~ 9 m depth (closure), rising to an elevation commensurate with SLC over the study period. The implicit contention of this result is that the volume of the nearshore acts as a minimum control on shoreline erosion as a function of SLC. This volume acts as a baseline to erosion regardless of the erosion losses to other sinks in the coastal system which are dependent on the storminess of the area.

Maurmeyer and Dean (1982) specifically consider a Bruun Rule type approach when considering profile development on barrier islands. They consider the likelihood of shoreline recession as a function of barrier beachface profile, as well as barrier width and bayside beach profile development. In the light of discrete SLC they indicate that recession is given by

$$R = s(L_O + W + L_B)/((B_O + h_O) - (B_B + h_B)) \quad \text{(Eq. 13.6)}$$

where the definitions are given in Figure 13.13. They note that if W, L_B, B_B and h_B are set to zero, then Eq.13.6 reduces to the Bruun Rule. A field test of this model and the Bruun model with respect to barrier island recession rates shows the Bruun model to deviate in prediction by 60 per cent while their formula has only 3 per cent deviation (Dean and Maurmeyer, 1983).

Figure 13.13: A cross-section of a USA east coast barrier to show the parameter definition of the model (Eq. 13.6) proposed by Maurmeyer and Dean (1982) to account for shoreline recession by sea-level change on sand barriers

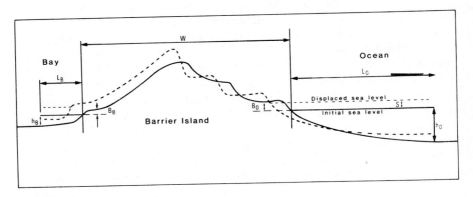

Source: After Maurmeyer & Dean (1982).

Continuous sea-level change

It is unlikely that the Bruun Rule will apply easily to the development of
the overall coastal response of the world's clastic shorelines in the light of
continuously varying SLC over the last 20 ka. Bruun (1983) does suggest
that multiple SLC, or discrete steps in sea-level change, may lead to a series
of stepped offshore profiles. However, continuous SLC and sediment
movement is unlikely to allow such morphology to persist. It is feasible that
only when sea-level rise is less than a critical threshold, somewhat greater
than current sea-level rise, will the effects of the Bruun Rule appear, in that
SLC is not fast enough to washout the reaction and response times of
beaches to wave processes.

Before considering beach profile response to continuous SLC some
understanding of the interaction effects of the controlling macro-scale
constraints is now required, in particular the relative fluctuations of sea
level and sediment supply. Curray (1964) shows that a shoreline can react
to a rising sea level by either advancing oceanwards (regressive offlap/
overlap), or retreating landwards (transgressive overlap), depending on the
rate of sediment deposition on the profile (Fig. 13.14) (see Devoy, this

Figure 13.14: The relationship between sea-level change (rise or fall) and
sediment supply to the beachface (erosion or deposition) in order to
define shoreline transgression or regression. Some possible depositional
environments are indicated particularly for mid-low latitude sandy beaches
within the transgressive-regressive domain. Timescale 100 ka

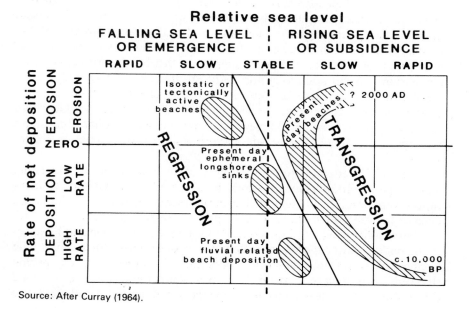

Source: After Curray (1964).

vol., Ch. 16). This process is irrespective of sediment source. Unless the wave environment is dominated by constructive swell, moving material into the nearshore-beach zone from offshore, the most likely source of extra sediment must be from alongshore. It is rare that longshore progradation is universal to any substantial coastal length during rising sea levels. If it does occur, then such areas are usually in the proximity of the deltas of 'big' rivers carrying sediment loads which include coarse $\gg$ sand-sized material. Most big rivers, like the Mississippi, Nile and Amazon, carry predominant silt loads, but a proportion of the total load may be of a size and volume to reside initially in river mouth distributary bars and shoals. Under wave attack these feed alongshore to adjacent beaches (Curray and Moore, 1964). Penland *et al.* (1981) give a good illustration of how the size and storm resistance of beaches peripheral to part of the Mississippi delta varies inversely with beach position downdrift from the feeding distributary mouth. Along most of the world's clastic shorelines it is more likely that progradation will be at the cost of updrift erosion. All of the major progradational, gravel-dominated forelands of the British Isles are the product of updrift erosion as sea level rose in the Holocene. An example of such is Magilligan foreland in Ulster (Carter, 1979).

Curray's work implies that the reaction of any clastic beach to allo-genetic change can only be studied in the joint context of sea-level rise and deposition. Over the last 10 ka most beaches have moved along a path of rapid to slow SLC. The problem has been to fix a similar path for sediment supply. This is clearly a regionally dependent variable and impossible to generalise as it depends on continental shelf gradient and prior sediment availability. Apart from fluvial point source inputs which must be localised, the rate of sediment supply per unit length of coast is some function of Δsl, in that slow shoreline recession, for whatever reason, will be allied to exhaustion of offshore sediment supply as the beach profile achieves equilibrium. At this point longshore sediment transport comes to dominate the beach transport system. This will have the effect of reducing beach deposition rate, except where beaches are adjacent to ephemeral longshore sediment sinks related to sediment cells, for example aeolian dunes and estuary shoals. This changing time pattern is graphically shown on Figure 13.14.

Swift (1975) attempts to take this type of analysis further by indicating which type of profile response would be observed given Δsl and sediment supply variation. He points out that beach profile adjustment reflects the balance between the build-up of the upper shoreface (beach) by construc-tive fairweather waves and the erosion of the upper shoreface and build-up of the middle/lower shoreface (nearshore) by destructive storm waves. The balance of storm to non-storm conditions, therefore, controls profile structure, as well as the coastal slope over which the translating shoreline drives the profile. Beach profile steepness is a function of sediment size

(Shepard, 1963) with gravel generating $<30°$ slopes, while sand forms $<5°$ slopes. If these beach response slopes are steeper or shallower than the coastal slope, then profile deviations from the equilibrium form cited in the Bruun Rule are bound to occur.

Variation in the curvature of the profile, as a function of grain size and storm/non-storm balance, plus variation in the angle of profile translation, a function of sea-level rise and continental shelf gradient, leads to a variety of profile translation models. Some of these models, developed in the context of barrier island sedimentation, are shown in Figure 13.15. Profiles A to C show the effects of increasing sediment supply to the profile and decreasing Δsl. In A the profile translates in a 'Bruun type' manner parallel

Figure 13.15: Swift's (1975) analysis of model beach profile conditions and structure as a function of sea-level change and sediment supply. Note that, over time, any given coastal profile may be a combination of such profile types as sediment supply and sea-level fluctuates

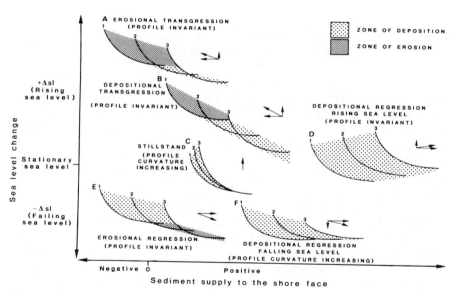

Source: After Swift (1975).

to the transgressed surface. Lack of profile sediment means the profile erodes through back-barrier deposition, so that lagoon and marsh sedimentation emerges on the shoreface as the barrier retreats. With increasing efficiency of the lagoon's tidal system to pass terrestrial derived sediment onto the profile, situation B may arise. Middle profile deposition together with decreasing sea-level rise allows the beach profile to transgress at a

441

divergent angle from the coastal slope and thus override back-barrier deposition. In time the rise in back-barrier depositional level will reduce the sediment load to the beachface and the profile will translate more like case A, parallel to the coastal slope, but over a thick layer of lagoon related deposition. If sea-level rise reduces to near zero, continuing sediment supply to the profile will cause upper profile progradation and an apparent regression ensues (model C). This model is considered by Swift (1975) to apply to the US barriers between 4,000 and 7,000 BP. Position D is related to high rates of sediment supply to the profile, so that the profile moves up and outward in an apparent regressive sedimentary sequence despite rising sea-level. This is similar to Curray's (1964) river deposition model. Categories E and F relate to falling sea levels with the profile moving down the coastal slope. Profile E shows the Bruun Rule in reverse with wave-based erosion of the lower profile and upper profile deposition, while F shows deposition across the whole profile. Here sufficient mobile fine sediments are available for lower profile and inner shelf deposition. As environmental controls alter, these profile morphologies (A–F) may be compounded.

To move towards a macro-scale understanding of shifting coastal morphologies by use of Swift's form of analysis, it is important now to consider the degree of sediment removal from the nearshore element of the profile by wave attack. This will control the nature of profile translation and recruitment of sediment for beachface supply when terrestrial sources are not available. On low-angle coasts rapid marine inundation erodes into deposits associated with former lower sea levels. In particular, barrier beaches retreat over lagoonal sands, clays and associated organic marsh sediments. The extent to which this material is eroded from the profile depends on the rate of shoreline migration and the storminess of the near-shore wave climate. Slow Δsl and storm conditions will result in deeper profile erosion than would occur under rapid Δsl and swell waves. Figure 13.16 shows schematically these approaches applied to a barrier coast (Fischer, 1961; Kraft, 1971). The resultant base of erosion known as a ravinement marks the boundary between back-barrier deposition and profile deposition (Swift, 1968). The depth of ravinement can sometimes be used as an index of Δsl if wave climate can be held as constant. Belknap and Kraft (1981) make use of this concept in their development of a model to account for differences in relict and modern coastal morphology at varying positions across the Delaware continental shelf. The potential for morphological preservation is reflected in the ravinement depth as a function of Δsl (Fig. 13.17). In the early Holocene, rapid transgression on the outer shelf allowed greater facies preservation potential than the present slow sea-level rise appears to be allowing in coastal Delaware.

Figure 13.17 stresses that shoreline morphology during marine inundation must vary due to the decreasing Δsl over time and the increasing tidal

Figure 13.16: The rate of sea-level change effects the depth to which the beachface will be eroded into underlying sediments. The preservation of facies and hence our ability to interpret past stratigraphies may be hampered by the differential erosion rates associated with variable rates of sea-level rise. However, such a difference in facies preservation may be used as an index of sea-level change rates

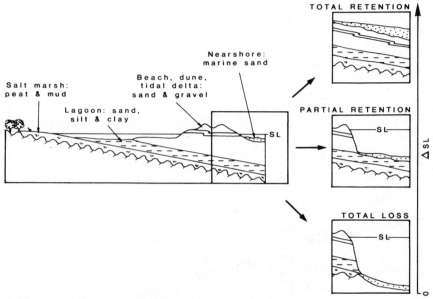

Source: After Kraft (1971).

amplitude at the shoreline as the continental shelf is flooded. Further, whilst incident wave energy was high at the outer shelf position, at the modern shoreline energy is reduced due to wave shoaling over the width of the continental shelf. The model is used to demonstrate three potential coastal morphological suites:

(A) on the outer shelf when Δsl was rapid, high wave energy and low tidal range existed. This generated wide but low barriers dominated by overwash. Shallow profile erosion depths ensure high preservation of depositional sequence.

(B) mid-shelf when Δsl was decreasing, moderate wave energy and high tidal range existed. This generated narrow and high barriers with some overwash. Moderate to deep profile erosion ensured reduced preservation potential of any depositional sequence.

(C) present shoreline with slow Δsl, moderate wave energy and a mesotidal range. This leads to high and narrow barriers dissected by inlets and intermittent overwash. Deep profile erosion ensures little depositional survival of back-barrier sedimentation in the front of modern barriers.

443

Figure 13.17: Belknap and Kraft (1981) made use of the facies preservation concept shown in Figure 13.16 to reconstruct the depositional environments at three points across the Delaware continental shelf. Coastal morphology and sediment preservation are shown to vary spatially as a function of decreasing rate of sea-level change over the Late Pleistocene and Holocene periods. See text for further explanation of conditions and morphology at points A, B and C

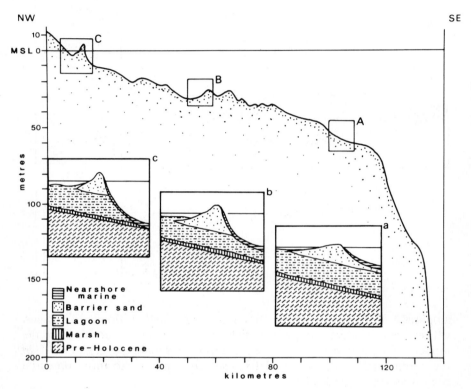

Source: After Belknap and Kraft (1981).

COASTAL RESPONSE TO SEA-LEVEL CHANGE: THREE EXAMPLES

It would be impossible in this chapter to consider all types of coastal response to sea-level change; instead three problem coastal responses have been selected which reflect recent and continuing research interest: (1) the migration mode of sand barriers, particularly off the USA eastern seaboard; (2) the development and migratory behaviour of gravel beaches and gravel-dominated barriers around the British Isles; (3) the cause of transgressive sand dune systems along the eastern Australian coast. The

444

principal question in these cases is how does sea-level change relate to the origin and form of the coastal phenomenon.

Sand barrier islands of the eastern USA seaboard

These features have been a major stimulus to coastal investigators over the last twenty years. Probably more has been written on these barrier islands and their genesis than any other coastal phenomenon. Such sand barriers are linear bodies lying above high water, approximately parallel to the main coastal trend, and separated from the mainland by a lagoon which is generally tidal and connected to the nearshore by tidal breaches/inlets across the barriers. Barriers usually occur in micro- to meso-tidal (<4 m) environments, forming under east coast swell (Davies 1972) and storms (hurricanes in the south, intense cyclonic depressions in the north). Leatherman (1979), Reinson (1979) and Otvos (1981) provide good reviews of the mixture of aeolian, wave and tidal processes active in shaping the barriers as well as discussing the stratigraphical implications of such deposits.

Barriers, and hence low energy back-barrier sedimentation, together with tidal inlet sedimentation, are initially controlled by the low-angle continental slope (<0.5°) over which the Atlantic Ocean transgressed, as well as by the high volume of sand available on the pre-transgressive surface. Nummedal (1982) shows that barrier origins are not singular. North of New York State, most barriers reflect reworking of glacigenic sediments, while south of New Jersey, sediments are of predominantly reworked fluvial or deltaic origin.

The origin of barriers, though contentious (Gilbert, 1885; Johnson, 1919; Swift, 1975), is not the point at issue here; rather interest is centred on how the barriers have reacted to positive SLC in the Holocene. Despite a rising sea level, there are substantial regressive sedimentary barriers with incremental beach/dune ridge plains showing older ridges landward of younger ones on this coastline (see Devoy, this vol., Ch. 10, for details). Most of the later regressive phases in barrier formation relate to fluctuations in sediment supply, though even a falling sea level coupled with a consistent sediment supply may trip a threshold to allow regressive sedimentation. Morton (1979) shows how regressive and transgressive barriers alternate along the Gulf of Mexico. Transgressive units are associated with subsiding relict delta-heads, while in the inter-delta areas trailing spits off of the flanks of the deltas converge to form a trap in which subsequent longshore drift builds a regressive beach sequence seawards. The commoner pattern in the northern states is for transgressive stratigraphy to form, with narrow and low profile barriers being pushed or rolled onshore (Kraft, 1971, Kraft *et al.*, 1979 and Kraft and John, 1979) (see Devoy, this vol., Ch. 10).

445

It has long been assumed that a continuously rising sea level leads to a continuous spatial migration of the shoreline (Fig. 13.18A). Dillon (1970) recognised that this process could be tempered by the amount of sediment, particularly in the longshore direction, available for barrier build-up. Assuming no major sediment supply, the barrier top would eventually be washed over by storm swash and beachface material transferred to the back barrier. A rising sea level would allow exhumation of old back-barrier sediment through the beachface, so recycling the barrier in step with SLC. Dillon makes the point that if the sediment supply at time T_1 is sufficient to build up the barrier above sea level S_1, the same is not valid at T_2 given an associated sea level S_2 (Fig. 13.18B and C). The area of the barrier expands geometrically as the barrier builds vertically and there comes a point when the barrier growth rate can not match SLC (S_3). At that point storm wash-over will reduce the barrier top and eventually the swash zone and wave base will plane the barrier crest and then 'jump' the lagoon to develop a surf zone near to the lagoon's mainland coast.

This idea of barrier drowning and spatial 'jumping' has become prominent in recent Holocene interpretations of transgressive USA barriers. In particular, Rampino and Sanders (1980) have identified, 5 km offshore on the Long Island shelf, what they describe as a remnant of a barrier formed before 7,000 BP (Rampino, 1979) with the next discernible barrier sedimentation at 2 km offshore. The lack of any tidal-inlet sediments between the two barrier sites is the criterion they use for evidence of barrier 'jumping', given that inlet sedimentation is an integral element of longshore barrier stratigraphic relationships. Despite Leatherman's (1983a) opposition, they believe that a barrier island was drowned by erosion of the 'superstructure' as a consequence of failure in the longshore sediment supply (Rampino and Sanders, 1983). A similar process of barrier 'jumping' has been identified by Staubblefield et al. (1984) for a lense-shaped sand unit, forming a distinctive scarp of 8 m relief, on the outer New Jersey continental shelf. They also identify a period of regressive barrier seaward development prior to a point when the barrier drowned. This combination of features could be explained by a diminishing sediment supply in conjunction with an initial sea-level stillstand (or negative sea movement) followed by an increasing Δsl (marine inundation). But such speculation must remain as such, when the morphological controls are multivariable and difficult to specify for any particular site. The possibility of drowned barriers elsewhere is supported by Amos and King (1984) who suggest that the mean height of a class of submarine ridges (27 m) differentiates them as old barriers, given that submarine processes of sand ridge building rarely form features bigger than 15 m in height.

Leatherman (1983a) specifically attacked Rampino and Sanders's idea of barrier drowning off Long Island. He suggests that the lack of inlet sedimentation offshore reflects a lack of inlets generally in the barriers,

Figure 13.18: Models of fine clastic barrier migration under sea-level rise, applied to eastern USA Holocene coastal stratigraphies. A. Continuous spatial migration of a barrier rolling onshore under a rising sea level. B. Discontinuous spatial migration of a barrier due to barrier drowning as a consequence of sea-level rise outstripping sediment supply. C. Barrier status related to constant sediment supply volume with a rising sea level

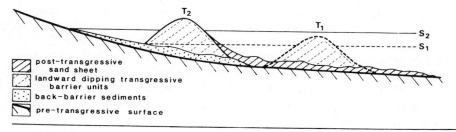

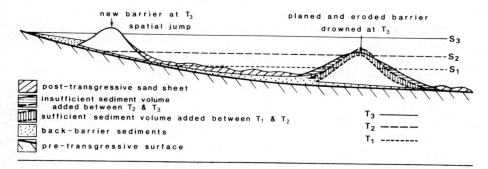

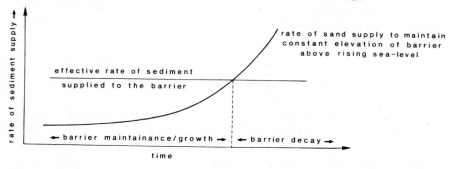

Source: After Dillon (1970).

447

given their micro-tidal regime (Hayes, 1979), rather than being due to an origin from barrier 'jumping', arguing that spatially discontinuous barrier stratigraphy can come from a spatially continuous transgressive barrier. Leatherman invokes the mechanism of shoreface retreat proposed by Swift (1975), in which variations of sediment supply, Δsl and coastal slope can cause major changes in the deposition left in the wake of a retreating beach profile, the result being that an offshore relict barrier can be related to a reduction in Δsl and increased sediment supply (Fig. 13.15B) and be a temporary blip on the longer-term conditions leading to a deeper erosional profile (Fig. 13.15A). Swift *et al.* (1984) likewise use this argument to account for the development of the offshore relict barrier identified by Staubblefield *et al.* (1984). They prefer to explain the lack of barrier sediments between the two known barrier positions as due to a 'smearing' out of back-barrier stratigraphy by shoreface retreat, with the presence of the mid-shelf scarp due to variation in modal beach profile as it retreats (Fig. 13.19A and B).

One possibility of testing the 'jumping' hypothesis is to examine the modern behaviour of barriers given the recent acceleration in Δsl. Leatherman (1983b) notes that few US barriers are presently both overwashing and migrating consistently onshore, with the result that barrier width remains constant. He notes that a number of barriers (Assateague, Hatteras and Fire Island) are narrowing as sea level rises. Narrowing barriers are identified by a seaward displacement of the back-barrier saltmarsh. This view on barrier narrowing is also supported by Jarrett (1983). Working on North Carolina barriers, he notes that the role of overwash is, as yet, only a minor process in recent barrier dynamics, while the only consistent trend in barrier structure (over a variety of barrier islands) is a tendency for barrier narrowing. This slimming process is reminiscent of Dillon's view of barrier failure with rising sea levels. Today Δsl is considerably lower than Δsl envisaged for past barrier starvation and drowning. However, the major reductions in longshore sediment supply observed along many urbanised barriers at present, due to coastal protection schemes, may be creating the sediment scarcity conditions whereby recently accelerating Δsl has been sufficient to produce barrier drowning along some parts of the eastern USA seaboard.

Gravel dominated beaches of the southern British Isles

These beaches contrast with the US barrier beaches in terms of sediment, wave climate, tidal range and coastal slope angle, plus the relative low level of sea-level rise since ~5,000 BP. The principal difference is that of sediment size, with gravel dominating over 70 per cent of south coast beaches, as well as being a major component of Atlantic beaches of

Figure 13.19: Swift *et al.* (1984) attempt to explain the structure of Holocene sedimentation on the Delaware continental shelf (B) by using a barrier migration model requiring a spatially continuous retreating beach profile. Relief variation could be explained in terms of varying Δsl and sediment supply to the shoreface (see Fig. 13.15), rather than requiring barrier jumping as proposed by Staubblefield *et al.* (1984)

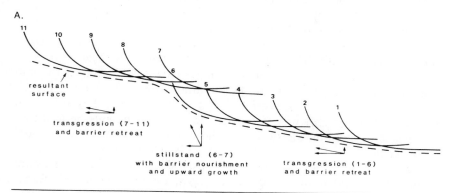

A.

resultant
surface

transgression (7-11)
and barrier retreat

stillstand (6-7)
with barrier nourishment
and upward growth

transgression (1-6)
and barrier retreat

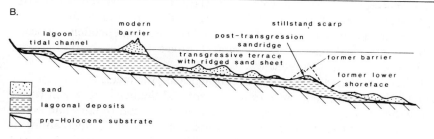

B.

lagoon
tidal channel

modern
barrier

stillstand scarp

post-transgression
sandridge

transgressive terrace
with ridged sand sheet

former barrier

former lower
shoreface

sand

lagoonal deposits

pre-Holocene substrate

Source: After Swift *et al.* (1984).

Northwest Europe (Orford, 1979). It is difficult to generalise about 'gravel beaches' in that few such beaches are free from sand as a subsidiary matrix infill, while most have a fringing low tide sand terrace. The high gravel content is a function of the 'Pleistocene Inheritence', being either glacigenic or periglacigenic in origin. This increased sediment size has repercussions for beach formation, steeper beach slopes generating reflective wave regimes in comparison to the dissipative regimes of most US barrier beaches, and beach drainage (substantial cross-beach seepage). These beaches are storm-wave dominated (Davies, 1972), driven by Atlantic westerly depressions, superimposed on a southwesterly swell wave component which is responsible (via refraction) for beach plan orientation. Present tidal ranges are mainly of a macro-scale (>4 m), such that barrier development is inhibited; though the steep and structurally variable continental shelf of the Western Approaches also appears non-conducive to

449

major longshore barrier formation. The present coast has a crenellate appearance with headland–bay units reinforcing the formation of long-shore coastal cells. These counter longshore continuity of beaches, as they frequently support reversals in sediment drift direction (Bowden and Orford, 1984).

The wave climate, sediment type and coastal slope combine under a rising sea level to form coastal depositional units tied to the shoreline, that is, there is a paucity of free barrier forms. Carter and Orford (1984) indicate a number of differences between fine and coarse clastic barriers, when the latter arise. The primary difference relates to beach porosity, where gravel's increased permeability enables terrestrial discharges to drain from the beachface without always requiring a distinct channel breach (Carter et al., 1984). This is crucial, in that such breaches enlarged by tidal flows would be the only way for coarse, beachface sediments, moved to a back-beach position during storms (onshore beach rolling), to be returned to the beachface and hence regulate onshore beach movement. It is rare to find gravel ridges with landward lagoons, other than where terrestrial valleys are truncated by transgressive ridges, for example Chesil Bank, England (Carr, 1969) and Tacumshin Barrier, southeast Ireland (Carter and Orford, 1980). Carter et al. (1984) show that the continuum of cross-beach drainage, or seepage to an open tidal channel, is dependent on beach permeability, making it rare for gravel ridges to lose their longshore coherence through such breaches.

Other differences related to gravel beaches are the manner in which they undergo both consistent onshore movement and the height to which wave-lain sediment occurs. This latter point makes it difficult to correlate beach crest height with sea-level position. In essence, small-scale fluctuations in gravel beach transgressions are not tied solely to SLC to the same degree as fine clastic barriers have been shown to be. Orford (1977, 1979) and Orford and Carter (1984) show that steep gravel beachfaces reinforce storm wave run-up, facilitating beach crest accretion during storms. The reflective wave regime of gravel beaches also accounts for longshore spatial consistency of beach crest build-up (Orford and Carter, 1984). Elevation of beach height by high-magnitude, low-frequency storms at one sea-level position may well be capable of absorbing the ridge-forming effects of subsequent smaller sea-level increases. The consistent onshore movement of crestal sediment during storms and lack of sediment return keeps the gravel beach tied or fixed to the terrestrial shoreline. This slow rolling onshore is reflected in the internal structure of gravel ridges with landward dipping units prominent (Carr and Blackley, 1974; Carter and Orford, 1980, 1984). Sediment provenance on some ridges also reflects onshore sediment movement at the expense of longshore movement (Hardy, 1964).

Control on beach ridge 'drowning' appears to depend on accelerated Δsl and failure in longshore sediment supply rate. Little can be said about Δsl

and beach ridge stability, as little firm evidence of offshore remnants congruent with drowned gravel features can be offered from UK waters, although Le Fournier has identified drowned sand barriers offshore of Picardy, eastern English Channel. There are gravel-based shoals, but these could be glacigenic lags or tidally induced (e.g. Hails, 1975), as well as being possible drowned features. Failure of longshore gravel supply often results not in ridge breakdown by storms, as observed in the sand counterpart, but rather the reverse, because gravel beaches tend to build up as a low bottom beach elevation allows larger waves to reach closer inshore. Such waves generate higher elevated swash and build up the crest. Clearly this process can not go on indefinitely, as extreme storms will overwhelm the crest with consequent crestal avalanches encouraging back ridge transgression (Orford and Carter, 1982). Although evidence of spatially intermittent overwashing can be found, the longshore consistency of ridge form shows that processes in the long term keep the ridge's retreat consistent with respect to longshore gradients of breaker height.

Although the majority of gravel forms along this coast appear as single, fixed, transgressive ridges, apparent regressive sedimentary features can be found, for example at Dungeness Foreland, Kent. However, these features tend to be regressive due to longshore sediment supply and represent the modern phase of gravel features once prominent updrift (Eddison, 1983). This longshore shift in gravel sedimentation is also a function of present-day stationary sea level. As Δsl declines to 0, gravel sources become greatly reduced or exhausted. At such times gravel beaches tend to be driven further onshore by wave action as longshore supply diminishes, while local headlands are able to breach the longshore sediment supply. The disturbed wave energy gradient of the new coastal cell then reworks the existing gravel distribution. Even mature beaches, if stranded against a headland, tend to reorganise within the coastal compartment (Orford, 1986). This results in an increasing emphasis on longshore sediment sorting dependent on different rates of transport as a function of material size. Longshore cells, with narrow gravel ridges passing downdrift into sand dunes, are common on beaches with long-term sea-level stability and dominant obliquely impinging waves.

Quaternary coastal sequences in eastern Australia

The Pleistocene and Holocene barrier-based morpho-stratigraphies along the eastern New South Wales (NSW) and southern Queensland coast offer marked differences to their eastern USA counterparts (Thom, 1983; Short and Hesp, 1984). These differences are imposed by the greater degree of initial coastal embayment, the narrower and deeper continental shelf, the restricted longshore sediment supply and higher wave energy found in

Australia compared to the USA.

The NSW coastline trends NNE–SSW and marks a stable plate edge of Cretaceous age. The continental shelf is very narrow ($<$35 km) and steep (80 per cent $>$50 m). The shelf tends to widen and decrease in slope northwards towards the Queensland boundary, where the protection of the Great Barrier Reef alters the nearshore wave climate considerably. Wave energy is dominated by southeast swell and wind wave with energy 5-10 times that of eastern USA. Tidal ranges are classified as micro-tidal ($<$2 m). Dominant winds in the north are onshore from the southeast, while in the south they are offshore from the west. Sea level has oscillated $\pm$ 1 m about the present-day datum which has not changed significantly since the end of the main postglacial transgression at ~6,000 BP (Thom and Chappell, 1975) (see Hopley, this vol., Ch. 12).

The present coastline is rock headland controlled, defined by a series of river mouth embayments and deep river estuaries. The coast is highly compartmentalised south of 33°S, with embayments becoming progressively more open northwards. It is the combination of embayed coast and steep shelf that dominates and controls shoreline deposition pattern. In a series of papers (Melville, 1984; Roy *et al.* 1980; Roy and Thom, 1981; Thom, 1978, 1983) the barrier stratigraphies found in these embayments have been identified. In short, during the Pleistocene the exposure of the flat inner continental shelf at lower sea levels reinforced a strong northwards littoral drift of terrestrially derived sediments (fluvial) towards the NSW–Queensland border. During marine inundation and high interglacial sea levels the headland boundaries of coastal compartments became more pronounced, reducing this northerly drift to negligible levels. At such times lack of longshore feeding meant that bay shorelines were predominantly fed from offshore sources.

The long-term reduction of sediment in the south compared to the north can be recognised in the different barrier structures found north and south. In the south pre-Holocene sedimentation is rare. Holocene barriers are thinner and predominantly transgressive, with three recognised types of barrier depending on the river embayment size and shape (Roy *et al.*, 1980). Most of these barriers are suffering shoreline erosion and appear to have been like this for several thousand years.

In the northern more open embayments, Pleistocene barrier sediments still remain with Holocene barriers, composed of major regressive beach ridge sequences, seawards of them. From a wide range of ^{14}C dates, it is now clear that most of the Holocene barrier sedimentation postdates ~ 6,000 BP and, therefore, is younger than the main postglacial transgression (Thom, 1978). The regressive sequences of the northern coast are seen as a result of reworking of inner shelf material, as the shoreface profile of equilibrium is reasserted during the stillstand following the last transgression. The excess of littoral sediment in the north gave rise to regressive barrier sequences (beach ridge plain) as well as major transgressive dune-sheets. This appears to be a sediment excess characteristic of pre-Holocene times as well, as Pleistocene dunes have been recognised below both current sea-level datum and Holocene dunes in the north. Roy and Thom (1981) refer to this thicker sedimentation record as 'depositional transgressive stratigraphy'. In the south, although the same concept of shoreface equilibrium profile has been recognised, there appears to be substantially less sediment available in the nearshore zone, such that a deepening of the existing profile allowed a more erosive phase to operate on the barrier shorelines. Dominant offshore winds plus sediment deficit also prevented major dune development on these barriers. Transgressive barrier migration in the south operates along the lines advocated by Swift (1975) through shoreface retreat, as discussed previously. Roy and Thom (1981) call this sedimentation pattern 'erosional transgressive stratigraphy'.

Thom (1978) points out that many of the Holocene deposits show episodic erosion and deposition regardless of sea-level 'stationarity'. Instability is shown by mixtures of beach ridge and foredune deposition and erosion, while older transgressive dunes exhibit podzolic soil profile development and nested dune structures, indicating intermittent periods of sediment mobility interspersed with vegetative stability. The causes of such variation are as yet unknown. Thom suggests that episodic events during stillstands could be associated with sediment supply variation, storminess variation and minor fluctuations of sea level.

The building of dunes opens up the still unresolved question of the specific relationship between dune development and sea-level change. Pye (1984) and Pye and Bowman (1984) recognise four models of coastal dune emplacement. All have the basic premises that dune sediment source is marine-controlled, onshore wind systems dominate and that vegetative matts are able to grow. Figure 13.20 outlines these four models. Figure 13.20A advocates dune building as a consequence of high sea levels sweeping material up, which is then stabilised by vegetation during lower sea-level periods (Bretz, 1960). Figure 13.20B shows the opposite view, that it is that when lower sea levels occur the wind is able to rework the marine sediments prism thus exposed above the shoreline. Following the advent of higher sea levels, the sediment supply is cut off and vegetative stabilisation occurs (Sayles, 1931). The third and fourth models are variants of the first two in that dunes are a consequence of rising sea-level periods (Fig. 13.20C) (Cooper 1958), while falling sea-level periods (Fig. 13.20D) have been suggested as the basis for dune development (Schofield, 1975, Marks and Nelson, 1979).

Thom (1978) is definite that the NSW–Queensland dunes are related to the phase of sea-level rise pre-6,000 BP. Pye and Bowman (1984) show ^{14}C evidence from six dune sites that indicates dune emplacement between 10,000 and 6,000 BP, prior to the stillstand. Yet Marks and Nelson (1979)

453

Figure 13.20: Models of transgressive dune-sheet development as a function of sea-level change. A. Dunes with high sea-levels. B. Dunes with low sea levels. C. Dunes with rising sea-levels. D. Dunes with falling sea levels

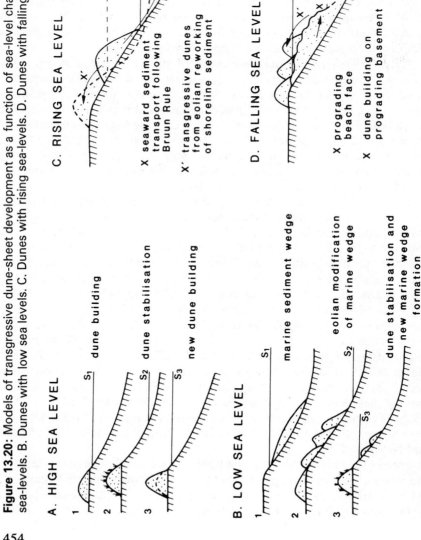

454

are positive that their New Zealand examples are related to falling sea levels and assert like Schofield (1975) that episodic periods of dune building are interspersed by erosion due to short periods of rising sea level.

Is it possible to reconcile these differences? There are two basic models, in that models A and C, and B and D, are reaching the same conclusion respectively. The difference between C and D resides in the process by which marine-controlled sediment is released for wind mobilisation. Neither model is invalid in what it proposes on this point. The dispersal of beachface and nearshore sediment depends on the storminess of the wave climate. Predominant swell will move sediment onto the beachface while dominant storms will strip the beachface. Given a falling sea level plus swell conditions, beachface sediments will be built up and more likely to be exposed intertidally to wind action. Given a falling sea level plus dominant storms, beachface sediments may not be available on a scale sufficient for transgressive dune sheets. Both the Δsl and the temporal extent of the sea-level change will control the sediment volume available for aeolian action. In a phase of marine inundation, the model of shoreface retreat by profile translation identifies the possibility of considerable beachface sediment being available, but it has to be mobilised upwards into an intertidal or supratidal position. Clearly the balance of storm versus swell is crucial, as is the Δsl. A slow Δsl may allow the removal of eroded equilibrium-related sediment to an offshore or longshore position, while a fast Δsl may be able to keep sufficient material within the intertidal zone for aeolian activity to rework it.

Obviously, these models of dune emplacement are only first order approximations, in that a full understanding of the constraints on the mechanisms of beach profile translation is required for each site. The necessity for multiple dune emplacement models underlines the final point of this chapter, that sea level *per se* is only one element of a highly inter-related system of littoral processes and responses. So far it has been difficult to define precise models of coastal response to sea-level change. In the future, given our growing recognition of other controls, it can only be harder!

CONCLUSIONS

Changes in sea level effect responses in coastal morphology; however, they are responses constrained by a number of other factors: wave and tidal regime, sediment composition, volume and supply rate, plus coastal slope angle. Variability of coastal response indicates that a zonal/regional approach to shoreline development is produced by the interaction of these controlling variables, regardless of the azonal nature of beach processes *per se*. The rate of sea-level changes appears to be crucial in the development of variable morphology within any one zonal coastal assemblage. Attempts to elucidate models of coastal response, based on modern examples under accelerating sea-level change, may not reflect past conditions of similar sea-level change, given the modern longshore constraints on sediment supply imposed by man's intervention into the littoral zone. In general, a rising sea level produces a single narrow transgressive beach unit, while a falling sea level leads to deposition in the form of multiple beach ridge units. There appears to be no general rule for dune development associated with the direction of sea-level change. Several models have been proposed which relate sea-level rise to coastal change; the Bruun Rule applies to small-scale rises, although this model's constraints are difficult to comply with in the real world. Swift's model proposes that larger sea-level changes can lead to beach profile spatial translation that depends on the rates of both sea-level change and sediment supply. Further specification of beach and barrier behaviour under sea-level change have been made, for example barrier drowning and aeolian dune development, but have proved contentious. The key to further understanding of sea-level change and coastal morphology lies in two areas: (1) increased and detailed exploration of the world's continental shelves for facies and stratigraphical data, to enable reconstruction of past shoreline chronologies, and (2) a better understanding of the processes and dynamics controlling modern coastal units, that can be used to explain the morpho-stratigraphic evidence existing in the nearshore and offshore zones of the world's shorelines.

ACKNOWLEDGEMENTS

I should like to thank both Bill Carter and Bob Devoy for their comments on this chapter, as well as Maura Pringle for cartographic assistance.

REFERENCES

Allison, H. (1980) 'Enigma of the Bruun Formula in shore erosion', in J.J. Fisher and M.L. Schwartz (eds), *Proc. Per Bruun Symp., IGU Comm. on the Coastal Environment*, Western Washington University, Bellingham, Wash., pp. 67-78.

Amos, C.L. and King, E.L. (1984) 'Bedforms of the Canadian Eastern seaboard: a comparison with global occurrences', *Mar. Geol.*, *57*, 167-205.

Barnett, T.P. (1983) 'Recent changes in sea level and their possible causes', *Climatic Change*, *5*, 15-38.

—— (1984) 'The estimation of "Global" sea level change: a problem of uniqueness', *J. Geophys. Res.*, *89C*, 7980-8.

Belknap, D.F. and Kraft, J.C. (1981) 'Preservation potential of transgressive coastal lithosomes on the US Atlantic shelf', *Mar. Geol.*, *42*, 429-42.

Bird, E.C.F. (1976) *Shoreline Change During the Past Century*, Int. Geogr. Union Work. Group, University of Melbourne.

Bloom, A.L. (1978) *Geomorphology*, Prentice-Hall, Englewood Cliffs, NJ.

Bowden, R. and Orford, J.D. (1984) 'Residual sediment cells on the morphologically irregular coastline of the Ards Peninsula, Northern Ireland', *Proc. Roy. Irish Acad., 84B*, 13-27.

Bretz, J.H. (1960) 'Bermuda, a partially drowned, late mature, Pleistocene karst', *Bull. Geol. Soc. Am., 71*, 1729-54.

Bruun, P. (1962) 'Sea level rise as a cause of shore erosion', *J. Water Ways, Harb. Div., Am. Soc. Civ. Eng., 88 (WW1)*, 117-30.

—— (1980) 'The "Bruun Rule": Discussion on boundary conditions', in *Proc. Per Bruun Symp., IGU Comm. on the Coastal Environment*, Western Washington University, Bellingham, Wash., pp. 79-83.

—— (1983) 'Review of conditions for uses of the Bruun Rule of erosion', *Coastal Engr., 7*, 77-89.

Cant, D. (1984) 'Development of shoreline–shelf sand bodies: in a Cretaceous epeiric sea deposit', *J. Sedim. Petrol., 54*, 541-56.

Carr, A.P. (1969) 'Size grading along a pebble beach: Chesil Beach, England', *J. Sedim. Petrol., 39*, 297-311.

—— and Blackley, M.W.L. (1974) 'Ideas on the origin and development of Chesil Beach, Dorset', *Proc. Nat. Hist. Arch. Soc., 95*, 9-17.

Carter, R.W.G. (1979) 'Recent progradation of the Magilligan Foreland, Co. Londonderry, Northern Ireland', *Les Côtes Atlantiques de l'Europe*, Centr. Nat. Exploit. Ocean. Brest, No. 9, 1-16.

—— (1983) 'Raised coastal landforms as a product of modern process variation and their relevance in eustatic sea level studies: examples from eastern Ireland', *Boreas, 12*, 167-82.

——, Johnston, T.W. and Orford, J.D. (1984) 'Stream outlets through mixed sand and gravel coastal barriers: examples from southeast Ireland', *Zeit. Geomorph., 28*, 427-42.

—— and Orford, J.D. (1980) 'Gravel barrier genesis and management: a contrast', *Coastal Zone '80, Am. Soc. Civ. Eng., 2*, 1304-20.

—— and Orford, J.D. (1984) 'Coarse clastic barrier beaches: a discussion of the distinctive dynamic and morphosedimentary characteristics', *Mar. Geol., 60*, 377-84.

Chapman, D.M. (1981) 'Coastal erosion and sediment budget with specific reference to the Gold Coast, Australia', *Coastal Engr., 4*, 207-27.

Chappell, J.C. (1983) 'Thresholds and lags in geomorphic changes', *Austr. Geogr., 15*, 357-66.

Cooper, W.S. (1958) *Coastal Sand Dunes of Oregon and Washington. Geol. Soc. Am. Mem., 104*.

Cram, J.M. (1979) 'The influence of continental shelf width on tidal range: paleoceanographic implications', *J. Geol., 87*, 117-30.

Cronin, T.M. (1983) 'Rapid sea level and climatic change: evidence from continental and island margins', *Quat. Sci. Rev., 1*, 215-44.

Curray, J.R. (1964) 'Transgressions and regressions', in R.L. Miller (ed), *Papers in Marine Geology: Shepard Commemorative Volume*, Macmillan, New York, pp. 175-203.

—— (1969) 'Shore zone sand bodies barriers, cheniers and beach ridges', in D.J. Stanley (ed), *The New Concept of Continental Margin Sedimentation*, American Geological Institute, Washington, DC, JC-II-1 to JC-II-19.

—— and Moore, D.G. (1964) 'Holocene regressive littoral sand, Costa de Nayarit,

Mexico', in L.N.J.U. van Straateen (ed), *Developments in Sedimentology, No. 1*, Elsevier, Amsterdam, pp. 76-82.

Davies, J.L. (1972) *Geographical Variation in Coastal Development*, Oliver & Boyd, Edinburgh.

Dawson, A.G. (1980) 'Shore erosion by frost: an example from the Scottish Late-glacial', in J.J. Lowe, J.M. Gray and J.E. Robinson (eds), *Studies in the Late-glacial of N.W. Europe*, Pergamon Press, Oxford, pp. 45-53.

Dean, R.G. (1982) *Models for Beach Profile Response*, Dept. of Civil Engineering, Technical Report No. 30, University of Delaware, Newark, Del.

―― and Maurmeyer, E.M. (1983) 'Models for beach profile response', in P.D. Komar (ed), *Handbook of Coastal Processes and Erosion*, CRC Press, Baca Rotan, Fla, pp. 151-65.

Dillon, W.P. (1970) 'Submergence effects on a Rhode Island barrier and lagoon and inference on barrier migration', *J. Geol.*, *78*, 94-106.

Dubois, R.N. (1975) 'Support and refinement of the Bruun Rule on beach erosion', *J. Geol.*, *83*, 651-7.

―― (1977) 'Predicting beach erosion as a function of rising water level', *J. Geol.*, *85*, 470-6.

―― (1980a) 'Nearshore evidence in support of the Bruun Rule on shore erosion', abstract in J.J. Fisher and M.L. Schwartz (eds), *Proc. Per Bruun Symp.*, *IGU Comm. the Coastal Environment*, Western Washington University, Bellingham, Wash.

―― (1980b) 'Hypothetical shore profiles in response to rising water level', in J.J. Fisher and M.L. Schwartz (eds), *Proc. Per Bruun Symp.*, *IGU Comm. on the Coastal Environment*, Western Washington University, Bellingham, Wash., pp. 13-32.

Eddison, J. (1983) 'The evolution of the barrier beaches between Fairlight and Hythe', *Geogr. J.*, *149*, 39-53.

Edelman, T. (1970) 'Dune erosion during storm conditions', in *Proc. 12th Int. Conf. Coastal Engr.*, Am. Soc. Civ. Eng., New York, pp. 719-29.

Emery, K.O. (1980a) 'Relative sea levels for tide-gauge records', *Proc. Nat. Acad. Sci.*, *77*, 6968-70.

―― (1980b) 'Continental margins — classification and petroleum prospects', *Am. Ass. Petrol. Geol. Bull.*, *64*, 297-315.

Faniran, A. and Jeje, L.K. (1983) *Humid Tropical Geomorphology*, Longman, London.

Fischer, A.G. (1961) 'Stratigraphic record of transgressive seas in light of sediment-ation on the Atlantic coast of New Jersey', *Am. Ass. Petrol. Geol. Bull.*, *45*, 1656-66.

Fisher, J.J. (1980a) 'Holocene sea level rise, shoreline erosion and the Bruun Rule overview', in J.J. Fisher and M.L. Schwartz (eds), *Proc. Per Bruun Symp.*, *IGU Comm. on the Coastal Environment*, Western Washington University, Bellingham, Wash., pp. 1-15.

―― (1980b) 'Shoreline erosion, Rhode Island and North Carolina coasts — test of Bruun Rule', in J.J. Fisher and M.L. Schwartz (eds), *Proc. Per Bruun Symp.*, *IGU Comm. on the Coastal Environment*, Western Washington University, Bellingham, Wash., pp. 32-54.

Flemming, N.C. (1965) 'Form and relation to present sea level of Pleistocene marine erosion features', *J. Geol.*, *73*, 799-811.

Forbes, D.L. and Taylor, R.B. (in press) 'Coarse-grained beach sedimentation under paraglacial conditions', in D. Fitzgerald and P. Rosen (eds), *Glaciated Coasts*, Academic Press, London and New York.

458

Galliers, J.A. (1968) 'Barrier beaches and lagoons of the Ghana coast', *Br. Geomorph. Res. Group, Paper No. 5*, 77-87.

Galvin, C.J., Jr (1983) 'Sea level rise and shoreline recession', *Coastal Zone '83, 3*, 2684-705.

Gilbert, G.K. (1885) 'The topographic features of lake shores', in *5th Ann. Rept. US Geol. Surv.: 1883-84*, Washington, DC, pp. 69-123.

Gornitz, V., Lebedeff, S. and Hansen, J. (1982) 'Global sea level trends in the past century', *Science, 215*, 1611-14.

Greenwood, B. and Davis, R.A., Jr (eds) (1984) *Hydrodynamics and Sedimentation in Wave-dominated Coastal Environments*, Elsevier, Amsterdam.

Guidish, T.M., Lerche, I., Kendall, C.G.St.C. and O'Brien, J.J. (1984) 'Relationship between eustatic sea level change and basement subsidence', *Am. Ass. Petrol. Geol. Bull., 68*, 164-77.

Guilcher, A. (1962) 'Morphologie de la Baie de Clew (Co. Mayo, Irlande)', *Bull. Ass. Geogr. France, 303/4*, 53-65.

Hails, J.R. (1975) 'Sediment distribution and Quaternary history', *J. Geol. Soc. (London), 131*, 19-36.

Hands, E.B. (1980) 'Bruun's concept applied to the Great Lakes', in J.J. Fisher and M.L. Schwartz (eds), *Proc. Per Bruun Symp., IGU Comm. on the Coastal Environment*, Western Washington University, Bellingham, Wash., pp. 63-6.

—— (1983) 'The Great Lakes as a test model for profile responses to sea level changes', in P.D. Komar (ed), *Handbook of Coastal Processes and Erosion*, CRC Press, Boca Rotan, Fla, pp. 167-89.

Hansen, J., Lacis, A. and Rind, D. (1983) 'Climatic trends due to increasing greenhouse gases', *Coastal Zone '83, 3*, 2796-810.

Hardy, J.R. (1964) 'Sources of some beach shingles in England', *20th Int. Geogr. Congr.*, Paper 1651 (abs.), p. 112.

Hayes, M.O. (1967) 'Relationship between coastal climate and bottom sediment type on the inner continental shelf', *Mar. Geol., 5*, 111-32.

—— (1979) 'Barrier island morphology as a function of tidal and wave regime', in S.P. Leatherman (ed), *Barrier Islands*, Academic Press, London and New York, pp. 1-28.

Hollin, J.T. (1977) 'Thames interglacial sites, Ipswichian sea level and Antarctic surges', *Boreas, 6*, 33-52.

Inman, D.L. and Nordstrom, C. (1971) 'On the tectonic and morphological classification of coasts', *J. Geol., 79*, 1-21.

Jardine, W.G. (1975) 'The determination of former sea levels in areas of large tidal range', in P.D. Suggate and M.M. Cresswell (eds), *Quaternary Studies*, Roy. Soc. New Zealand, Wellington, pp. 163-8.

Jarrett, J.T. (1983) 'Changes of some North Carolina barrier islands since the mid-19th century', *Coastal Zone '83, 1*, 641-61.

Johnson, D.W. (1919), *Shore Processes and Shoreline Development*, Wiley, New York.

Johnson, J.W. (1949) 'Scale effects on hydraulic models involving wave motion', *Trans. Am. Geophys. Union, 30*, 517-25.

Kidson, C. (1982) 'Sea level changes in the Holocene', *Quat. Sci. Rev., 1*, 121-51.

King, C.A.M. (1972) *Beaches and Coasts*, Edward Arnold, London.

Komar, P.D. (1976), *Beach Processes and Sedimentation*, Prentice-Hall, Englewood Cliffs, NJ.

—— (1983) *Handbook of Coastal Processes and Erosion*, CRC Press, Boca Raton, Fla.

Kraft, J.C. (1971) 'Sedimentary environment, facies patterns and geologic history of

459

a Holocene marine transgression', *Bull. Geol. Soc. Am.*, *82*, 2131-58.

——, Allen, E.A., Belknap, D.F., John, C.J. and Maurmeyer, E.M. (1979) 'Process and morphologic evolution of an estuarine and coastal barrier system', in S.P. Leatherman (ed) *Barrier Islands*, Academic Press, London and New York, pp. 149-84.

—— and John, C.J. (1979) 'Lateral and vertical facies relations of transgressive barriers', *Am. Ass. Petrol. Geol. Bull.*, *63*, 2145-63.

Leatherman, S.P. (1979) *Barrier Islands*, Academic Press, London and New York.

—— (1983a) 'Barrier island evolution in response to sea-level rise: a discussion', *J. Sedim. Petrol.*, *53*, 1026-31.

—— (1983b) 'Barrier dynamics and landward migration with Holocene sea level rise', *Nature*, *301*, 415-17.

Le Fournier, J. (1980) 'Modern analogue of transgressive sand bodies off Eastern English Channel', *Bull. Cent. Rech. Explor.–Prod. Elf-Acquitaine*, *4*, 99-118.

Ludwig, G., Muller, H. and Streif, H. (1981) 'New dates on Holocene sea level changes in the German Bight', *Int. Ass. Sedim.*, *Spec. Publ. No. 5*, 211-19.

Marks, G.P. and Nelson, C.S. (1979) 'Sedimentology and evolution of Omaro Spit, Coromandel Peninsula', *NZ J. Mar. Freshw. Res.*, *13*, 347-72.

Maurmeyer, E.M. and Dean, R.G. (1982) 'Sea level rise and barrier migration: an extension of the Bruun Rule to account for landward sediment transport', (Abst), *Meeting NE and SE Sect. Geol. Soc. Am.*, Washington, DC, March 1982.

May, J.P. and Tanner, W.F. (1973) 'The littoral power gradient and shoreline changes', in D.R. Coates (ed), *Coastal Geomorphology*, State University of New York, Binghampton, NY, pp. 43-60.

Melville, G. (1984) 'Headland and offshore islands as dominant controlling factors during Late Quaternary barrier formation in the Forster-Tuncurry area, NSW, Australia', *Sed. Geol.*, *39*, 243-71.

Morton, R.A. (1979) 'Temporal and spatial variations in shoreline changes and their implications, examples from the Texas Gulf Coast', *J. Sedim. Petrol.*, *49*, 1101-12.

Nummedal, D. (1982) *Barrier Islands*, Coastal Res. Group. Tech. Rept. No. 82-3, Dept. of Geology, Louisiana State University.

Nummedal, P., Cuomo, R.F. and Penland, S. (1984) 'Shoreline evolution along the northern coast of the Gulf of Mexico', *Shore and Beach*, *52*, 11-17.

Orford, J.D. (1977) 'A proposed mechanism for storm beach sedimentation', *Earth Surf. Proc.*, *2*, 381-400.

—— (1979) 'Some aspects of beach ridge development on a fringing gravel beach, Dyfed, West Wales', *Les Côtes Atlantiques de l'Europe*, Centr. Nat. Exploit. Ocean. Brest, No. 9, 35-49.

—— (1986) 'Coasts: environment and landforms', in P. Fookes and P. Vaughan (eds) *Handbook of Engineering Geomorphology*, Surrey University Press, Guildford, pp. 203-17.

—— and Carter, R.W.G. (1982) 'Crestal overtop and washover sedimentation on a fringing sandy gravel barrier coast, Carnsore Point, southeast Ireland', *J. Sedim. Petrol.*, *52*, 265-78.

—— and Carter, R.W.G. (1984) 'Mechanisms to account for the longshore spacing of overwash throats on a coarse clastic barrier in southeast Ireland', *Mar. Geol.*, *56*, 207-26.

Otvos, E.G. (1981) 'Barrier island formation through nearshore aggradation — stratigraphic field evidence', *Mar. Geol.*, *43*, 195-243.

Penland, S., Boyd, R., Nummedal, D. and Roberts, H. (1981) 'Deltaic barrier development on the Louisiana coast', *Gulf Coast Ass. Geol. Soc. Trans.*, *31*, 471-6.

Pugh, D.T. and Faull, H.E. (1983), 'Tides, surges and mean sea level trend', in Inst. Civ. Eng. (ed), *Shoreline Protection*, T. Telford, London, pp. 59-70.

Pye, K. (1984) 'Models of transgressive coastal dune building episodes and their relationship to Quaternary sea-level change: a discussion with reference to evidence from eastern Australia', in M.W. Clark (ed), *Coastal Research: UK Perspectives*, Geo-Books, Norwich, pp. 81-104.

—— and Bowman, G.M. (1984) 'The Holocene marine transgression as a forcing function in episodic dune activity on the eastern Australian coast', in B.G. Thom (ed), *Australian Coastal Geomorphology*, Academic Press, Sydney, pp. 179-96.

Rampino, M.R. (1979) 'Holocene submergence of southern Long Island, New York', *Nature*, *280*, 132-4.

—— and Sanders, J.E. (1980) 'Holocene transgression in south-central Long Island, New York', *J. Sedim. Petrol.*, *50*, 1063-80.

—— and Sanders, J.E. (1983) 'Barrier island evolution in response to sea level rise: reply', *J. Sedim. Petrol.*, *53*, 1031-3.

Reinson, G.E. (1979) 'Barrier island systems', in R.G. Walker (ed), *Facies Models*, Geoscience Canada, Reprint 1, pp. 57-74.

Rosen, P. (1978) 'A regional test of the Bruun Rule on shoreline erosion', *Mar. Geol.*, *26*, M7-M16.

—— (1980) 'An application of the Bruun Rule in Chesapeake Bay', in J.J. Fisher and M.L. Schwartz (eds), *Proc. Per Bruun Symp., IGU Comm. on the Coastal Environment*, Western Washington University, Bellingham, Wash., pp. 55-62.

Roy, P.S. and Thom, B.G. (1981) 'Late Quaternary marine deposits in New South Wales and southern Queensland — an evolutionary model', *J. Geol. Soc. Australia*, *28*, 471-89.

——, Thom, B.G. and Wright, L.D. (1980) 'Holocene sequences on an embayed high energy coast: an evolutionary model', *Sed. Geol.*, *26*, 1-19.

Sayles, R.W. (1931) 'Bermuda during the Ice Age', *Proc. Am. Acad. Arts Sci.*, *66*, 381-468.

Schofield, J.C. (1975) 'Sea-level fluctuations cause periodic postglacial progradation, south Kaipara barrier, North Island, New Zealand', *NZ J. Geol. Geophys.*, *18*, 295-316.

Schwartz, M.L. (1965) 'Laboratory study of sea level rise as a cause of shore erosion', *J. Geol.*, *73*, 528-34.

—— (1967) 'The Bruun theory of sea level rise as a cause of shoreline erosion', *J. Geol.*, *75*, 76-92.

—— and Milicic, V. (1980) 'The Bruun Rule: a historical perspective', in J.J. Fisher and M.L. Schwartz (eds), *Proc. Per Bruun Symp., IGU Comm. on the Coastal Environment*, Western Washington University, Bellingham, Wash., pp. 6-12.

Schwartz, R.K., Hobson, R. and Musialowski, F.R. (1981) 'Subsurface facies of a modern barrier island shoreface and relationship to the active nearshore profile', *Northeastern Geol.*, *3*, 283-96.

Scott, D.B. and Greenberg, D.A. (1983) 'Relative sea level rise and tidal development in the Fundy tidal system', *Can. J. Earth Sci.*, *20*, 1554-64.

—— and Medioli, F.S. (1978) 'Vertical zonation of marsh foraminifera as accurate indicators of former sea levels', *Nature*, *272*, 528-31.

Shepard, F.P. (1963) *Submarine Geology*, Harper & Row, New York.

Short, A.D. and Hesp, P.A. (1984) *Beach and Dune Morphodynamics of the Southeast Coast of South Australia*, Coastal Studies Tech. Unit Rept. No. 84/1,

461

Dept. of Geography, University of Sydney.

Sissons, J.B. (1967) *The Evolution of Scotland's Scenery*, Oliver & Boyd, Edinburgh.

Smith, D.E. and Dawson, A.G. (1983) *Shorelines and Isostasy*, Academic Press, London and New York.

Staubblefield, W.L., McGrail, D.W. and Kersey, D.G. (1984) 'Recognition of transgressive and post-transgressive sand ridges on the New Jersey continental shelf', in R.W. Tillman and C.T. Siemers (eds), *Siliclastic Shelf Sediments*, Soc. Econ. Pal. Min., Spec. Publ. No. 34, 1-24.

Straaten, L.M.J.U. van, (1965) 'Coastal barrier deposits in south and north Holland', *Med. Geol. Sticht., N.S. 17*, 41-75.

Sunamura, T. and Horikawa, K. (1974) 'Two-dimensional beach transformations due to waves', in *Proc. 14th Conf. Coast. Engr.*, pp. 920-37.

Swift, D.J. (1968) 'Coastal erosion and transgressive stratigraphy', *J. Geol., 76*, 449-56.

—— (1970) 'Quaternary shelves and their return to grade', *Mar. Geol., 8*, 5-30.

—— (1975) 'Barrier island genesis: evidence from the Central Atlantic Shelf of eastern USA', *Sed. Geol., 14*, 1-43.

Swift, D.P. (1980) 'Bruun's Rule and continental shelf sedimentation: an oceanographer's viewpoint', abstract in J.J. Fisher and M.L. Schwartz (eds), *Proc. Per Bruun Symp., IGU Comm. on the Coastal Environment*, Western Washington University, Bellingham, Wash.

——, McKinney, T.F. and Stahl, L. (1984) 'Recognition of transgressive and post-transgressive sand ridges on the New Jersey continental shelf: a discussion', in R.W. Tillman and C.J. Siemers (eds), *Siliclastic Shelf Sediments*, Soc. Econ. Pal. Min., Spec. Publ. No. 34, 25-36.

Synge, F. (1980) 'Raised beaches in Ireland', *Bull. Ass. Français Etude Quat., 17*, 77-9.

—— (1981) 'Quaternary glaciation and change of sea-level in the south of Ireland', *Geol. Minjb., 60*, 305-15.

Takeda, I. and Sunamura, T. (1982) 'Formation and height of berms', *Trans. Jap. Geomorph. Union, 3*, 145-57 (in Japanese).

Thom, B.G. (1978) 'Coastal sand deposition in southeast Australia during the Holocene', in J.L. Davies and M.A.S. Williams (eds), *Landform Evolution in Australasia*, ANU Press, Canberra, 197-214.

—— (1983) 'Transgressive and regressive stratigraphies of coastal sand barriers in southeast Australia', *Mar. Geol., 56*, 137-58.

—— and Chappell, J. (1975) 'Holocene sea-levels relative to Australia', *Search, 6*, 90-3.

Tooley, M.J. (1974) 'Sea level changes during the last 9000 years in N.W. England', *Geogr. J., 140*, 18-42.

Trenhaile, A.S. (1983) 'The development of shore platforms in high latitudes', in D.E. Smith and A.G. Dawson (eds) *Shorelines and Isostasy*, Academic Press, London and New York, pp. 77-93.

Tricart, J. (1972) *Landforms of the Humid Tropics, Forests and Savannas*, Longman, London, pp. 86-123.

Usoroh, E.J. (1977) 'Coastal development in the Lagos area', PhD dissertation, University of Ibadan. (Cited in A. Faniren and L.K. Jeje (1983), *Humid Tropical Geomorphology*, Longman, London.)

Vail, P.R., Mitchum, R.M., Jr and Thompson, S., III (1977) 'Global cycles of relative change of sea level', in C.B. Payton (ed), *Seismic Stratigraphy — Application to Hydrocarbon Exploration, Mem. Am. Ass. Petrol. Geol., 26*, 83-97.

Vanney, J.R. and Stanley, D.J. (1983) 'Shelfbreak physiography: an overview', in D.J. Stanley and G.T. Moore (eds), *The Shelfbreak: Critical Interface on Continental Margins*, Soc. Econ. Pal. Min., Spec. Publ. No. 33, 1-24.

Vellinga, P. (1983) 'Predictive computational models for beach and dune erosion during storm surges', *Delft Hydraulic Lab., Rept.*, *294*, 1-14.

Walton, T.J., Jr (1978) 'Coastal erosion — some causes and some consequences', *Mar. Tech. Soc. J.*, *12*, 28-33.

Worsley, T.R., Nance, D. and Moody, J.B. (1984) 'Global tectonics and eustacy for the past two billion years', *Mar. Geol.*, *58*, 373-400.

Zeuner, F.W. (1946) *Dating the Past: An Introduction to Geochronology*, Methuen, London.

14

Man's Response to Sea-level Change

R.W.G. Carter

INTRODUCTION

The surface of the undisturbed sea is called the geoid. However, this surface is neither level nor still. Even if gravity waves are absent, it is unusual if the sea surface is not undergoing some change. Man has an uneasy relationship with sea level. In many respects it is convenient to live as near the sea as possible, yet such encroachment brings hazards, particularly to the unwary. The seaside has many attractions. Not only is the sea an important resource — for food, power, minerals and recreation — but also an agency for communication and defence and a provider of flat land for agriculture. For all combinations of these reasons many states are marinocentric. Capital cities, unless sited by political whim, are invariably coastal in location. In the nineteenth century, with the advent of cheap mass transport, it became fashionable to take holidays by the sea. Resorts developed in Britain, France and along the eastern US seaboard. This pattern was later repeated in Spain, Greece and Mexico and is now devouring the coastlines of previously 'exotic' countries like Gambia, Sri Lanka and Bali (Indonesia). Such a taste for coast dwelling has spilled over into retirement homes, industrial relocation and all that goes with it. The population of Florida, America's most 'coastal' state, has been increasing at the rate of 4 per cent per year for over twenty years (Graff and Wiseman, 1978). Almost all these people reside near the coast — only a tiny fraction have any comprehension of the risk. Similar, although admittedly not so spectacular trends, may be discerned in almost every developed country. In less developed nations coastal cities are growing spectacularly; prime examples are São Paulo in northeast Brazil, which is expected to have over ten million inhabitants, mostly poor, by 1990, and Kowloon in Hong Kong.

In recent years the problems and relationships between man and sea level have been highlighted in a number of articles and books. Although well meaning, the thrust of many of these articles has been the denigration of various peer groups, particularly civil engineers, for their failure to act or react responsibly to sea-level hazards. The response from these groups has often been not to answer the criticism, but, with the acumen born of a successful politician, to dispute the validity of the data and the credentials of the questioner.

The aim of this chapter is to examine the interlaced theme of sea level, the coast and man; first in terms of the nature of the problem and, second, in terms of the engineering and non-engineering solutions. Much of the information on which this chapter is based is either from secondary 'grey' literature or from personal communications. A substantial amount of this material does not relate to sea level *sensu stricto* but focuses on tangible manifestations of sea-level rise — storms, floods and avulsions. In contrast, the responses of man to sea-level fall are very poorly documented.

THE NATURE OF SEA-LEVEL CHANGE

It is impossible to quote a universally appropriate rate of sea-level change, although various predictions have been attempted (see Titus, this vol., Ch. 15). Secular deformation of the geoid is caused by differential lithospheric loading, variations in regional water balances (glaciations, or evaporation of sequestered seas) and tectonic changes in ocean basin shape (Mörner, 1976; Donovan and Jones, 1979; Carey, 1980). It is possible that geoidal variations migrate (Mörner, 1976; Newman *et al.*, 1980), inducing low frequency oscillations. Furthermore, shorter-term sea-level variations may be forced by astrophysical and geophysical phenomena (Fairbridge, 1984), perhaps manifest through subtle climatic changes. These may be in the form of predictable tides, both short (12-500 hours) and long period (up to 18.6 years), or unpredictable surges (Murty, 1984). These factors determining sea-level changes are discussed in detail in earlier chapters.

Despite difficulties in collecting and analysing sea-level data, there is a general consensus that global sea-level is rising. The most recent estimates of Emery (1980), Gornitz *et al.* (1982) and Barnett (1983) put the global rise at between 1.1 and 3 mm a^{-1}. However, in many countries more concern is expressed about short-term supra-elevations of level or storm surges than over long-term secular rises in the sea surface. Although storm surges recur only infrequently along many coasts, they wreak considerable havoc, both in terms of loss of life and damage to property. Two types of site are most vulnerable: (1) those low-latitude coasts subject to intense tropical storms, typhoons or hurricanes (Simpson and Reidl, 1981) (Fig. 14.1) and (2) coasts bordering shallow enclosed water bodies (sea and lakes) where meteorological forcing, due to extreme high or low atmospheric pressure, may combine with high tides (not lakes) to raise coastal water levels (Murty, 1984) (Fig. 14.1). At any one time and place a surge may involve several components, including tide, wave set-up, current set-up, wave type and form, direct wind stress, terrestrial ponding and inlet jets.

465

Figure 14.1: Global incidence of tropical cyclones coincides with coasts liable to infrequent, and largely unpredictable, short period increases in the sea surface. In addition the shallow enclosed or semi-enclosed seas and lakes indicated are often subject to meteorological forcing, causing surges (negative or positive deviations from predicted water levels)

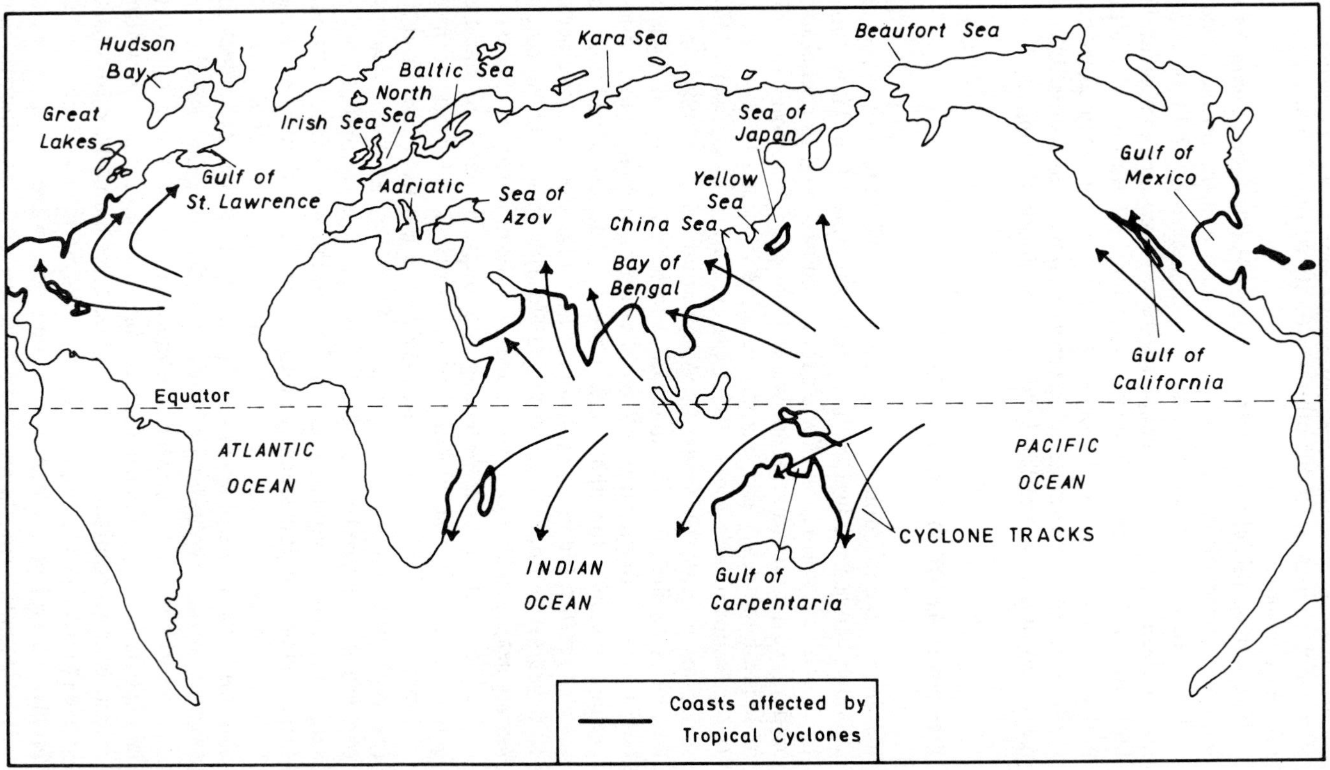

466

The relationship between changes in mean sea level and changes in the frequency and magnitude of surges is not simple. A disparity in interpretation is apparent in recent research around the margins of the North Sea. Rohde's (1980) view is that the frequency of marine flooding in the German Bight has mirrored rises in sea level since the sixteenth century. Thus the two are directly connected. Alternatively, work by Graff (1978, 1981) on extreme tides at British ports does not support fully Rohde's view, as the magnitude–frequency curves for many east coast locations are decidedly non-linear, in some cases becoming time-independent. Intuitively, the frequency of extreme levels must be related not only to mean sea level, but also to shifts in tidal constituents and amphidromic points (Pugh, 1982), changes in storminess (Resio and Hayden, 1975; Thom, 1978; Hayden, 1981), shifts in storm tracks (Lamb, 1981; Hayden, 1981) and local conditions of wave type, shoaling slope and tidal stage. However, the degree to which any one or any combination can force a destructive rise in sea level is still unknown.

STRATEGIES TO OFFSET THE EFFECTS OF SEA-LEVEL RISE

In many undeveloped regions societies view the coast as a hazard (White, 1974; Kates, 1978), accepting loss and adjustment as part of their lifestyle. Socio-economic structures are often based around the unpredictability of risks, viewed, as Islam (1974) describes for coastal Bangladesh, as 'God's Will'. However, such an extreme teleological position is not often tenable in developed societies although Baumann and Sims (1974) show that it still exists. In many developing nations burgeoning coastal populations (largely due to in-migration) and increasing investment in property and technology have reorientated perceptions and expectations of many coastal communities and states. The most sophisticated responses to sea-level change hazards are found, not unexpectedly, in highly developed countries like the United States, Australia and Japan. In some countries the threats of sea-level rise to coastal communities have attained something of a 'doomsday' status to judge from many recent prognostications (Kaufman and Pilkey, 1980; McCleish, 1980; Jackson and Reische, 1981). While a timely warning can help to alert public awareness, and also generate research funds, over-despondent forecasts may lead both to anxiety and even fear among individuals. It may also result indirectly in depressing coastal economies, for instance through falling property values, and disrupting social structure. It is clear that a wholly effective public relations strategy for espousal of coastal hazards has yet to be devised. These themes are taken up in more detail by Titus in Chapter 15.

Despite these reservations, it is clear that the twentieth century concentration of coast dwellers has brought with it serious problems of hazard mitigation. There are five main ways in which these problems may be tackled:

(1) Through construction of sea defences using civil engineering techniques.
(2) The use of biophysical methods to strengthen the shoreline.
(3) Implementation of coastal plans, particularly those advocating the judicious use and allocation of coastal land resources.
(4) Improvement in building design and technique for coastal structures and the enforcement of building codes to achieve better storm/flood proofing.
(5) Through social 'engineering', via education, legislation and manipulation of coastal populations.

All five have been tried with varying degrees of success and failure, although rarely in concert under the auspices of a workable coastal management policy. Some of the most advanced statements are to be found in various US state coastal plans, especially those for California (1975) and Florida (1982). Each of the strategies listed above will be discussed in turn.

Construction of sea defences

> For 'tis the sport to have the engineer
> Hoist with his own petard.
>
> *(Hamlet, III, iv)*

This section concentrates on the rationale and justification for sea-defence works. Traditionally coastal engineers have concentrated on fixing shoreline position regardless of environmental consequences. Construction of seawalls, often in conjunction with harbour or river mouth 'improvements' may be traced back several thousand years to various Middle Eastern civilisations (Straub, 1952; Pannell, 1964). Modern coastal defence works date largely from the middle nineteenth century, the heyday of the Victorian civil engineer. In this period coastal protection assumed an almost lyrical air; Sir John Rennie (1845) in his Presidential Address to the Institute of Civil Engineers in London said 'Where can man find nobler or more elevated pursuits ... than to interpose a barrier against the raging ocean?' Such confidence was translated through energy and enterprise into many huge breakwaters, piers, groynes, bulkheads and seawalls, dotted around the coasts of Europe and the Americas and throughout the various monarchical empires, for example India.

Numerous published examples affirm the efficacy of coastal engineering works in holding the shoreline and preventing storm damage (e.g. O'Brien and Johnston, 1980; Berkeley-Thorn and Roberts, 1981), although there have been some celebrated failures (Sines Breakwater, Portugal). Sadly such static solutions are by and large inappropriate in the long term to dynamic problems. 'Holding the shoreline' may lead to rapid depletion of the beach, initiate terminal scour, cut off downdrift supplies of sediment and disrupt the normal exchange of sediment between nearshore, beach and dune. Furthermore, structures do not adjust to sea-level changes. This latter point has led to a heated argument between 'concerned' geologists (mainly based on the eastern US seaboard) and coastal engineers. The focus of the debate has been the management of the US barrier islands, although many of the points raised apply equally well elsewhere. The 'shoreline debate' is epitomised in an article by Pilkey and others which appeared in *Geotimes* 1981, entitled 'Old solutions fail to solve beach problems'. The article is a thinly disguised attack on the coastal engineering profession, claiming it has a vested financial interest in ignoring the geological evidence for sea-level change. The paper goes on to maintain that current 'protection' measures are inherently obsolescent in such circumstances and questions the objectivity of many engineers. Such accusations have put engineers on the defensive. There are two basic responses (see O'Brien, 1982): first, that the critical data for assessing sea-level rise are ambiguous and thus ignoring them is justifiable, and second, that even if sea-level is rising 'so what?'. By this O'Brien implies that the design life of many coastal structures is shorter than many of the observed secular variations in sea level and so are unaffected by them.

These divergences in opinion are not easily reconciled. Perhaps it is fair to point out that many engineers do recognise the problems posed by changing sea levels, especially those of short duration, and many design criteria take these into account. An engineer, Per Bruun, is commonly credited as the originator of the 'Bruun Rule' (Schwartz, 1980) which relates certain types of shoreline recession to sea-level rise (see Orford, this vol., Ch. 13). On the obverse side, some geologists (Wanless, 1982) remain sceptical of the importance of sea-level change to current erosion problems.

Over the last forty years many of the engineering 'tactics' for treating coast erosion have changed. Some changes have come about through increased understanding of coastal dynamics, others from improved design or materials. Still more have been forced through financial expediency. One major advance is the realisation that coastlines should not be protected in piecemeal fashion. Boundaries between protected and unprotected coasts often act as foci for serious problems. The ends of seawalls commonly show terminal scour, relating to the diffraction of waves around the end of the wall. Jetties may induce shoaling, groynes intercept or deflect longshore drift and cause sediment starvation downdrift (Hails,

1977). Improved designs have led to more sophisticated structures, for example 'permeable' revetments or low relief, permeable groynes. In the latter example, one is perhaps tempted to question the need for such a structure at all, other than to boost public confidence that 'something is being done'. Often it would appear that seawalls are built solely to assuage local desires, with critical decisions being made in a political rather than an environmental arena. Similarly, beach nourishment techniques are often designed to recover sediment lost through earlier inadequate protection schemes — see the example from Florida outlined on page 481. Thus one may reach the somewhat ludicrous position that a naturally stable beach, destroyed by protection measures, is having to be restabilised through the artificial introduction of sand. The long-term funding of such projects is questionable.

In some places where sea level is indisputably rising, very complex and sophisticated solutions have been implemented to alleviate marine flooding hazards. Two examples occur on either side of the southern North Sea, along the coast of Holland and in the Thames Estuary, southern England, and another in the Lagoon of Venice in northeast Italy. Taking these in turn:

The Dutch coast

Protection of the low-lying Dutch coast has long been a paramount objective of the government. The disastrous 1953 storm surge led to a renewed national effort (Koestner, 1984), which has resulted in the construction of storm surge barriers around the entire coast. The lynch-pins of this protection policy are the storm surge barriers across the Rhine–Meuse–Scheldt delta (Fig. 14.2). Following the 1953 storm, closure of the estuaries was authorised by the 1958 Delta Act. Progressive sealing of the estuaries will culminate in the closure of the Eastern Scheldt in 1986. Enclosed by a continuous sea defence line comprising earth embankments, stone revetments, natural and artificial dunes, the Netherlands is 'designed' to withstand a 1 in 5,000 year surge, although known sea-level imbalances over such a time period renders such statistics meaningless.

The Thames Estuary

London on the upper reaches of the Thames Estuary (Fig. 14.3) has been threatened by flooding for centuries. However, the extension of the urban area onto lower-lying land, especially to the east of the city, plus construction of many subterranean communications and waste disposal systems, has magnified the possible effects of marine flooding many times over. Many British Government offices and archives, as well as major commercial institutions, are located in this area. Added to the flooding problem, sea level in the estuary is rising. This is partly due to the continuing isostatically controlled subsidence of the area (Devoy, 1982), but is also related to

470

Figure 14.2: Most of the Dutch coast is protected by storm surge barriers. The most elaborate network is found in southwest Holland (Zeeland) where a series of barrages, dykes, revetments and artificial dunes have been constructed to withstand postulated 1 in 5000 year supra-elevations of sea level. The 'Delta Plan' had led to major physical, chemical and biological changes in the Zeeland environment; for example, many of the former estuaries are now brackish or freshwater impoundments

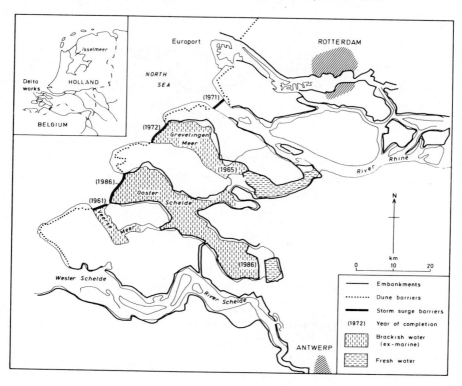

man's activities. At Southend, near the estuary mouth, sea level is rising about 1.4 mm a^{-1} (Pugh and Faull, 1983). At London Bridge, however, the rise is at least twice the rate (Rossiter and Lennon, 1965). Since 1790 high tides in London have risen by almost a metre. This disparity would seem to be related to man-induced modifications in the tidal wave (Prandle and Wolf, 1978), by, for example, deepening the estuary through dredging or narrowing it by marginal land reclamation.

Despite the construction of 112 km of estuary margin embankments in the mid-nineteenth century, serious flooding took place in 1881, 1928 and 1953 (Devoy, 1980; Gilbert and Horner, 1984). Although repeated raising of the embankments has provided temporary relief, it was decided in 1976 to construct a tidal barrage at Silverton, about 12 km below London

471

Figure 14.3: The Thames Barrage, opened in 1983, is located at Silverton and designed to protect much of central London from not only occasional storm surges, but also the steadily rising sea level (see graph)

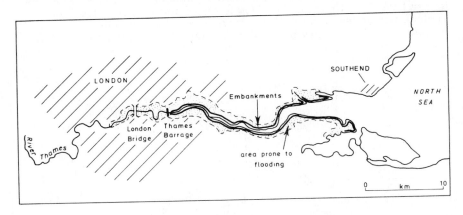

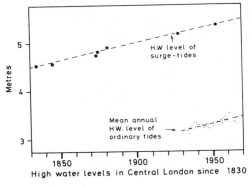

High water levels in Central London since 1830

Source: After Bowen (1972).

Bridge. This structure, costing £850 million and opened in 1983, is designed to cope with the 1 in 1,000 year high tide flood, about 1.2 m above the highest hitherto recorded (in 1978) at London Bridge. It has been estimated (see Gilbert and Horner, 1984) that a single major flood could cause at least £4000 million worth of damage. In addition to the main barrage, numerous other small improvements have been undertaken, including floodproof upgrading of lockgates, flapvalves and estuary walls, designed both to stop marine incursions and allow terrestrial drainage. Obviously the cost benefits of such a scheme are very favourable. A somewhat similar barrage scheme protects the Japanese city of Osaka (Whipple, 1984).

Venice Lagoon

Considerable international concern has been expressed over the fate of Venice's unique historical and cultural heritage as the city is faced with a rising sea level. The city is built on an island within a partitioned tidal lagoon (Fig. 14.4) and has a long history of periodic tidal inundation (Pirazolli, 1973, 1983). The evidence put forward by Pirazolli suggests that sea level remained relatively stationary until around 1900, after which it has risen by 0.2-0.3 m ($\sim$ 3.5 mm a^{-1}). The problem is compounded by the foundation settlement of individual structures; some have subsided 0.5 to 0.8 m since construction in the seventeenth or eighteenth century. In addition groundwater abstraction has led to subsidence. A ban on local industries pumping water from the coastal aquifer appears to have slowed the subsidence rate in the last twenty years.

The lagoon and barrier coast of the Venice region forms part of the Po Delta margin, the sort of area liable to subsidence through increasing lithospheric loading by sediment accumulation. However, the post-1900 sea-level rise in the lagoon appears to have been triggered by man. Passino and Todisco (1984) demonstrate that extensive infilling of the lagoon edge (over 10 per cent of the water area has been lost), especially in the central Malamocco and eastern Lido basins, together with extensive dredging of the lagoon and the tidal passes to improve navigation, have led to a major adjustment in the tidal curve. (Spring tidal range is 0.9 m.) Prior to 1900 the Adriatic tidal wave was severely damped and retarded on entering the lagoon, but removal of the marginal shallows (high tidal friction areas) and improved access has resulted in the progressive merging of the external and internal tide regimes. Although the shallow lagoon has always been susceptible to meteorologically forced surges, these are now far more damaging when associated with enhanced tidal levels. The increase in flood occurrences is shown on the lower inset in Figure 14.4.

The disastrous flood of 1.95 m above mean sea level in 1966 alerted both conservationists and the Italian government to the plight of Venice. In the last decade, a major international effort has been mounted to preserve and restore many of the art treasures and historic buildings. Consequently, the Italian government has been pursuing the objective of protecting the lagoon from high water levels. It is proposed to build four mobile flood barriers across the three passes (the Lido pass will have two barriers — see Figure 14.4), serving both to attenuate the tidal wave and to reduce the tidal volume inflow, so depressing the water level within the lagoon. In addition a number of short *en échelon*, fixed transverse bulkheads are to be built within the pass throats. These measures allow for both continuing navigation and free exchange of water between the sea and the lagoon — the latter being an essential prerequisite if water quality within the lagoon is to be improved.

473

Figure 14.4: A rising sea level has posed numerous problems for Venice, especially in the last thirty years (see upper inset). Much of the attention has focused on the progressive deterioration of the historic buildings and their contents. Current proposals envisage construction of surge barriers across the tidal passes to depress extreme tide heights

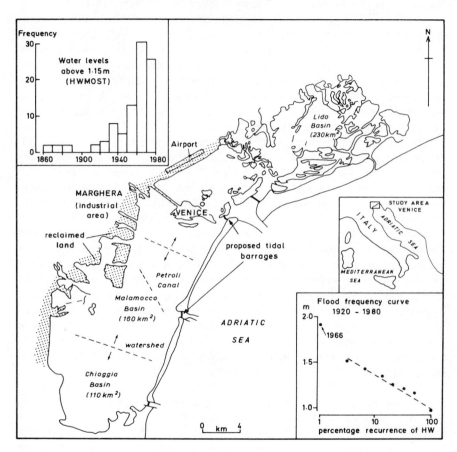

The environmental ramifications of such major coastal engineering works are many. Barrages tend to disrupt or destroy the encircled estuarine ecosystems (Saejis, 1982), through reduction of tidal flow, leading to losses in diversity and occasionally productivity. Also a barrage will result in modification of the external tidal waves and as a consequence sediment will be redistributed. Many of these physical and biological topics are covered in the environmental impact statements associated with tidal barrage proposals, for example the Severn (Shaw, 1980) or the Bay of Fundy (Daborn, 1977; Gordon and Dadswell, 1984).

474

In recent years it has become evident that while engineering solutions for combating rising sea level may work in isolation, they tend to create more problems than they solve within the whole environment. The response of some engineers has been to increase the complexity and the cost of these 'solutions', so that many coastal sites boast a history of increasingly grandiose and expensive structures. A case in point is Miami Beach in southeast Florida. Heie rising sea level, around 2.3 mm a^{-1} (Hicks, 1981), acting on a barrier island with a known geological history of instability (Chardon, 1977), has caused severe erosion — up to 3 m a^{-1}. The problem has been exacerbated by the development of the island since 1920, which destroyed some of the foredunes. The first bulkheads and groynes appeared on the beach in 1925 and 1927 respectively (Hansen, 1947) and backshore armouring in the 1930s (Bruun et al., 1962). During the same period (1920-40) the rest of the dunes were removed to allow development right to the water's edge. Miami Beach's erosion problems were made worse through lack of co-operation between riparian owners. Most people simply protected their own property in the best way they could, often at the expense of their neighbours'. In later years this inconsideration led to a number of legal disputes over maintenance, repair and damage (Mills et al., 1976). Further back-barrier developments following World War II led to quantities of dredge spoil being used as fill on the oceanfront, partly for protection but also as it provided a convenient dumping ground. By 1955, the entire 13.5 km was groyned in an attempt to arrest longshore drift material. In the early 1960s, beach nourishment was being promoted as 'the answer' (Bruun and Manohar, 1963; p. 19), although early experiments at nearby Key Biscayne and Virginia Key were a failure (Park, 1969). Despite these setbacks, the beach was eventually restored between 1978 and 1980 through the importation of 10.7 million m^3 of offshore sand at a cost of $80 million. Perpetual maintenance is required to maintain this artificial beach. A final irony is that the dunes wilfully destroyed in the 1920s and 1930s have now been rebuilt as a hurricane surge protection line (Fig. 14.5).

The Miami Beach experience has been repeated elsewhere. Ocean City, Maryland, shows very much the same pattern (Leatherman, 1981), as does much of the northern New Jersey shore (Nordstrom, 1977). In Britain coast protection at Bournemouth (Willmington, 1983) has followed a broadly similar trend, culminating in a nourishment scheme in the late 1970s. (It must be pointed out that sea-level rise has not been a major issue here, although it is known to be rising slowly (Pugh and Faull, 1983).) The efficacy of these and other similar schemes must be set against the expenditure and full consideration of alternatives. The fact remains that in some cases it might be more prudent to remove completely the offending engineering structure, in the light of sea-level rise.

475

Figure 14.5: Bal Harbour at the north end of Miami Beach is part of a major tourist resort built on a geologically unstable barrier island. A combination of rising sea level and wilful environmental destruction had led to severe erosion all along this coast. The photograph shows the 'new' beach, restored through importation of offshore sand between 1978 and 1980, and the reinstated (and irrigated!) duneline, designed to withstand hurricane surges. Unfortunately the duneline gaps, provided for access, may negate this latter function by channelling overwash onto the back barrier

Biophysical methods

While engineering structures might be said to work *against* coastal processes, biophysical processes are generally thought to work *with* them. Yet as we shall see this is not always true.

As noted before (and also in Chapter 13 by Orford), both long- and short-term changes in sea level induce morphological response. Rises in sea level may cause beach and cliff erosion, overwashing, embayment/estuarine infilling and terrestrial flooding. In all these cases there is a re-distribution of materials (Hails and Carr, 1975).

Biophysical methods, largely involving the planting and/or encouragement of vegetation, have been advocated for defence on 'soft' coastlines, for example on dune or marsh coasts, for centuries. Planting techniques have been in use in Denmark since at least the thirteenth century (Sorensen, 1983) and some principles are enshrined in English Common Law as prescriptive (common) rights. Widespread dune reclamation and stabilisation by planting marram *Ammophila arenaria* were practised in the eighteenth and nineteenth centuries in western France (Bremontier, 1833), and the original Dutch, Danish and French methods seem to have been globally disseminated (Brookes, 1979; Woodhouse, 1982). Similarly propagation and transplanting of coastal wetland and marsh species, notably varieties of *Spartina*, is a common technique in North America (e.g. Knutson, 1974; Race and Christie, 1982; Zedler, 1983). *Spartina anglica* was planted widely in the British Isles both for land reclamation and as a wave-stilling technique to reduce shoreline wave energy (Hubbard and Stebbings, 1967; Ranwell, 1967), but is now treated as an invasive weed, requiring chemical or physical eradication.

Encouragement of peritidal or supratidal vegetation has drawbacks. Artificially nurtured monoculture stands are susceptible to disease or disturbance, which may lead to a catastrophic collapse of the ecosystem. Failure to provide for succession by maintaining a suitably diverse sward may also lead to difficulties. The die-back in the late 1950s of *Spartina anglica* along the coast of southern England (Goodman *et al.*, 1959) is an example. *Ammophila brevigulata* plantings in the mid-Atlantic states of the US often fail after 3-4 years due to the susceptibility of phenotypes to insect-borne pathogens.

Perhaps more seriously, vegetation plantings cause morphological disequilibrium. Almost all vegetation acts to accumulate sediment, which in turn provides nutrients and trace elements essential for sustaining growth. Where the sediment supply is finite, abstraction by vegetation may lead to morphological or sedimentological changes elsewhere in the system. The best-documented example of this type of response is found on the barrier island coasts of the eastern US seaboard. Under natural conditions these barriers are migrating upward and landward in response to the slowly rising sea-level (~ 3 mm a^{-1}). However, from the 1930s onward the US National Park Service policy was to encourage, through the planting of dune vegetation, a high, secure backshore ridge against storm washover. In the early 1970s, Dolan (1972, 1973) highlighted the weaknesses in this policy, which in effect was promoting the accumulation of sand in high dunes instead of on the beach or back barrier, and, in so doing, ousting the overwash processes vital to the maintenance of the barrier system. This alteration forced both morphodynamic and sedimentological changes (Fig. 14.6). In the former the beach shifts from naturally dissipative (flat profile, spilling waves) towards artificially reflective (steep profile, plunging waves), leading to an increase in beachface slope, perhaps accompanied by changes in sediment texture (Dolan, 1972; Leatherman, 1979). The appearance of a reflective beach may favour the creation of subharmonic or synchronous

Figure 14.6: Idealised response of a barrier system to artificial construction of high 'barrier' foredune ridges

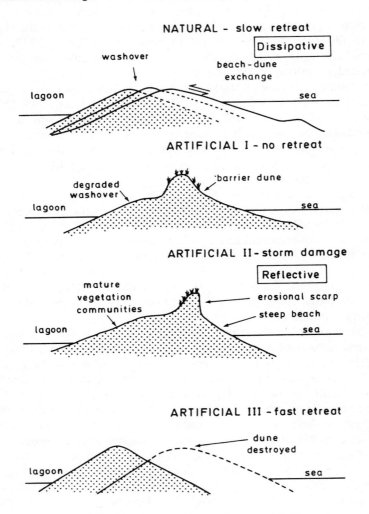

edge waves which in turn may cause erosion (Holman, 1983). Ecological consequences of encouraging crestal dunes follow the same pattern. Under natural conditions there is a strong balance between morphology, particularly of dunes and washovers, and ecology (Godfrey *et al.*, 1979). A wide floristic diversity is useful in modulating response of the barrier to sea-level change. When a range of species is present, adjustments to morphological change are facilitated, suitable plants colonising new surfaces. Formation of

478

a barrier dune causes the inner (landward) communities to mature. For example, *Ammophila* spp. become depauperate and are replaced by less sediment-tolerant species. Sudden removal of a foredune finds the back-barrier vegetation 'unprepared'; the dominant species are unable to cope with engulfment and there is often insufficient *Ammophila* spp. root stock, or rhizomes, to respond to the requirements for large-scale revegetation. Accordingly the barrier becomes highly susceptible to further erosion.

It would appear that somewhat similar processes occur on all dunes and are not restricted to barrier islands. Excessive planting of foredune slopes in Ireland engenders an abnormal sequence of slope failures (Carter, 1980; Carter and Stone, in press) which extends the sediment retention time in the dunes from 3 to 27 years in the examples studied by Carter (1980). These dunes display an abnormal, intermittent failure response pattern to short-term sea-level variations.

The overall impression from such studies is that while biophysical methods are a useful adjunct to shoreline stabilisation programmes, great care must be taken to marry techniques with anticipated variability in the system. Biophysical 'overstrengthening' by the planting and encouragement of vegetation in the face of known trends in sea level, simply shifts the system response from a slow, paced inundation, where morphological and ecological changes are in tune, to one where response is erratic or even catastrophic. Morphology and ecology become 'detuned' and effective management is severely handicapped. The construction of artificial coastal ecologies is probably a far more sensitive process than many of the early advocates would have envisaged.

Planning for sea-level change

The usual 5-25 year timescales employed in planning science are too brief to consider secular sea-level change. Similarly, many structural design lives are short (~ 100 years) and the normal decision-making processes are an amalgam of short-term political and financial contigencies. Moreover, few plans or planners emphasise the distinctive nature of the coastline, being content to treat the coast as part of a general landmass (see Carter, 1987). Only in the most rapidly rising or falling land zones would sea level be a major factor and many of these are sparsely populated and have few planning controls. Most planning policies and designs associated with sea level are aimed at mitigating short-term flood hazards. In many countries planners are content to vest such considerations with coastal (or river) engineers, assuming that sea defences can be built to sufficient criteria to obviate marine flooding. This is particularly true of Britain where regional water authorities are responsible for sea defences. In these circumstances no special land use zoning may exist to safeguard against sea-level rise. In the USA, city, county and state planning varies greatly from place to place, so that a number of different approaches exist for dealing with sea-level changes. These include determination of costs and benefits for development proposals (often both from economic and environmental viewpoints), regulation of taxes and subsidies for coastal properties, shoreland zoning on the basis of hazards, building control regulations and access provision. Some of these will be discussed below, whilst further consideration of the theme and the related concepts of social 'engineering' are given by Titus in Chapter 15.

United States land use planning can be regulatory or non-regulatory. Often there is a permit system for development, but no overall strategic plan, so that providing the requisite permission is forthcoming, development may take place anywhere. In other areas only planning guidelines are issued, but most authorities have some kind of plan and permit system. Selective subsidies may determine who builds what, where.

The US Coastal Zone Management Act (CZMA) (1972) was authorised to help states draft and implement coastal management plans (Eliopoulos, 1982). So far all coastal states except Texas, Georgia, Illinois, Ohio, Indiana and Minnesota have participated fully in the programme. Inevitably the integrated approach favoured by the CZMA initiative includes planning, especially in the siting of particular facilities such as power plants, roads, accesses and marinas, as well as dealing with more nebulous resources like scenic beauty and potential hazards. Continuing Federal support for the CZM office is now (1984) in some doubt (see Galloway, 1982), although progress is likely to be maintained at a State level.

Although there is no Federal control of coastal planning in the US, the National Flood Insurance Program (NFIP) started in 1968 is widely recognised as having promoted development in low-lying coastal areas (Platt, 1978). Under the NFIP the US Congress legally obligated the Federal Government to reimburse flood victims for damage and loss. Heavy subsidies (up to 80 per cent) have made the NFIP extremely attractive to coastal property owners and acted as an incentive to developers. Four coastal areas — Galveston, New Orleans, Miami and Tampa — have 37 per cent of the total national indemnities under the NFIP (Monday, 1984). The Reagan Administration have decided, purely as a cost-cutting exercise rather than an environmental one, not to extend the NFIP to new developments on barrier islands — a decision warmly applauded by the US coastal conservation lobby as an aid to deterring future expansion in these vulnerable areas.

The only US state, besides Hawaii, entirely within the 'coastal zone' is Florida, an area particularly prone to the rising sea-level hazard because of its wide, low-lying shoreline. Until recently, environmental planning in Florida has taken something of a back seat to political aims and commercial exploitation (Carter, 1973; Blake, 1980; Adams, 1982), with the result that many buildings are too close to the sea. In an attempt to rectify matters, the Florida legislature passed a series of laws designed to preserve their coastlands (Guy, 1983). A major step forward was the establishment of a coastal set-back line (CSL) (Purpura, 1972) behind which all development should remain. Subsequently set-back lines have been adopted elsewhere, for example in New Zealand (Gibb, 1981). At first the set-back line was defined simply as a linear distance (10 m) from the 'present dynamic beachface' or (6.5 m) from the seawardmost dune crest. However, since 1972 the definition of the Florida CSL has been improved greatly by extensive studies into wave processes and morphology at regular intervals longshore (Balsillie et al., 1983), so that in places the CSL may be several hundred metres inland. Unfortunately, many variances to established set-back lines have been granted. As well as CSL studies, hurricane surge penetration investigations have been, or are being, undertaken for all urban shorelines in Florida (e.g. Dean and Chiu, 1981) and the susceptibility of soft coasts to storm damage is being examined systematically (Stone and Morgan, 1984). The reports by Dean and Chiu present simulations of storm damage across specific urban coasts, based on meteorological, wave and morphological data, for recurrence intervals from 10 to 500 years. No account is taken of long-term sea-level change. The ideal end product of all these investigations will be a comprehensive land use plan, designed to minimise damage from sudden water level supra-elevations.

Regrettably, most coastal zones have become inhabited before any serious consideration is given to the problems of sea-level rise, or for that matter any other environmental hazard. Thus some type of storm warning system is desirable. The commonest procedure is to prepare for evacuation from low-lying coastal sites. There are two main types of evacuation — horizontal and vertical. The commonest, horizontal evacuation, involves moving large numbers of people inland. This can be difficult especially where escape route capacities are low so that the tend to jam with traffic, and/or warning systems are inadequate. Sanibel Island in southwest Florida is unusual in that it has a self-imposed (by city ordinance) population 'cap' among its environmental controls (Clark, 1977). The cap is set at a level commensurate with the hurricane evacuation capacity of the access causeway to the island.

Vertical evacuation requires people to shelter in high-rise (often reinforced) buildings. This method is advocated where evacuation routes are either of low capacity or tortuous, suitable buildings are available, and/ or the population, for one reason or another, cannot respond quickly. A good example of this is Miami Beach where causeways to the mainland are narrow, the population has a high proportion of elderly and infirm and there are many suitable high-rise buildings (Carter and Orford, 1982). In cases like this vertical evacuation is the most sensible option.

Almost all tropical or subtropical countries affected by tropical cyclones, typhoons and hurricanes operate some form of storm warning service, based on weather forecasting (World Meteorological Organisation, 1983). Some of these services, for example the one in India, have existed for well over a hundred years. Unfortunately, the narrowness and extreme unpredictability of many tropical storm tracks makes reliable and effective forecasting more than 12 to 24 hours ahead difficult. The best preparation for a major coastal storm is undoubtedly to move inland, although this is not always easy to accomplish. For example, even a 24-hour storm warning for the Florida Keys would not be enough time to evacuate everybody. The most complete and sophisticated warning systems are those in the USA, the Philippines and in Australia. These include detailed plans of evacuation routes, their capacity, their problem areas and reception zones. Communities are issued with detailed instructions about when they are to leave, what they may take with them, how they are to escape and where they are going to congregate. Some human aspects of storm evacuation will be considered later.

A cognate area of interest is in tsunami (long-period impulsive waves) warnings, particularly on Pacific margin and island coasts (Houston and Tsai, 1983). Tsunamis may cause rapid local fluctuations in sea level, often several metres above msl. In California 1 in 500 year tsunami heights exceed +3 m at San Francisco and +2.2 m at Los Angeles. On the Hawaiian islands extreme sea surface supra-elevation may reach over 15 m at the most vulnerable localities (Houston *et al.*, 1977) — an example is shown in Figure 14.7.

Structural protection

It is unlikely that buildings and other structures will be specifically designed to withstand long-term changes in sea level, although gradual encroachment would most likely result in *in situ* drowning, as is apparent in Venice. Somewhat ironically, drowned buildings in sheltered localities may stand a far better chance of preservation underwater than above it. More realistic is the need to design buildings to withstand temporary rises in sea level during storms, without incurring prohibitive costs. One should always be prepared to abandon a site to the sea! However there are problems with capital-intensive, long-life projects like power plants, which are built without due regard to potential sea-level change. Tooley (1979) voices concern over the construction of coastal nuclear power stations in Britain, many of which appear to be vulnerable to sea-level change. Despite numerous and rigorous site selection criteria (e.g. Ducsik, 1979) the possibilities of sea-level change seem rarely to be considered in planning nuclear installations on the coast.

Figure 14.7: Predicted tsunami heights on the island of Hawaii in the eastern Pacific Ocean. Predictions are affected by coastal configuration, shelf morphology and width, wave approach directions and shelter behind nearby islands (to the northwest in this case). Hazard maps of this kind are of inestimable value in planning set-back lines for development

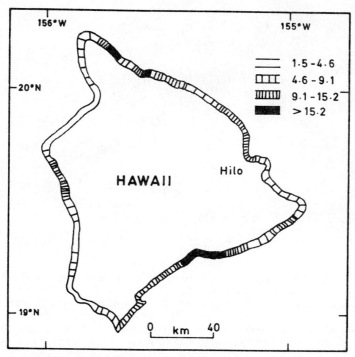

Combined tsunami and highest astronomical tide (HAT) 50 year exceedence probabilities (metres)

Source: After Houston and Tsai (1983).

Problems with coastal houses are most acute where regulations have allowed construction at or near the water's edge, particularly on fragile barrier islands (Nordstrom and McCluskey, 1984), in foredunes, or along cliff edges (Kuhn and Shepard, 1983) (Fig. 14.8). In many places coastal residents have attempted to safeguard their houses by building bulkheads, groynes and seawalls (Fig. 14.9), although in the main these tend to exacerbate or shift the problem. One alternative is to stormproof individual dwellings, in such a way that damage is minimised. Stormproofing involves a combination of correct siting and design. Ideally siting should be inland

Figure 14.8a: A cliff edge house in southern California. The structure is designed, at considerable cost, to withstand storms, surges and probably long-term sea-level rise; the adjacent cliff is not. The aesthetics of the scene appear to have received a low priority

Figure 14.8b: Dune/beach edge houses on Fire Island, New York. These buildings are founded on pilings in the anticipation that storm water will pass beneath, leaving the main structure intact. In addition this type of construction has, until recently, qualified for NFIP support, while more conventional designs have not. These particular houses were damaged by a storm in the Spring of 1984, and many repairs have still to be carried out

Figure 14.9: Bulkheads, revetments and groynes have all been used in attempts to offset the effects of sea-level rise on this part of the New Jersey coast (Ocean City). A more immediate consequence has been the loss of beach sand allowing increased exposure to wave attack

of the seawardmost dune ridge or cliff top, at a distance unlikely to be affected by normal beach/back-beach sediment exchange. Access to the shore should be by elevated walkway, or via a set of paths opened in rotation. Such procedures are normal in Holland and Germany and are being copied widely elsewhere.

Design criteria are more personal. A quick perusal of coastal styles will lead to the inevitable conclusion that 'anything goes'. However, some designs are better than others, for example hip roofs are better at withstanding wind than the more common gable roof. One way of illustrating some principles of design is to chart the effects of rising sea level through a storm on an individual structure; in this example a wooden platform house constructed on a barrier (Fig. 14.10). Stage I is the Washover Stage. Sea level rises sufficiently to allow overwashing of the barrier crest. Conventionally ground-level houses will be badly damaged or even collapse at this point. Some with shallow foundations may float away. Houses built on pilings are least vulnerable as the washover passes below. Stage II, the Wave Peaking Stage, occurs as the sea level rises to the point where waves are rapidly shoaling, but not breaking, across the barrier crest. Upward-directed stresses may destroy horizontal structures (floors, piers,

485

Figure 14.10: A schematic view of the effects of rising and falling sea levels through a storm surge, on a barrier-located dwelling. See text for full explanation

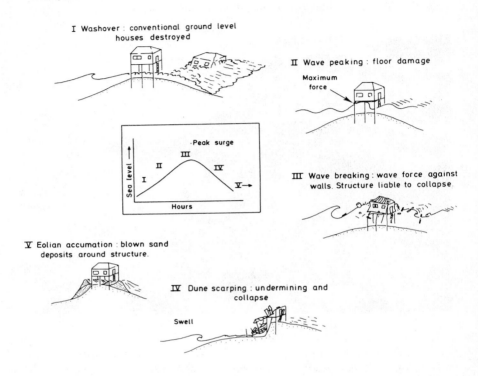

boardwalks). Houses on stilts are most at risk when waves are peaking under them. Continuing sea-level rise introduces Stage III, the Wave Breaking or Impedence Stage, in which the wave force is directed against the vertical seaward elements of the structures. Rapid variations in dynamic pressure cause massive failure. In addition waves may be armed with loose floating debris. The Undercutting Stage (IV) occurs as the beach is eroded. While this may occur during storm as waves attack the dune front, it is equally likely to take place during storm decay, when long-period swells may excite edge waves capable of dominating the run-up and causing erosion (Holman, 1983). Undercutting may undermine pilings, resulting in collapse. Such events can be avoided by adequate set-back lines and piling to depths well below the known beach cut. Finally in Stage V storm recovery processes are likely to lead aeolian transfer from the beach to the dunes. At times this may be very rapid, perhaps 0.5-1.0 m of deposition in a few days. Beachfront constructions intercept this blown sand and induce sedimentation. Nordstrom and McCluskey (1985) have calculated that such properties on Fire Island, New York, control up to 95 per cent of leeside sand accumulation. The sequence outlined above is idealistic, and often not executed in its entirety. In many cases Stages I, II and III are missing, with IV and V remaining as the most likely.

In recent years a number of publications have sought to provide guidelines for shorefront construction. These include Collier *et al.* (1977), Federal Emergency Management Agency (FEMA) (1984) and Pilkey *et al.* (1983). Much of the advice centres on the construction methods most likely to withstand the special stresses encountered during storms, although the Pilkey volume covers broader issues of siting and planning.

Social 'engineering'

Short- and long-term changes in sea level represent a hazard to individuals and communities. Perceptions of the threat posed by sea-level change vary from place to place and person to person. Despite a degree of residual awareness, many people are prepared to disregard the evidence and 'invasion' of hazard zones continues unabated.

While procedures to offset temporary supra-elevations in sea level associated with storms are well established in developed countries like the USA and Australia, the picture is very different in many less developed nations. In these a myriad of additional problems tends to extend the risks of personal injury or death. The following discussion examines this disparity, taking developed and developing nations in turn.

Developed nations

In 1972 the US Army Corps of Engineers published new guidelines for coastal management within the framework of the National Coastline Study. For the first time this recognised a policy of social and community measures to reinforce engineering practices. These new measures included both legal and fiscal provisions, such as outlawing coast damaging activities, taxing and coastal zoning, as well as more subtle ones, including disincentive propaganda and adult and juvenile education programmes. Many of these latter concepts have been tried, often in conjunction with storm warning services. Some have been monitored by sociologists and psychologists (e.g. Baumann and Sims, 1974).

Figure 14.11 illustrates the main decision-making pathways that should be encountered when assessing risks from sea-level change. It is unlikely that information on long-term rises or falls in sea level would be well known below the scientific–government level, although such data are becoming more widely debated in the US and have obviously affected the decision-making process in examples like the Thames Estuary flood alleviation scheme and the Delta Plan.

Figure 14.11: Institutional decision making pathways used in assessing risks from sea-level change. See text for details

DECISION MAKING ROUTES FOR SEA-LEVEL CHANGE HAZARDS

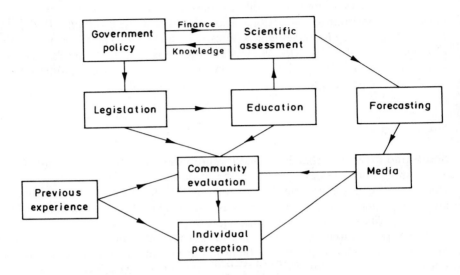

The decision-making structure in Figure 14.11 highlights the community evaluation, particularly in determining individual response. In his seminal study on perception of the coast erosion hazard, Mitchell (1974) noted, within his sample, quite different community-level perceptions along the eastern US seaboard. The individual perception, while embedded within that prevailing in the community, is often crucial in managing the shoreline. This is perhaps somewhat idiosyncratic of the situation in the US, where so much of the beachfront is privately owned. None the less, in most places, persistent individual views are often picked up by elected or nominated representatives and come to dominate any debate. In his example of coast erosion, Mitchell (p. 176) suggests that increasing hazard awareness may not be needed so much as access to objective information. However, shoreline erosion is a regular process which may be related to specific distances and/or objects, so that hard data are meaningful to most people. With sealevel rise there is a credibility gap; people simply cannot imagine the effects of rises of only a few millimetres per year. In addition the relative infrequency of large storms and marine flooding mitigates against objective assessment and undistorted perception. Pollard's (1978) mainly anecdotal account of the 1953 North Sea storm surge in eastern England is fascinating, not only for the eye-witness tales, but also for the overwhelming realisation that the coastal communities were totally unprepared, both mentally and physically, for such an event, despite the fact that they lived on a coast where flooding and avulsion had been commonplace throughout recorded history. It may have been that knowledge and appreciation of the coastal hazard had been lost within a few generations following the construction of extensive sea defences in the nineteenth and early twentieth centuries.

There is an obvious attraction in being able to influence individual and community views, improving awareness of and response to hazards. The main difficulty in applying such techniques would appear to be the paucity and unevenness of the market research. Potent prescriptions cannot be based on weak descriptions. The only sizeable body of work on this topic appears to be studies into hurricane warning response among coastal towns in the US. The potential destructiveness of these events has spurred research into the effectiveness of hazard preparation programmes.

Major storms obviously stick in the memory, although time can warp the facts. Clark and Carter (1979) found that people who had had direct experience of major coastal storms were more receptive to later information. Conversely, Baker (1979a) suggest that passage of time leads to complacency among the survivors and a general underestimation of future threats. Frank (1980) makes the same point, which he ascribes to the lack of a really destructive storm (in terms of casualties) since 1900. There have been several studies relating storm warnings to public response. Four US examples are reassessed by Baker (1979b). In these examples most people had heard about the storm on TV or radio broadcasts. Personal decisions whether or not to evacuate seem to be based on several nebulous criteria such as TV/radio reports, friends' or neighbours' advice and warnings from officials. Surprisingly, evacuation is not really related to awareness, intelligence or interest, possibly because such more alert cohorts are keen to remain to see the storm, or believe they know better. Baker found almost no strong predictors of evacuation behaviour, although there was a weak relationship between the forecast 'strength' of the storm and people's confidence in the day-to-day weather bulletins. Such indefinable responses make the potential for management very difficult. It is unlikely that a single interventionist strategy, for example, by telling people to leave once a siren was sounded, would illicit more than a patchy response.

These conclusions are discouraging. Attempts to influence individuals and communities come through better education and legislation. In the US the Hurricane Season (August–November) brings a cascade of 'promotions', ranging from free storm-tracking charts, literature and TV programmes to warnings printed on supermarket bags and the backs of parking tickets. Despite this 'caring' image some communities are now shying away from such activities, for fear that, in the absence of any storm, the publicity will put off potential newcomers and investors.

The somewhat loose US system may be contrasted with that existing in

Holland. Here there is compulsory education about storm hazards plus a highly organised emergency service, integrating scientific (forecasting — Stormwaarschnwing Dienst) with local reconnaissance (sea defence observations — Dykebewaking) and rescue (Bescherming Bevolking) services. The local activities are organised by water, fire and police authorities, but staffed, in part, by volunteers until military help arrives.

Developing nations

Poor communications, high population densities, low levels of literacy and income, poor land use planning and building standards and restrictive social organisations all combine to act against rational hazard management in developing countries. White (1974) stresses the point, re-emphasised in her Table 30-4, that storms of similar magnitude have markedly different effects depending on the developmental status of the affected coastal community. In developed countries damage costs are high but loss of life low, exactly the opposite of the situation in developing countries.

Developing countries most at risk include India, Pakistan, Bangladesh, Mozambique, Malaya, Borneo and Indonesia. Throughout these countries, Southern (1979) estimates that, on average, 15,000 lives are lost and damages amounting to over $200 million are caused each year by coastal storms. Although a break-down of these figures is not available, a sizeable proportion of the losses is due to temporary rises in sea level.

The problem is perhaps most acute in Bangladesh (Fig. 14.12). Here the conjunction of a densely populated coastal lowland (often more than 500 people km^{-2}) and a broad shallow shelf ideal for the generation of surges constitute an area of exceptional risk (Frank and Husain, 1971; Das, 1972). At least seven surges resulting in the loss of over 5,000 lives each have taken place in the last 200 years. Three of the surges, in 1876 1897 and 1970, caused over 100,000 deaths.

The devastation caused by the 1970 storm, not in itself particularly severe, was due to the coincidence of high tide and a rapidly moving cyclone, creating water levels up to 6 m above normal. In addition to the horrific loss of life, some estimates putting it as high as 300,000, the economic and social structures of all Bangladesh were placed under great strain for many months. This was largely because the important fishing industry was decimated, leading to an acute food shortage. Some of the reasons for the devastation are discussed by Islam (1974). His study suggested that although most of the population were aware of storm dangers, they were generally accepted as 'God's Will'. Despite warnings, many families preferred to stay on their small land holdings, sheltering in the hopelessly inadequate buildings. Many felt secure behind earth embankments and forests located by the Government as protective measures. Both types of protection were ineffective. Very few people were evacuated, only 5 per cent who survived the 1970 surge had gone to pre-

Figure 14.12: Map of coastal Bangladesh, scene of numerous major disasters due to temporary sea-level rises. The combination of wide shelf, subsiding delta margin and occasional fierce onshore cyclones makes the low-lying and heavily populated coastlands exceptionally vulnerable to disaster. The figure marks the track of the 1970 cyclone, believed to have killed up to 300,000 people

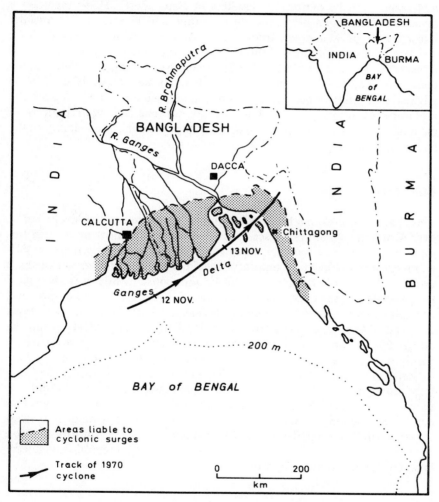

specified shelters; far more (38 per cent) had climbed trees. The decision to remain put is complex. It represents in part a need to safeguard belongings from looters, in part a reaction to social and religious conditions (for example many women were in purdah, thus it was against established convention to go outside), and in part a desire to remain together rather than

491

risk evacuation and possible separation. Further, provision of emergency international relief aid throws up paradoxes in attitude. For instance, Islam (1974) feels that many people will suffer the cyclone or surge in anticipation of the relief aid to come.

Since the 1970 disaster a variety of steps has been suggested and taken to lessen the risks of further disasters in Bangladesh. These included a review of disaster warning procedures (Frank and Husain, 1971) and the introduction of a simpler Information–Alert–Warning system (WMO, 1983). In addition, numerical models for surge prediction have been tested (Das *et al.*, 1974; Johns *et al.*, 1983) to help forecasting. Notwithstanding these sophisticated advances, many difficulties still exist. Islam (1974) advises that only major land reforms, better communications and the rephasing of crops, so that harvests do not coincide with periods of maximum risk, will engender a significant improvement in the ability of the coastal population of Bangladesh to survive surges.

CONCLUSION

This chapter has taken a broad view of man's response to sea-level change, over a number of temporal scales. It is evident that most interest in this topic has, until very recently, been focused on the hazards presented by relatively infrequent but temporary rises in sea level. The impact of these events varies greatly from place to place and, consequently, it is often hard to generalise. However, the influence of long-term secular sea-level changes cannot be ignored. Such projects as the Thames Barrier have been designed both to alleviate storm surge flooding and to protect, within the design life of the engineering structure, against a rising sea-level trend. Similarly, the recent shoreline 'debate' over the progressive erosion of the eastern US seaboard has served to highlight the need to consider the long-term sea levels as well as the more immediate storm surge penetrations. It seems likely that the next twenty years will see a major reappraisal of coastal engineering solutions to shoreline 'protection' (in its narrowest sense as 'defence against the sea'), involving a move towards *a priori* preventative measures rather than *a posteriori* 'cures', integrating engineering, ecological and social approaches.

ACKNOWLEDGEMENTS

My thanks to Julian Orford and Bob Devoy for commenting on the original draft. Julian, in particular, directed me to several important references that I had neglected. It is a pleasure to thank Killian McDaid and Nigel McDowell for their technical help.

REFERENCES

Adams, J.W.R. (1982) 'The politics of beach nourishment', *Shore. Beach, 50 (1)*, 3-5.

Baker, E.J. (1979a) 'Predicting response to hurricane warnings: a reanalysis of data from four studies', *Mass Emergencies, 4*, 9-24.

—— (1979b) 'Geographical variations in hurricane risk and legislative response', *Coast. Zone Mang. J. 5*, 263-83.

—— (ed), (1980), *Hurricanes and Coastal Storms*, Florida Sea Grant Report No. 33. Gainesville, Fla.

Balsillie, J.H., Athos, D.E., Bean, R., Clark, R.R. and Ryder, L.L. (1983) 'Florida's program of beach and coast preservation', in J. Mondy (ed), *Preventing Coastal Flood Disasters*, Natural Hazards Research Center, Boulder, Col., 109-22.

Barnett, T.P. (1983) 'Recent changes in sea level and their possible causes', *Climatic Change, 5*, 15-38.

Baumann, D.D. and Sims, J.H. (1974) 'Human response to the hurricane', in G.F. White (ed), *Natural Hazards: Local National and Global*, Oxford University Press, New York, 25-30.

Berkeley-Thorn, R. and Roberts, A.G. (1981) *Sea Defence and Coast Protection Works*, Thomas Telford, London.

Blake, N.M. (1980) *Land into Water — Water into Land*, Florida State University, Tallahassee, Fla.

Bowen, A.J. (1972) 'The tidal regime of the river Thames; long-term trends and their possible causes', *Phil. Trans. R. Soc. Lond. A.*, *272*, 187-99.

Bremontier, N.T. (1833) 'Mémoire sur les dunes', *Ann. Pont. Chausée, 69*, 145-242.

Brookes, A. (1979) *Coastlands: A Practical Conservation Handbook*, British Trust for Conservation Volunteers, Wallingford, Oxford.

Bruun, P., Chui, T-Y, Gerritsen, F. and Morgan, W.H. (1962) 'Storm tides in Florida in relation to coastal topography', *Florida Eng. Ind. Expt. Stat. Bull.*, *109*, 1-76.

Bruun, P. and Manohar, M. (1963) 'Coast protection for Florida development and design', *Florida Eng. Ind. Expt. Stat. Bull.*, *118*, 1-16.

Carey, S.W. (1980) 'Causes of sea-level oscillations', *Proc. Roy. Soc. Vict., 92*, 13-17.

Carter, L.J. (1973), *The Florida Experience: Land and Water Policies in a Growth State*, Johns Hopkins University Press, Baltimore, Md.

Carter, R.W.G. (1980) 'Slope failure and eroding sand dunes', *Biol. Conserv., 18*, 117-22.

—— (1987) *Coastal Environments*, Academic Press, London and Orlando, Florida.

—— and Orford, J.D. (1982) 'Miami Beach and hurricanes', *Geogr. Mag., 52*, 442-8.

—— and Stone, G.W. (in press) 'Mechanics of sand dune cliff retreat: Magilligan Foreland, Northern Ireland', *Earth Surf. Proc. Landform.*

Chardon, R. (1977) 'Cartographic analysis of coastal change', *Geo. Sci. Man., 18*, 257-67.

Clark, J. (1977) *The Sanibel Report*, Conservation Foundation, Washington, DC.

—— and Carter, T.M. (1979) *Response to Hurricane Warnings as a Process*, Natl. Haz. Warn. Rept. Ser. No. 79-08, University of Minnesota, Minneapolis.

Collier, C.A., Eshaghi, K., Cooper, G. and Wolfe, R.S. (1977) *Guidelines for Beachfront Construction with Special Reference to the Coastal Construction Set-back Line*, Florida Sea Grant Report No. 20, Gainesville, Fla.

Daborn, G. (ed) (1977) *Fundy Tidal Power and the Environment*, Report No. 28, Acadia University, Wolfville, Nova Scotia.

Das, P.K. (1972) 'Prediction model for storm surges in the Bay of Bengal', *Nature*, *239*, 211-13.

—— Sinha, M.C. and Balasubramanyam, V. (1974) 'Storm surges in the Bay of Bengal', *Quart. J. Roy. Met. Soc.*, *100*, 437-49.

Dean, R.G. and Chiu, T.Y. (1981) *Hurricane Tide Frequency Analysis for Broward County, Florida*, Rept. COEL-81/001, University of Florida, Gainesville, Fla.

Devoy, R.J. (1980) 'Postglacial environmental change and man in the Thames Estuary', in F.H. Thompson (ed), *Archaeology and Coastal Change*, Occasional Paper No. 1, Society of Antiquaries of London, London, 134-48.

—— (1982) 'Analysis of the geological evidence for Holocene sea-level movements in southeast England', *Proc. Geol. Ass. Lond.*, *93*, 65-90.

Dolan, R. (1972) 'Barrier dune system along the outer banks of North Carolina: a reappraisal', *Science*, *176*, 286-8.

—— (1973) 'Barrier islands: natural and controlled', in D.R. Coates (ed), *Coastal Geomorphology*, State University of New York, Binghampton, NY, 263-78.

Donovan, D.T. and Jones, E.J.W. (1979) 'Causes of world-wide changes in sea level', *J. Geol. Soc. Lond.*, *136*, 187-92.

Ducsik, D.W. (1979) 'Power plants in the coastal zone? — an analysis of utility siting practices and their implications for coastal zone management', in *Proc. Fifth Ann. Conf. Coastal Soc.*, 366-80.

Eliopoulos, P.A. (1982) 'Coastal zone management: program at a crossroads', *Environmental Reporter Monogr. 30*, 1-48.

Emery, K.O. (1980), 'Relative sea levels from tide-gauge records', *Proc. Nat. Acad. Sci. USA*, *77*, 6968-72.

Fairbridge, R.W. (1984) 'Planetary periodicities of the Sun, Earth and Moon', *Int. Symp. on Milankovitch and Climate*, Washington, DC (to be published).

FEMA (1984) *Design and Construction Manual for Residential Buildings in Coastal High Hazard Areas*, US Dept. of Housing and Urban Development, Washington, DC.

Frank, R.A. (1980) 'Living with coastal storms: seeking an accommodation', in E.J. Baker (ed), *Hurricanes and Coastal Storms*, Florida Sea Grant Report No. 33, Gainesville, Fla, 4-10.

—— and Husain, S.A. (1971) 'The deadliest tropical cyclone in history?', *Bull. Am. Met. Soc.*, *52*, 438-44.

Galloway, T.D. (1982) *The Newest Federalism: A New Framework for Coastal Issues*, Times Press, Wakefield, Rhode Is.

Gibb, J.G. (1981) *Coastal Hazard Mapping as a Planning Technique for Waiapu County, East Coast, North Island, New Zealand*, New Zealand Water and Soil Technical Publ., *21*, 1-63.

Gilbert, S. and Horner, R. (1984) *The Thames Barrier*, Thomas Telford, London.

Godfrey, P.J., Leatherman, S.P. and Zaremba, R. (1979) 'A geobotanical approach to classification of barrier beach systems', in S.P. Leatherman, (ed), *Barrier Islands*, Academic Press, New York, 99-126.

Goodman, P.J., Braybrooks, E.M. and Lambert, J.M. (1959) 'Investigations into "die-back" in *Spartina townsendii* Agg. I The present status of *Spartina townsendii* in Britain', *J. Ecol.*, *47*, 651-77.

Gordon, D.C., Jr and Dadswell, M.J. (1984) *Update on the Marine Environmental Consequences of Tidal Power in the Upper Reaches of the Bay of Fundy*, Canadian Technical Report on Fisheries and Aquatic Sciences, No. 1256, Ottawa.

Gornitz, V., Lebedeff, S. and Hansen, J. (1982) 'Global sea level trend in the past century', *Science, 215*, 1611-14.

Graff, J. (1978) 'Abnormal sea levels in the northwest of England', *Dock. Har. Auth. J., 58*, 366-71.

—— (1981) 'An investigation of the frequency distributions of annual sea level maxima at ports around Great Britain', *Est. Coastal Mar. Sci., 12*, 389-449.

Graff, T.O. and Wiseman, R.F. (1978) 'Changing concentrations of older Americans', *Geogr. Rev., 68*, 379-93.

Guy, W.E., Jr (1983) 'Florida's coastal management program: a critical analysis', *Coastal Zone Mang. J., 11*, 219-48.

Hails, J.R. (1977) 'Applied geomorphology in coastal zone planning and management', in J.R. Hails (ed), *Applied Geomorphology: A Perspective of the Contribution of Geomorphology to Interdisciplinary Studies and Environmental Management*, Elsevier, Amsterdam, 317-68.

—— and Carr, A. (eds) (1975) *Nearshore Sediment Dynamics and Sedimentation: An Interdisciplinary Review*, Wiley, London and New York.

Hansen, H.J. (1947), 'Beach Erosion Studies in Florida', *Florida Eng. Ind. Expt. Stat. Bull. 16*, 1-68.

Hayden, B.P. (1981) 'Secular variation in Atlantic coast extratropical cyclones', *Mon. Weather Rev., 109*, 159-67.

Hicks, S.D. (1981), 'Long-period sea level variations for the United States through 1978', *Shore. Beach, 49*, 26-9.

Holman, R.A. (1983) 'Edge waves and the configuration of the shoreline', in Komar, P. (ed), *Handbook of Coastal Processes and Erosion*, CRC Press, Boca Raton, Fla, 21-34.

Houston, J.R., Carver, R.D. and Markle, D.G. (1977), *Tsunami-wave Elevation Frequency of Occurrence for the Hawaiian Islands*, Technical Report H-77-16, US Army Waterways Experiment Station, Vicksburg, Miss.

—— and Tsai, F.Y. (1983) 'Tsunami hazards in coastal zones', in *Proc. Coastal Zone '83*, Am. Soc. Civ. Eng., New York, 2024-36.

Hubbard, J.C.E. and Stebbings, R.E. (1967) 'Distribution, date of origin and acreage of *Spartina townsendii* (S.L.) marshes in Great Britain', *Proc. Bot. Soc. Br. Is., 7*, 1-7.

Islam, M.A. (1974) 'Tropical cyclones: coastal Bangladesh', in G.F. White (ed), *Natural Hazards: Local National and Global*, Oxford University Press, New York, 19-25.

Jackson, T.C. and Reische, D. (1981) *Coast Alert: Scientists Speak Out*, Coast Alliance/Friends of the Earth, Washington, DC.

Johns, B., Sinha, P.C., Dube, S.K., Mohanty, U.C. and Rao, A.D. (1983) 'Simulation of storm surges using a three-dimensional numerical model: an application to the 1977 Andhra cyclone', *Quart. J. Roy. Met. Soc., 109*, 211-24.

Kates, R.W. (1978) *Risk Assessment of Environmental Hazard*, SCOPE/Wiley, Chichester and New York.

Kaufman, W. and Pilkey, O.H. (1980) *The Beaches are Moving*, Anchor Press, Doubleday, New York.

Knutson, P. (1977), *Planting Guidelines for Marsh Development and Bank Stabilization*, US Army Corps Eng., Fort Belvoir, Washington, DC.

Koestner, M. (1984) 'Introduction to the Delta case studies', *Wat. Sci. Tech., 16*, 1-9.

Kuhn, G.G. and Shepard, F.P. (1983) 'Beach processes and sea cliff erosion in San Diego County, California', in P. Komar (ed), *Handbook of Coastal Processes and Erosion*, CRC Press, Boca Raton, Fla, 267-84.

Lamb, H.H. (1982), *Climate, History and the Modern World*, Methuen, London.

495

Leatherman, S.P. (1979) 'Barrier dune systems: a reassessment', *Sed. Geol.*, 24, 1-16.

—— (1981) 'Barrier beach development: a perspective on the problem', *Shore. Beach*, 49, 2-9.

MacLeish, W.H. (1980) 'Our barrier islands are the key issue in 1980, the "Year of the Coast"', *Smithsonian*, 11, 46-59.

Maloney, F.E. and O'Donnell, A.J., Jr (1976) 'Drawing the line at the ocean front: the role of coastal construction setback lines in regulating development in the coastal zone', *Florida Law Rev.*, 30, 383-404.

Mills, J.L., Woodson, R.D. and Solomons, E.K. (1976) *Compilation of Laws Relating to Florida Coastal Zone Management*, Department of Natural Resources, Tallahasee, Fla.

Mitchell, J.K. (1974) *Community Response to Coastal Erosion*, Department of Geography Research Paper No. 156, University of Chicago, Chicago, Ill.

Monday, J. (ed) (1984) *Preventing Coastal Flood Disasters*, Nat. Haz. Res. Centre, Denver, Col.

Mörner, N.-A. (1976) 'Eustacy and geoid changes', *J. Geol.*, 84, 123-52.

Murty, T.S. (1984) *Storm Surges — Ocean Meterological Tides*, Bulletin No. 212, Dept. of Fisheries and Oceans, Ottawa.

Newman, W.S., Marcus, L.F., Pardi, R.R., Paccione, J.A. and Tomecek, S.M. (1980) 'Eustasy and deformation of the geoid; 100-6000 radiocarbon years BP', in N.-A. Mörner, (ed), *Earth Rheology, Isostasy and Eustasy*, Wiley, Chichester and New York, 555-67.

Nordstrom, K.F. (1977) *The Coastal Geomorphology of New Jersey I. Management Techniques and Management Strategies*, Center for Coastal and Environmental Studies, Technical Report No. 1-77, Rutgers University, New Brunswick, NJ.

—— and McCluskey, J.M. (1984) 'Considerations for the control of house construction in coastal dunes', *Coast. Zone Mang. J.*, 12, 385-402.

—— and McCluskey, J.M. (1985) 'The effects of houses, and sand fences on the eolian sediment budget at Fire Island, New York', *J. Coastal Res.*, 1, 39-46.

O'Brien, M.P. (1982) 'Saving the American Beach', *Shore. Beach*, 50 (2), p. 5.

—— and Johnson, J.W. (1980) 'Structures and sandy beaches', in *Proc. Coastal Zone '80*, Am. Soc. Civ. Eng., 2718-40.

Pannell, J.P.M. (1964) *An Illustrated History of Civil Engineering*, Nelson, London.

Park, F.D.R. (1969) 'Virginia Key — Key Biscayne beach nourishment project', *Shore. Beach*, 37, 32-5.

Passino, R. and Todisco, A. (1984) 'Environmental aspects connected with "high water" protection of Venice', *Wat. Sci. Tech.*, 16, 319-36.

Pilkey, O.H. (1981) 'Geologists, engineers and a rising sea-level', *Northeastern Geol.*, 3, 150-8.

—— and Evans, M. (1981) 'Rising sea: shifting shore', in T.C. Jackson and D. Reische, *Coast Alert: Scientists Speak Out*, Coast Alliance/Friends of the Earth, Washington, DC, 13-47.

Pilkey, O.H., Sp, Pilkey, W.D., Pilkey, O.H., Jr and Neal, W.J. (1983) *Coastal Design*, Van Nostrand Rheinhold, New York.

Pirazzoli, P. (1973) 'Inondations et niveaux marins à Venice', *Mem. Lab. Géomorph. EPHE, Dinard*, 22, 1-284.

—— (1983), 'Flooding ('Acqua Alta'), in Venice (Italy): a worsening phenomenon', in E.C.F. Bird and P. Fabbri (eds), *Coastal Problems of the Mediterranean Sea*, IGU-CCE, Bologna, 23-31.

Platt, R.H. (1978) 'Coastal hazards and national policy: a jury-rig approach', *Am. J. Plan.*, 44, 170-80.

Pollard, M. (1978) *North Sea Surge: The Story of the East Coast Floods of 1953*,

Daltons, Lavenham, Suffolk.

Prandle, D. and Wolf, J. (1978) 'The interaction of surge and tide in the North Sea and River Thames', *Geophys. J. Roy. Astron. Soc., 55*, 203-16.

Pugh, D.T. (1982) 'A comparison of recent and historical tides and mean sea-levels off Ireland', *Geophys. J. Roy. Astron. Soc., 71*, 809-15.

—— and Faull, H.E. (1983) 'Tides, surges and mean sea level trends', in *Shoreline Protection*, Thomas Telford, London, 59-69.

Purpura, J.A. (1972) 'Establishment of a coastal setback line in Florida', in *Proc. Thirteenth Inst. Eng. Conf., Florida*, 1599-615.

Race, M.F. and Christie, D.R. (1982) 'Artificial creation of marsh ecosystems', *Env. Management, 6*, 317-28.

Ranwell, D.S. (1967) 'World resources of *Spartina townsendii* (Sensu lato), and economic use of *Spartina* marshland', *J. Appl. Ecol., 4*, 239-56.

Rennie, J. (1845) Presidential address, *Min. Proc. Inst. Civ. Eng. Lond., 4*, 23-5.

Resio, D. and Hayden, B.P. (1975) 'Recent secular variations in mid-Atlantic winter extratropical storm climate', *J. Appl. Meteor., 14*, 1223-34.

Rohde, H. (1980) 'Changes in sea level in the German Bight', *Geophys. J. Roy. Astron. Soc., 62*, 291-302.

Rossiter, J.R. and Lennon, G.W. (1965) 'Computation of tidal conditions in the Thames Estuary by the initial value method', *Proc. Inst. Civ. Eng. Lond., 31*, 25-56.

Saejis, H.L.F. (1982) *Changing Estuaries*, Rijkswaterstaat Communications No. 32, The Hague.

Schwartz, M.L. and Milicic, V. (1980) 'The Bruun Rule: an historical perspective', in J.J. Fisher and M.L. Schwartz (eds), *Proc. Per Bruun Symp., IGU Comm. on the Coastal Environment*, Western Washington University, Bellingham, Wash., 6-12.

Shaw, T.L. (ed) (1980) *An Environmental Analysis of Tidal Power Stations: With Particular Reference to the Severn Barrage*, Pitman, London.

Simpson, R.H. and Reidl, H. (1981) *The Hurricane and Its Impact*, Blackwell, Oxford.

Sorensen, H.B. (1983) 'Administrative practices concerning shoreline protection in Denmark', in *Shoreline Protection*, Thomas Telford, London, 21-5.

Southern, R.L. (1979) 'The global socio-economic impact of tropical cyclones', *Aust. Meteor. Mag., 27*, 175-95.

Stone, G.W. and Morgan, J.P. (1984) 'The concept of storm wave susceptibility and its application to four Florida coastal areas', in L.S. Tait (ed), *The New Threat to Beach Preservation*, Tallahassee, Florida Shore and Beach Preservation Association, Tallahassee, Fla, 19-36.

Straub, H. (1952) *A History of Civil Engineering*, Leonard Hill, London.

Thom, B.G. (1978) 'Coastal sand deposition in southeast Australia during the Holocene', in J.L. Davies and M.A.J. Williams (eds), *Landform Evolution in Australia*, Australian National University Press, Canberra, 197-214.

Trafford, B.D. and Braybrooks, R.J.E. (1983) 'The background to shoreline protection in Great Britain', in *Shoreline Protection*, Thomas Telford, London, 1-8.

Tooley, M.J. (1979) 'Sea-level change during the Flandrian Stage and the implications for coastal development', in K. Suguio, T.R. Fairchild, L. Martin and J.-M. Flexor (eds), *1978 Int. Symp. on Coastal Evol. in the Quat.*, São Paulo, Brazil, 502-33.

Verhulst, A. and Gottschalk, M.K.E. (1980) *Transgressies en Occupatiegeschiedenis in de Kustgebeieden Van Nederland en Belgie*, Pub. No. 66, Cent. Land. Gesch., Ghent.

Wanless, H.R. (1982) 'Sea-level is rising, so what?', *J. Sed. Petrol., 25*, 1051-4.

Whipple, A.B.C. (1984) *Storm,* Time-Life Inc., Amsterdam.

White, A. (1974) 'Global summary of human response to natural hazards: tropical cyclones', in G.F. White (ed), *Natural Hazards: Local National and Global,* Oxford University Press, New York, 255-65.

Willmington, R.H. (1983) 'The renourishment of Bournemouth beaches 1974-1975', in *Shoreline Protection,* Thomas Telford, London, 157-62.

Woodhouse, W.W., Jr (1982) 'Coastal sand dunes of the U.S.', in R.R. Lewis (ed) *Creation and Restoration of Coastal Plant Communities,* CRC Press, Boca Raton, Fla, 31-60.

World Meteorological Organisation (WMO) (1983) *Human Response to Tropical Cyclone Warnings and their Content,* Report No. 12, WMO, Geneva.

Zedler, J.B. (1983) 'Salt marsh restoration: the experimental approach', in *Proc. Coastal Zone '83,* Am. Soc. Civ. Eng., 2578-86.

15

The Greenhouse Effect, Rising Sea Level and Society's Response

J.G. Titus

INTRODUCTION

The previous chapters have described various studies of current sea-surface processes. We now apply some of our knowledge in these fields to paint a picture of the likely consequences of an 'experiment' that humanity is conducting on the earth's climate.

Increasing concentrations of atmospheric carbon dioxide, methane, nitrous oxide, chlorofluorocarbons and other gases are expected to warm our planet a few degrees in the next century by a mechanism commonly known as the 'greenhouse effect'. Such a warming would alter atmospheric and oceanic circulation, and raise sea level about one metre in the next century and perhaps several metres in the next two centuries.

Although the most significant consequences are unlikely to occur before the middle of the twenty-first century, decisions in the next few decades will largely determine both the magnitude of the 'greenhouse warming' and how prepared society is to cope with it. These decisions will depend on the adequacy of our understanding of the basic processes involved and on how well scientists convey to decision makers and the general public what they learn.

This chapter examines the basis for expecting a global warming and accelerating rise in sea level, the likely impacts, possible responses and the time constraints society faces.

THE GREENHOUSE WARMING

Since the latter part of the nineteenth century, scientists have known that the carbon dioxide and water vapour in the atmosphere keeps the earth much warmer than it would otherwise be (Arrhenius, 1896, 1908). These gases allow sunlight to penetrate, but absorb outgoing infrared radiation, warming the atmosphere. Svante Arrhenius originally coined the term 'greenhouse effect' at the turn of the century. Technically, the glass panels of a greenhouse prevent convectional cooling while the 'greenhouse gases' prevent radiational cooling; nevertheless, the analogy is close enough to convey the basic idea.

In spite of the general acceptance of the greenhouse effect, the possibility that humanity would alter the worldwide concentration of greenhouse gases enough to cause significant consequences was little more than speculation during the first half of the twentieth century. Arrhenius noted that the combustion of fossil fuels might one day lead to a doubling of atmospheric carbon dioxide (CO_2), which he estimated would raise the earth's average surface temperature 4 to 6° C. However, Barrell et al. (1919) expressed concern that depletion of natural sources might eventually reduce the CO_2 concentration enough to cool the earth substantially.

Given the unusually cold weather that had characterised the nineteenth century, it is not surprising that the possibility of an anthropogenic global warming did not generate substantial interest. When global temperatures warmed in the 1930s (Fig. 15.1), with noticeable adverse impacts on agriculture such as the 'dust bowl' in the United States, some people again suggested that rising CO_2 could be responsible (Callendar, 1940). However, no one had set up monitoring stations in areas sufficiently remote to approximate background concentrations of carbon dioxide. The prevailing view was that the oceans, which contain sixty times as much CO_2 as the atmosphere, would absorb anthropogenic CO_2 as rapidly as human activities released it.

In 1957, however, Roger Revelle and Hans Seuss demonstrated that the upper layers of the ocean would not take up CO_2 as rapidly as had been thought. Rather, mankind was embarking unconsciously on a grand geophysical experiment whose results could be profound and largely irreversible. Shortly thereafter, as part of the International Geophysical Year, monitoring stations for CO_2 were set up at Mauna Loa, Hawaii and the South Pole. We now know that CO_2 concentrations have increased from 315 parts per million (ppm) in 1958 to 340 ppm in 1980 (Keeling et al., 1982). Studies of ice cores and tree rings have revealed that the concentration was probably between 260 and 280 ppm in the midde of the nineteenth century (Keeling, 1978; Neftel et al., 1982).

Carbon cycle modellers and energy economists generally expect atmospheric CO_2 to double around the middle of the twenty-first century (Nordhaus and Yohe, 1983). These projections imply that annual emissions will rise 1 or 2 per cent per year, compared with the 4 per cent annual increase that took place until the OPEC oil embargo of 1973. While mature economies are expected to use energy more efficiently and derive an increasing share from non-fossil fuel sources, most developing nations are expected to increase per capita consumption as their economies more closely approximate the patterns of the developed world. None of the

Figure 15.1: Global temperature trend for (a) the past century, (b) the millennium and (c) 25,000 years

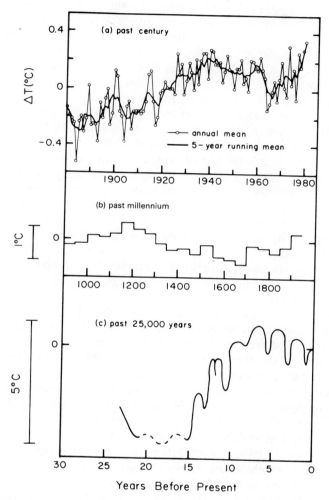

Source: Barth and Titus (1984).

assessments assume that the nations of the world take measures deliberately to reduce use of fossil fuels to avert a greenhouse warming.

In the last decade we have learned that humanity is increasing the concentrations of other gases that absorb infrared radiation. Although these gases have far lower atmospheric concentrations than carbon dioxide, they are often more efficient absorbers of infrared radiation than CO_2, and absorb energy in different parts of the spectrum. Ramanathan *et al.* (1985)

evaluated dozens of minor greenhouse gases and concluded that their combined impact is likely to be as significant as carbon dioxide.

Chlorofluorocarbons (CFCs) received substantial attention in the 1970s because of their potential impacts on stratospheric ozone. These gases are used widely in refrigeration, insulation and packaging foam, dry cleaning, and electronics (Palmer *et al.*, 1980). Their biggest use has been as aerosol propellants (Tukey *et al.*, 1979), a use that the United States and a few other countries have banned. Japan, the Soviet Union, and most European countries, however, continue to use CFCs as propellants.

Methane concentrations are also increasing, but the sources are less clear. Wetlands, rice paddies, feedlots, and some industrial processes release significant quantities (Rasmussen and Khalil, 1981). Little has been done to estimate the relative contributions of each source. Moreover, it is possible that these sources are not the main cause of increased concentrations. Unlike carbon dioxide, methane degrades photochemically in a matter of decades. The rate of methane's oxidation, however, depends in part on the abundance of OH^- in the atmosphere. Emissions of carbon monoxide appear to have reduced the atmospheric abundance of OH^- (Levine *et al.*, 1985).

As the consensus emerged that concentrations of greenhouse gases are increasing, scientific inquiry shifted to examining the likely impacts on climate. Such assessments have fallen broadly into two categories: palaeoclimatology (the study of past climates) and mathematical climate modelling. Both of these fields have important limitations. It is not clear whether the concentration of greenhouse gases has ever increased as rapidly as the next century may see; there may be no direct historical analogue. Moreover, if one could demonstrate such an increase, there would be difficulties in separating the cause from the effect and in distinguishing short- and long-term impacts. Climate models are limited in that they can only approximate the climate and consider only the impacts that are known. Moreover, today's models generally use grid cells 500 to 1,000 kilometres wide, preventing a refined specification of even the processes that are understood. Nevertheless, the combined insights from physics, palaeoclimatology and climate modelling do permit some important inferences. Hansen *et al.* (1984) showed that a doubling of atmospheric carbon dioxide would warm the earth 1.2°C if nothing else changed. They showed that their model adequately explains the difference in temperatures between Earth, Venus (whose atmosphere is 97 per cent CO_2 but whose cloudy atmosphere gives it a high albedo) and Mars (whose thin atmosphere has a negligible greenhouse effect).

Although climatologists generally agree on the direct impact of a doubling of CO_2, the resulting warming could be expected to alter other aspects of the climate. Most assessments have concluded that these 'climatic feedbacks' would amplify the direct warming from increased CO_2 (Nierenberg *et al.*, 1983). Figure 15.2 illustrates estimates by Hansen *et al.* (1984) of the direct impact and scale of these feedbacks. A warmer atmosphere would hold more water vapour, which is also a greenhouse gas, and would thus cause additional warming. Retreating snow and sea ice coverage would reduce the reflectivity of the earth's surface, causing additional warming. The impacts on clouds are poorly understood. Although some assessments suggest that a reduction in cloud cover would be likely (Hansen *et al.*, 1984), this is a tentative finding. Changes in cloud height could also be a factor. One can not rule out the possibility that the clouds would at least partly offset the greenhouse warming.

Figure 15.2: Estimated global warming due to a doubling of greenhouse gases: direct effects and climate feedbacks. Although Hansen *et al.* (1984) estimate a positive feedback from the clouds, a negative feedback cannot be ruled out

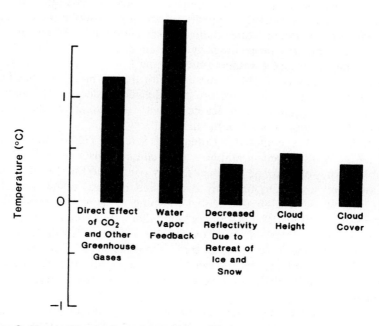

Source: Goddard Institute for Space Studies, New York.

The geological record provides some evidence that climatic feedbacks in the past have tended to amplify the direct impact of radiative changes. It is widely believed that cyclic changes in the shape of the earth's orbit and in the tilt of its axis have resulted in cyclic fluctuations in the amount of solar radiation absorbed by the earth's surface. Hayes *et al.* (1976) suggested that

503

these fluctuations are the driving force behind the cycles of ice expansion and retreat (i.e. ice ages) that have characterised the earth during the last two to three million years.

Fluctuations in sunlight may be the closest analogue to externally generated increases in greenhouse gases, in that both alter the earth's radiative balance. This evidence appears to support the belief that on balance, the climate amplifies global temperature changes caused by other factors. The changes in sunlight have been small, yet the changes in temperatures have been about 5°C (NAS, 1975; Hansen *et al.*, 1984). Because most of the glaciers have vanished, the reflectivity feedback would clearly be less in the future than it has been in the past. Unfortunately, there is little prospect for securing accurate data on past cloud cover. However, Hansen *et al.* (1984) show that the increase in temperatures resulting from the retreating ice cover in the last 18,000 years is consistent with the feedback estimated by their model, which shows clouds to have a positive feedback.

Although our ability to project climate or explain past changes is limited, the current state of knowledge appears sufficient to describe the evolution of global average temperature changes over moderate periods of time. Figure 15.3 compares projections by Hansen *et al.* (1981) with observations of the mean global temperature for the last century. The warming expected from CO_2 is roughly consistent with the warming that has taken place, although it hardly explains the decade-to-decade fluctuations. However, when changes in volcanic activity and solar irradiation are considered, the model fits the data fairly well.

In 1979, the US National Academy of Sciences (NAS) convened a panel of geologists, oceanographers and climatologists to evaluate the available studies on climatic feedbacks. They concluded that 'we have tried but have been unable to find any overlooked physical effect that could reduce the currently estimated global warming to negligible proportions' (Charney *et al.*, 1979). The NAS concluded that a doubling of atmospheric carbon dioxide would result in an eventual warming of 1.5 to 4.5°C, with the warming at the poles two to three times as great. Because of the thermal delay introduced by the oceans they suggested that the warming would lag a couple of decades behind the increases in atmospheric gas concentrations.

Given the many uncertainties regarding future economic growth, energy sources and consumption patterns, emissions of other greenhouse gases, the gases' residence times in the atmosphere, the various climatic feedbacks and the extent to which oceanic heat absorption will delay temperature increase, it is impossible to state with any certainty how rapidly the earth will warm in the next century. Moreover, unpredictable fluctuations in volcanic activity and solar irradiations could temporarily offset or amplify the warming. Nevertheless, Hoffman *et al.* (1983) and Hoffman *et al.* (1985) have formulated scenarios of future temperature increases. These

504

Figure 15.3: Global temperature trend. See text for explanation

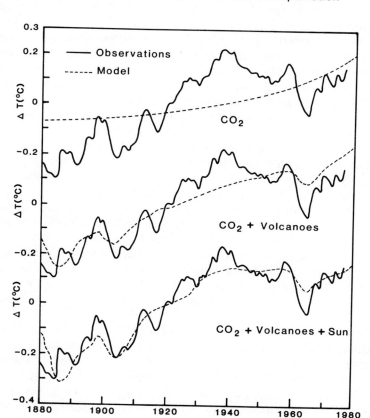

Source: Hansen *et al.* (1981).

scenarios have been developed by determining upper and lower bounds for all of the major uncertainties and calculating the implications. On the basis of these assumptions Hoffman *et al.* (1985) estimate that the global warming (over 1980 levels) will be 0.6 to 1.2°C by 2025; 1:2 to 2.6°C by 2050; and 2.3 to 7.0°C by 2100. They also estimate '*unrealised warming*', the eventual warming that would result from gases emitted by a particular year once the oceans have come into equilibrium. By 2050, they estimate unrealised warming of 0.8 to 4.6°C, implying that by 2050 the earth is likely to be locked into a warming of 2.5 to 9.0°C.

The estimates of unrealised warming generally suggest that society can only keep the global warming below a particular level by curtailing emissions long before temperatures reach this threshold. The need for anticipatory measures is even greater when one realises that there is also social inertia. Even a commitment to curtail emissions of greenhouse gases would take decades to implement and would reduce but not eliminate emissions. Seidel and Keyes (1983), for example, suggest that even if society decided to reduce greenhouse gas emissions today, the most that could be hoped for would be equivalent to banning shale oil, synthetic fuels and coal, i.e. cutting total emissions in half, by the year 2000. Even such a drastic policy would only delay the warming expected through to 2040 by 25 years and would result in a warming of over three degrees by 2100.

Society will probably wait for observations to confirm the greenhouse warming before implementing policies to slow emissions. Hansen *et al.* (1984) predict that the temperature signal will emerge from the noise by the year 2000. Other than bans on aerosol CFCs, limiting emissions would be so costly that it would take at least ten years to reach a worldwide agreement to do so. Unless future technological breakthroughs occur, drastic reductions in emissions would not be politically feasible until global warming before implementing policies to slow emission. Hansen *et al.* Even then, the poorest nations and those that benefit from a warmer climate may be reluctant to curtail emissions drastically without compensation.

Perhaps the most important impact of a global warming would result from changes in atmospheric and oceanic circulation. Because a warmer atmosphere can hold more water vapour, an increase in rainfall worldwide is generally expected. Nevertheless, this increase in rainfall may be more than offset by increased evaporation. Reductions in ice cover at high latitudes would enable polar areas to warm more than the equator, reducing latitudinal temperature differences. Because atmospheric and oceanic circulation are largely driven by these differences, an overall reduction in circulation could be expected. Climatologists generally expect increased desertfication in interior areas to result (Manabe *et al.*, 1981). However, it is not possible to predict whether particular areas will have increased or decreased water availability.

SEA-LEVEL RISE

One impact that can be projected is the rise in sea level. Unlike precipitation, sea level in a particular area is largely determined by global and latitudinal averages. In particular, increases in the average ocean temperature and in the mass of the oceans would raise sea level everywhere (see chapters in Part One).

Although sea level can be measured relative to the coast in particular areas, there is no absolute measure of worldwide sea level (Peltier, this vol., Ch. 3). The best proxy thus far has been to take weighted averages from all the tide gauges in the world. Numerous researchers have concluded that this measure of global sea level has risen 10 to 15 cm in the last century (Barnett, 1984; Gornitz *et al.*, 1982). Figure 15.4 illustrates the estimate of worldwide sea level by Gornitz *et al.*, who also concluded that a 5 cm rise in the last century can be attributed to thermal expansion of the upper layers of the ocean resulting from the 0.4°C average global warming over that period (Hansen *et al.*, 1981). Meier (1984) estimates that mountain glaciers have retreated enough to raise sea level 2-7 cm per century since 1900 and could add 30 cm to sea level in the next century. The contribution of Greenland to sea level has been small but could be 15 to 40 cm by the year 2100 (Bindschadler, 1985). Antarctic contributions of water mass to the oceans could add as much as one metre to sea level in the next century, although some believe that the ice sheets have recently gained mass (Polar Research Board, 1985).

Figure 15.4: Sea-level trends and estimated trends due (1) solely to thermal expansion for various values of climate sensitivity to (2) a CO_2 doubling

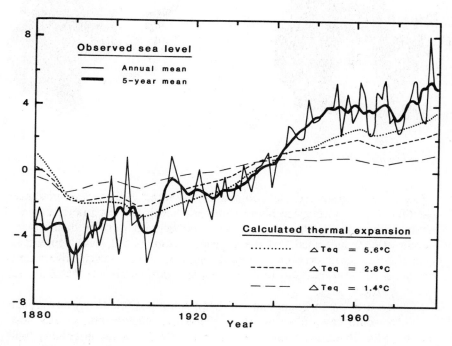

Source: Gornitz *et al.* (1982).

In the long run, perhaps the most ominous possibility is a disintegration of the West Antarctic Ice Sheet (Mercer, 1970; 1978). Although West Antarctica has only 10 per cent of the continent's ice, it is more vulnerable to a global warming than East Antarctica because the West Antarctic Ice Sheet rests on the bottom of the ocean, not on land (Fig. 15.5). This ice sheet is buttressed by floating ice shelves, without which the ice sheet would slide into the ocean. Glaciologists have estimated that the ice shelves could melt in about a century with the projected global warming (Thomas *et al.*, 1979) and that the ice sheet would take another two (Hughes, 1983) to five (Bentley, 1983) centuries to completely disintegrate, which would raise global sea level about 6 metres. Differentially higher figures for sea level rise would be recorded in many regions in the short term (Clark and Lingle, 1977).

Figure 15.5: (a) Antarctic ice cover today and (b) after a 5–10°C warming

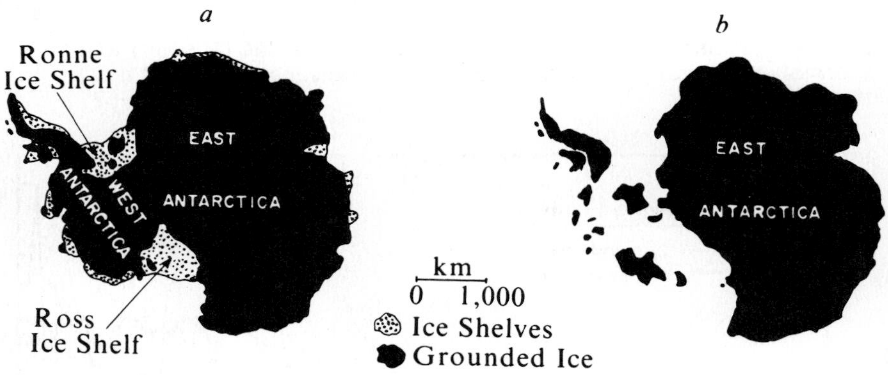

Source: Mercer (1978).

Coastal geologists and glaciologists have offered independent evidence that the West Antarctic Ice Sheet vanished during the last Interglacial (~ 120,000 years ago), which was one or two degrees warmer than today. Mercer (1970) suggested that the ice sheet disintegrated, based on fossils from areas under its present location. Several coastal geologists have found shorelines 100,000 years old about 6 m above today's sea level (Marshall and Thom, 1976).

Glaciers are so poorly understood that it is not possible to say how much the earth would have to warm for an irreversible deglaciation of the West Antarctic Ice Sheet to commence; one can not rule out the possibility that even the one degree warming expected by 2030 would initiate such a deglaciation. Fortunately, it is generally agreed that even if the process is set in motion soon, it will take several centuries (Thomas *et al.*, 1979). Thermal expansion of ocean water, the melting of Greenlandic and mountain glaciers and perhaps some meltwater around the fringes of Antarctic glaciers, are likely to be the main mechanisms by which the greenhouse warming raises the sea in the next century.

To provide decision makers with rough estimates of future sea-level rise, Hoffman *et al.* (1983) developed a variety of scenarios of how much the sea would rise for particular assumptions, illustrated in Figure 15.6. Unfortunately, they were only able adequately to assess thermal expansion and had to adjust the resulting estimate in a crude fashion to account for glacial contributions. Figure 15.7 and Table 15.1 illustrate the projections of Hoffman *et al.* (1983) and a subsequent estimate by Revelle (1983). The former estimated that thermal expansion is likely to raise sea level 28 to 115 cm by 2100, with 72 cm most likely. Considering all contributions to sea level, they estimated that global sea level will rise 13 to 55 cm over 1980 levels by the year 2025 and 56 to 345 cm by 2100. Revelle estimated

Figure 15.6: Factors considered in projecting future sea-level rise

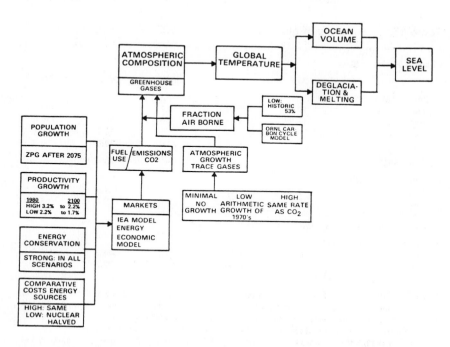

Source: US Environmental Protection Agency.

Figure 15.7: Global sea-level rise scenarios: low, mid-range low, mid-range high, and high

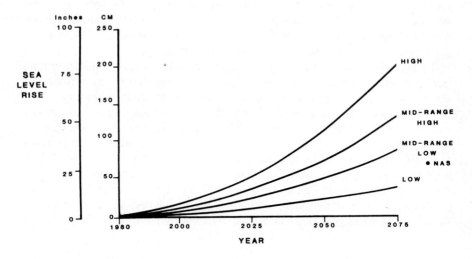

Source: Hoffman *et al.* (1983).

Table 15.1: Estimates of global sea-level rise 1980–2100 (centimetres)

Absolute rise over 1980

	2000	2025	2050	2075	2100
Hoffman *et al.* (1983)[a]					
high	17.1	54.9	116.7	212.7	345.3
mid–high	13.2	39.3	78.6	136.8	216.6
mid–low	8.8	26.2	52.3	91.2	144.4
low	4.8	13.0	23.8	38.0	56.2
Revelle (1983)	—	—	—	70[b]	—
Hoffman *et al.* (1985)					
high	5.5	20.7	54.7	191.9	367.2
low	3.5	10.1	20.5	36.4	58.7

Rate of sea-level rise (centimetres per century)

	2000	2025	2050	2075	2100
Hoffman *et al.* (1985)					
high	37.4	85.0	266.0	646.7	740.0
low	20.5	32.6	49.7	77.2	102.6

Notes: [a]Hoffman *et al.* (1983) concluded that the high and low scenarios are unlikely but cannot be ruled out. The low and mid–low scenarios estimate thermal expansion and assume the total rise will be twice that estimate; the high and mid–high scenarios estimate thermal expansion and assume the total rise would be three times as great.
[b]Revelle (1983) estimates a 70 cm rise by 2080 assuming no Antarctic ice discharge. He suggests that Antarctica could add 2 m a^{-100} after 2050.

that thermal expansion would raise sea level 30 cm, that melting of Greenlandic and mountain glaciers would each add 12 cm and that factors responsible for current trends would continue, accounting for a total rise of 70 cm.

A major deficiency of both reports was that glacial process models had not been developed to estimate the contribution of particular ice fields to sea-level rise. Recently, the NAS Polar Research Board (1985) has provided assessments for the contribution of particular glaciers. Using their results and other new information Hoffman *et al.* (1985) formulated new scenarios of future sea-level rise. As Table 15.1 shows, the new information on glaciers revealed that the crude assumptions in their 1983 report had overstated ice contributions in the next fifty years, but was consistent with the expected contribution through to 2100. The new scenarios imply a rise of 10 to 21 cm by 2025 and 59 to 367 cm by 2100.

Although global sea-level rise is most closely related to climate, decision makers must consider the rise relative to a particular location (see Part One). Assessments of the impacts of a rise in sea level due to the greenhouse effect have generally been added to current trends in land subsidence or uplift, generally 15-20 cm of subsidence in the US. Changes in currents, prevailing winds and storminess, as well as crustal warping due to redistribution of mass away from polar areas, may also cause the local impact in some areas to differ significantly from the global average. These factors have not, however, been estimated as yet in a year-by-year form, necessary for decision making.

IMPACTS OF FUTURE SEA-LEVEL RISE ON HUMAN ACTIVITIES AND THE ENVIRONMENT

A rise in sea level of one or two metres would inundate low-lying areas and drown coastal wetlands, erode shorelines by hundreds of metres, increase coastal flooding and raise the salinity of rivers, bays, and aquifers in coastal areas (Barth and Titus, 1984; see discussion in Orford, Ch. 13, Carter, Ch. 14 and Lewis, Ch. 18. From work on Pauanui Beach, New Zealand, on coastal erosion, Gibb and Aburn (1986) have estimated that a future CO_2 induced rise in sea level would turn a beach now in dynamic equilibrium into a zone of long-term erosion. They estimated a rise in global sea level of 70 cm in the next 100 years would cause loss of property and damage amounting to NZ$17 M. On a world scale, if the rise in sea level took place overnight, it would undoubtedly cause severe dislocations for both ecosystems and human infrastructures. If a planned retreat took place, those dislocations could be avoided, although everything would be displaced inland somewhat. Coastal communities will have to weigh the costs of a retreat against the costs of maintaining current locations.

511

This section summarises previous studies of the potential impacts of sea-level rise with an eye toward identifying decisions made today whose outcomes could be influenced by consideration of this issue. The next section will examine some of the implications of those studies. Although the examples will be based on conditions in the United States, the general principles are widely applicable.

Wetlands

By definition, the elevation of coastal wetlands is generally less than the highest tide in a given month. Because the tidal range in most US estuaries is of the order of one metre, a rise in sea level of one metre would probably destroy about three-quarters of the coastal wetlands in that country. Although many nations are in areas with higher average tidal ranges (see Devoy, Fig. 10.1), their wetlands would be similarly threatened by such a rise in sea level. The marshes and swamps of Louisiana (USA) are already converting to open water as a result of a combination of human modification of the Mississippi River and the deltaic plain's current subsidence of 1 m per century (Gagliano *et al.*, 1981). Many of the world's other river deltas are rapidly eroding because the rivers that created them have been dammed or leveed (Milliman *et al.*, 1984). A rise in sea level of 1 m would dramatically accelerate erosion processes in Louisiana, Bangladesh, Egypt, Columbia, and other deltas (see Carter, this vol., Ch. 14).

For the last several thousand years, sea level has risen slowly enough for the vertical growth of marshes to keep pace throughout most of the United States, as shown in Figure 15.8. As the sea rose new marsh was formed inland, but the seaward edge of the marsh did not generally retreat as rapidly and often advanced. Thus, in most coastal areas, there is far more land at marsh elevation than immediately above the marsh. However, if the sea rises more rapidly than the marsh's ability to keep pace, the marsh would retreat at the seaward boundary more than it encroaches inland, resulting in 50-90 per cent of the marsh being lost in the typical US estuary (Titus, 1985; Kana *et al.*, 1984). Moreover, the loss could be 100 per cent if there is no vacant land for new marsh to form. In many areas of the United States, environmental laws prevent development on coastal marshes; but areas just inland of the marsh are being developed extensively. Ensuring that there is sufficient land for new marsh to form will require either placing some areas off-limits to development, initiating long-term plans soon to retreat from these areas if the sea rises a certain amount, or initiating short-term abandonment plans later.

Adopting any of these coastal management approaches would be difficult. To deprive property owners of the right to build because of environmental problems predicted 50-100 years hence would effectively require

512

Figure 15.8: Coastal marshes have kept pace with the slow rate of sea-level rise that has characterised the last several thousand years. Thus, the area of marsh has expanded over time as new lands were inundated. If in the future, sea level rises faster than the ability of the marsh to keep pace, the marsh area will contract. Construction of bulkheads to protect economic development may prevent new marsh from forming and result in a total loss of marsh in some areas

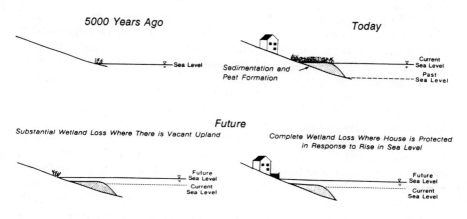

Source: US Environmental Protection Agency.

purchasing the land in most cases, which would be costly. Long-term plans for abandonment would not impose substantial costs in the short term, but they would require unprecedented foresight to enact and would require eternal vigilance to enforce. If the issue is not addressed until, say, 2025 when the sea has risen 30 cm and a 1 m rise is on the horizon, the impact of such a plan on property values could be prohibitive. Nevertheless, there may be strategic opportunities for making use of information on the possibility of a rise in sea level. For example, some developers might agree to a requirement that particular areas revert to wetland one hundred years hence, if the sea rises a metre or two, in return for the right to develop now a few acres of existing wetlands for marinas — particularly developers who do not believe the sea will rise or are unconcerned about the distant future! Environmentalists who expect the sea to rise and are more interested in the long-term survival of particular ecosystems than with the short-term maintenance of a particular number of acres would also view such a trade-off as acceptable (Titus, 1984).

513

Beach erosion and coastal resorts

Some geologists have suggested that rising sea level is the driving force for erosion along most of the world's coasts (Bird, 1976; Pilkey *et al.*, 1981). The Danish engineer Per Bruun first showed that a rise in sea level would cause shorelines to retreat by considerably more than the area that was directly inundated (Bruun 1962; Bruun 1983). Bruun showed that rising sea level upsets the balance between erosion from storms and the natural rebuilding of the beach that generally takes place during calm periods. If there was no erosion, shoreline retreat would be inversely proportional to the slope immediately above the original water level. The 'Bruun Rule' shows that when erosion from sea-level rise is considered, shoreline retreat is inversely proportional to the slope of the entire beach profile. For further detail readers are referred to the in-depth discussion by Orford (Chapter 13) of this and related shoreline concepts.

Using the Bruun Rule, the erosion from a 1 cm rise in sea level would be 0.5-1 m along the New Jersey (Kyper, 1985) and Maryland (Everts, 1985) coasts, 1-2 m along the South Carolina coast (Kana *et al.*, 1984), 2-4 m along the San Francisco coast (US Army Corps of Engineers, 1979) and up to 10 m along the Florida coast (Bruun, 1962). Because most resorts have housing within 30 m of the shore, even the 30 cm rise projected for the next 40 years along the US coast could be important.

As discussed in Chapter 14 by Carter, coastal communities could respond to a one metre rise in sea level by building dykes and seawalls, raising the land surface, or retreating (Sorensen *et al.*, 1984). Because levees and seawalls do not prevent a beach from eroding, this option would not be viable for resort communities. The decision whether to retreat or raise the land surface would be largely a matter of economics. Retreat would be most likely in lightly developed areas. Two states with lightly developed shores, North Carolina and Maine, already have several policies that encourage retreat. Both have outlawed structural erosion control measures. North Carolina requires immovable structures to be set back from the dune line by 60 years of erosion and smaller houses to be set back by 30 years of erosion, with the assumption that the house will be moved back in the future. Maine requires developers to demonstrate that a house will still be standing one hundred years hence.

States with heavily developed coasts base land use decisions on the assumption that the shore will be stable. Many coastal cities have relied primarily upon dykes, bulkheads and seawalls for protection. The Dutch have long found it worthwhile to protect even agricultural lands from the sea using such techniques. In the United States it will continue to be cost-effective to protect many urban areas, such as New Orleans ($\sim$ 2-3 m below sea level) in this fashion (Sorensen *et al.*, 1984). However, for much of the Atlantic coast of both North and South America, the coastal economy is based on recreational beaches, so seawalls are unacceptable. In the past, resorts have generally relied on both beach nourishment and the construction of groynes for protection. However, with a more rapidly rising sea level in the future, it will be necessary to add increasing amounts of sand to the beach and the costs will escalate. Moreover, it will not be sufficient to nourish the beach: entire barrier islands will have to be raised in place. Nevertheless, it will continue to be cost-effective to defend the shore in many of the most intensely developed resorts. Ocean City, Maryland, has seven million visitors per year. Their projected $2 million annual cost for coastal protection schemes would only amount to about 25 cents (US) for every visitor (Titus et al., 1985).

Ocean City provides a rare example of the risk of erosion from an accelerated rate of sea-level rise influencing practical decision making. Because the beach is currently eroding at 0.5 m a^{-1}, due to sand moving south along the shore, the state decided in the 1970s to construct a series of groynes to trap sand and slow erosion. However, a study of the implications of an accelerated sea-level rise (Titus et al., 1985) and a series of newspaper articles (Pilkey, 1985; Peters, 1985) pointed out that the groynes would not curtail the erosion caused by sea-level rise. Shortly thereafter, the state's governor decided that the state would instead rebuild the beach by pumping in sand from offshore, which in the short term offsets erosion caused by sea-level rise as well as by alongshore transport (Associated Press, 1985).

Flooding

Although the impacts of sea-level rise on shorelines has received the most attention, the impact on flooding could be the most economically important (Barth and Titus, 1984). The higher sea level would provide a higher base for storm surges to build upon (see Carter, this vol., Ch. 14). Erosion would leave some properties more vulnerable to direct attack from ocean waves. Finally, higher sea level would slow natural and artificial drainage during rainstorms. Relatively little work has been done to estimate the impact such sea-level rise would have on flooding in particular areas. At current rates of sea-level rise, this is understandable. An extra centimetre of flooding will increase damage, but usually not in a way that warrants conceptually different treatment from the current flood alleviation protection activities. An important exception is the Thames River Barrier, whose design includes an allowance for sea-level rise.

The more substantial rates of sea-level rise expected in the future and discussed earlier would warrant more widespread attention. Kana et al. (1984) estimated that a 1.6 m rise would convert the 100-year flood plain in the area of Charleston, South Carolina to a 10-year flood plain. Gibbs

(1984) estimated the resulting economic impacts through to 2075 at $2 billion. However, he also estimated that the impact could be reduced by 60 per cent if anticipatory measures were implemented such as building on higher ground and designing coastal protection structures for higher sea levels. Leatherman (1984) estimated that a 1.6 m rise in sea level would enable a 100-year storm to overtop the Galveston seawall and otherwise disrupt the Texas coast. Gibbs suggested that the area would realise substantial savings by extending the ocean side seawall to completely encircle Galveston Island.

Communities are not likely to begin erecting levees or altering land use plans until the rate of sea-level rise has accelerated notably; major public works and changes in land use are controversial enough. It may be possible, however, for those who design coastal structures to factor the risk of accelerated sea-level rise in design decisions. La Roche and Webb (in press) examined the impact of a 30 cm sea-level rise on the drainage system for a watershed in Charleston. The city plans to overhaul this system, which is over one hundred years old. The cost of bringing the system up to standard would be $4.81 million for today's sea level and $5.07 million for a 30 cm rise. If the system is built for today's sea level and the sea rises 30 cm, the system would have to be overhauled at a cost of $2.41 million. Thus, for an additional expenditure of $260,000, the city could insure itself against the risk of a second overhaul at nine times the cost. The wisdom of such insurance depends, of course, on how one assesses the probability of future sea-level rise and the importance of the system functioning in the future (see Barth and Titus, 1984).

This approach to risk management may not be appropriate where potentially catastrophic or irreversible impacts are possible. Flynn et al. (1984) examined implications of sea-level rise for hazardous waste sites in the United States. Operating chemical disposal sites in flood plains are subject to strict regulations that would be tightened automatically as the sea, and consequently the groundwater table, rises. However, such sites can be closed without consideration of the possible impacts of future sea-level rise. Fortunately, nuclear waste disposal sites are generally more than 70 metres above sea level.

Saltwater intrusion

A rise in sea level would enable saltwater to advance inland in coastal aquifers and upstream in estuaries (Sorensen et al., 1984). This impact of a rise in sea level has been the least researched. The impact on coastal aquifers in the United States is probably dwarfed by the fact that many have been overpumped to the point where they are becoming salty at a much more rapid rate than would occur as a result of sea-level rise (Kana *et al.*, 1984; Luzier, 1980).

The only water resource agency in the United States to examine its vulnerability to a rise in sea level has been the Delaware River Basin Commission. In an analysis supporting the Commission's comprehensive plan, Hull and Tortoriello (1979) estimated that the 13 cm rise in sea level expected in the region for the 1965-2000 period would enable salt concentrations to migrate upstream 2-4 km in the Delaware River and Bay. More recently, Hull *et al.* (1986) estimated that a 73 cm rise in sea level would approximately double salt concentrations during droughts in parts of the river, including Philadelphia's drinking water intake. Lennon *et al.* (1986) suggested that such increases in river salinity would contaminate parts of the Potomac–Raritan–Magothy aquifer. This is recharged by the river and provides water for much of New Jersey, but has now been pumped well below sea level (Fig. 15.9). A number of measures can be taken in response to such increased river salinity. Intakes can be moved upstream; communities can shift to alternative water supplies; the aquifers' hydraulic connection with the river can be impeded with injection barriers. In the Delaware Basin, reservoirs are the chief tool for counteracting salt-water intrusion.

High salinity is only a problem during droughts. Thus, it is possible to store water during the rainy season and release it during periods of low flow. The Delaware River Basin Commission estimates that a large reservoir could offset the saltwater intrusion resulting from a 73 cm rise in sea level (Hull and Titus, 1986). However, other impacts of the 'greenhouse effect' could offset or exacerbate the consequences of sea-level rise. A study by the NASA Goddard Institute for Space Studies for the Environmental Protection Agency suggested that areas like the Delaware Basin could experience more annual rainfall, up to ten times as many droughts, or possibly both (Rind and Lebedeff, 1984).

The difficulties posed by uncertain future sea-level rise are even greater for climate change. Yet there are constructive responses to this uncertainty. Steven Schneider of the National Center for Atmospheric Research emphasises the need for 'robust' approaches: instead of optimising for present conditions, which may be a very poor guide for the future, we should identify strategies that are not likely to be vulnerable to possible changes (Chen *et al.*, 1983). For an area like the Delaware River Basin, a robust approach might be to identify possible reservoir sites and ban development in them. If the reservoirs prove to be unnecessary, the areas can be developed later; if they are necessary, the sites will be available. Such an approach would be more robust than building reservoirs now, which would be a waste of money if they are unnecessary, or failing to identify reservoir sites, which could leave the area without any suitable sites.

Figure 15.9: Delaware estuary, USA. Freshwater aquifer depth (in feet) below sea level

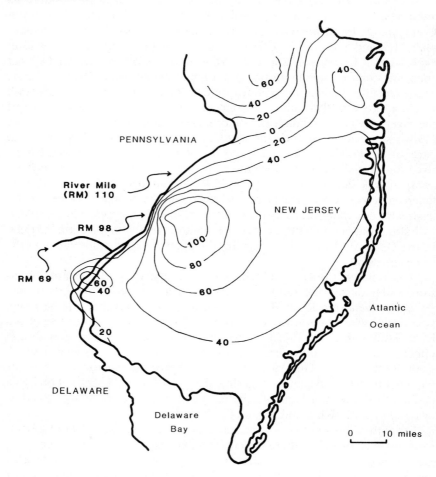

Source: After Luzier (1980).

WHEN IS A SOCIAL RESPONSE NECESSARY?

Although assessments of potential impacts of sea-level rise have received considerable attention, there has been less focus on determining precisely when the accelerated sea-level rise from the 'greenhouse effect' really requires concrete actions to be taken. The need to assess the consequences of future sea-level rise depends on the likely timing of decisions that use

518

the information and the time lag between completion of research and implementation of policy. One can only imagine how long it would take for society to phase out the use of fossil fuels even if there was a scientific consensus on the need to do so. It may already be too late to avert a substantial warming. Nevertheless, the decision process will not begin until there are more conclusive findings. Research to determine whether it will be necessary to curtail greenhouse gases is generally recognised as important.

There is less of a consensus, however, on the need to undertake research and policy analysis on how to adapt to whatever impacts do take place. Many observers maintain that it will not be possible or necessary to address the consequences of sea-level rise until the impacts become noticeable. If they are correct and if the necessary research could be completed in about twenty years, it might be safe to hold off on these investigations until the year 2000.

We agree with this perspective to the extent that people respond to the accelerated rise in sea level by erecting coastal protection structures. Seawalls, beach restoration and levees can all be implemented rapidly when and if the rise occurs. However, a policy to place major areas off-limits to development, or to require development to retreat from the shore, could take many decades to develop and implement. If we fail to prepare until problems emerge we may leave the next generation with only one option: hold back the sea.

In spite of the potential benefits of planning for the future, most observers believe that society is unlikely to address sea-level rise until it becomes a problem. People with something to lose will be sceptical and the near-term losers will outshout the long-term winners, even if the winners outnumber them. But institutional problems may make it more difficult to address the problem later than to address it now. Once the sea rise accelerates property owners will want the shore to be protected. By this time, a requirement to retreat would require compensation, which could impede such a plan. As a result, property owners along estuaries would protect their lots from becoming a marsh and the marsh would be lost. Barrier island communities could be forced to pay for beach nourishment even where a retreat would be more economical.

A planning example

To illustrate the issues involved in determining when it might be necessary to take concrete measures to adapt to the 'greenhouse warming', we provide a numerical economic illustration of a moderately developed hypothetical barrier island in the United States (Fig. 15.10). Although our focus is on erosion related decisions, the principle may also apply to land use

519

Figure 15.10: Planning for a retreating barrier island: plan view of a 20 ×
330 m strip. Case A: An ocean front house moved to the bay side. Case B:
Bay front owner takes over new bay front lot; ocean front house moved to
the lot previously occupied by the bay front owner

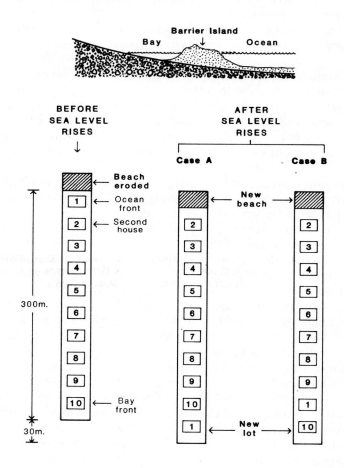

decisions related to water availability and flood vulnerability.

Table 15.2 describes the economics of a 20 metre-long section of a
barrier island 330 metres wide. This section has ten 20 × 30 m sized lots
and a 20 × 30 m section of beach. We assume that the ocean front lot sells
for $200,000, the bay front lot for $100,000 and the remaining lots for
$50,000. Furthermore, there is either a $100,000 structure on the ocean
front lot or no structure at all, and there are $50,000 structures on the
other nine lots.

If the slope of the beach profile is 1 per cent, a 30 cm rise in sea level

Table 15.2: Example tract on a barrier island resort with single family houses

Assumptions

		$
1.	Ocean front house	100,000
2.	Other houses	50,000
3.	Ocean front lot	200,000
4.	Bay front lot	100,000
5.	Other lots	50,000
6.	Creating a bayside lot with fill	1,000
7.	Moving ocean front house	20,000
8.	Moving other houses	10,000
9.	Cost of defending shore for 30 cm rise	
	(a) at $6m^{-3}$	45,000
	(b) at $15m^{-3}$	112,500

Cost of retreating	*Without plan*	*With plan*
	$	$
Loss to ocean front property owner	181,000	31,000
New lot created	1,000	1,000
Bayside house moved onto new lot	10,000	10,000
Ocean front house moved to old bay lot	20,000	20,000
Ocean front premium	150,000	0
Gain to second-from-the-ocean owner of ocean front premium	150,000	0
Social cost	31,000	31,000
Cost to ocean front owner	181,000	31,000
Less: Gain to second-from-the-ocean owner	150,000	0

Source: Hypothetical values based on typical moderate density beach resort property on a barrier island along the Atlantic Coast of the United States.

would erode the shore 30 m, completely consuming the ocean front lot. Assuming that the vertical and horizontal dimensions of the beach profile are 1,250 m and 12.5 m respectively, the amount of sand necessary to counteract a one metre rise in sea level would be 25,000 m³. The cost of holding back the sea can be expected to increase over time because the least expensive sand would be used first. Only a fraction of the sand placed on the beach has sufficient grain size not to wash away. Thus beachfill projects are designed as the highest quality deposits are exhausted. We assume that the first 7,500 m³ would cost $6 m⁻³, the next 7,500 would cost $9 m⁻³. Successive 30 cm increments of sea level rise would thus cost $45,000, $67,500, $90,000, and $112,500. Because both the rate of sea-level rise and the unit cost of sand can be expected to rise, the annual cost of holding back the sea is likely to accelerate considerably.

The economic impact of a given rise in sea level would depend on whether the ocean front lot is vacant and on whether institutional impediments prevent the least-cost solution. If there is a structure on the ocean

521

front lot, then the $45,000 required to hold back the sea for the first 30 cm rise is clearly justified by the $300,000 property value. If the land is vacant, the owner would be willing to pay $45,000 to protect his $200,000 lot. However, the community at large would be almost indifferent. Although the ocean front owner would eventually lose his $200,000 investment, the owner of the second house from the ocean would realise a $150,000 premium as his property became ocean front. For the community the net loss is $50,000, the value of an interior lot.

Now consider the fourth 30 cm increment of sea-level rise, for which it would cost $112,500 to stabilise the shore. Again, the $300,000 ocean front property would appear to warrant protection. However, if the ocean front property owner were to buy the adjacent house, he could reap the $150,000 capital gain resulting from it becoming ocean front, for a net loss of only $150,000. Moreover, if the $100,000 house could be moved to another site at a cost of $20,000, an additional $80,000 would be salvaged, for a net loss of only $70,000, far less than the cost of holding back the sea.

One possible destination for the ocean front house would be a newly created lot on the bay side, as shown in Figure 15.10, Case A. A 30 m × 20 m lot created by filling a shallow bay would only cost about $1,000. It would also be necessary to compensate the bay side owner for taking away his access to the water. This problem might be resolved by moving his house to the new bay lot (Fig. 15.10, Case B), or by paying him a bay front premium. If the cost of moving the bay front house was another $10,000, the ocean front owner's net loss would be reduced to $31,000, equal to the cost of moving the two houses plus the cost of creating the new lot.

A retreat will be economical once the cost of saving the ocean front lot exceeds the cost of moving two houses and creating a new lot. If a storm destroyed the ocean front house, the cost of retreat would be further reduced because it would not be necessary to move the ocean front house! However, against this apparent economic superiority one must weigh the institutional costs of achieving the common good. The first major hurdle would be to establish the principle of creating land on the bay side as a response to ocean side erosion. The idea of moving houses around and giving bay front land to the one-time ocean front owner seems a bit extravagant at first. But it is far cheaper to build new land on the bay side than to defend the ocean side, and the barrier would erode away if new land on the bay side is not created.

Perhaps the most difficult aspect of a retreat is that it would cause the ocean front premium to change hands. Although an ocean front owner might purchase the second house from the ocean to hedge against sea-level rise, this would not always be possible. Even though the net cost to the community of retreating might be only $31,000, the ocean front owner is facing an additional loss of at least $150,000 to agree to a retreat, while seeing that premium accrue to his neighbour. Similarly, the bay front owner might insist on a cash payment in addition to having his house moved to the new bay front lot. These transfers of wealth could cause people to make decisions that increase everyone's costs. If the ocean front house is built to withstand a major storm, it may remain long after the sunbathing portion of the beach has migrated onto the property. To maintain the recreational use of the beach, the entire community may have to pay to defend the shore, even though it would be cheaper to move the house back.

One way to avoid this institutional bottleneck would be to plan for it in advance. The problem need not occur if by the time it is no longer cost-effective to hold back the sea, the ocean front property owner has been required to move his house to a newly created bayside lot. If implemented today, such a plan would probably not be construed as an unconstitutional 'taking' in the United States, since the impact would be so far in the future and the owner would receive ownership of another parcel. But even if ocean front owners had to be compensated, the cost would be small.

Table 15.3 shows the approximate impact of such provisions on the value of the ocean front premium, compared with a policy of stabilising the shore at no cost to the ocean front taxpayer. The impact on property values is only a few hundred dollars per house if a plan is enacted in 1990, but thousands of dollars if a plan is enacted in 2020 and more if a plan is further delayed.

Against the savings from being prepared we must weigh the cost of planning for a rise that does not subsequently materialise. We believe that the latter costs are not likely to be great, as long as a definitive assessment

Table 15.3: Wealth transfer in a particular year from ocean front owner to second-from-the-ocean owner from adopting long-term plan, effective 2050

Year	Market estimate of probability of significant rise by 2050		Value of premium ($) 5 and 10% interest			
	low	high	low/5%	high/5%	low/10%	high/10%
1990	0.2	0.4	1,606	3,212	99	198
2000	0.3	0.6	3,924	7,848	383	766
2010	0.4	0.8	8,522	17,045	1,325	2,651
2020	0.5	0.9	17,352	31,235	4,298	7,736
2030	0.6	1.0	33,920	56,533	13,377	22,295
2050	0.8	1.0	73,669	92,086	46,265	57,831
2055	1.0	1.0	117,528	117,428	93,138	93,138

Note: If long-term retreat covenant exists, ocean front premium will decline by a factor of $1 - p(1+r)^{(-t)}$, where p is the probability the sea will rise enough to require retreat, r is the interest rate, and t is the number of years until the effective date of the retreat.
Source: Hypothetical values used to test sensitivity of the typical ocean front property value to long-term planning.

were possible before 2010. The major role of long-term planning is not to require expensive investments that will only be fruitful if and when the sea rises; rather it is to help keep open possibilities that would otherwise be foreclosed. Provisions that require actions to be taken only when the sea rises will not cost much if the sea does not rise.

CONCLUSION

A growing body of evidence suggests that mankind is modifying the atmosphere in a fashion that will warm our planet about four degrees ($^{\circ}$C) in the next century and raise sea level a metre or two. Such a rise in sea level would inundate low-lying areas, erode beaches, increase coastal flooding and increase the salinity of rivers, bays and aquifers.

Will the consequences of the 'greenhouse warming' unfold more or less rapidly than our ability to address them? Thus far, the scientists and policy makers are ahead of the experiment. If the sea rises thirty centimetres or a new desert is created, we will not be surprised. It is less clear, however, whether we will be prepared to act and take the necessary measures in a timely fashion to deal with the problem. We will never know whether society can be far-sighted until practical responses are developed and proposed. If the 'greenhouse effect' is viewed as a *problem* there is a tendency to search for a collective *solution*, such as curtailing the use of fossil fuels, which is beyond anyone's capability. If it is viewed as a *phenomenon*, then anyone whose activities or descendants would be affected can examine ways to avoid adverse impacts and capitalise on opportunities. Here the value of better understanding does not lie in convincing a majority but in educating specific individuals with the greatest incentive to use the information.

The decision whether or when to plan for the consequences ultimately depends on the trade-off between what will happen if we plan for a rise that does not take place and what will happen if we fail to plan for a rise that does occur. If by planning for a rise in sea level one means banning all development along the coast, or immediately building seawalls that will not be necessary for several decades, the costs of planning for a rise that does not materialise would be substantial. However, there are anticipatory measures that would not in retrospect impose unwarranted costs even if the rise does not take place.

Additional research into the climate change–sea level linkage and the repercussion on shoreline position is warranted and will continue. Research funded by national governments, nevertheless, logically focuses on basic applied research issues. There is a gap between the results of this research and the practical understanding by individuals and communities on how it affects them. Most of this gap can be filled inexpensively by supplementing ongoing studies of coastal sites with an evaluation of the implications of sea-level rise.

There is a possibility, although hopefully an unlikely one, that mankind will continue emitting greenhouse gases even after the earth has warmed three or four degrees (C), and in so doing will heat the planet to intolerable temperatures in the next few centuries. If there is anything we could do to make this less likely, it would be desirable. International research and public understanding of the consequences will be required. Whatever your field of endeavour, we hope that you will consider the potential consequences of the 'greenhouse warming', so that we can keep science ahead of this 'experiment'.

REFERENCES

Arrhenius, Svante (1896) 'On the influence of carbonic acid in the air upon the temperature of the ground', *Phil. Mag., 41*, 237.

—— (1908), *Worlds in the Making*, Harper, New York.

Associated Press (1985) 'Doubled erosion seen for Ocean City', *Washington Post*, 14 November 1984, Maryland Weekly Section.

Barnett, T.P. (1984) 'The estimation of "global" sea level change, a problem of uniqueness', *J. Geophys. Res., 89 (C5)*, 7980-8.

Barrell, J., Schuchert, C., Woodruff, C., Lull, R. and Huntington, E. (eds) (1919) *The Evolution of the Earth and Its Inhabitants*, Yale University Press, New Haven, Conn.

Barth, M.C. and Titus, J.G. (eds) (1984) *Greenhouse Effect and Sea Level Rise*, Van Nostrand Reinhold, New York.

Bentley, C.R. (1983) 'West Antarctic ice sheet: diagnosis and prognosis', in Department of Energy, *Proceedings: Carbon Dioxide Research Conference: Carbon Dioxide, Science and Consensus. Conference 820970*, National Technical Information Service, Springfield, Va., IV. 3-IV.50.

Bindschadler, R. (1985) 'Contribution of the Greenland ice cap to changing sea level', in M.F. Meier (ed), *Glaciers, Ice Sheets and Sea Level*, National Academy Press, Washington, DC, 80-91.

Bird, E.C.F. (1976) *Shoreline Changes During the Past Century*, IGU Working Group, University of Melbourne.

Bruun, P. (1962) 'Sea level rise as a cause of shore erosion', *Journal of Waterways and Harbors Division, Am. Soc. Civ. Eng., 1*, 116-30.

—— (1983) 'Review of conditions for uses of the Bruun rules of erosion', *Coastal Engineering, 7*, 77-89.

Callendar, G.S. (1940) 'Variations in the amount of carbon dioxide in different air currents', *Quart. J. Roy. Met. Soc., 64*, p. 223.

—— (1949) 'Can carbon dioxide influence climate?', *Weather, 4*, p. 310.

Charney, J. (Chairman, Climate Res. Board) (1979) *Carbon Dioxide and Climate: A Scientific Assessment*, National Academy Press, Washington, DC.

Chen, R.S., Boulding, E. and Schneider, S.H. (eds) (1983) *Social Science Research and Climate Change*, Reidel, Dordrecht.

Clark, J.A. and Lingle, C.S. (1977) 'Future sea-level changes due to West Antarctic ice sheet fluctuations', *Nature, 269*, 206-9.

Everts, C.H. (1985) 'Effect of sea level rise and net volume change on shoreline position', in J.G. Titus, S.P. Leatherman, C.H. Everts, D.L. Kriebel and R.G. Dean, *Potential Impacts of Sea Level Rise on the Beach at Ocean City, Maryland*, US Environmental Protection Agency, Washington, DC, 67-97.

Flynn, T.J., Walesh, S.G., Titus, J.G. and Barth, M.C. (1984) 'Implications of sea level rise for hazardous waste sites in coastal floodplains', in M.C. Barth and J.G. Titus (eds), *Greenhouse Effect and Sea Level Rise*, Van Nostrand Reinhold, New York, 271-94.

Gagliano, S.M., Meyer-Arendt, K.J. and Wicker, K.M. (1981) 'Land loss in the Mississippi deltaic plain', in *Transactions of the 31st Annual Meeting of the Gulf Coast Association of Geological Societies*, Gulf Coast Associates of Geological Societies, Corpus Christi, Tex., 181-207.

Gibb, J.G. and Aburn, J.H. (1986) *Shoreline Fluctuations and an Assessment of a Coastal Hazard Zone along Pauanui Beach, Eastern Coromandel Peninsula, New Zealand*, Water and Soil Technical Publication No. 27, Wellington, New Zealand.

Gibbs, M.J. (1984) 'Economic analysis of sea level rise: methods and results', in M.C. Barth and J.G. Titus (eds), *Greenhouse Effect and Sea Level Rise*, Van Nostrand Reinhold, New York, 215-51.

Gornitz, V., Lebedeff, S. and Hansen, J. (1982) 'Global sea level trend in the past century', *Science, 215*, 1611-14.

Hansen, J., Johnson, D., Lacis, A., Lebedeff, S., Lee, P., Rind, D. and Russell, G. (1981) 'Climate impact of increasing atmospheric carbon dioxide, *Science, 213*, 957-66.

——, Lacis, A.A., Rind, D.H. and Russell, G.L. (1984) 'Climate sensitivity to increasing greenhouse gases', in M.C. Barth and J.G. Titus (eds), *Greenhouse Effect and Sea Level Rise*, Van Nostrand Reinhold, New York, 57-77.

Hayes, J.D., Imbris, J. and Shackelton, N.J. (1976), 'Variations in the Earth's orbit: pacemaker of the Ice Ages', *Science, 194*, 1121-31.

Hoffman, J.S., Keyes, D. and Titus, J.G. (1983) *Projecting Future Sea Level Rise*, 2nd rev edn, Government Printing Office, Washington, DC.

——, Wells, J.B. and Titus, J.G. (1985) 'Future global warming and sea level rise', in P. Bruun (ed), *Iceland Symposium September 1985*, National Energy Authority, Reykjavik, 53-71.

Hughes, T. (1983) 'The stability of the west Antarctic ice sheet: what has happened and what will happen', in Department of Energy, *Proceedings: Carbon Dioxide Research Conference: Carbon Dioxide. Science and Consensus*, DOE Conference 820970, Dept. of Energy, Washington, DC, IV.51-IV.73.

Hull, C.H.J. and Tortoriello, R.C. (1979) *Sea-Level Trend and Salinity in the Delaware Estuary*, Staff Paper, Delaware River Basin Commission, West Trenton, NJ.

——, Thatcher, M.L. and Tortoriello, R.C. (1986) 'Salinity in the Delaware estuary', in C.H.J. Hull and J.G. Titus (eds), *Greenhouse Effect, Sea Level Rise, and Salinity in the Delaware Estuary*, US Environmental Protection Agency, Washington, DC, 15-39.

—— and Titus, J.G., (1986) 'Responses to salinity increases', in C.H.J. Hull and J.G. Titus (eds), *Greenhouse Effect, Sea Level Rise, and Salinity in the Delaware Estuary*, US Environmental Protection Agency, Washington, DC, 55-63.

Kana, T.W., Michel, J.M., Hayes, M.O. and J.R. Jensen (1984) 'The physical impact of sea level rise in the area of Charleston, South Carolina', in M.C. Barth and J.G. Titus (eds), *Greenhouse Effect and Sea Level Rise*, Van Nostrand Reinhold, New York, 105-50.

526

Keeling, C.D. (1978) 'Atmospheric carbon dioxide in the 19th century', *Science,* 202, p. 1109.
—— Bacastow, R. and Whorf, T. (1982) 'Measurements of the concentrations of carbon dioxide at Mauna Loa Observatory, Hawaii', in W. Clark (ed) *Carbon Dioxide Review: 1982,* Clarendon Press, New York, 377-84.
Kyper, T. (1985) *Coastal Zone '85, Am. Soc. Civ. Eng.,* New York.
La Roche, T. and Webb, K. (in press), 'Impact of sea level rise on coastal drainage systems' in Charleston, South Carolina', in *Impact of Sea Level Rise on Coastal Drainage Systems,* US Environmental Protection Agency, Washington, DC.
Leatherman, S.P. (1982) *Barrier Island Handbook,* University of Maryland, College Park, Md.
—— (1984) 'Coastal geomorphic responses to sea level rise: Galveston Bay, Texas', in M.C. Barth and J.G. Titus (eds), *Greenhouse Effect and Sea Level Rise,* Van Nostrand Reinhold, New York, 151-78.
—— (1985) 'Geomorphic effects of accelerated sea level rise on Ocean City, Maryland', in J.G. Titus, S.P. Leatherman, C.H. Everts, D.L. Kriebel and R.G. Dean, *Potential Impacts of Sea Level Rise on the Beach at Ocean City, Maryland,* US Environmental Protection Agency, Washington, DC, 33-65.
Lennon, G.P., Wisniewski, G.M. and Yoshioka, G.A. (1986) 'Impact of increased river salinity on New Jersey aquifers', in C.H.J. Hull and J.G. Titus (eds), *Greenhouse Effect, Sea Level Rise, and Salinity in the Delaware Estuary,* US Environmental Protection Agency, Washington, DC, 40-54.
Levine, J.S., Rinsland, C.P. and Tennille, G.M. (1985) 'The photochemistry of methane and carbon dioxide in the troposphere in 1950 and 1985', *Nature, 318,* 254-7.
Luzier, J.E. (1980) *Digital Simulation and Projection of Head Changes in the Potomac–Raritan–Magothy Aquifer System, Coastal Plain, New Jersey,* US Geological Survey, Reston, Va.
Manabe, S., Wetherald, R.T. and Stouffer, R.J. (1981), 'Summer dryness due to an increase of atmospheric CO_2 concentrations', *Climatic Change, 3,* 347-86.
Marshall, J.F. and Thom, B.G. (1976) 'The sea level in the last Interglacial', *Nature, 263,* 120-1.
Meier, M.F. (1984) 'Contribution of small glaciers to global sea level, *Science, 226,* 1418-21.
Mercer, J.H. (1970) 'Antarctic ice and interglacial high sea levels, *Science, 168,* 1605-6.
—— (1978) 'West Antarctic ice sheet and CO_2 greenhouse effect: a threat of disaster', *Nature, 271,* 321-5.
Milliman, J.D., Quraishee, G.S. and Beg, M.A.A. (1984) 'Sediment discharge from the Indus River to the ocean, past present and future', in B.U. Haq and J.D. Milliman (eds) *Marine Geology and Oceanography of Arabian Sea and Coastal Pakistan,* Van Nostrand Reinhold, New York, 65-70.
National Academy of Sciences (NAS) (1975) *Understanding Climate Change,* National Academy of Sciences, Washington, DC.
Neftel, A. Oeschger, H., Schwander, J., Stauffer, B. and Zumbrunn, R. (1982) 'New measurements on ice core samples to determine the CO_2 content of the atmosphere during the last 40,000 years', *Nature, 295,* 220-3.
Nierenberg, W.A., *et al.* (1983), 'Synthesis', in Carbon Dioxide Assessment Committee, *Changing Climate,* National Academy Press, Washington, DC, 483-93.
Nordhaus, W.D. and Yohe, G.W. (1983) 'Future carbon dioxide emissions from fossil fuels', in Carbon Dioxide Assessment Committee, *Changing Climate,*

527

National Academy Press, Washington, DC, 287-302.

Palmer, A., Moose, W., Quinn, T. and Wolf, K. (1980) *Economic Implications of Regulating Chlorofluorocarbon Emissions for Non-Aerosol Applications*, Rand Corporation, Santa Monica, Calif.

Peters, J. (1985) 'War of the waves', *Baltimore Sun*, reprint of articles published 3, 10, 17 and 24 June.

Pilkey, O. (1985) 'The twilight of Ocean City', *Washington Post*, 26 May.

——, Howard, J., Brenninkmeyer, B., Frey, R., Hine, A., Kraft, J., Morton, R., Nummedal, D. and Wanless, H. (1981) *Saving the American Beach: A Position Paper by Concerned Coastal Geologists*, Results of the Skidaway Institute of Oceanography Conference on America's Eroding Shoreline, Skidaway Institute of Oceanography, Savannah, Ga.

Polar Research Board, US National Academy of Sciences (1985), *Glaciers, Ice Sheets and Sea Level: Effect of a CO_2-Induced Climatic Change*, National Technical Information Service, Springfield, Va.

Ramanathan, V., Singh, H.B., Cicerone, R.J. and Kiehl, J.T. (1985), 'Trace gas trends and their potential role in climate change', *J. Geophys. Res.*, *90(D3)*, 5547-66.

Rasmussen, R.A., and Khalil, M.A.K. 'Atmospheric methane: trends and seasonal cycles', *J. Geophys. Res.*, *86*, 9826-32.

Revelle, R. (1983) 'Probable future changes in sea level resulting from increasing atmospheric carbon dioxide', in Carbon Dioxide Assessment Committee, *Changing Climate*, National Academy Press, Washington, DC.

—— and Seuss, H. (1957) 'Carbon dioxide exchange between atmosphere and ocean and the question of an increase of atmospheric CO_2 in the past decades', *Tellus*, *9*, 18.

Rind, D. and Lebedeff, S., NASA Goddard Space Flight Center (1984) *Potential Climatic Impacts of Increasing Atmospheric CO_2 with Emphasis on Water Availability and Hydrology in the United States*, Government Printing Office, Washington, DC.

Seidel, S. and Keyes, D. (1983) *Can We Delay a Greenhouse Warming?*, Government Printing Office, Washington, DC.

Sorensen, R.M., Weisman, R.N. and Lennon, G.P. (1984) 'Control of erosion, inundation and salinity intrusion caused by sea level rise', in M.C. Barth and J.G. Titus (eds) *Greenhouse Effect and Sea Level Rise*, Van Nostrand Reinhold, New York, 178-214.

Thomas, R.H., Sanderson, T.J.O. and Rose, K.E. (1979) 'Effect of climatic warming on the West Antarctic ice sheet', *Nature*, *277*, 355-8.

Titus, J.G. (1984) 'Planning for sea level rise before and after a coastal disaster', in M.C. Barth and J.G. Titus (eds), *Greenhouse Effect and Sea Level Rise*, Van Nostrand Reinhold, New York, 253-69.

—— (1985) 'Sea Level Rise and Wetlands Loss', in *Coastal Zone '85*, American Society of Civil Engineers, New York.

——, Leatherman, S.P., Everts, C.H., Kriebel, D.L. and Dean, R.G. (1985) *Potential Impacts of Sea Level Rise on the Beach at Ocean City, Maryland*, US Environmental Protection Agency, Washington, DC.

Tukey, J.W. (Chairman) (1979) *Protection Against Depletion of Stratospheric Ozone by Chlorofluorocarbons*, National Academy Press, Washington, DC.

US Army Corps of Engineers (1979), *Ocean Beach Study: Feasibility Report*, San Francisco District, San Francisco.

Part Five

Sea-surface (Sea-level) Changes: Some Wider Implications

16

Hydrocarbon Exploration and Biostratigraphy: The Application of Sea-level Studies

R.J.N. Devoy

INTRODUCTION

Hydrocarbons, in the form of the fossil fuels, oil, gas and coal, are a vital source of energy for man and their discovery and exploitation has become a multi-billion dollar industry (Anderson, 1985; Sampson, 1975). In short-term geological time (10 Ma) these resources are non-renewable and by definition, therefore, are becoming scarcer as man uses them. Probability estimates indicate that ~ 80 per cent of total petroleum and coal resources will have been used up within approximately the next 100 years and 300-400 years respectively (Fischer, 1974). Such figures are inevitably open to dispute on the basis of the forecaster's politics and economic philosophy, the vagaries of market demand and technological change. Nevertheless, the figures do serve to emphasise the present dependency and likely continued short-term importance to the world of hydrocarbons as a primary source of energy.

The attempts and good resolutions, at both national and international levels since the oil crises in the West of the 1970s, to diversify the energy base of industrial economies have, at best, been only moderately successful. Yet despite their importance in the life of man, ownership of hydrocarbon resources is concentrated in the hands of relatively few countries. Whatever the political and economic factors at work here, the factor of spatial distribution, with the resources derived from a restricted zone of the earth's crust, must be of primary importance. The principal hydrocarbons are located in the thick sedimentary rock sequences of the continents and structurally passive adjacent continental margins. The vast areas of the ocean floors, composed of thin oceanic crust (Stewart, 1977), and the ocean ridges are unlikely to contain any such resources. The original depositional environments for sediments rich in the organic matter necessary for subsequent hydrocarbon development, are essentially ones of the shallow marine–coastal zone. Initial accumulation of the source rocks, therefore, took place in environments where relative sea-level changes would also have been registered and, in many instances, preserved in the sedimentary rock record. Tissot (1979) suggests that this environmental association is more than just one of chance. Thus relative sea-level changes are seen as playing an important role in determining, firstly, the type of organic matter produced, the building blocks of hydrocarbons, and in turn conditioning the nature of the fuel formed; secondly, in influencing the provision of the conditions necessary for its preservation within the depositional environment.

These factors of initial origin, coupled with the vagaries of subsequent hydrocarbon generation and preservation through geological time, have ensured that these fossil fuels are restricted in extent, even within the provenance of sedimentary rocks themselves. Geological knowledge of the likely location of oil and gas shows that only 2 per cent of sedimentary structures are certain to contain such resources and that at least 42 per cent of such rocks have no hydrocarbon potential (Warman, 1978; Birks, 1978). Such probability statistics, nevertheless, make the task of exploration no easier as it is not certain as to exactly where the resources lie. In the case of oil and gas, these may migrate hundreds of kilometres through the source rocks to their eventual location in reservoir structures (Tissot and Welte, 1978). Major areas of the continental margins and even of the world's landmasses, as in Africa, Asia and Antarctica, remain unexplored in any detail. However, the scale of the task and the technical difficulty of working beneath a cover of water, in the case of the continental margins, means that the exploration for new hydrocarbon resources is both time consuming and expensive. However, improvements in exploration–retrieval technology, coupled with the impetus of increasing world demand in the twentieth century and exhaustion of reserves from these accessible locations, have led to a movement offshore. In the petroleum industry this has occurred most notably in the areas of the North Sea, the Gulf of Mexico and the East and South China Seas. This progressive movement into deeper water for both petroleum production and exploration (Fig. 16.1) has been matched by an escalation in costs, with the risk investment in developing an exploration well running at $8.5 M or more for each well (Birks, 1978; Anderson, 1985).

This pattern illustrates clearly the constant need within the industry to develop (1) Reliable, efficient methods of identifying hydrocarbon locations, and (2) Cost–time saving techniques in exploration. In regard to both these goals the development of ever-improving stratigraphic and geophysical methods of examining extensively the form and age of the continental margins have been of fundamental value. Use of coring and seismic survey techniques has led to advancements in depositional system concepts and an improved understanding of the sedimentary environments in which hydrocarbons originated. In particular, the new study of seismic stratigraphy (Payton, 1977; Brown and Fisher, 1980) has resulted in interest in

Figure 16.1: Trend of progressive movement into deeper water offshore for oil exploration and production

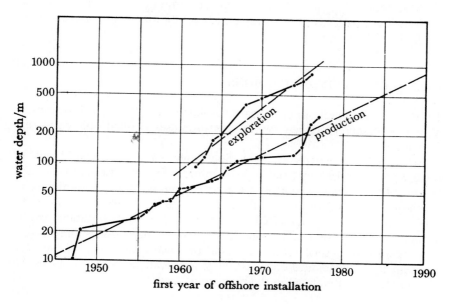

first year of offshore installation

Source: Goldman (1978).

the relationship–application of relative sea-level change patterns to the location and identification of hydrocarbon (oil and gas) source rocks. However, Vail *et al.*'s (1977; Vail and Todd, 1981) interpretation of the origin of many geological unconformities, recorded by seismic data from continental margin sedimentary sequences, as the result of long-term 'cycles' of relative sea-level change, has been widely criticised (Hallam, 1981; Watts, 1982; Brown and Fisher, 1980; see Chappell, this vol., Ch. 2 and Mörner, this vol., Ch. 8). Nevertheless, the new data produced by seismic stratigraphy has provided an important stimulus to studying the mechanisms and long-term form of sea-level changes at the global scale. For commercial interests, whether the sea-level control mechanism is the right one as an explanation of the recorded stratigraphic patterns is a secondary concern. If the seismic model works, who cares!

The aims of this chapter are, therefore, threefold:

(1) To examine the nature of the relationship between hydrocarbon generation and relative sea-level change.
(2) Examination of the relative sea-level/coastal onlap–seismic model and the related underpinning biostratigraphic techniques in hydrocarbon

533

exploration, together with their linkage to the concepts of global relative sea-level changes.

(3) The application of such models and techniques to understanding (a) long-term patterns of environmental evolution and (b) the location of hydrocarbon resources.

THE SOURCE OF HYDROCARBONS

To understand the relevance of sea-level change to hydrocarbon location, it is first necessary to review briefly the processes of coal and petroleum formation.

Both coal and petroleum formation begin with the accumulation of organic matter, primarily from large quantities of plant material (Tissot and Welte, 1978). Development occurs generally within slowly subsiding sedimentary basins and both are subject during their formation to the same geological processes of initial bacterial action, incorporation within sediments, compaction and geothermal heating. Increases in temperature and pressure over time following burial produce a sequence of physical, biochemical and chemical changes (diagenesis–catagenesis) in the original plant material. Alterations in the type-ratio of component H, O, C, N and S compounds from the original hydrocarbons present in the fossil material take place, leading to a progressive increase in carbon content with maturity. In petroleum formation a further stage (metagenesis) occurs with a continued rise in temperature and pressure, causing complete reorganisation of the hydrocarbon source material and resulting in the major phase of gas production (Tissot *et al.*, 1974; Cooper, 1978; Philippi, 1965).

Despite these similarities in process, basic differences in the modes of coal and petroleum formation remain. Of primary importance are the type and environmental source of plant material in the sediments. Coals form as a solid mass in the sediment–rock sequence. They are predominantly the remains of terrestrial higher plants, accumulating either *in situ* or from inwashing over relatively short distances into semi-terrestrial–coastal lagoon, deltaic, swamp environments. Optical and physico-chemical examination of the resultant coal shows that the different types of constituent plant remains give rise to three main groups of component materials, each with different physical-chemical properties (Stack *et al.*, 1975; Tissot and Welte, 1978) (Table 16.1). A similar examination of petroleum compounds shows that these form from predominantly the liptinite-exinite group (Hunt, 1979). Here the source material of algal remains, resin, spores, pollen and plant cuticles are rich in the lipids (waxes, fat substances) especially important in petroleum formation. This material, derived mainly from lower aquatic plants and bacteria, accumulated largely within marine environments as finely divided particulate and dissolved organic matter,

Table 16.1: Interrelationships and terminologies used in kerogen and coal petrographic description

ENVIRONMENTAL SOURCE	TERMINOLOGIES			MAIN PRODUCE
	Constituent materials	*Maceral group (coal)*	*Kerogen*	
Aquatic	ALGAL Amorphous organic matter (<1μm), lipid rich remains of plants, algal fragments, spores, etc.	Liptinite (or Exinite)	Type 1 Type II	Oil and Gas
Sub-aerial (Terrestrial)	Herbaceous (Fibrous-lignin) Woody-lignin rich plant structures Coaly (carbon rich, angular to sub-angular brittle particles — formed from woody material oxidatively degraded before burial)	Vitrinite (or Huminite) Inertinite	Type III	Coal and Gas

Source: Based on Tissot and Welte, 1978; Cooper, 1978.

mixed with fine sand-sized inorganic sediments. The insoluble component of this material (kerogen) accounts for ~80 per cent of the total organic content and forms the principal base material for petroleum compounds. Following the release of liquid/mobile petroleum from these source rocks, migration of the oil and gas takes place under the pressure of burial. Eventual collection occurs in permeable, porous reservoir rocks, generally sandstones, siltstones and carbonates (Chapman, 1973). In addition to the existence of a suitable reservoir rock, a further condition for collection is the presence of a seal, or trap to prevent further escape and dispersal of the petroleum from the reservoir. These traps are usually of two types, either stratigraphic, formed by depositional–post depositional processes, or structural (tectonic) in form (Brenner, 1984).

535

From the above discussion it is evident that a number of environmental boundary conditions exist which determine both the conditions necessary for initial accumulation of organic material and the type of hydrocarbon resource produced; chief among these are the types of component plant material and the nature of the depositional environment. Other both related and independent geological limiting conditions may also be identified. These are:

(1) The first appearance in time of the plant associations necessary for generation of the resource type.
(2) Sufficient production of plant biomass.
(3) The existence of a chemically reducing environment, necessary for the preservation of plant material.
(4) A balance of energy levels in the depositional environment; low enough to allow preservation of organic matter and high enough to ensure its supply−accumulation in sufficient quantity.
(5) The inorganic sedimentation rate.

Detailed discussion of these factors is given in Tissot and Welte (1978).

THE INFLUENCE OF SEA LEVEL IN HYDROCARBON LOCATION

Studies of palaeogeography, plate tectonics and earth surface processes show that depositional environments suitable for hydrocarbon formation are dynamic in geological time. This results both in major changes in their regional−global location and consequently in the types of hydrocarbon resource formed. It is thus necessary to consider what large-scale mechanism(s) control both the distribution of these environments and the occurrence, therefore, of the limiting conditions in hydrocarbon formation. The operation of global changes in relative sea level may provide one such mechanism. As defined by the boundary conditions, the majority of suitable hydrocarbon source environments are water dominated and most are situated on the continental margins under the influence of marine processes. Factors of continental drift, mountain building−tectonism, climatic change and biological evolution, for example, must influence changes in these environments through time. Yet, with the possible exception of evolutionary processes, many of these factors operate either wholly or partly through changes in the position of land and sea. The study of long-term patterns of relative sea-level change show that the world's oceans have probably experienced a series of cyclic rises and falls in level relative to their present position (see Chappell, this vol., Ch. 2) and that these fluctuations have taken place throughout Phanerozoic time at a number of different scales of resolution. At the first order level, Pitman

(1978) and others (Vail *et al.*, 1977; Hays and Pitman, 1973; Hallam, 1977, 1981) indicate two major phases of sea-level rise since the Cambrian. The last rise reached ~ +350 m above present levels (Pitman, 1978; Sleep, 1976) before falling to ~ +60 m in the Middle Miocene and subsequently oscillating widely around its present position (Fig. 16.2).

In a different context examination of changes in the estimated abundance of marine phytoplankton, the main source material for type 1 kerogen, similarly show two maxima (Tappan and Loeblich, 1970)(Fig. 16.2). These peaks in global organic productivity correspond in time to the two phases of highest sea level and a causal relationship is inferred by Tissot (1979). These times of high sea level would have resulted in a major expansion in areas of shallow continental margin waters, consequently stimulating an increase in marine organic productivity and the subsequent preservation of this biogenic material. In the intervening phase of lower sea

Figure 16.2: Global cycles of first (1) and second (2) order relative changes in coastal onlap (sea level), together with variations in time in abundance of phytoplankton productivity/composition. It is important to note that changes in the coastal onlap curve cannot be related exactly to the pattern of sea-level change (see Vail *et al.*, 1983). The two curves, however, would be similar and the pattern of coastal onlap may be taken as the broad trend/scale of associated sea-level changes

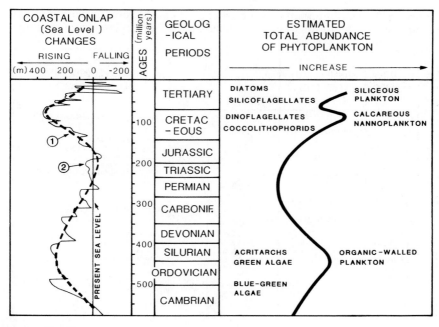

Source: After Tissot (1979).

537

level, exposure of the continental shelves would have led to an expansion of semi-terrestrial environments, favouring the growth of swampland plant communities and the development of coal. Analysis of the distribution of major world coal and petroleum resources lends support to this suggested correlation (Strakov, 1962; Bois *et al.*, 1975). Rocks of Jurassic and Cretaceous age (times of the last first-order sea-level maximum) account for ~ 70 per cent of the major oil resource rocks of the world. By comparison ~ 70 per cent of coal resources are concentrated in rocks of the Carboniferous, Permian, Jurassic and Cretaceous, formed during the time of relative sea-level fall and subsequent rise (Fig. 16.3).

Recognition of second and third-order cycles of relative sea-level change have also been made (Vail *et al.*, 1977, 1980; Hallam, 1978; Hancock and Kauffman, 1979; Rona, 1973; May *et al.*, 1984), operating at timescales of 10 Ma (Fig. 16.2). At these levels changes in the direction of relative sea-level movement are seen as particularly important in influencing the environmental boundary conditions that determine both the type of organic matter produced and its subsequent preservation (Bitterli, 1963; Tissot, 1979). As discussed by Tissot and Welte (1978), major factors in the destruction of organic matter are chemical oxidation and the action of aerobic–heterotrophic organisms. Thus environmental conditions that cause the depletion of oxygen in water and sedimentary environments

Figure 16.3: Distribution of major coal and oil source rocks in time by comparison with phases of global relative coastal onlap (sea-level) change

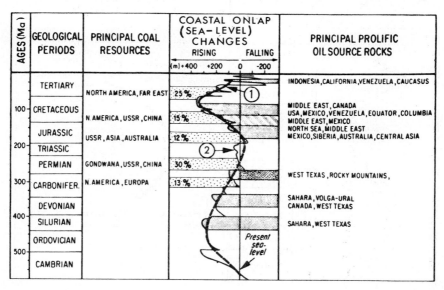

Source: After Tissot (1979).

538

encourage the preservation of organic matter.

At times of relative sea-level rise shallow epicontinental seas seem commonly to have been formed through the flooding of former continental depressions (Laughton, 1975; Ziegler, 1981). In these land-locked seas conditions would have been ideal for the deposition of organic rich sediment. The shallow water and relatively warm conditions of these marine environments would have favoured high organic productivity (Tappan and Loeblich, 1970). High influx of nutrients from the surrounding emergent landmasses may often have been responsible for prolific algal blooms and deoxygenation of the waters. Such blooms would temporarily create anoxic bottom conditions suitable for the preservation of organic matter. Operation of this process, through repeated coccolith–dinoflagelate blooms, may explain the origin of the Kimmeridge Clay oil shales (Jurassic) and oil source rocks of the North Sea (Gallois, 1976). Restricted circulation and production of stagnant–partially renewed water, creating anaerobic bottom conditions within such shallow seas, may provide an alternative method for organic preservation (Hallam, 1967, 1975). Conversely, during phases of relative sea-level fall, a regressive sedimentary sequence would have been initiated (Vail *et al.*, 1977; Hallam, 1981) with progradation of coastal lowland areas. The resulting development of widespread marginal coastal swamp environments, far from the main zones of coarse sediment influx, would have favoured the accumulation of terrestrial plant matter and coal formation.

Together these second- and third-order alternations of relative rise and fall in sea level provide a possible mechanism for conditioning the variable distribution and abundance in time of the major coal and petroleum resources. Comparison of the shorter timescale occurrences of petroleum and coal source rocks with these second-order changes in relative sea level again show a strong degree of correlation (Fig. 16.3).

STRATIGRAPHIC TECHNIQUES IN HYDROCARBON EXPLORATION

Stratigraphy forms the study of rock successions and their interpretation as sequences of events in the earth's geological history (Lowe and Walker, 1984). This study has been achieved through a variety of methods: division of the sediment–rock sequence into discrete units on the basis of rock composition (lithostratigraphy), fossil content (biostratigraphy) and classification by age (chronostratigraphy). The traditional tools employed in such studies, the analyses of observed rock type and micro-/macrofossil content, have more recently been added to by new quantitative techniques, including radiometric dating, electrical resistivity analyses, geochemical tracing and, more controversially, seismic profiling. From the earliest days of stratigraphy (Hancock, 1977) one of its primary aims has been to correlate and establish the synchroneity of the rock strata in different localities. 'The very basis of stratigraphy is correlation. Standard methods of correlation are concerned with seeking and identifying identical points or levels in comparable stratigraphical sequences' (Cubitt and Reyment, 1982). Shaw (1982) further defines the process of correlation itself as 'the determination of geometric relationships between rocks, fossils, or sequences of geologic data for interpretation and inclusion in facies models, palaeogeographic reconstructions, or structural models'. These views on the role of stratigraphy—correlation serve to indicate an important and growing application of such studies, namely, geohistory analysis. Here quantitative stratigraphic techniques are used to characterise the nature of geological-environmental change through time (van Hinte, 1978). One of the main beneficiaries and at the same time instigators of this development have been the hydrocarbon industries. In petroleum exploration and resource evaluation particularly these stratigraphic techniques have been of value in three main areas:

(1) Geochronology — the dating of rock sequences, for use in the location of potential source rocks.
(2) Correlation/matching of source rocks.
(3) Palaeogeographic reconstruction, for determination of the likely location and size of a hydrocarbon resource.

Normally more than one stratigraphic technique would be employed in the study of these areas. However, biostratigraphy has perhaps become one of the most widely applied and valuable of the techniques used (Poag, 1977). As such biostratigraphic zonation has been employed in each of the three above stratigraphic spheres outlined as of use to the petroleum industry. Any method of analysing earth history will, however, be subject to a variety of limitations, inherent in the nature and assumptions of the technique itself and imposed by the vagaries of data preservation in the geological record. Thus the examination of rock sequences, especially when undertaken for purposes of environmental reconstruction, is best carried out using a complex of approaches, providing data support for, and a check on the validity of interpretation from any one source. The following discussion centres upon two such interrelated techniques, those of biostratigraphy and seismic-stratigraphy, the latter forming the area of primary interest in terms of the information and applications for sea-level changes.

Biostratigraphy

Whilst biostratigraphy is in its own right a key tool in stratigraphic correlation, its role in establishing local—regional scale chronologies ensures it a focal position in other forms of stratigraphic correlation. In this context seismic-stratigraphy and the establishment of a global pattern of coastal onlap/relative sea-level change are dependent upon calibration by an independent chronology. As such, they are intimately linked to biostratigraphy. Hence a consideration of this technique forms an integral part of examining the new developments in these other fields.

Biostratigraphy is a large and still expanding area of geo-biological science and readers are referred to Kauffman and Hazel (1977) for a more detailed discussion of the topic. The biostratigraphic concept (fossil zonation) is based upon the widespread occurrence in sediments–sedimentary rocks of fossil organisms, usually found concentrated in rocks derived from former marine environments. In particular, the study of microfossils (micropalaeontology) plays a key role in the technique and its commercial application (Poag, 1977). Biostratigraphy, as such, is concerned with the organisation of the fossils contained within rock strata into discrete units, based upon their variety and abundance. These units (biozones) represent an ordinal sequence of events, or progression in time, and provide the basis for a geological timescale.

Biozones used in stratigraphic classification generally fall into four main types, assemblage, range, acme and interval zones; readers are referred to Berggren (1978) for their definition. Prior to the Quaternary the biostratigraphic record in the Phanerozoic is divided up mainly on the basis of evolutionary changes in organisms and hence concerns the use of forms of range zone and, to a limited extent, acme zones. These represent parts of the geological record during which an organism first appeared, developed and became extinct. Most microscopic organisms found fossil in the sedimentary record can be used with differing degrees of success in stratigraphic zonation. Micro-/macrofossils most commonly used include those of foraminifera, diatoms, coccoliths, palynomorphs, radiolaria and ostracods (Haq and Boersma, 1978). Apart from their use as time–stratigraphic markers and hence in correlation, knowledge of changes in the habitat and ecology of these organisms, for example differences between planktonic and benthic (vagile, sessile) forms, makes them of particular value as well in palaeoenvironmental reconstruction. Planktonic, free-floating organisms such as silico-flagellates, radiolaria, calcareous nannoplankton, together with some foraminifera and diatoms, are of especial value as oceanic temperature indicators. Benthic organisms, including ostracods and again some foraminifera and diatoms, are good indicators of physico-chemico parameters of sea water composition, changes in water depth and sedimentary environment.

Palynomorphs differ from the other fossils mentioned in that they form a composite group of organic-walled micro-organisms. These consist of pollen/spores derived from land plant communities and components of aquatic algae in the form of acritarchs and dinonflagellates (Owens, 1981;

541

Zaitzeff and Cross, 1970; Evitt, 1964). Use of palynomorphs as biostratigraphic tools in the petroleum industry has grown in importance since 1960 (Cross, 1964; Kuyl *et al.*, 1955; Muller, 1959; Hopping, 1967; Stanley, 1969; Tschudy and Scott, 1969). Often found in abundance in favourable depositional environments ($\geqslant 1,000$ pollen/spore grains g^{-1}), this group has a number of advantages as a biostratigraphic tool over other fossil indicators. Representatives of this group occur throughout the geological record from the Precambrian onward and present a rapidly evolving series of forms for use as time–stratigraphic markers (Griggs, 1970; Doyle, 1977; Parry *et al.*, 1981). Further, these fossils are found in sediments of both marine and terrestrial environments and may, therefore, be used for direct correlation between continental and oceanic rock strata. However, variations in the modes of dispersal, post-depositional reworking and source of the fossils, between local and regional components, may make such correlation difficult (Doyle, 1977; Heusser, 1978; Owens, 1981). The concentration in lithofacies derived from sources of coal and petroleum generation, namely the continental shelf margins, does, however, give this group particular value in palaeoenvironmental reconstruction for use in location of potential hydrocarbon source rocks. More recently, recognition of time–pressure–temperature related changes in the organic wall of palynomorphs (Owens, 1981; Staplin, 1977) has led to their application in determining the maturity of sediments and indication of their potential as a hydrocarbon source.

By comparison, the development of a stratigraphic timescale has been dominated by the study of planktonic foraminifera, though many other microfossils have been used in a similar fashion (Haq and Boersma 1978; Boersma, 1978). In each case the length of the timescale developed depends upon factors of first appearance and duration of occurrence, or the abundance of the organism in the geological record. Chronostratigraphic units based on foraminifera and other groups have now been linked with numerical timescales derived from the record of earth magnetic polarity reversals or direct from radiometric techniques (Berggren, 1969, 1972; van Hinte, 1976a, b). Consequently the age of biostratigraphic units can be expressed accurately in terms of 'millions of years', units in the Cenozoic commonly representing a range of age of ~ 1 Ma and ~ 2 Ma in the Mesozoic (van Hinte, 1978). In the time–stratigraphic record greater accuracy and resolution can now come from the combination of the individual fossil zonation schemes.

As in any technique biostratigraphy is subject to a range of limitations dependent primarily upon the biological understanding of its component fossil forms and their geological interpretation–correlation in space and time. Of critical importance is the definition of the fossil species, based upon recognition of external variations in morphology (Berggren, 1978). This morphology may be subject to a variety of changes, leading possibly to confusion in species identification and their charting over time (in the fossil record). Asexually reproducing organisms (covering most of those mentioned here) may present a wide variation in morphology of the same species, affecting their classification, whilst others undergo fundamental changes during their life cycle. Factors of evolutionary–environment based changes, such as intra specific variation (phenotypic–genotypic changes), adaptive radiation or autogenetic development may also cause errors in species recognition and palaeoenvironment reconstruction. Progressing from these specific considerations, definition of the biozone itself in space and time presents a problem. There may be regional variations in the occurrence of taxa and species diversity, leading to differences in the application of zonation schemes and correlation of strata.

In the Cenozoic the abundance and short stratigraphic ranges of low-latitude planktonic foraminifera allow the recognition in these areas of 35-40 biozones, covering a time span of ~ 65 Ma (Gradstein and Agterberg, 1982). Moving into the mid latitudes the absence of many low-latitude forms, reduction in species diversity and increasing stratigraphic range leads to a progressive reduction in stratigraphic resolution, whilst standard zonation becomes inapplicable in high-latitude areas. Diachroneity also forms a major problem in biozone correlation, particularly where fossil assemblage zones are used. In long-term geological time and in the deep ocean, the occurrence of a fossil form or fossil assemblage may be uniform over large areas. Under these conditions and particularly where sedimentation rates are slow biozones are often effectively synchronous. Closer to the continental margins, or elsewhere where changing environmental conditions stimulate variations in species composition, new biozones will have to be constructed. The time and spatial relationship of these with those of deeper ocean may be complex. In the recognition of range zones the influences of organism migration patterns, speciation factors, local competition, adverse ecological conditions, sea-level changes (van Couvering and Berggren, 1977) and expansion into unfilled niches may areally and temporally distort the first and last occurrence of fossil indicators (Poag, 1977; Eldredge and Gould, 1977; Kauffman and Hazel, 1977). Further, Hazel (1977) argues that the construction of finite regional biostratigraphic units ('standard sections') derived from a composite of overlapping local sections is based upon a self-supporting circular argument. Use of such sections in long-distance correlation may lead to significant chronostratigraphic errors. These problems of correlation over long distances throw the technique back upon the need for linkage with an absolute timescale and close interval definition of biostratigraphic units. In either case this may not be possible, due to lack of data availability, money or opportunity for new coring work, or non-preservation of data in the geological record.

Sea-level change and biostratigraphy

The direct relationship between biostratigraphy and sea-level studies lies in two main fields. Firstly, in the use of fossil data as indicators of former water depth changes through palaeobathymetric reconstruction and, secondly, in the provision of a timescale. As Hallam (1981) and Tappan and Loeblich (1970) point out, marine organisms are in themselves generally not good bathymetric indicators. For water depth, except perhaps in the littoral zone, is not an environmental parameter as such (Fig. 16.4). Changes in organism and species composition result more from variations

Figure 16.4: Depth zonation in oceans of major invertebrate and algal groups

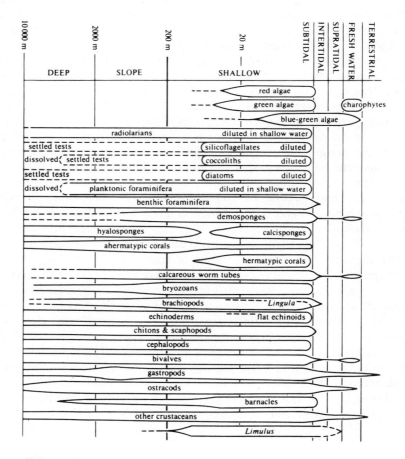

Source: Hallam (1981).

544

in light, water chemistry and food supply, which in turn may be affected by or result from bathymetric changes. The use of the observation that the ratio of planktonic to benthic forms generally increases with water depth to infer sea-level changes is misleading. This relationship results more from changes in habitat and the decrease in abundance of benthic forms offshore (Funnell, 1967). Despite these drawbacks many fossils may be used to infer changing water depths. Different organism groups or even individual species do live in different depth locations. For example, dinoflagellates are predominantly neritic in distribution whilst coccoliths and many marine diatoms occupy open ocean areas. Studies by Barr and Berggren (1980), Berggren and Aubert (1980) have used such location differences in foraminifera to infer major changes in water depth with falling relative sea level, at sites in Libya and California respectively during the Eocene from 49-50 Ma. Similarly, other studies have used differences in the distribution/ abundance relationship of pollen and spores with distance from the shore qualitatively to define marine to non-marine environments and also possibly changes in water depth (Upshaw, 1964; Zaitzeff and Cross, 1970). Here the ratio of terrestrial organisms (pollen and spores) decreases in proportion to marine microplankton with increasing distance from the shore, although influences of reworking and surface currents may be a distorting factor in this pattern (Heusser, 1978). Biostratigraphic information of these types, based on inferences from the fossil record, provide an important source for identification of long-term changes in relative sea level. For example, Hallam (1978) suggests a long-term cyclic pattern of sea-level change, with a general rising trend in relative sea level through the Jurassic (see Fig. 16.13), based largely upon fossil faunal criteria. Other studies for the Ordovician-Silurian (McKerrow, 1979) Carboniferous (Ramsbottom, 1979) and Cretaceous times (Cooper, 1977; Hancock and Kauffman, 1979) use different fossil groups similarly to show cyclic variations of sea level. The contribution of sea-level models from these sources provides not only an important data base in themselves for characterising the global pattern of relative sea-level change, but also a basis for testing other sea-level models, such as those derived from seismic stratigraphy.

SEISMIC STRATIGRAPHY AND GLOBAL RELATIVE SEA-LEVEL CHANGE

The development of seismic stratigraphy as a major technique in the examination of thick offshore sedimentary sequences (Payton, 1977; Brown and Fisher, 1980) has become controversially linked, particularly through the work of Vail *et al.* (1977), with the development of a long-term record of global sea-level change (see Mörner, this vol., Ch. 8). In the view of these

545

workers, the environmental mechanism of sea-level change forms a dominant depositional control upon sediments forming on continental margins. The record of sea-level changes preserved in such sediments, therefore, in conjunction with the stratigraphic interpretation of the sediments through seismic profiling (seismic stratigraphy) is thus seen as an important tool in hydrocarbon exploration. Vail and Hardenbol (1979) write,

> Studies of global changes in sea level throughout geological history are important to hydrocarbon exploration as an instrument of stratigraphy and geochronology, especially in areas that lack well or outcrop inform- ation. Sedimentation responding to changes in sea level, has produced a unique stratigraphic record by shaping sedimentary sequences that can be studied by seismic reflection.

The methodology

Details of the methodology underpinning the approach have been given in a number of papers, Vail *et al.* (1977, 1980 1983), Vail and Hardenbol (1973) and Vail and Todd (1981). In brief, it is argued that the position of relative sea level forms the base level to which processes of subaerial and submarine sediment deposition build. The balance between rates of appar- ent water volume-level (sea-level) change, sediment influx and land– crustal movement are reflected in lateral and vertical shifts over the conti- nental shelf/slope of the position of sedimentary facies. A trend of relative rise in sea level is broadly equated with the onlap of littoral–coastal deposits and offlap of marine sediments over the continental margin (Fig. 16.5). This change may result from:

(1) Actual sea-level rise, while the sea floor remains stable, or rises at a slower rate;
(2) Sea level remains stationary while the underlying depositional surface subsides;
(3) Sea level falls while the sea floor subsides at a faster rate.

Conversely, a trend of falling relative sea level is linked to the onlap of marine sediments and seaward progradation–downward shift of the coastal onlap model (Fig. 16.6). In this case changes may result from:

(1) A real fall of sea level whilst the sea floor remains stable, or subsides at a slower rate;
(2) Sea level remaining stationary, whilst the sea floor is rising, *or*
(3) A continued rise in sea level, whilst the sea floor rises at a faster rate.

546

Figure 16.5: Marine offlap–coastal onlap model of sedimentation equated with a relative sea-level rise

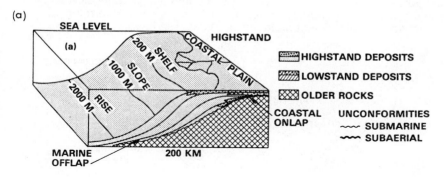

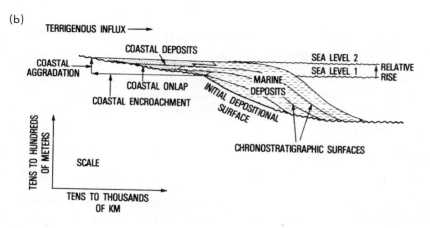

Source: Vail *et al.* (1977).

Recognition of the changes recorded in the sedimentary sequence is achieved through seismic profiling. Here primary seismic reflections resulting from velocity–density contrasts in the rock sequence are generated by physical surfaces in the rock. These are composed primarily of bedding surfaces and unconformities; it is the latter that form the main focus for the seismic stratigraphic approach.

Unconformities are characterised by a lack of continuity in deposition and are interpreted as resulting mainly from the non-deposition or erosion of sediments during times of rapid relative sea-level rise and fall respectively (Vail *et al.*, 1977). These surfaces represent significant time gaps separating older from younger rocks everywhere along their line of occurrence and their presence–identification in the stratigraphic sequence is

547

Figure 16.6: Downward shift of the coastal onlap sequence and the marine onlap of sediments is equated with a trend of relative sea-level fall. In (b) the downward shift occurs between Units 5 and 6 in sequences A and B, sequence A forming during a relative sea-level rise with B forming in response to a subsequent rapid fall in relative sea level

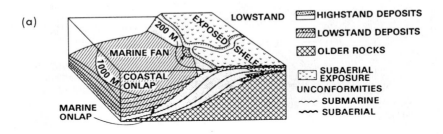

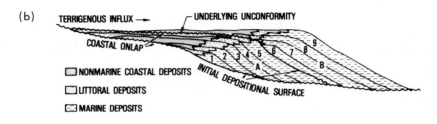

Source: After Vail *et al.* (1977).

essential to the methodology. Important assumptions made are that unconformities, or their correlative conformities in the rock sequence (Fig. 16.7) (Vail *et al.*, 1977), are not diachronous and may be assigned a specific geological age. They define the top and base of the base stratigraphic unit, or depositional sedimentary sequence, which is composed of a relatively conformable succession of genetically related strata (Fig. 16.7 and see also Fig. 16.8). Further, unconformities are seen as forming the initial depositional surface for subsequent sediment accumulation and it is with respect to this surface that the direction of relative sea-level change is interpreted from the sedimentary sequence. Consequently, Vail *et al.* (1977, 1983) argue that unconformities constitute major chronostratigraphic boundaries and their recognition provides the basis for setting up a chronological framework—correlation of coastal onlap sequences from different continental margins.

Initial dating of a coastal onlap sequence is determined from biostratigraphic data derived from available deep borehole records and these in turn are tied to geochronometric timescales (Vail *et al.*, 1983; van Hinte,

Figure 16.7: Concepts of depositional sequence, unconformities and hiatus and their chronostratigraphic interpretation

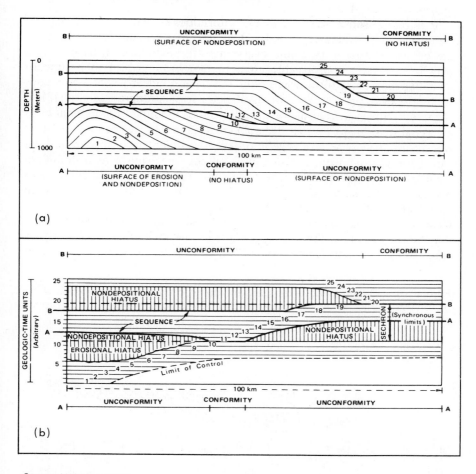

(a)

(b)

Source: Vail *et al.* (1977).

1978). Having established the local–regional pattern of unconformities — coastal onlap sequence the correlation, or more strictly 'matching' (Shaw, 1982), of three or more such patterns from different global locations is used as the basis for constructing a global sequence of coastal onlap. Further discussion of this procedure is given in Vail *et al.* (1983) and Vail and Hardenbol (1979), whilst details of the quantification of the seismic stratigraphic record into the pattern of coastal onlap and the direction–scale of relative sea-level change may be found in Vail *et al.* (1977: Part 3). Subsequent comparison of the global model with other regional curves for

549

coastal onlap is made both to help further define the established global model and to identify any anomalies that may be related to regional tectonic or local structural events.

It is recognised that although unconformities of the same age may be found in many widely separated sedimentary basins, such unconformities are not necessarily to be found everywhere throughout each basin. There may be extensive conformable areas in that basin, whilst other unconformities may be of local origin only and have no clear relation to a controlling sea-level mechanism, or significance for a stratigraphic model. It is suggested, however, that major unconformities identified on a global basis are generated through primarily rapid changes in relative sea level and that unconformities of the same age in different sedimentary basins around the world must result from the operation of this mechanism as a dominant environmental control. The rate of tectonic subsidence along continental margins is seen as too slow, as decaying in time and as varying between basins and is, therefore, viewed as unsatisfactory as an explanation of the observed synchroneity in unconformities Vail *et al.* (1983). An example may be seen in the sedimentary record from the shelf edge of West Africa (Fig. 16.8) (Vail and Hardenbol, 1979). Here geohistory analysis suggests

Figure 16.8: Seismic line taken perpendicular to the shelf edge offshore from West Africa, showing erosional and depositional patterns of Tertiary strata

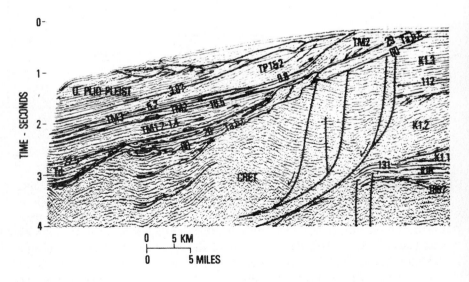

Source: Vail and Hardenbol (1979).

550

that subsidence ceased and the shelf became stable by the Mid-Cretaceous. The identification of subsequent major Tertiary unconformities at 60, 29, 22.5 and 6.6 Ma in the seismic record are thus interpreted as the result of rapid falls in relative sea level, interspersed with phases of normal deposition.

Application of the model to hydrocarbon location

On the basis of a composite of such regional sequences a model for global coastal onlap has been constructed for the Phanerozoic (Fig. 16.9). The application of such a model is seen as of importance to hydrocarbon exploration in four main areas (Vail *et al.*, 1977):

(1) The linkage of seismic stratigraphy to the sediment/sea-level model provides an improved basis for stratigraphic and structural analyses within sedimentary basins.
(2) In areas where detailed drill hole records are sparse or non-existent, the model allows a reasonable estimation of the age of strata, and thus

Figure 16.9: Generalised global cycles of relative coastal onlap during Phanerozoic time, coupled with a more detailed plot of data for the Mesozoic and an estimate of associated changes in global relative sea level

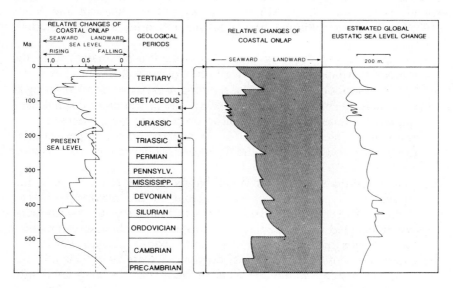

Source: Derived from Vail *et al.* (1977) and Vail and Todd (1981).

551

likelihood of hydrocarbon presence, before need for recourse to expensive drilling operations.

(3) Development of a global system of geochronology, which, unlike related biostratigraphic techniques, allows the Phanerozoic to be divided up into time units on a common criterion.

(4) The model of sedimentary facies change and underlying causative process, if correct, should aid in reconstruction of the environment represented by the sedimentary sequence and help in the location of oil and gas traps.

Long-term changes in sea level — support for the seismic stratigraphic model

The existence of long-term changes in relative global ocean water volume during Phanerozoic time seems to be accepted as both theoretically possible and a necessary response to geophysical changes in the earth's crust (Pitman, 1978; Donovan and Jones, 1979). It is unclear, however, as to the cause(s) of such changes and whether they were recorded as global or regional events. These doubts have great significance for both the feasibility and interpretation of the seismic coastal overlap model established by Vail *et al*.

Many causes have been proposed in explanation of assumed global changes in sea level and discussion of these may be found both in this volume (Chappell, Ch. 2; Mörner, Ch. 8) and in Donovan and Jones (1979) and Hays and Pitman (1973). Six main theories may be isolated:

(1) The growth and decay of major hemispherical ice masses form the most effective mechanism for variation of ocean water volume at short-term timescales of 100 ka-1 Ma, resulting in rapid sea-level lowering at rates of 1000 cm a^{-1000} (Pitman, 1978). The operation of this mechanism would cause relative sea level to drop below the continental shelf and would effectively account for the widespread oscillations of relative sea level observed from the Oligocene/Miocene onwards (Tanner, 1968; Rona, 1973).

(2) Changes in the shape of the ocean basins through sediment infill and consequent displacement of water across the landmasses (Hays and Pitman, 1973). In particular, changes in volume of the ocean ridge system coupled with variations in the rates of sea-floor spreading may form a major long-term cause of sea-level change (Pitman, 1978).

(3) Alternations of 'high' and 'low' orogenic activity or associated variations in melting within the asthenosphere may be associated with trends of falling and rising global sea level respectively (Hays and Pitman, 1973; Sloss and Speed, 1974).

552

(4) Decantation of epicontinental seas, or alternatively new flooding of deep continental basins, may produce global sea-level events.

(5) Change in the volume of the hydrosphere through the addition of juvenile water (see Chapter 2).

(6) Expansion in time of earth size (Hallam, 1971).

Of these potential causes, Pitman's model (1978) has achieved wide support as a primary explanation of first-order changes in global sea-level (Fig. 16.2) and has been employed by Vail *et al.* in the calibration of their independently derived seismic stratigraphic model of relative coastal onlap/sea-level change. Growth in ocean ridge volume consequent upon a rapid increase in sea-floor spreading rates, given stability in factors (5) and (6), would result in a long-term trend of relative sea-level rise. Conversely, reduction in the rate of sea-floor spreading would lead to a decrease in ridge volume and consequently a trend of relative sea-level fall (Fig. 16.10). Computations based upon geophysical data of changes in sea-floor spreading show a rapid pulse of spreading between 110.5 to 84.5 Ma, resulting in ridge growth and a relative sea-level rise of ~350 m above present levels by ~85 Ma (Larson and Pitman, 1972). This estimate appears to agree with independent geological data for an Upper Cretaceous sea-level maximum (Sleep, 1976). Decrease in spreading rates after this date caused the ocean ridges to contract with a consequent long-term fall of relative sea level. Subsequent variations in the rate of sea-floor spreading, superimposed on this broad pattern, may further provide a partial explanation of the second order cycles of sea-level change (Fig. 16.2).

Record of changes in sea level are often derived from the thick sedimentary sequences accumulating on passive (Atlantic type) continental margins, the focus for oil and gas exploration. These zones effectively form hinge lines for ocean basin subsidence and, as such, would be sensitive to changes in ocean volume and thus relative sea-level movement, as argued by Vail *et al.* (1977). Pitman (1978) proposes, in a simplified 'passive margin' model, a method of calculating shoreline changes over time on such a margin. Such changes would occur as a function of a balance between ocean ridge volume–fluctuating spreading rates, tectonic ocean basin subsidence, sediment influx and shelf margin loading (Fig. 16.11). Important boundary conditions here, based on geological data, are that total subsidence rates do not exceed 2-4 cm a^{-1000}, subsidence is always greater than sea-level rise, or fall, and that this does not normally exceed a rate of 1 cm a^{-1000}. (As noted by Orford, this vol., Ch. 13, Worsley *et al.* (1984) indicate that Vail *et al.* type changes in sea level are of the order of 1-2 mm Ma^{-1} during phases of continental movement of 25-50 Ma.) Further, it is assumed that sediment supply is able to keep pace with any combination of subsidence and sea-level rise, thus maintaining an approximate equilibrium

553

Figure 16.10: (a) Model of the repercussions for relative sea-level change and other elements of the environment caused by an increase and subsequent decrease in the velocity of crustal plate motion since the Upper Cretaceous. (b) Calculated relative sea-level curve based on ocean spreading rates

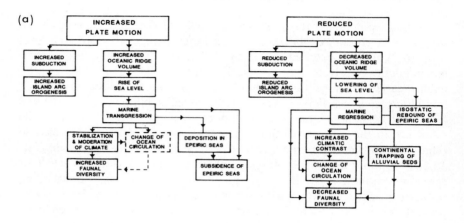

(a)

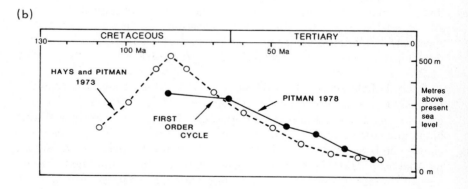

(b)

Source: (a) Hays and Pitman (1973). (b) After Vail *et al.* (1977).

profile with the sea surface. In this situation a shoreline position (X_L) relative to the hinge line may be found from the expression:

$$X_L = (R_{SL} + S)_{RSS}$$

where S = O, i.e. sedimentation is uniform over the continental shelf. In short the model shows that a relative rise in sea level may result simply from short-term changes in the rates of sea-floor spreading—ridge volume

554

Figure 16.11: Generalised model of an Atlanctic-type (passive) continental margin, modelled as a platform subsiding at a constant rate about a fixed hinge line. Thus, the subsidence rate decreases linearly from maximum values at the shelf edge to zero at the hinge line. It is assumed that sedimentation (and erosion) rates are distributed so as to maintain a constant slope. D = distance from hinge line to shelf edge; X_L = distance from hinge line to shoreline; X = distance from hinge line to any point on the shelf or coastal plain; S_L = shelf and coastal plain slope; R_{SS} = rate of subsidence at shelf edge of basement platform relative to an horizontal plane that extends through the hinge line; R_{SL} = rate of sea-level change (positive downward) relative to the same horizontal plane; dy_{ss}/dt = rate of vertical movement of the shelf surface relative to the same horizontal plane; dy_{sw}/dt = rate of vertical movement of the sea-level surface relative to the shelf surface; dx/dt = rate of movement of the shoreline relative to the hinge line

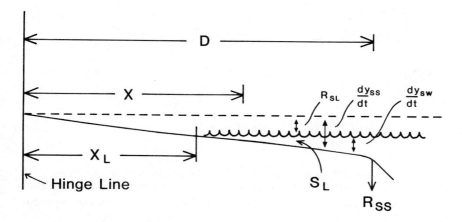

Source: Pitman (1978).

change. Thus a relative fall of sea level may occur when the general trend is one of rise because of a decrease in the rate of rise. Alternatively, a relative sea-level rise may occur with a falling sea-level trend because of a decrease in the rate of fall. Examination of the post Cretaceous produced decline in relative sea-level position using this model successfully predicts oscillations from the trend (Fig. 16.12), a pattern confirmed in broad detail from geological sources.

Despite the feasibility of this type of ridge volume change–sea-floor spreading model as a cause of sea-level movement, the concept can only account for a much slower and vertically smaller scale of sea-level change (<0.7 cm a^{-1000}) (Pitman, 1978) than that of the glaciation mechanism. Thus explanation of shorter frequency global changes of sea level prior to the Miocene remain an enigma. Mörner's proposal (see Chapter 8) of

Figure 16.12: Pattern of Late Mesozoic–Tertiary phases of relative sea-level change calculated from (i) ridge volume changes and the simplified model of margin subsidence and (ii) geological record observations (scale is arbitrary). Curve A (solid line) shows the calculated relative sea-level rise/fall sequence (the dashed line indicates possible curve changes if 'Deep Sea Drilling Project' results are used to recalibrate the magnetic timescale used). Curves B and C show the sea-level records from North America and Africa respectively, using stratigraphic data

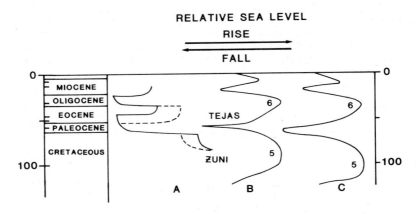

Source: Pitman (1978).

palaeogeoid changes as an explanation remains open to question and the results of further study. As yet, therefore, no universal mechanism, other than that of glaciation, is considered feasible in accounting for the apparent geologically rapid falls of relative sea-level below the shelf edge recognised by Vail *et al.* or, in a related context, for the development of unconformities on continental margins (Watts, 1982). Yet if their records are correct such large-scale oscillations of sea level at times of non-glaciation were commonplace. Support for this viewpoint may be found in the palaeontologic record. Studies of Tertiary foraminifera by Barr and Berggren (1980), in cores taken offshore from California, and by Berggren and Aubert (1980), from Libya, independently show a sudden switch, between 50-49 Ma, from assemblages indicative of bathyl conditions to those of neritic environments. This is taken to indicate the occurrence of a widespread and rapid fall of relative sea level at this time, amounting to 300-400 m change in water depth. However, this estimate is probably too large, being based on the fossil data alone.

Similar biostratigraphic evidence of repeated rapid changes in relative sea level during geological time comes from a variety of other sources (Hallam, 1981; McKerrow, 1979; Ramsbottom, 1979; Matthews and Cowie, 1979). Study of the Jurassic record (Hallam, 1975, 1978) through sedimentary, faunal and geochemical techniques has allowed recognition, in a wide range of facies, of repeated series of upward deepening and shallowing sedimentary sequences. These can be traced and correlated accurately over large areas ($10^{5/6}$ km^2) within northwest Europe and even as far as South America. Due to the widespread nature and synchroneity of these changes it is argued that global sea-level change provides the most plausible mechanism in explanation of the pattern. Although Hallam (1981) admits that quantification of the vertical extent of the sea-level changes is difficult, the observed record is best represented by (1), a rapid rise and fall, interspersed with a longer phase of stillstand, or alternatively by a combination of this model with (2), a rapid rise followed immediately by a slow fall (Fig. 16.13). Comparison of this tentative pattern of sea-level change for the Jurassic with that of Vail *et al.* (1983), Vail and Todd (1981), shows a strong similarity between the two models. Both show a long-term trend of rising relative sea level coupled with a series of apparent rapid, short-term oscillations. Differences do occur; in particular Hallam

Figure 16.13: Comparison of estimated curves for global eustatic sea-level change during the Jurassic

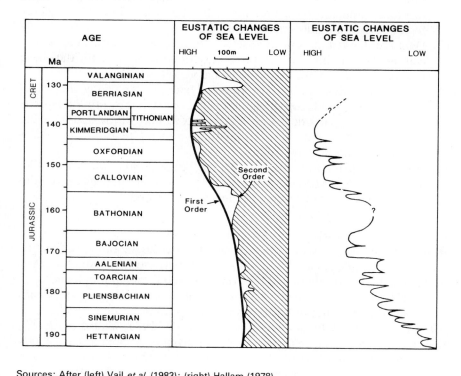

Sources: After (left) Vail *et al.* (1983); (right) Hallam (1978).

557

(1981) points to the opposing trends shown between the two curves in the Sinemurian/Hettangian ages. This may be a function of the different types of data base — approach taken, accuracy in inferring the degree and direction of change dependent on the respective methodologies, the source areas and size of data base used in representing apparent regional–global trends.

Problems with the seismic stratigraphic model of sea-level change

The validity of the Vail *et al.* seismic stratigraphic model has been widely questioned. Are the rapid short-term (second, third order) falls of relative sea level/coastal onlap an artifact of the seismic technique, the plotting and interpretation of the inferred sedimentary record, or are they real? Criticisms by Hallam (1981), Pitman (1978), Thorne and Watts (1984), Watts (1982), Pitman and Golovchenko (1983) and Brown and Fisher (1980) centre to some extent on the data sources, their credibility and universality:

(1) Much of the primary seismic and borehole information is not on public record but remains on confidential file with Exxon Ltd and is, therefore, not open to independent scrutiny.
(2) Data are derived, almost exclusively, from sediment sequences on the continental margins of interest to oil exploration. As such the data base may be biased and not have true global validity.
(3) Discovery of differences in record detail between the Vail *et al.* 'global' patterns and more focused, higher resolution studies by other researchers on specific parts of the geological record may throw doubt on the accuracy of other sections of the seismic stratigraphic record for which independent checks are lacking.
(4) Variations in relative sea level on the scale of 300-400 m (see Fig. 16.14) over ⩽ 1 Ma seem unreasonable, or even impossible, given the mechanisms of sea-level change currently proposed.

Caution of Vail *et al.*'s results along these lines are, in effect, more 'a feeling in the bones' by other researchers than accurate and specific criticism of the techniques used and the results themselves. Nevertheless, they do highlight areas of doubt and lead onto more substantial problems.

Since the original publication of the global sea-level record (Vail *et al.*, 1977) some important modifications in concept have become necessary. It has been realised by the authors (Vail and Todd, 1981; Vail *et al.*, 1983) that the recording of relative sea-level movements cannot be directly related to the seismic record of coastal onlap in its upper stages. In reality, a sea-level/shoreline position should approximate (in the stratigraphic record) to the facies boundary between coastal and alluvial plain sediments. A lack of sufficient resolution in the seismic technique makes it impossible to distinguish between these two facies. Consequently alluvial plain sediments, with their greater dip, and those of the littoral zone are combined in the coastal onlap record. This source of error causes a steepening of the 'curve' for coastal onlap, resulting in the showing of more abrupt changes in the direction of movement than are real. Variations in the rate of ocean basin sedimentation and depth over geological time may also influence the reconstruction of the apparent rate of sea-level change (Hays and Pitman, 1973). A more accurate plot of changing relative sea level direction should show more gradual shifts in direction. Thus the global charts developed represent relative changes in coastal onlap and not sea level *per se*, although the plots are similar. However, changes of the relative sea-level curve resulting from this modification, to show more gradual shifts in direction of movements over time, may now be at variance with the palaeontological data. As discussed, this often shows rapid palaeo-bathymetric changes, although local factors and interpretational error may again be the cause.

Other imponderables also exist, for example, in the method of calculating the scale of vertical relative sea-level change/coastal onlap, particularly in interpreting the rate of sea-level fall from an erosional surface. Although the age of unconformities, and thus the timing of sea-level fall, have been dated to ≤ 1 Ma in some locations (Vail *et al.*, 1977), this may not be the pattern everywhere. Development of other surfaces may have taken longer and been achieved through erosion interspersed with stillstand phases spread over a longer time span. Further, it has been realised that the maxima and minima in relative sea-level rise and fall do not necessarily coincide with the sediment cycle of coastal onlap. Changes in sediment budget or subsidence rates may mean that a peak in sea-level rise occurs in the middle of a coastal onlap cycle, whilst the basal position of sea-level fall occurs after the downshift in coastal onlap and thus above the associated unconformity. These modifications to the original model are not seen as altering either its underlying global sea-level change cause or its validity, nor the usefulness of the technique to hydrocarbon exploration.

In the view of others many questions remain to be answered (Brown and Fisher, 1980: pp. 101 and 104). Recognition of the existence of global cycles of sea-level change depends much upon the accuracy of the dating techniques used and these in large part are derived from biostratigraphic sources. In spite of improvements in biostratigraphic timescales through correlation with geochronometric data (van Hinte, 1978), major areas of disagreement remain. Pitman (1978) in his calculation of post-Cretaceous relative sea-level movements, recognised that possible errors in the chronology used would result in variations both in the timing and calculation of the vertical scale of sea-level change. Further, the results of Pitman's work, used by Vail *et al*. to calibrate their data, is based upon the subsidence–sedimentation model (Fig. 16.11). Here the vertical extent of sea-level change events is seen as limited to positions above the continental shelf edge. However, the first- and second-order cycles shown by Vail *et al*. as occurring throughout the Phanerozoic are vertically large and rapid in form and exceed this boundary condition. These changes are more akin to the changes in sea level associated with glaciation. Yet, as Vail *et al*. admit, the glacial mechanism does not seem feasible as an explanation of the seismic stratigraphic record of global sea-level change prior to the Oligocene/ Miocene. Even for this time evidence of glaciations affecting the earth on a global scale is tenuous (see Mörner, this vol., Ch. 8).

Doubts also exist as to whether the data underpinning the model truly fulfil the criteria set out by Vail *et al*. as necessary for determining the global significance of unconformities. The process of comparing regional cycles of change as a basis for constructing the global model is open to bias from sites showing the most pronounced pattern. Hallam (1981) evidences here the major Oligocene drop in the coastal onlap curve. It is suggested that this is based largely upon data from the North Sea, other regions showing less significant departures (Fig. 16.14). Further, one may question the validity and degree of accuracy in comparing zones like the North Sea, for example, with others used such as the Gippsland Basin (see Chappel, this vol., Ch. 2) or the San Joaquin Basin. It may be, therefore, that elements of the global curve are weighted in favour of one or two sites only and that these are of no more than regional significance. This problem of representivity becomes worse with increasing geological age. Little seismic data is available from the oceans–continental margins prior to ~200 Ma and the Palaeozoic curve appears to rely heavily upon evidence from North

Figure 16.14: Comparison of regional cycles of relative change of coastal onlap from four continents and averaging to construct global cycles

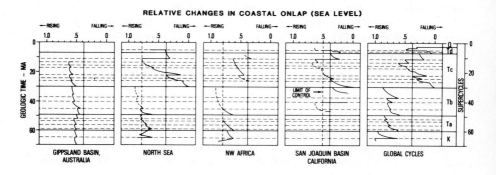

Source: Vail *et al*. (1977).

America alone. For many, publication and research of a much broader data base is necessary to validate the global correlations shown in the seismic stratigraphic model, although this may be contrary to commercial interests at present. On this basis Brown and Fisher (1980) also question the causal role of 'global sea-level change' itself in determining the sedimentary record on the continental margins that have been examined.[1] It is possible that many of the margins studied underwent similar ¯sedimentary and tectonic histories (i.e. North Sea, Northwest Africa, Gulf of Mexico). Support for this view comes from Watts (1982), who suggests that the controls of tectonism and continental margin flexure may have conditioned the sedimentary record, rather than a mechanism of global sea-level change. In many situations, however, whether sea level or other factors formed the dominant environmental control in shaping the stratigraphic record would be difficult to determine from the available data base.

A final major area of criticism centres upon the sedimentological interpretation of the seismic profile itself. Disagreement (Brown and Fisher, 1980) focuses on the reality of the Vail *et al.* model for downward shift in coastal onlap, showing sediment movement across the continental shelf and the transfer of land-river sediments to shelf edge deltas. Here, further transfer downslope would occur via erosional submarine canyons to produce onlapping submarine fans, derived directly from river-coastal sources. In Brown and Fisher's experience they have only rarely been able to demonstrate in borehole-supported seismic data the existence of such a sediment sequence. Their interpretation of the relevant seismic records suggests the alternation of prograding clinoforms and onlapping deep marine facies (Fig. 16.15), produced from the erosion of relict shelf-slope sediments through turbidity and related autosuspension mechanisms. The importance of this disagreement in interpretation is relevant in two main areas:

(1) The existence and need for a downward shift of coastal onlap below the shelf edge as a universal explanation of continental margin deposition in response to a sea-level change mechanism.
(2) Application of the seismic-control onlap model to the location of hydrocarbon resources. Brown and Fisher state that the most productive hydrocarbon traps and best-quality reservoirs occur not in prograding river-fed onlap systems (as per Vail *et al.*), but in submarine onlapping fans derived from relict sediment sources.

Clearly the arguments are ones for sedimentologists and require improved stratigraphic/sediment models for continental margin deposition, based upon accurate ground truthing of seismic records through borehole and related field data.

Figure 16.15: A. Nature of offlap and onlap slope deposition, and B. Processes of deposition. Models by Vail *et al.* (1977) infer that sediment deposited in deep water is introduced by direct river/delta deposition. Brown and Fisher (1980) suggest that most slope sediment in onlap systems is derived by submarine erosion processes from slope/shelf areas. The submarine fans shift landward with sediment supply in response to such erosion/shelf retreat

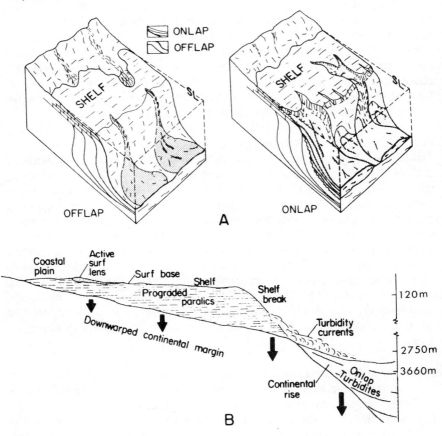

Source: Brown and Fisher (1980).

LONG-TERM SEA-LEVEL CHANGE — APPLICATIONS AND CONCLUSIONS

The applications of the seismic stratigraphic sea-level model to the hydro-carbons industry, as suggested by Vail and co-workers, are clear and do not need restating in detail. They lie primarily in helping both to locate such

562

resources and to reduce the cost of resource exploration. For one major oil company at least the value of such an approach seems to have been established. Direct spin-offs from the model may lie, if it becomes more widely validated, in providing a global system of ocean basin geochronology. This could be used as a cross-check on biostratigraphic timescales, although the problems of circular argument resulting from the present dependence of the published seismic records on biostratigraphic data are obvious.

In the broader sphere of environmental evolution, the development of the coastal onlap/sea-level model has provided great stimulus to the quantitative study and modelling of long-term global sea-level changes. Many environmental scientists have speculated on the importance of such sea-level movements as a dominant control on earth environmental change. The interrelated factors of changing ocean basin shape and direction of sea-level movement may have had profound repercussions on global climate, ocean circulation and even upon biological evolution (Hays and Pitman, 1973). In terms of climate and ocean circulation the ubiquitous shallow epicontinental seas and ocean areas of the Cretaceous probably encouraged and stabilised the existence of the moderate, equable earth climate recorded at this time. Subsequent relative falls in sea level, consequent upon ocean ridge spreading and ocean basin growth, may be linked with the subsequent onset of climate deterioration, ice cap growth, eventual continental glaciations and the setting-up of strong temperature, environmental gradients. The environmental stresses caused by such changes may have been an important factor in the subsequent faunal–floral crises of the Mesozoic–Cenozoic. Whatever the balance of truth and speculation in such scenarios the detailed study of long-term sea-level changes, encouraged by the debate over the 'seismic sea-level' model, cannot but help to add to our knowledge of earth environmental processes.

Finally, at the other end of the time spectrum, the study of sea level may have relevance to the location of the industries and people dependent on the hydrocarbon resource. In the same way as sea-level change factors influence the initial location, type and discovery of hydrocarbon resources, so they may help to determine areas where such resources may best be refined or used. Commercial and economic interests may dictate exactly where in the world these activities occur. Nevertheless, most petroleum refineries and many ancillary industries are necessarily located in the coastal zone. As the chapters on modern coastal processes showed (Part Four, this volume) this environment is both dynamic and dangerous! Studies of sea-level change mechanisms as factors in determining coastal position, vertical water motion and consequent flood hazard zoning are, therefore, important. In this context both long-term and short-term sea-level studies are an essential part of sensible environmental planning and management.

REFERENCES

Anderson, R.O. (1985) *Fundamentals of the Petroleum Industry*, University of Oklahoma Press and Weidenfeld & Nicolson, Norman, Okla., and London.

Barnard, P.C. and Cooper, B.S. (1981) 'Oils and source rocks of the North Sea area', in L.V. Illing and G.D. Hobson, *Petroleum Geology of the Continental Shelf of Northwest Europe*, Inst. Civ. Eng., Heydon & Son, London, pp. 169-75.

Barr, F.T. and Berggren, W.A. (1980) 'Lower Tertiary biostratigraphy and tectonics of northeastern Libya', in *Second Symp. on the Geology of Libya*, Academic Press, London and New York, pp. 48-61.

Berggren, W.A. (1969) 'Cenozoic chronostratigraphy, planktonic foraminiferal zonation and the radiometric time scale', *Nature*, *224*, 1072-5.

—— (1972) 'A Cenozoic time scale — some implications for regional geology and palaeogeography', *Lethaia*, *5*, 195-215.

—— (1978) 'Marine micropaleontology: an introduction', in Haq, B.U., and Boersma, A. (eds) *Introduction to Marine Micropaleontology*, Elsevier, New York, pp. 1-17.

—— and Aubert, J. (1980) *Paleogene Benthonic Foraminiferal Biostratigraphy and Bathymetry of the Central Coast Ranges of California*, USGS Prof. Paper No. 26.

Birks, J. (1978) 'Introduction to aspects of economics and logistics', in 'Sea floor development: moving into deep water', *Phil. Trans. Roy. Soc. Lond., A.*, *290*, 3-19.

Bitterli, P. (1963) 'Aspects of the genesis of bitumous rock sequences', *Geol. Mijnb.*, *42*, 183-201.

Boersma, A. (1978), 'Foraminifera', in B.U. Haq and A. Boersma (eds). *Introduction to Marine Micropaleontology*, Elsevier, New York, pp. 19-77.

Bois, C., Gess, G., Perrodan, A. and Pommier, G. (1975) *Ninth World Petroleum Congr. Proc.*, 2.

Brenner, R.L. (1984) *Petroleum Stratigraphy: A Guide for Non-Geologists*, Reidel-IHRDC, Dordrecht and Boston.

Brown, L.F. and Fisher, W.L. (1980) *Seismic Stratigraphic Interpretation and Petroleum Exploration*, Am. Ass. Petrol. Geol., Tulsa, Okla.

Chapman, R.E. (1973) *Petroleum Geology*, Elsevier, Amsterdam and New York.

Cooper, B.S. (1978) 'Estimation of the maximum temperature attained in sedimentary rocks', in G.D. Hobson (ed.), *Developments in Petroleum Geology*, Applied Science Publications, London, pp. 127-46.

Cooper, M.R. (1977) 'Eustasy during the Cretaceous: its implications and importance', *Palaeogeography, Palaeoclimatol., Palaeoecol.*, *22*, 1-60.

Couvering, J.A. van and Berggren, W.A. (1977) 'Biostratigraphical basis of the Neogene timescale', in E.G. Kaufman and J.E. Hazel (eds), *Concepts and Methods of Biostratigraphy*, Dowden, Hutchinson & Ross, Stroudsberg, Pa, pp. 283-302.

Cross, A.T. (1964) *Palynology in Oil Exploration, A Symposium*, Spec. Publ. No. 11, Soc. Econ. Pal. Min., Tulsa, Okla.

Cubitt, J.M. and Reyment, R.A. (1982) *Quantitative Stratigraphic Correlation*, Wiley, Chichester and New York.

Dikkens, A.J. (1985) *Geology in Petroleum Production*, Elsevier, Amsterdam.

Doyle, J.A. (1977) 'Spores and pollen: the Potomac Group (Cretaceous) angiosperm sequence', in E.G. Kauffman and J.E. Hazel, (eds), *Concepts and Methods of Biostratigraphy*, Dowden, Hutchinson & Ross, Stroudsberg, Pa., pp. 339-63.

Donovan, D.T. and Jones, E.J.W. (1979) 'Causes of world-wide changes in sea level', *J. Geol. Soc. Lond.*, *136*, 187-92.

Eldredge, N. and Gould, S.J. (1977) 'Evolutionary models and biostratigraphic strategies', in E.G. Kauffman and J.E. Hazel (eds), *Concepts and Methods of Biostratigraphy*, Dowden, Hutchinson & Ross, Stroudsberg, Pa., pp. 25-40.

Evitt, W.R. (1964) 'Dinoflagellates and their use in petroleum geology', in A.T. Cross (ed.), *Palynology in Oil Exploration, A Symposium*, Spec. Publ. No. 11, Soc. Econ. Pal. Min., Tulsa, Okla., pp. 65-72.

Fischer, S.C. (1974) *Energy Crisis in Perspective*, Wiley, New York.

Funnell, B.M. (1967) 'Foraminifera and radiolaria as depth indicators in the marine environment', *Mar. Geol.*, *5*, 333-47.

Gallois, R.W. (1976) 'Coccolith blooms in the Kimmeridge clay and origin of North Sea oil', *Nature*, *259*, 473-5.

Goldman, E.C. (1978) 'Offshore subsea engineering', *Phil. Trans. Roy. Soc. Lond.*, *A*, *290*, 99-111.

Gradstein, F.M. and Agterberg, F.P. (1982) 'Models of Cenozoic foraminiferal stratigraphy: northwestern Atlantic margin', in J.M. Cubitt, and R.A. Reyment, (eds), *Quantitative Stratigraphic Correlation*, Wiley, Chichester and New York, pp. 119-73.

Griggs, P.H. (1970) 'Palynological interpretation of the type section, Chuckanut Formation, northwestern Washington', in A.T. Cross (ed.) *Palynology of the Late Cretaceous and Early Tertiary*, Spec. Publ. 127, Geol. Soc. Am., pp.169-212.

Hallam, A. (1967) 'The depth significance of shales with bitumous laminae', *Mar. Geol.*, *5*, 481-93.

—— (1971) 'Re-evaluation of the palaeogeographic argument for an expanding earth', *Nature*, *232*, 180-2.

—— (1975) *Jurassic Environments*, Cambridge University Press, Cambridge.

—— (1977) 'Secular changes in marine inundation of USSR and North America through the Phanerozoic', *Nature*, *269*, 762-72.

—— (1978) 'Eustatic cycles in the Jurassic', *Palaeogeography, Palaeoclimatol., Palaeoecol.*, *23*, 1-32.

—— (1981) *Facies Interpretation and the Stratigraphic Record*, W.H. Freeman, Oxford.

Hancock, J.M. (1977) 'The historic development of concepts of biostratigraphic correlation', in E.G. Kauffman and J.E. Hazel (eds). *Concepts and Methods of Biostratigraphy*, Dowden, Hutchinson & Ross, Stroudsberg, Pa., pp. 3-22.

—— and Kauffman, E.G. (1979) 'The great transgression of the Late Cretaceous', *J. Geol. Soc. Lond.*, *136*, 175-86.

Haq, B.U. and Boersma, A. (1978), *Introduction to Marine Micropaleontology*, Elsevier, New York.

Hays, J.D. and Pitman W.C., III (1973) 'Lithospheric plate motion, sea-level changes and climatic and ecological consequences', *Nature*, *246*, 18-22.

Hazel, J.E. (1977) 'Use of certain multivariate and other techniques in Assemblage Zone biostratigraphy: examples using Cambrian, Cretaceous and Tertiary benthic invertebrates', in E.G. Kauffman and J.E. Hazel (eds), *Concepts and Methods of Biostratigraphy*, Dowden, Hutchinson & Ross, Stroudsberg, Pa, pp. 187-212.

Heusser, L. (1978) 'Spores and pollen in the marine realm', in B.U. Haq and A. Boersma (eds), *Introduction to Marine Micropaleontology*, Elsevier, New York, pp. 327-39.

Hinte, J.E. van (1976a) 'A Jurassic time scale', *Am. Ass. Petrol. Geol. Bull.*, *60*, 489-97.

—— (1976b) 'A Cretaceous time scale', *Am. Ass. Petrol. Geol. Bull.*, *60*, 498-516.

—— (1978) 'Geohistory analysis: application of micropalaeontology in exploration geology', *Am. Ass. Petrol. Geol. Bull.*, *62*, 201-22.

Hopping, C.A. (1967) 'Palynology and the oil industry', *Rev. Palaeobot. Palynol.*, *2*, 23-48.

Hunt, J.M. (1979) *Petroleum Geochemistry and Geology*, W.H. Freeman, San Francisco.

Kauffman, E.G. and Hazel, J.E. (1977) *Concepts and Methods of Biostratigraphy*, Dowden, Hutchinson & Ross, Stroudsberg, Pa.

Kuyl, O.S., Muller, J. and Waterbolk, H.T. (1955) 'The application of palynology to oil geology, with special reference to western Venezuela', *Geol. Mijnb., N.S.*, *17*, 49-76.

Larson, R.L. and Pitman III, W.C. (1972) 'World-wide correlation of Mesozoic magnetic anomalies and its implications', *Bull. Geol. Soc. Am.*, 83, 3645-62.

Laughton, A.S. (1975) 'Tectonic evolution of the northeast Atlantic ocean: a review', *Nor. Geol. Unders. Publ. 316*, 169-93.

Lowe, J. and Walker, M. (1984) *Reconstructing Quaternary Environments*, Longman, London.

McIver, R.D. (1975) 'Hydrocarbon sources from JOIDES deep sea drilling project', in *Panel Discussion 5, Petroleum Prospects in the Deep Ocean Regions*, preprint, Proc. 9th World Petrol. Cong. 1975, Applied Science Publications, London, pp. 1-13.

McKerrow, W.S. (1979) 'Ordovician and Silurian changes in sea level', *J, Geol. Soc. Lond.*, *136*, 137-45.

Matthews, S.C. and Cowie, J.W. (1979) 'Early Cambrian transgression', *J. Geol. Soc. Lond.*, *136*, 133-5.

May, J.A., Yeo, R.K. and Warme, J.E. (1984) 'Eustatic control on synchronous stratigraphic development: Cretaceous and Eocene coastal basins along an active margin', *Sed. Geol.*, *40*, 131-49.

Muller, J. (1959) 'Palynology of recent Orinoco delta and shelf sediments', *Micropaleontology*, *5*, 1-32.

Owens, B.C. (1981) 'Palynology, its biostratigraphic and environmental potential', in L.V. Illing and G.D. Hobson (eds), *Petroleum Geology of the Continental Shelf of Northwest Europe*, Inst. Petrol. Geol., Heyden & Son, London, pp. 162-8.

Parry, C.C., Whitley, P.K.J. and Simpson, R.D.H. (1981) 'Integration of palynological and sedimentological methods in facies analysis of the Brent Formation', in L.V. Illing and G.D. Hobson (eds), *Petroleum Geology of the Continental Shelf of Northwest Europe*, Inst. Petrol. Geol., Heyden & Son, London, pp. 205-14.

Payton, C.E. (ed.) (1977) *Seismic Stratigraphy: Applications to Hydrocarbon Exploration*, Am. Ass. Petrol. Geol., Tulsa, Okla.

Philippi, G.T. (1965) 'On the depth, time and mechanism of petroleum generation', *Geochim. Cosmochim. Acta*, *29*, 1021-49.

Pitman, W.C., III (1978) 'Relationship between eustasy and stratigraphic sequences of passive margins', *Bull. Geol. Soc. Am.*, *89*, 1389-403.

—— and Golovchenko, X. (1983), 'The effect of sea-level change on the shelf-edge and slope of passive margins'. *Soc. Econ. Palaeotol. and Mineral., Spec. Publ.*, *33*, 41-58.

Poag, C.W. (1977) 'Biostratigraphy in Gulf coast petroleum exploration', in E.G. Kauffman and J.E. Hazel (eds), *Concepts and Methods of Biostratigraphy*,

Dowden, Hutchinson & Ross, Stroudsberg, Pa., pp. 213-33.

Ramsbottom, W.H.C. (1979) 'Rates of transgression and regression in the Carboniferous of Northwest Europe', *J. Geol. Soc. Lond.*, *136*, 147-53.

Rona, P.A. (1973) 'Relations between rates of sediment accumulation on continental shelves, sea-floor spreading and eustasy inferred from the central North Atlantic', *Bull. Geol. Soc. Am.*, *84*, 2851-72.

Sampson, A. (1975) *The Seven Sisters: The Great Oil Companies and The World They Shaped*, Hodder & Stoughton, London.

Shaw, B.R. (1982) 'The correlation of geologic sequences', in J.M. Cubitt and R.A. Reyment (eds), *Quantitative Stratigraphic Correlation*, Wiley, Chichester and New York, pp. 7-18.

Sleep, N.H. (1976) 'Platform subsidence mechanisms and eustatic sea-level change', *Tectonophys.*, *36*, 45-56.

Sloss, L.L. and Speed, R.C. (1974) 'Relationship of cratonic and continental margin tectonic episode', in W.R. Dickinson (ed.), *Tectonics and Sedimentation*, Spec. Publ. No. 22, Soc. Econ. Pal. Min., pp. 89-119.

Stack, E., Mackowsky, M.Th., Teichmuller, M., Taylor, G.H., Chandra, D. and Teichmuller, R. (1975) *Textbook of Coal Petrology*, Gebruder Borntraeger, Berlin and Stuttgart.

Stanley, E.A. (1969) 'Marine palynology', *Oceanogr. Mar. Biol. Ann. Rev.*, *7*, 277-92.

Staplin, F.L. (1977) 'Interpretation of thermal history from colour of particulate organic matter — a review', *Palynology*, *1*, 9-18.

Stewart, J. (ed.) (1977) *Oceanography: Introduction to the Oceans*, Open University Press, Milton Keynes.

Strakov, N.M. (1962), 'Stages in the development of external geospheres and the formation of deposits in the history of the earth'. *Izvestiya Akademii Nauk SSR, Seriya Geologicheskaya*, *12*, 3-22.

Tanner, W.F. (1968) 'Tertiary Sea-level Symposium', *Palaeogeography, Palaeoclimatol., Palaeoecol.*, *5*.

Tappan, H. and Loeblich, A.R., Jr (1970) 'Geobiologic implications of fossil phytoplankton evolution and time–space distribution', in A.T. Cross (ed.), *Palynology of the Late Cretaceous and Early Tertiary*, Spec. Publ. 127, Geol. Soc. Am., pp. 247-340.

Thorne, J. and Watts, A.B. (1984), 'Seismic reflectors and unconformities at passive continental margins', *Nature, 311*, 365-8.

Tissot, B. (1979) 'Effects on prolific petroleum source rocks and major coal deposits caused by sea-level changes', *Nature, 277*, 463-5.

——— Durand, B., Espitalie, J. and Combaz, A. (1974) 'Influence of nature and diagenesis of organic matter in formation of petroleum', *Am. Ass. Petrol. Geol.*, *58*, 499-506.

——— And Welte, D.H. (1978) *Petroleum Formation and Occurrence*, Springer Verlag, New York.

Tschudy, R.H. and Scott, R.A. (eds) (1969) *Aspects of Palynology*, Wiley, New York.

Upshaw, C.F. (1964), 'Palynological zonation of the Upper Cretaceous Frontier Formation near Dubois, Wyoming', in A.T. Cross (ed.), *Palynology in Oil Exploration, A Symposium*, Spec. Publ. No. 11, Soc. Econ. Pal. Min., pp. 153-68.

Vail, P.R. and Hardenbol, J. (1979) 'Sea-level changes during the Tertiary', *Oceanus*, *22*, 71-9.

——— Hardenbol, J. and Todd, R.G. (1983), 'Jurassic unconformities, chronostrati-

567

graphy and sea-level changes from seismic stratigraphy and biostratigraphy', preprint to Am. Ass. Petrol. Geol. Mem. on 'Interregional Unconformities'.

—— Mitchum, R.M., Jr, Shipley, T.H. and Buffler, R.T. (1980) 'Unconformities of the North Atlantic', *Phil. Trans. Roy. Soc. Lond., A, 294,* 137-55.

—— Mitchum, R.M., Jr, Todd, R.J., Widmier, J.M., Thompson III, S., Sangree, J.B., Bubb, J.N. and Hatlelid, W.G. (1977) 'Seismic stratigraphy and global changes of sea-level', in C.E. Payton (ed.), *Seismic Stratigraphy: Applications to Hydrocarbon Exploration,* Mem. 26, *Am. Ass.* Petrol. Geol., Tulsa, Okla., pp. 49-212.

—— and Todd, R.G. (1981) 'Northern North Sea Jurassic unconformities, chronostratigraphy and sea-level changes from seismic stratigraphy', in L.V. Illing and G.D. Hobson (eds), *Petroleum Geology of the Continental Shelf of Northwest Europe,* Inst. Petrol. Geol., Heyden & Son, London, pp. 216-35.

Warman, H.R. (1978) 'Hydrocarbon potential of deep water', *Phil. Trans. Roy. Soc. Lond., A., 290,* 33-42.

Watts, A.B. (1982) 'Tectonic subsidence, flexure and global changes of sea level', *Nature, 297,* 469-74.

Worsley, T.R., Nance, D. and Moody, J.B. (1984) 'Global tectonics and eustasy for the past two billion years', *Mar. Geol., 58,* 373-400.

Zaitzeff, J.B. and Cross, A.T. (1970) 'The use of dinoflagellates and acritarchs for zonation and correlation of the Navarro Group (Maestrichtian) of Texas', in A.T. Cross (ed.), *Palynology of Late Cretaceous and Early Tertiary,* Spec. Publ. No. 127, Geol. Soc. Am., pp. 341-77.

Ziegler, P.A. (1981) 'Evolution of sedimentary basins in Northwest Europe', in L.V. Illing and G.D. Hobson (eds), *Petroleum Geology of the Continental Shelf of Northwest Europe,* Inst. Petrol. Geol., Heyden & Son, London, pp. 3-39.

Zimmerman, W. (1969) *Geschichte der Pfhanzen, eine Ubersicht,* Thieme, Stuttgart.

NOTE

1. Recent work by Schwarzacher, W. and Schwarzacher, W. (1986, 'The effect of sea-level fluctuations in subsiding basins'. *Computers and Geosciences, 12,* 225-7) shows that an equilibrium in sediment accumulation may be achieved independently of fluctuations in sea-level position.

17

Placer Deposits of the Nearshore and Coastal Zones: The Role of Littoral Processes and Sea-level Changes in Their Formation

Donald G. Sutherland

INTRODUCTION

Placer deposits are economically valuable mineral accumulations that have been formed principally by mechanical surface or near-surface processes. Littoral placers contain minerals of relatively low whole-rock abundance that, due to their physical and/or chemical stability and the operation of fluvial, marine and/or aeolian processes, have been transported to and concentrated in the coastal or nearshore zone. It follows that in this chapter no consideration is given to chemical precipitates such as manganese nodules or biochemical accumulations such as phosphorites. In addition, deposits of sand and gravel, although abundant in coastal areas and on the continental shelves, and of considerable value, are not discussed; for the processes that give rise to large accumulations of sand and gravel are frequently inimical to the concentration of low abundance heavy minerals.

Littoral placer deposits occur widely in both present-day and raised beaches around the world's coasts and have produced significant quantities of diamonds, gold, rutile, ilmenite, zircon and monazite as well as other minerals. Submerged river channels extending onto the shelf areas have also produced very large quantities of cassiterite. A variety of factors have been influential in their occurrence. Hinterland factors are important in the localisation of littoral placers, governing the supply of minerals to the coastal zone. Short-term sea-surface movements and related coastal processes are also important. For example, littoral processes result in the segregation of heavy minerals producing sorting both normal to the shore and alongshore. The interaction of dominant wave direction, coastal configuration and offshore profile also defines the extent of littoral circulation cells that have unique heavy mineral populations and within which placers develop. Finally, longer-term sea-level changes have resulted in the formation of raised beaches and submerged river channels in both of which placers may exist. The successive sea-level rises and falls of the Quaternary have had an important process function that has favoured formation of placers following sea-level rises, whilst reducing the possibilities of placer formation on the shelves.

Discussion in the chapter will now focus in turn upon these themes and through the discussion will attempt to show the interaction between such coastal processes–sea-level changes in the location and possible resource appraisal of placer deposits. A case study from the gold-bearing beaches at Nome, Alaska, is used to illustrate the complex interaction of sediment supply and sea-level change that is typically involved in the formation of many littoral placers.

LITTORAL PLACER DISTRIBUTION

Around the coasts of the world very many heavy mineral concentrations are known in raised marine deposits, the present littoral zone and in the immediate offshore areas (for reviews see, for example, Hails, 1976; Eliseev, 1980). A large proportion of these deposits are sub-economic and serve to confirm only the universality of the concentrating processes. They leave open the question as to the factors responsible for the creation of a deposit of suitable size and concentration of mineral to be economically workable. A selection of relevant deposits relating to gold, tin, diamonds, titanium minerals and zircon is shown in Figure 17.1, indicating their widespread nature. A number of mineral associations can be identified that helps explain the distribution of these placers.

Perhaps the most widely distributed littoral placers are the 'beach sand' heavy mineral accumulations consisting mainly of varying proportions of ilmenite, rutile, zircon and monazite as well as other more minor constituents. The bedrock sources for these minerals are dominantly in the metamorphic terrains of the shields or old, now quiescent mountain zones. Littoral concentrations are best developed along coasts flanking such areas where overall sediment budgets are low (Sutherland, 1985). Typical examples of this association are along the coasts of eastern and southwestern Australia, southern India and southeastern United States (Fig. 17.1). Where sediment supply is high as in the deltas of large rivers, some heavy mineral concentrations occur alongshore from delta distributaries (e.g. the Nile Delta, Wassef, 1981). Such areas, however, have not produced deposits of similar size and degree of mineral concentration to those along coasts with low sediment budgets.

The beach sand heavy minerals are of relatively low specific gravity (ranging from 4.0 to 5.5) and typically are very well sorted with mean particle sizes normally in the range of 0.1 to 0.3 mm. They are frequently reworked by aeolian processes into dune systems to the landward and such dunes are normally considered part of the littoral placer system. Beach sand placers have been dated from the Tertiary to the present and occur in altitude from over 100 m above to below present sea level. Exploration for

Figure 17.1: Distribution of certain gold (Au), cassiterite (Sn), diamond (D), ilmenite and rutile (Ti) and zircon (Zr) placers in coastal and nearshore areas

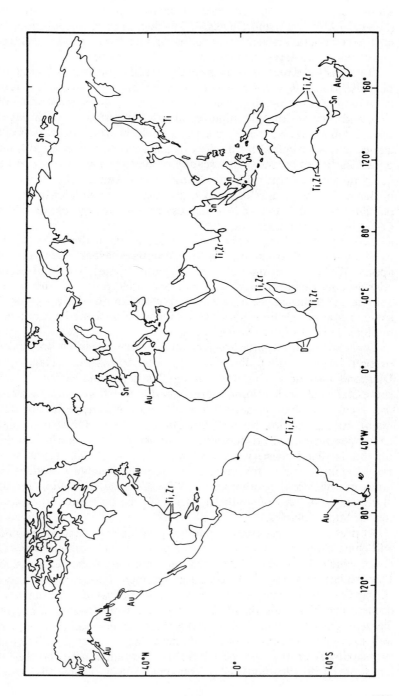

similar deposits on the continental shelves has met with little success (Schott, 1976; Jones and Davies, 1979) due, in part, to conditions favouring heavy mineral concentration following major rises in sea level (see later discussion of sea-level change factors).

Gold has been recovered from many beach deposits. As a large number of these deposits are the result of local marine erosion of auriferous bedrock, the resulting placers are typically small, although occasionally very rich, and they are of only minor economic interest. There is a larger-scale association between glaciated coasts and gold-bearing littoral deposits. This association is a result of the ability of a glacier to transport large volumes of poorly sorted sediment that is readily eroded by marine processes and its heavy mineral content being accordingly concentrated. The classic example of this is the Nome area of Alaska, discussed in detail later in the chapter, but the association extends to the eastern and western coasts of Canada, Chile, New Zealand and the USSR.

Gold beach placers, unlike those of diamonds or the beach sand heavy minerals, are typically limited in their longshore extent, the maximum size apparently being less than 30 km in length (Nome). The gold occurs in a considerable range of particle sizes from small gravel to fine sand. Most gold is recorded as being recovered close to the base of the littoral sediments although finer gold particles may be deposited with other heavy minerals in lenses in the beach sands.

Diamond is an important placer mineral that occurs in both raised and present beaches along the coast of southern Africa (Hallam, 1964). Diamond also occurs offshore along this coast and although there are successful small-scale mining ventures in the nearshore zone, farther out on the shelf there has been no successful economic exploitation despite a major attempt during the 1960s (Gurney, 1983). Occasional diamonds have been recovered offshore elsewhere, for example, during tin dredging off the Thailand coast (Garson et al., 1975), whilst diamonds have been traced along rivers to the margin of coastal sedimentary wedges in other areas of the world (Sutherland, 1982). To date, however, the deposits along the west coast of southern Africa are the only established diamondiferous littoral placers.

At present the west coast of southern Africa experiences a semi-arid to arid climate and the only major source of terrestrial sediment is the Orange River which drains the diamondiferous interior of the Kaapvaal craton. The present coast therefore has a low sediment budget and is also a high energy swell wave environment in which considerable sediment reworking occurs. The diamonds range in average size from over 2 carats ($\geqslant 6$ mm diameter particle) to under 0.01 carats ($\leqslant 1$ mm diameter particle), are very well size-sorted and progressively diminish in size away from river mouths (principally from the Orange River) in accordance with the direction of longshore drift (Stocken, 1962; Hallam, 1964). Diamonds have been recovered from over 1000 km length of coastline.

Cassiterite is a placer mineral of considerable importance in the near-shore area. It occurs in beach and nearshore sediments in various parts of the world (Fig. 17.1) but the most important area is Southeast Asia, particularly on the west coast of Thailand and Burma and around the 'tin islands' of Indonesia. Cassiterite has been recovered in relatively small quantities from the present-day beaches of this area (Hosking, 1971) but the major placer deposits are all former river channels which have been traced from the onland primary sources onto the shelf. The cassiterite in these river channels has only been transported for limited distances (van Overeem, 1960; Aleva, 1985) and is poorly sorted. More than one period of river incision is known, such periods having been related to low Quaternary sea levels (Aleva *et al.*, 1973). During periods of high sea level there has been limited reworking of cassiterite and barren or low-grade, fine-grained sediments have been deposited in the coastal areas. These cassiterite placers are the most extensive and valuable placers known to occur in the offshore areas of the world but they are not primarily the product of littoral or nearshore processes. Rather, they result from tectonic subsidence of the Sunda shelf allied to Quaternary sea-level change that has resulted in repeated marine invasion close to or across the primary sources of cassiterite.

Viewed broadly, the most extensive littoral placers are developed along swell dominated, trailing-edge coasts with low sediment budgets. Collision coasts (Inman and Nordstrom, 1971) and storm dominated coasts have many localities in which littoral placers are developed but the interactions of coastal configuration, wave climate and sediment supply favour the development of placers in the first broad category, as is now to be discussed.

LITTORAL SEDIMENT TRANSPORT: DETERMINING FACTORS IN PLACER DEVELOPMENT

Shore and beach forming processes, determining particularly the sorting and transport of placer minerals, result essentially from short duration movements of the sea surface. The dynamic nature of the shore and near-shore zone in effecting longshore transport of sediment gives rise to the conceptual sub-division of the coast into three units: (1) sources, where there is a net addition of sediment to the coast, (2) transport zones, where sediment is moved alongshore and a dynamic equilibrium between supply and loss of sediment is maintained, and (3) accumulation areas or sediment sinks where there is a net loss of sediment to longshore drift (Zenkovitch, 1967). This concept makes no reference to scale and the sizes of the units are the result of the interaction of coastal configuration, nature and angle of wave attack and, to a lesser extent, the offshore profile (see Orford, this vol., Ch. 13, for discussion of related coastal/offshore–sedimentation process concepts).

For placer deposits, a threefold sub-division of heavy mineral sources can be identified. First are point sources, these being either mouths of rivers carrying particular heavy mineral suites or the intersections of primary ores by marine erosion. Second are multiple sources where a series of river mouths along a section of coast all supply the relevant minerals or where marine erosion intersects a number of ore deposits. This latter category may include erosion of mineralised raised marine sediments with reworking into the contemporary littoral zone. The third type of source may be termed disseminated. Here the relevant mineral is supplied to the coast rather evenly over a considerable distance because of marine erosion of host rocks, in which the relevant heavy minerals are uniformly distributed. Examples of this would be erosion of glacial tills or of low-grade sandstone strata.

The sediment supplied to the coast from these different sources contains varying absolute amounts of heavy minerals. Absolute abundance of heavy minerals is of less importance to the development of beach placers than are the size distributions of both the heavy minerals and the remainder of the sediment and, in particular, the relative proportions of the two sediment fractions in the hydraulically equivalent sizes that accumulate on the beach. Thus a supply of sediment may have a high total percentage of heavy minerals but if these are in the size classes that are either removed offshore or are overwhelmed by the equivalent size class of the bulk of the sediment, no placer will form. Conversely, a relatively low heavy mineral frequency can give rise to a placer if the size classes of the heavy mineral coincide with a low abundance in the equivalent size classes in the bulk of the sediment and are also suitable for incorporation in the beach. This factor of sediment supply is a complex function of original size distribution, sorting during fluvial transport as well as the climate and tectonic regimes of the hinterland.

The length of littoral transport zones, given uniform wave conditions, is largely governed by coastal configuration. Where coasts have a relatively smooth outline and wave attack is at a constant angle, material can be transported for many hundreds of kilometres. This is demonstrated for the diamond deposits along the Namibian coast north of the Orange River where a progressive reduction in average diamond size demonstrates transport and sorting of diamonds over a distance in excess of 250 km (Stocken, 1962; Sutherland, 1982). Elsewhere, however, indented coastlines inhibit the development of long transport paths and such coasts are characterised by individual sections or embayments with unique heavy mineral populations. Thus Rice *et al.* (1976) and Luepke (1980) have demonstrated that littoral circulation cells on the Californian and southwest Oregon coasts have unique heavy mineral populations, and Chauris (1982) has shown that many small but quite rich placers have developed independently in bays along the southern Brittany coast.

During transport attrition is a significant process and minerals such as cassiterite, gold and poorer quality diamonds are likely to be reduced to size fractions that cannot be recovered economically. Similarly, the dominantly siliceous associated sediments also suffer severe attrition and are removed from the higher energy nearshore environment, thus producing a relative enrichment in mechanically resistant minerals such as high-quality diamond, rutile and ilmenite.

Transport paths terminate in areas of accumulation, such as major spit complexes, depositional forelands or large offshore shoals, or in sediment traps where sediment is effectively removed from the beach–nearshore system by, for example, spillage down steep offshore breaks of slope (Zenkovitch, 1967) or into submarine canyons (Emery, 1960). Significant quantities of sediment may also be lost to the system by aeolian action. This latter factor may produce placer deposits directly by differential transport of sand-sized heavy minerals as has been widely recorded around the Australian coast (Gardner, 1955; Baxter, 1977). Alternatively, the loss of sand-sized sediment by aeolian action can produce a relative enrichment in the remaining coarser minerals, a factor that has operated along the diamondiferous west coast of southern Africa, particularly at times of low world sea level.

Except in areas of tectonic downwarping, accumulation forms develop until they reach equilibrium after which they are integrated into the transport zone. Large accumulations of sediment may contain high total volumes of heavy minerals but frequently these are at too low concentration to be exploitable. Certain accumulation forms may, however, act as focuses for heavy mineral concentration. Examples of increased proportions of heavy minerals towards the ends of spits have been given by Zenkovitch (1967) and a similar increase in the content of detrital zircon has been demonstrated on Spurn Head spit in eastern England by Hill and Parker (1984).

Areas of removal of sediment from the active zone by diversion down submarine canyons are likely to have been more widespread during periods of low world sea level than at present (see later discussion). Studies along the Californian coast have shown that the lighter minerals may be preferentially lost down submarine canyons with a consequent increase in the proportion of heavy minerals near the canyon intake (Emery and Noakes, 1968).

The above dynamic units make up littoral circulation cells (Shepard and Inman, 1950; Komar, 1976) and unique heavy mineral placers develop within given cells. The identification of such cells is therefore important in the exploration for placer deposits. Within individual cells, however, there are particular locations where heavy minerals are preferentially concentrated. On the basis of the distribution of wave energy along the coast, two broad categories of locality of heavy mineral concentration can be identified. The first relates to locations of net deposition that receive maximum amounts of wave energy. Such stretches of coast have high sediment fluxes, but particles that are stable in all but the highest energy events will be retained once deposited whilst finer particles are removed. There will accordingly be a net gain in the coarsest (or hydraulically-equivalent higher density) particles that are undergoing alongshore transport. This effect will be augmented by high rates of attrition of less stable minerals in such areas.

The second category relates to areas where there is a reduction in energy available for transport such that heavy minerals and coarse sediment that has been in motion are preferentially deposited whilst the finer or less dense particles continue alongshore.

In identifying specific localities along a coast where such general processes operate, the role of headlands is important. Headlands are obstructions to longshore drift and sediment tends to be deposited on their up-drift side, the beach becoming oriented to face the dominant wave direction. Such a beach grows until in equilibrium and sediment is passed around the headland, although heavy minerals are retained preferentially in the high energy zone immediately up-drift of the headland. This effect is well exemplified by the titaniferous beach sands along the eastern coast of Australia (Gardner, 1955) and has been noted in the auriferous raised beaches at Nome (Moffitt, 1913) and the diamondiferous beaches of Namibia (Hallam, 1964).

Material that passes a headland, however, enters a low energy area in its lee and here in both the beach and in the centre of the bay heavy minerals again accumulate (Hallam, 1964). Figure 17.2 illustrates this for an example from a diamond bearing raised beach to the south of the Orange River in South Africa (Keyser, 1972). The bay illustrated conforms to the ideal of a zeta-form bay in which the plan outline develops in equilibrium with the distribution of wave energy resulting from refraction and diffraction around the resistant headland (Silvester, 1974). A further locality illustrated in Figure 17.2 that is favourable to concentration of heavy minerals is the high energy tail of the bay where there is a maximum flux of sediment.

SEDIMENT SORTING PROCESSES

Waves begin to move sediment on the sea bed at depths approximately equal to half their wavelength (King, 1972), but significant movement of sediment is not experienced until depths typically less than 20-25 m and the greatest movement of material is in the area of the beach, dominated

Figure 17.2: (a) Diamond concentration in a former bay on the 25 m raised shoreline near Alexanders Bay, South Africa. (1) former land area; (2) 0–19 diamonds 1,000 ft^{-2} (92.9 m^{-2}); (3) 20-49 diamonds 1,000 ft^{-2}; (4) 50-99 diamonds 1,000 ft^{-2}; (5) $\geqslant$ 100 diamonds 1,000 ft^{-2}; (6) direction of dominant wave attack. (b) Idealised model of a zeta-form bay and dominant wave paths due to refraction and diffraction around a resistant headland (cf. Buchuberg in (a))

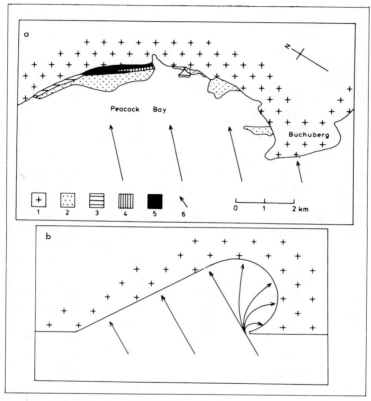

Sources: (a) After Keyser (1972); (b) Silvester (1974).

here by processes related to wave collapse. The break-point of a wave is a fundamental position in beach dynamics. Landwards of the break-point are the run-up and swash zones in which energy supply to the beach is greatest. In this area considerable volumes of sediment are put into suspension as well as larger particles being swept across the beach surface or, in extreme cases, being thrown up on the beach. Longshore movement of sediment is generally greatest in this zone both by transport of suspended particles by longshore currents and, in the coarser grades, by the resultant lateral movement on the beach face from oblique swash and normal backwash trajectories (see Komar, 1976; Hails and Carr, 1975).

577

Sediment put in suspension in the breaker zone can be moved beyond the break-point by return wave currents or, in certain sea states, by rip currents. This sediment can either be moved alongshore in wave induced, tidal or oceanic currents or be deposited, perhaps to be moved onshore again with a change in wave regime. Where waves outside the break-point interact with the bottom sediment they are capable of moving particles either shorewards or offshore, depending on the particle size and the wave characteristics (Grant, 1943). In certain circumstances it is possible for differential movement to occur with coarser particles being moved onshore and finer particles offshore under the influence of the same wave regime (Silvester, 1974).

As has been discussed for longshore movement, the waves interact with the beach and nearshore sediment which is sorted such that the different particle sizes will tend towards those zones in which they are in equilibrium with the energy supplied. A variety of processes related to selective particle entrainment, transport and deposition operate on the beachface (Komar and Wang, 1984; Slingerland, 1984) to produce concentrations of heavy minerals or coarse non-heavy-mineral particles on the upper part of the beachface and in such features as break-point bars.

It is possible to differentiate between two broad types of wave regime: long-period swell waves and short-period storm waves. Swell waves effect a net shoreward transport of material and are essentially constructional. The beach receives sediment during swell wave activity and is built up of lenses a few to several tens of particles thick with an elongate and cuspate form in plan. Size sorting in both the horizontal and vertical planes occurs during formation of these lenses. Under the influence of swash, sorting due to differential entrainment and transport (Slingerland, 1984) principally occurs. Backwash normal to the beachface mobilises a thin layer of sediment in which vertical dispersive sorting may occur (Sallenger, 1979), producing a characteristic inversely graded beach lamination (Clifton, 1969). Thus a constructional beach in which heavy minerals are concentrated consists of a series of stacked elongate lenses, the heavy minerals occurring preferentially at the base of the laminations and in those parts of the arcuate swash paths where coarser particles are located. Such lenses are widely reported in the beach placer literature and a typical example of the distribution of tin in a section of the beach in St Ives Bay in southwest England (Hosking and Ong, 1963) is shown in Figure 17.3.

Storm waves approach the shore with much shorter periods and greater variability of direction than do swell waves. They are typically steep and approach close to the beach prior to breaking which, together with their high frequency, results in the destruction of the beach with much sediment being removed to the offshore zone (King, 1972). The variability of direction and relatively short duration of storm wave attack diminishes the possibilities of longshore transport during storm events. Their principal role

Figure 17.3: Distribution of tin in a portion of the beach at the head of St Ives Bay, Southwest England. (a) Location of sampled section: 1. Intertidal sand beach; 2. Intertidal bedrock. Note dominant wave attack is from the southwest. (b) Distribution of tin: 1. Sample line; 2. 3,000-7,000 ppm; 3. > 7,000 ppm

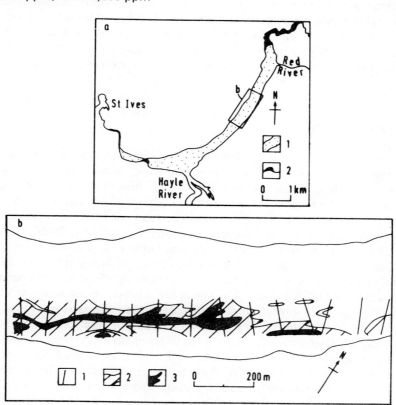

Source: After Hosking and Ong (1963).

is therefore to move sediment away from the upper beach. During such destruction of the beach coarser particles and heavy minerals tend to be left in the area of the upper beach zone (Rao, 1957; Komar and Wang, 1984), producing a heavy-mineral rich lag deposit. If the storm does not erode the whole beach then the lag deposit exists as a lens that is thicker than the beach laminations formed during constructive wave activity (Hosking and Ong, 1963). In severe storms the beach may be bared to bedrock or a lag gravel layer resting on bedrock and hence, with time, a basal layer enriched in heavy minerals accumulates. This process is particularly true for those minerals that are part of the traction (as distinct from saltation or suspension) population and such basal layers have been noted with auriferous (e.g. Moffitt, 1913) and diamondiferous (e.g. Hallam, 1964) beaches. In such examples the distribution of the heavy minerals is largely controlled by the details of the bedrock surface (e.g. Wright, 1964) or the particle size distribution of the basal sediments (e.g. Vorob'yev, 1979). These basal sediments may remain for long periods covered by constructional beach sediments in which more minor concentrations of heavy minerals are once again built up. It is therefore the alternation between destructive and constructive conditions, which often reflects seasonal climatic changes, that is responsible for the overall development of the richest layers in beach placers.

SEA-LEVEL CHANGE

Changes in the level of the sea in the relatively recent geological past have been important in both the formation of and ability to mine littoral placers. In various parts of the world, positive tectonic movements of the coastal zone have lifted marine deposits to altitudes that may be in excess of 100 m above present sea level. Placer accumulations in such locations are relatively easily prospected and mined. Elsewhere, negative tectonic movements have resulted in the submergence and burial of former river systems and their accompanying placer minerals, hence rendering these deposits amenable to low-cost bucket dredging. Of greater significance, however, to the formation of littoral placers have been the relatively rapid series of sea-level falls and rises experienced worldwide during the Quaternary as a result of the growth and shrinkage of northern hemispheric ice sheets.

During full glacial episodes the sea fell to levels that have been variously recorded around the world of between −100 and −150 m psl. Such depths are towards or at the margins of the continental shelves and hence very large areas of the shelves became dry land, across which rivers extended their channels and eroded to the lower base levels. At the same time, marked changes of climate occurred in the tropics and sub-tropics as well as in the higher latitudes, resulting in changes in the rates of supply of sediment to the coastal zone from those of the same areas today (Sutherland, 1985). Moreover, as sea level fell, estuaries were flushed out of sediment that had accumulated in them during the preceding periods of high relative sea level (see later discussion). As the sources of the majority of placer minerals are in the crystalline basement or associated igneous rocks, the lengthening of river courses increases the distance of transport for placer minerals to the coastal zone and hence increases the likelihood of dilution. Furthermore, as has previously been noted, at times of low sea level increased volumes of sediment were removed from the contemporaneous nearshore areas down submarine canyons. All these factors indicate that the sediment budget of glacial low shorelines would differ from that of neighbouring interglacial high shorelines. Such placer accumulations as may occur on the outer shelves, therefore, need not necessarily reflect the presence or absence of placers along the neighbouring present-day coasts. Any such deposits, however, on the outer shelves must be of only theoretical interest for the immediate future, for the technology does not yet exist to prospect or mine them.

The 'saw-toothed' nature of glacial cycles (Broecker and van Donk, 1970) has meant that rises of sea level at the terminations of glacial periods have occurred relatively rapidly. Any halts in sea-level rise that may have produced distinct shorelines would have been short-lived, with a consequent reduction in the possibilities for heavy mineral enrichment. Moreover, the unconsolidated sediments that cover most of the continental shelves were subjected to active high energy littoral processes during sea-level rise and much of the sediment was reworked, hence destroying any shoreline features that may have formed during the preceding fall in sea level. The landward migration of the high energy littoral zone would also tend to incorporate and transport landwards the coarser sediments and heavy minerals (Emery and Noakes, 1968), thus leaving behind a thinner lag of sandy sediments. These sands undergo reworking by the lower energy wave and tidal regimes of the shelves (Swift, 1974), the contained heavy minerals apparently being distributed uniformly rather than being concentrated (Jones and Davies, 1979; Field, 1980). On shelves where there are coarse sediments, a lag deposit of gravel may form in which heavy minerals are trapped and concentrated. This has been observed in the area offshore from the gold bearing beaches at Nome (Nelson and Hopkins, 1972) and off the diamondiferous coast of southern Africa (Murray et al., 1970).

The generally low relief and low gradient character of the shelves is a further factor that reduces the likelihood of significant placers being formed during low sea-level periods. The role of headlands in providing locations favourable for mineral enrichment has been discussed previously and such features would be much less frequent on the shelves. Thus heavy minerals would become more widely dispersed along the contemporaneous shorelines than along interglacial high shorelines, as has been demonstrated for the Oregon shelf by Scheidegger et al. (1971).

The interaction of glacio-eustatic, geoidal eustatic, glacio-isostatic and hydro-isostatic as well as tectonic factors has meant that different areas of the world have had different relative sea-level change histories since the last glacial maximum (Kidson, 1982; see also Peltier, this vol., Ch. 3, and chapters in Part Three). Thus in areas of intense ice-sheet glaciation relative sea level has generally shown a net sea-level fall since deglaciation. In extra-glacial areas, however, rises of sea level have occurred, in some areas the sea attaining its present level several thousand years ago whilst in others sea-level rise is still continuing (cf. Clark et al., 1978).

581

These variable sea-level histories have implications for placer development. The often geologically rapid net sea-level falls of the formerly glaciated areas are a further reason to add to those of wave climate and hinterland factors (Sutherland, 1985) for limited development of littoral placers in such areas. Sea-level rise, as discussed earlier, is likely to be a favourable factor in the production of placers along shorelines formed at the top of the rise, but the length of time during which such shorelines were occupied is also of relevance to the degree of sediment reworking and the adjustment of the coastal zone to the new sedimentary regime. Following sea-level rises, large estuaries act as sinks for coarse sediment, trapping sand-sized and larger material (Meade, 1969) and hence restricting the supply of heavy minerals to the coastal zone. This factor is most likely to have declined in importance on those stretches of coast where sea-level attained its present position several thousand years ago.

NOME ALASKA: A CASE STUDY

Current models of continental shelf sedimentation (Kennett, 1982; see also Orford, this vol., Ch. 13) emphasise the interrelationships between present-day processes, the legacy of Quaternary sea-level changes and variation in sediment budgets. Placer deposits are no less products of the same interactions and this may be illustrated by considering a specific example that clearly shows these factors operating: the Nome, Alaska gold placers. Gold was first discovered in river deposits in the Nome area in 1898 (Cobb, 1973) and the present beach was found to be gold-bearing in the following year. Subsequently, a complex sequence of sea-level change and periods of gold transport, reworking and concentration in several distinct beaches has been revealed.

In the Nome area a gently curving coastal foreland underlain by Tertiary and younger sediments described below is backed by highlands that rise to over 300 m less than 8 km inland. Bedrock is composed of Palaeozoic schists and limestones which have been intensively folded and faulted and traversed by a series of auriferous quartz veins and lenses particularly in the area around Anvil, Glacier and Dexter creeks (Fig. 17.4).

The earliest known littoral sediments are those of the Pliocene (at ~ −21 m msl) and Submarine Beaches (Tagg and Greene, 1973), the latter having two distinct phases, the Outer and Inner, at −10 m and −6 m respectively (Fig. 17.4). Gold reported to grade between 8.1 and 20.3 g m^{-3} was mined from the Submarine Beaches near the present mouth of the Snake River (Moffitt, 1913). Marine fossils from the Submarine Beaches have allowed the correlation of these sediments with the 'Beringian Transgression' (Hopkins et al., 1960; Hopkins, 1967; see also Mörner, this vol., Ch. 8) which has been dated to >2.1 Ma BP (Péwé,

Figure 17.4: (a) Location and map of gold placers around Nome, Alaska:
1. Former shorelines; 2. Possible former shorelines offshore; 3. Fluvial
placers; 4. Possible former channels of the Snake River; 5. Gold-bearing
'bench' gravels; 6. Limit of glaciation. (b) Cross-section marked A–B on (a):
1. Gold-bearing beach gravels; 2. Nome River Glaciation deposits; 3. Iron
Creek Glaciation deposits; 4. Fine grained marine sediments; 5. Gold
concentrations in g m^{-3}. See text for sources

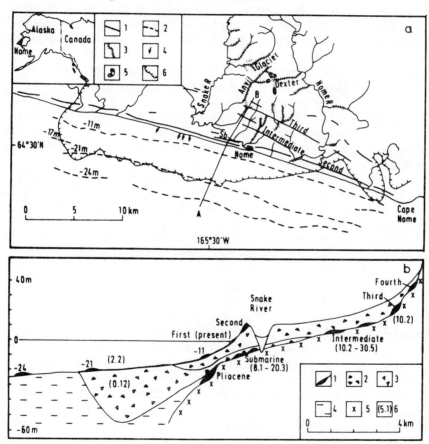

1975). The gold in these beaches most probably was derived by river
transport from the sources around Glacier and Anvil Creeks in the Snake
River basin, perhaps utilising the former channels of the Snake River cut
into bedrock to the west of its present mouth (Greene, 1970).

During the Early Pleistocene, the Nome area was glaciated at least once
(the Iron Creek Glaciation), the pre-existing auriferous deposits being
partly buried and partly eroded and incorporated into the glacial sediments. These sediments accordingly have a gold content which has been

583

estimated at ~0.12 g m^{-3} (Nelson and Hopkins, 1972). After the Iron Creek Glaciation, the 'Anvilian Transgression', dated to between 0.7 and 0.9 Ma BP (Péwé, 1975), resulted in the erosion of the glacial deposits and the formation of the Intermediate (+6 m), Monroeville (+10 m) and the Third (+21 m) beaches (Hopkins et al., 1960; 1974). The beach sediments are enriched in gold which may have been largely derived from reworking of the glacial deposits, although contribution from contemporaneous streams is indicated by the intercalation of rich gold-bearing fluvial and marine sediments near the mouth of Anvil Creek on the Third Beach (Moffitt, 1913).

The highest level reached by the sea in the Nome area is that represented by the Fourth Beach (+37 m msl) but as this beach has proved non-fossiliferous its age is unknown. Hopkins et al. (1960) record that the cliff at the rear of the Third Beach truncates the front edge of the Fourth Beach, thus implying an Early Pleistocene age (at least) for these upper deposits and suggesting that some of the gold recovered from the Third Beach was reworked from the higher level.

The next recorded major event was renewed glaciation (the Nome River Glaciation) the deposits of which overlie the Anvilian marine sediments and are themselves truncated by the Second Beach (Hopkins et al., 1960). This glacial event is assigned to the Illinoian Glaciation (~ 130,000-180,000 BP). As with the Iron Creek Glaciation, pre-existing gold-bearing sediments were reworked into the Nome River glacial deposits.

The Second Beach (+12 m) is the most continuous of the raised shorelines. It has been assigned to the last interglacial (~ 125,000 BP) and is the culmination of the 'Pelukian Transgression' (Hopkins et al., 1960; Hopkins, 1967; 1973). Elsewhere around the Alaskan coast, this relative sea-level rise has been shown to have had two phases (Hopkins, 1973), but only one period of marine deposition has been identified at Nome. It seems probable that much of the gold in the Second Beach has been the product of reworking of the deposits of the Nome River Glaciation.

During the last (Wisconsin) cold phase, glaciers in the hills behind Nome did not develop to such an extent that they covered the coastal foreland. This was a period of marine removal, which has been followed over the last ~15,000 years by the 'Krusensternian Transgression' culminating at ~6,000 BP in the present shoreline position (Péwé 1975). During this latter time of relative sea-level rise, periods of more stable sea level resulted in the construction of beaches at −24 m, −21 m and −11 m msl and possibly also at ~−17 m msl (Nelson and Hopkins, 1972). The gold content of these submerged beaches has not been reported, but the likely relatively brief periods of sea-level stillstand that produced the shoreline features at these depths would suggest lower concentrations than on the beaches formed at the peaks of major phases of marine inundation. In addition to the submerged beaches, lag gravels were produced during this phase of relative sea-level rise by reworking of glacial deposits and these thin deposits show significant enrichment in gold (~ 2.2 g m^{-3}) (Nelson and Hopkins, 1972).

The present beach has formed at the maximum of the 'Krusensternian Transgression'. Gold has been recovered from more than 32 km along the coast where it has been found to be concentrated close to the base of the beach gravels and in lenses of garnet rich sands. Concentrations have been reported to vary between 0.2 and 300 g m^{-3} (Moffitt, 1913). Net longshore drift is in an eastwards direction (Nelson and Hopkins, 1972) and the gold is reported to diminish in size away from the area around Nome at the mouth of the Snake River. An eastwards transport of gold is also indicated for the Third Beach by the decrease in size of gold in that direction from the mouth of Anvil Creek and the position of gold concentration locations around headlands (Moffitt, 1913). Over 1,500 kg of gold were recorded as having been recovered from the present beach (Moffitt, 1913).

The Nome deposits illustrate a complex history of mineral transport, concentration and reworking by fluvial, glacial and marine processes. They are instructive in that they imply that in other areas similarly complex histories of placer development are likely to be found. It is also notable that no deposits comparable in size and concentration to the beaches formed after sea-level rises have been found in the offshore area, despite a wide distribution of gold during periods of lowered sea level (Nelson and Hopkins, 1972). (See Appendix II.)

CONCLUSION

Placers formed in the littoral or nearshore zones are the products of a complex of factors interacting at a variety of scales and they have been termed 'polygenetic' by Hails (1969). Small-scale beach processes that produce sediment sorting and heavy mineral segregation operate universally and result in heavy mineral concentrations. The development of sufficiently large concentrations to form placers of economic interest is, however, the product of medium- to large-scale processes. Principal among these are (1) the hinterland factors governing both the rate of supply and the size distribution of heavy minerals and the associated bulk of the sediment, (2) the coastal configuration and wave climate that determine the sizes of littoral transport cells and the position of zones of preferential mineral entrapment as around headlands, and (3) the history of sea-level change which modifies the above and also favours production of heavy mineral concentrations following major sea-level rises.

Placers related to the present beaches or nearshore areas as well as raised beaches have had a long history of mining and will continue to contribute significantly to the mineral industries for the foreseeable future.

During the 1960s in particular, considerable exploration activity was directed at the offshore areas and deposits analogous to those of the present coastal zones were sought. Although mineral recoveries were made, no sustained placer mining venture was established as a result of this exploration. Severe technical problems related both to prospecting and mining were encountered and the adequacy of the conceptual prospecting models that were applied was only rarely tested. Renewed interest is now being shown in the potential mineral resources of the shelves (e.g. Ballard and Bischoff, 1984). The present chapter has discussed a number of factors that apparently favour littoral placer development following major sea-level rises and are adverse to placer formation on the shelf areas. However, there may be areas in which submerged littoral placers occur and knowledge of factors that are favourable as well as those that are adverse to shelf placer development and preservation is important in identifying the areas in which exploration should be carried out.

REFERENCES

Aleva, G.J.J. (1985) 'Indonesian fluvial cassiterite placers and their genetic environment'. *J. Geol. Soc. Lond.*, *142*, 815-36.
—— , Fick, L.J. and Krol, G.L. (1973) 'Some remarks on the environmental influence on secondary tin deposits', *Bull. Bur. Min. Res. Austr.*, *141*, 163-72.
Ballard, R.D. and Bischoff, J.L. (1984) 'Assessment and scientific understanding of hard mineral resources in the EEZ', *US Geol. Surv. Circ.*, *929*, 185-208.
Baxter, J.L. (1977) 'Heavy mineral sand deposits of Western Australia', *Geol. Surv. Western Aust. Min. Res. Bull.*, *10*.
Boyle, R.W. (1979), 'The geochemistry of gold and its deposits', *Bull. Geol. Surv. Canada*, *280*.
Broecker, W.S. and van Donk, J. (1970), 'Insolation changes, ice volumes and the 0^{18} record in deep-sea cores', *Rev. Geophys. Space Phys.*, *8*, 169-98.
Chauris, L. (1982), 'Les sables lourds des plages du Mor Bras — introduction à l'étude des placers littoraux en Bretagne méridionale', *Bull. Soc. Sc. Nat. Ouest de la France*, *N.S.*, *4*, 1-58.
Clark, J.A., Farrell, W.E. and Peltier, W.R. (1978) 'Global changes in postglacial sea level: a numerical calculation', *Quat. Res.*, *9*, 265-87.
Clifton, H.E. (1969) 'Beach lamination: nature and origin', *Mar. Geol.*, *7*, 553-9.
Cobb, E.H. (1973) 'Placer deposits of Alaska', *US Geol. Surv. Bull.* 1374.
Eliseev, V.I. (1980) 'Placers of the coastal areas outside the USSR and their genetic types', *Litol. Polez. Iskop.*, *4*, 33-43. Trans. Plenum Publishing Corp., New York, 1981.
Emery, K.O. (1960) *The Sea off Southern California*, Wiley, New York.
—— and Noakes, L.C. (1968) 'Economic placer deposits of the continental shelf', *United Nations ECAFE, CCOP Tech. Bull.*, *1*, 95-111.
Field, M.E. (1980) 'Sand bodies on coastal plain shelves: Holocene record of the US Atlantic inner shelf off Maryland', *J. Sed. Petrol.*, *50*, 505-28.
Gardner, D.E. (1955), 'Beach-sand heavy-mineral deposits of Eastern Australia', *Austr. Bur. Min. Res. Geol. Geophys. Bull.*, *28*.
Garson, M.S., Young, B., Mitchell, A.H.G. and Tait, B.A.R. (1975) 'The geology

of the tin belt in Peninsular Thailand around Phuket, Phangnga and Takua Pa.', *UK Inst. Geol. Sci. Overs. Mem., 1*.

Grant, U.S. (1943) 'Waves as a sand-transporting agent', *Am. J. Sci., 241*, 117-23.

Greene, H.G. (1970) 'A portable refraction seismograph survey of gold placer areas near Nome, Alaska', *US Geol. Surv. Bull., 1312B*.

Gurney, J.J. (1983) 'Sea diamond developments', *Indiaqua, 36(3)*, 13-17.

Hails, J.R. (1969) 'The nature and occurrence of heavy minerals in three coastal areas of New South Wales', *J. Proc. Roy. Soc. New South Wales, 102*, 21-39.

—— (1976) 'Placer deposits', in K.H. Wolf, (ed.) *Handbook of Strata-bound and Stratiform Ore Deposits. Vol. 3: Supergene and Superficial Ore Deposits; Textures and Fabrics*, Elsevier, New York, pp. 213-44.

—— and Carr, A. (eds) (1975) *Nearshore Sediment Dynamics and Sedimentation: An Interdisciplinary Review*, Wiley, London.

Hallam, C.D. (1964) 'The geology of the coastal diamond deposits of southern Africa (1959)', in S.H. Haughton (ed.), *Some Ore Deposits of Southern Africa, Vol. II*, Geol. Soc. S. Afr., Johannesburg, pp. 671-728.

Hill, P.A. and Parker, A. (1984) 'Detrital zircon of Spurn Head, Yorkshire, England: the economic potential', *Trans. Inst. Min. Metall. (Sect. B: Appl. earth sci.), 93*, B35-B38.

Hopkins, D.M. (1967) 'Quaternary marine transgressions in Alaska', in D.M. Hopkins (ed.) *The Bering Land Bridge*, Stanford University Press, Stanford, Calif., pp. 47-90.

—— (1973) 'Sea level history in Beringia during the past 250,000 years', *Quat. Res., 3*, 520-40.

——, MacNeil, F.S. and Leopold, E.B. (1960) 'The coastal plain at Nome, Alaska — a Late Cenozoic type section for the Bering Strait region', in *Proc. 21st Int. Geol. Cong., Copenhagen 1960, Part 4*, pp. 46-57.

——, Rowland, R.W., Echols, R.E. and Valentine, P.C. (1974) 'An Anvilian (Early Pleistocene) marine fauna from western Seward Peninsula, Alaska', *Quat. Res., 4*, 441-70.

Hosking, K.F.G. (1971) 'The offshore tin deposits of southeast Asia', *United Nations ECAFE, CCOP Tech. Bull., 5*, 112-29.

—— and Ong, P.M. (1963) 'The distribution of tin and certain other "heavy" metals in the superficial portions of the Gwithian/Hayle beach of West Cornwall', *Trans. Roy. Geol. Soc. Cornwall, 19*, 351-92.

Inman, D.L. and Nordstrom, C.E. (1971) 'On the tectonic and morphologic classification of coasts', *J. Geol., 79*, 1-21.

Jones, H.A. and Davies, P.J. (1979) 'Preliminary studies of offshore placer deposits, eastern Australia', *Mar. Geol., 30*, 243-68.

Kennett, J.P. (1982), *Marine Geology*, Prentice-Hall Englewood Cliffs, NJ.

Keyser, U. (1972) 'The occurrence of diamonds along the coast between the Orange River estuary and the Port Nolloth reserve', *Bull. Geol. Surv. S. Afr., 54*, 1-23

Kidson, C. (1982) 'Sea-level changes in the Holocene', *Quat. Sci. Rev., 1*, 121-51.

King, C.A.M. (1972) *Beaches and Coasts*, (2nd edn,) Edward Arnold, London.

Komar, P.D. (1976) *Beach Processes and Sedimentation*, Prentice-Hall, Englewood Cliffs, NJ.

—— and Wang, C. (1984) 'Processes of selective grain transport and the formation of placers on beaches', *J. Geol., 92*, 637-55.

Luepke, G. (1980) 'Opaque minerals as aids in distinguishing between source and sorting effects on beach-sand mineralogy in southwestern Oregon', *J. Sed. Petrol., 50*, 489-96.

Meade, R.H. (1969) 'Landward transport of bottom sediments in estuaries of Atlantic coastal plain', *J. Sed. Petrol., 39*, 229-34.

Moffitt, F.H. (1913) 'Geology of the Nome and Grand Central quadrangles, Alaska', *US Geol. Surv. Bull.*, *533*.

Murray, L.G., Joynt, R.H., O'Shea, D. O'C., Foster, R.W. and Kleinjan, L. (1970) 'The geological environment of some diamond deposits off the coast of South West Africa', *Inst. Geol. Sci. Rept.*, *70/13*, 119-41.

Nelson, C.H. and Hopkins, D.M. (1972), 'Sedimentary processes and distribution of particulate gold in the northern Bering Sea', *US Geol. Surv. Prof. Paper 689*.

Overeem, A.J.A. van (1960) 'The geology of the cassiterite placers of Billiton, Indonesia', *Geol. Mijnb.*, *39*, 444-57.

Péwé, T.L. (1975), 'Quaternary geology of Alaska', *US Geol. Surv. Prof. Paper 835*.

Rao, C.B. (1957) 'Beach erosion and concentration of heavy mineral sands', *J. Sed. Petrol.*, *27*, 143-7.

Rice, R.M., Gorsline, D.S. and Osborne, R.H. (1976) 'Relationships between sand input from rivers and the composition of sands from the beaches of Southern California', *Sedimentology*, *23*, 689-703.

Sallenger, A.H. (1979) 'Inverse grading and hydraulic equivalence in grain-flow deposits', *J. Sed. Petrol.*, *49*, 553-62.

Scheidegger, K.F., Kulm, L.D. and Runge, E.J. (1971) 'Sediment sources and dispersal patterns of Oregon continental shelf sands', *J. Sed. Petrol.*, *41*, 1112-20.

Schott, W. (1976) 'Mineral (inorganic) resources of the oceans and ocean floors: a general review', in K.F. Wolf (ed.), *Handbook of Stratabound and Stratiform Ore Deposits, Vol. 3: Supergene and Surficial Ore Deposits: Textures and Fabrics*, Elsevier, New York, pp. 245-94.

Shepard, F.P. and Inman, D.L. (1950) 'Nearshore water circulation related to bottom topography and wave refraction', *Trans. Am. Geophys. Union*, *31*, 196-212.

Silvester, R. (1974) *Coastal Engineering*, vol. II. Elsevier, Amsterdam.

Slingerland, R. (1984) 'The role of hydraulic sorting in the origin of fluvial placers', *J. Sed. Petrol.*, *54*, 137-50.

Stocken, C.G. (1962) 'The diamond deposits of the Sperrgebiet, South West Africa', in *Field Excursion Guide, 5th Ann. Congr. Geol. Soc. S. Afr.*, pp. 1-16.

Sutherland, D.G. (1982) 'The transport and sorting of diamonds by fluvial and marine processes', *Econ. Geol.*, *77*, 1613-20.

—— (1985) 'Geomorphological controls on the distribution of placer deposits', *J. Geol. Soc. Lond.*, *142*, 727-37.

Swift, D.J.P. (1974) 'Continental shelf sedimentation', in C.A. Burke and C.L. Drake (eds), *The Geology of Continental Margins*, Springer Verlag, Berlin, pp. 117-35.

Tagg, A.R. and Greene, H.G. (1973) 'High-resolution seismic survey of an offshore area near Nome, Alaska', *US Geol. Surv. Prof. Paper 759-A*, A1-A23.

Vorob'yev, V.B. (1979) 'Mineral distribution in a gold beach placer', *Int. Geol. Rev.*, *21*, 72-8.

Wassef, S.N. (1981) 'Distribution and properties of placer ilmenite in East Rosetta beach sands, Egypt', *Mineral. Deposita*, *16*, 259-67.

Wright, J.A. (1964) 'Gully pattern and development in wave-cut bedrock shelves north of the Orange River mouth, South West Africa', *Trans. Geol. Soc. S. Afr.*, *67*, 163-71.

Zenkovitch, V.P. (1967) *Processes of Coastal Development*, Oliver & Boyd, Edinburgh.

18

Sea-surface Variations and Energy: Tidal and Wave Power

A.W. Lewis

INTRODUCTION

The movements of the sea surface are a result of the various external or 'body' forces acting upon the mass of the ocean (see this vol., Chappell, Ch. 2 and Peltier, Ch. 3). These movements (waves and tides) may be characterised by the periodicities associated with them which can vary from a few seconds to several days. Vertical surface-water movements and the associated subsurface particle motions represent a reservoir of stored energy. This energy has a lifetime equal to the lifetime of the solar system and is inexhaustible. Various estimates have been made as to the size of this energy resource, and Isaacs (1979) shows that the quantities of energy available from all sea-surface movements would amount to the earth's total energy requirements in the year 2000. A number of schemes have been proposed for the harnessing of this vast power resource but many show little understanding of the engineering problems involved. It is proposed to discuss this resource here under two main headings based on the two major groups of cyclic sea-surface movements. First are the long period motions associated with *Tides* which have a dominant periodicity of around 12.5 hours and secondly the *Wind Generated Waves* having periods varying between four and twenty seconds. Within each section a brief introduction to the basic physics will be presented to give an understanding of the processes involved. There will not be any detailed mathematics as only the principles are important. Only a small number of formulae will be given and the reader is directed to appropriate texts for proof.

TIDES

Tidal motion in the ocean and the resulting variations in sea-surface level is caused by the gravitational forces of the bodies in our solar system. Early mariners had little understanding as to their cause and speculated as to the

589

existence of submerged giant whales breathing or angels around the periphery of the ocean dipping their feet (see Darwin (1898)). Pliny ascribes the moon to be the source of tides by inference from observations of Pytheas, a Greek merchant who visited tin mines in Cornwall, England. Little progress was made on the understanding of tides until Galileo published a treatise which was later suppressed by Papal decree during his persecution. Gravitational attraction was first postulated by Newton, and Laplace (1799) outlined a tidal theory based on this. Airy (1861) and Darwin (1898) published the basis for modern tidal theory and recent analyses of the equations of motion for the real ocean are reviewed by Hendershott (1977).

Observations of the tidal level at a particular location over the period of at least one month yield some interesting phenomena. Figure 18.1 shows tidal curves for typical coastal locations in European and other ocean waters. The sea surface rises and falls regularly twice per day and this type of tide is therefore referred to as 'semi-diurnal'. There are locations where the tidal motion is diurnal with only one oscillation per day and others where a mixture of the two occur (Defant, 1961). The time when the level is falling is called the *ebb tide* and the rising level is called the *flood tide*. The vertical level difference between high and low water is called the tidal range.

Tide generating forces arise from the net gravitational attraction of the sun and moon. These forces obey Newton's Law of Gravitation and thus the magnitude of the force due to the moon is about twice that of the sun. The components of these forces tangential to the earth's surface give rise to the variations in sea-surface level. A full mathematical treatment is given in Neumann and Pierson (1966), and a more general account by McMillan (1966). The time between successive high water levels in Figure 18.1 is just under 12.5 hours. This is a result of the predominance of the moon's gravitational forces following the lunar day which is 24 hours and 50 minutes long. The range of the tide varies throughout the month with two maxima and two minima. This can be explained using the concept of equilibrium tide theory, which assumes the earth to be completely covered by ocean. It can be shown that water will be drawn out to an ellipsoid shape by the moon's tide generating force (Harvey, 1976) as shown in Figure 18.2a. Treating the solar tide generating force in a similar way another ellipsoid shape is generated. Figure 18.2b shows the moon's orbit around the earth which takes one synodic month (29 days 12 hours). When the sun and the moon are co-linear, at points a and c, their forces combine to make an ellipsoid of high eccentricity. At this time the maximum tidal ranges are experienced and are called *spring tides*. When the sun and moon are orthogonal, at points b and d, the resulting ellipsoid has low eccentricity. Minimum tidal ranges are experienced and are called *neap tides*.

In the open ocean the vertical tidal is usually less than about one metre.

Figure 18.1: Typical tidal curves

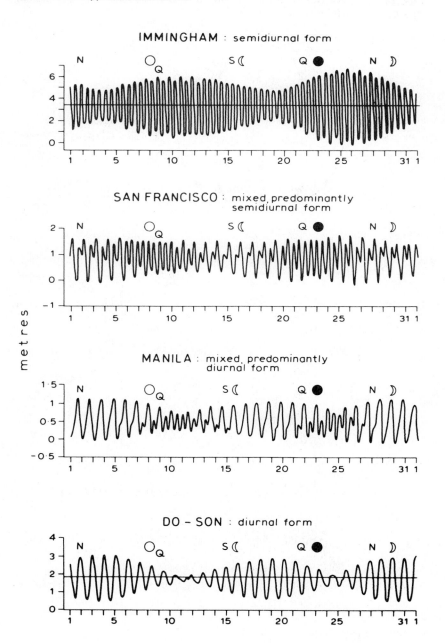

Source: Pethick (1984) after Defant (1961).

Figure 18.2: (a) Formation of the equilibrium tide ellipsoid. (b) Springs–Neaps cycle in tidal forces

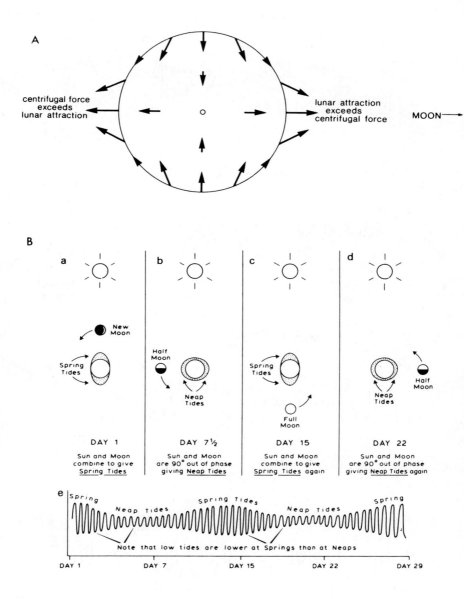

Sources: (a) Harvey (1976); (b) Pethick (1984).

Large tidal ranges, however, occur in certain coastal locations as a result of topographic amplification (see Devoy, this vol., Ch. 1 and Fig. 10.1). The propagation of the tidal wave into shoaling water decreases its wavelength and hence its amplitude increases. The Bay of Fundy on Canada's eastern seaboard experiences the world's largest tidal ranges, in excess of 15 m (Scott and Greenberg, 1983). Cook Inlet, Alaska, has a tidal range in excess of 7 m and the Valdez Peninsula in Argentina has a range up to 8 m. In Europe the largest tidal ranges are experienced on the northern coasts of Brittany, France (13.5 m) and the Severn Estuary, UK ($\geqslant$ 12 m). Other locations with large tidal ranges include the Barents Sea and Sea of Okotsk, USSR, Port Darwin, N.W. Australia, in China in the Yang-tsien-kiang region and on the northern shores of the Indian Ocean in the Gulf of Kutch.

ENERGY UTILISATION FROM THE TIDAL MOTION

The total amount of energy in the tides is estimated to be 7×10^{17} joules (Hendershott, 1977). Mechanisms for conversion of this tidal energy into useful power fall into two categories: (1) devices which utilise the kinetic energy available from the flow of water induced by tidal-level differences, and (2) devices which utilise the potential energy available in the rise and fall of tidal level.

Type (1) devices have been used in Europe since the early part of the present millennium. The Domesday Book mentions many on the south coast of England. These tidal mills were mainly used for grinding corn, although there are records of pumps being used in estuary situations. The principle of operation is based on the flow being directed past a water wheel built on the shoreline. The obvious disadvantage of this is that as the level falls, the wheel will lose contact with the flow. In order to try to overcome this, wheels attached to floating pontoons were utilised in estuary or river channel locations (Wailes, 1941).

Modern concepts are all based on the device type (2) and consist of a system to prevent the normal return of sea-surface level from high water to low water. The resulting potential energy is converted to electrical power by directing the return flow through suitable machinery. The simplest form of this system is to construct a barrier across an embayment or estuary (Fig. 18.3a). This barrier must be fitted with sluice gates for filling and water turbines for electricity production. Figure 18.3b illustrates the operation of this system. On the flood tide the sluice gates are open and the basin is filling. At high water the sluice gates are closed to trap the water inside the basin. As the water level falls outside on the ebb tide, the trapped water gains potential energy. When sufficient level difference has been built up the turbine gates are opened and the stored water flows out, generating

Figure 18.3: Single basin tidal scheme (ebb generation only). (a) Plan view of layout. (b) Water-level variations (hours). (c) Power output from turbines (hours)

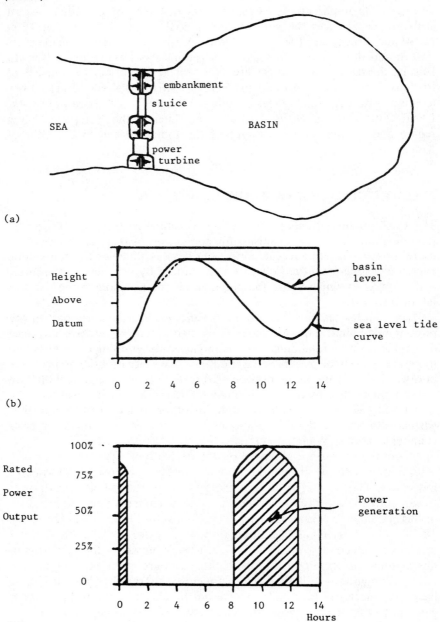

(a)

(b)

(c)

electricity. Generation continues until just before low water when the turbine gates are closed and the sluice gates opened for refilling the basin. A typical example of such a system is an early corn-grinding mill near Pembroke in South Wales, UK which is still operating on this principle (Fig. 18.4). The blocking of the estuary and alteration of the tidal regime by such devices can have important environmental implications where ecosystems are dependent on intertidal areas. Shaw (1974) discusses these wider implications with specific reference to the Severn Estuary, UK.

One of the difficulties with a scheme such as this is that power output from the system is only available for a portion of the tidal cycle, as shown in Figure 18.3c. This factor, coupled with the fact that the tidal period is longer than the solar day, means that generation does not coincide well with the demand for power. A number of possibilities have been proposed to try to overcome these problems. It is possible to utilise turbine generators which operate with flow in two directions and thus generate power on the flood tide. Other possibilities involve more than one basin interconnected in some way to circulate the water and extend the generation time. Power generated by tidal action at times when it is not needed is used to pump water from one basin to another. A full description of these multi-basin schemes and their operation is given by Simeons (1980). Alternative proposals for re-timing the power produced involve integration with conventional pump-storage units such as those at Dinorwig in the UK or Vianden Centrale in Luxembourg. Gibson and Wilson (1978) show that significant increases in output result from this integration.

In a single-basin scheme two-way operation is possible with appropriate machinery. One advantage of this is that the output can be increased by using the machinery in a pumping mode. The energy used to pump water into the pool around high tide is relatively low as its potential energy has only to be increased by a small amount. When this is released on the following tide the energy content is apparently increased as shown in Figure 18.5. The shaded volume added by pumping above the high water level has a potential energy equal to its weight multiplied by $h/2$. If this volume is retained until low water then it will have a potential energy equal to $(h+R)/2$ which is a significant increase.

TIDAL POWER GENERATION: PRACTICAL SYSTEMS

Modern development of tidal energy potential has been centred in three locations. France, the United Kingdom and Canada. Extensive studies have been carried out of potential sites but only two working schemes have been constructed. This section will discuss examples of potential development together with the equipment used for power generation.

All tidal energy schemes have the following elements: (1) basin

Figure 18.4: Tidal mill in Wales. (a) Barrier, showing the flood tide filling through a sluice. (b) Overall view of the corn grinding mill

(a)

(b)

Figure 18.5: Potential energy gain by pumping on inflow. (a) Water pumped into the basin near time of High Water. (b) same water volume released near Low Water

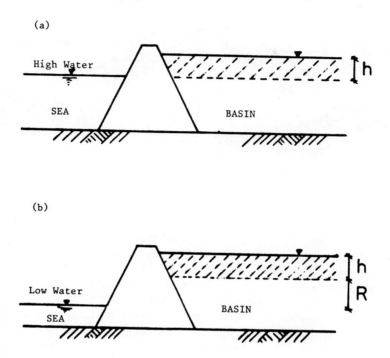

impounding embankment, (2) sluice gates to allow filling or emptying, and (3) turbine systems for electricity generation.

(1) Basin impoundment is usually achieved with an embankment and a typical cross-section is shown in Figure 18.6. Low cost materials such as dredged aggregate or imported rocks will be used. The impermeable core is an essential feature of construction which will consist of a central impermeable diaphragm wall constructed in concrete and flanked by a filter layer. Obviously sites in natural bays with narrow entrances become attractive in reducing costs for the impoundment. In these and other situations the source of the aggregate material is important. Abstraction of material from coastal or offshore sources may have serious repercussions on the coastal sediment and hydrodynamic systems, as may the structure itself (Hails, 1977, Derbyshire *et al.*, 1979).

(2) Sluice gates must be provided to allow for filling or emptying the basin and these can either be of the simple rise/fall vertical gate design or of a

Figure 18.6: Typical impounding embankment cross-section

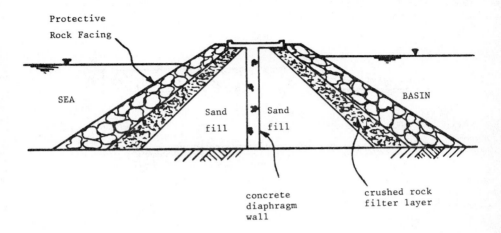

sector design (Bruun, 1980). These would be incorporated into prefabricated caisson units and floated on to a site as suggested by Clare and Oakley (1981) and Severn and Campbell (1978). Figure 18.7 shows a cross-section of a typical unit of the rise and fall sluice gate design.

(3) Turbine generation systems have been progressively developed to convert the kinetic energy of flow into mechanical energy and for electricity generation. Turbines for use in tidal energy schemes must operate efficiently at head differences which are much lower than conventional machinery. Slow speed propeller type turbines are the most suitable as they operate well in low head regimes. Figure 18.8 shows alternate configurations for the turbines, all of which are horizontal axial flow systems. Conventional installations are not feasible because of the increased cost of civil engineering works. The bulb turbine (Fig. 18.8a) has all of the generator housed within the body in a watertight casing. These units were developed for the Rance tidal power scheme and despite initial problems with sealing they now function satisfactorily. Figure 18.8b shows a simpler option for power take-off with the generating machinery above water level. This tube turbine or slant axis machine has the disadvantage of its associated large structure. Figure 18.8c shows a recent development which offers the advantages of the generator out of contact with the water and a compact machine. The Straflo turbine developed by Escher Wyss has a rim-generator where electricity is generated by electromagnetic induction between the rotor (propeller) and the coils around the turbine housing.

Figure 18.7: Typical sluice structure for flow control into a tidal basin

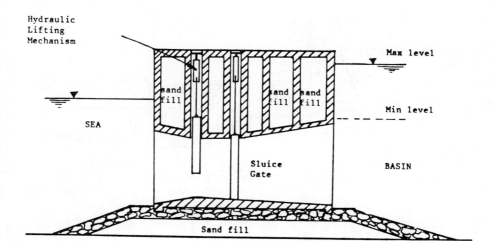

TIDAL POWER CASE HISTORIES

There are only three tidal power stations in operation worldwide although studies have been carried out for a large number of potential sites (Charlier, 1978). A selection of these will be discussed here.

The Rance Estuary, France

French engineers have traditionally been interested in the potential for energy production using the large tides experienced along their northern coasts. Plans for power dams were drawn up and in 1928 an abortive start was made on the construction of a pilot plant near Aber-Wrach which is north of Brest (Ley 1954). Gibrat (1966) rekindled interest which led to the construction of the first tidal power plant in the world at the Rance Estuary near St. Malo. The structure bridges a 750 m-long gap near the mouth of the estuary. Figure 18.9 shows the location and plan of the closure works. Figure 18.8a shows the cross-section of the bulb turbines used for power generation. Construction took a total of six years to complete with the first power generated on 19 April 1966 and the full commissioning in December 1967. The total capacity of the scheme is 240 MW

599

Figure 18.8: Tidal energy turbine generators suitable for low head use

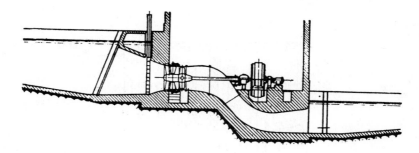

(a) Cross-section through power house for bulb turbine

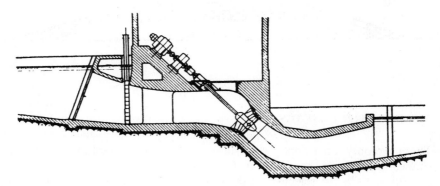

(b) Cross-section through power house for slant axis turbine

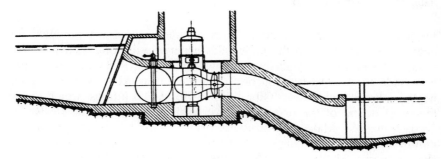

(c) Cross-section through power house for straight flow turbine

Source: Leliavsky (1982).

Figure 18.9: Layout of La Rance tidal scheme, France. 1. Access building to large rooms, floor at ~ 16.65 m level (marine charts). 2. Shaft for descent to rooms at level −7 m, diameter 12 m. 3. Access galery at 7 m, passing under the lock, about 80 m long. 4. Navigation lock: 65 × 13 m. 5. Administrative building and main access to the plant. 6. Equipment disassembly bays and maintenance shop. 7. Twenty-four bulb-unit bays. 8. Control bay. 9. Wall at plant end. 10. Rock-fill dyke. 11. Six sluices equipped with 15 × 10 m gates. 12. Line departure unit, three 225,000 v lines.

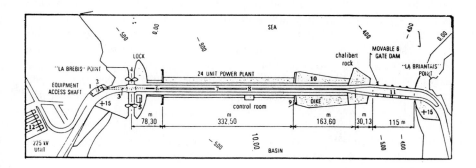

Source: EDF, Usine Marémotrice de la Rance.

and the maximum tidal range is 13.5 m. An area of 2200 ha is enclosed at the high water level and the basin volume is 184×10^{16} m³. A good résumé of the operation to date, together with the problems, is given by Banal and Bichon (1981).

Kislaya Bay, USSR

A small evaluation system was developed in the USSR which utilised the bulb turbine similar to the Rance Scheme. It has an installed capacity of only 400 kW and was built to evaluate the use of tidal power at a remote location. The system is unique in that all of the construction was completed as a single floating caisson 36 × 18 m in size and weighing 5,200t. All of the generating plant and turbine was installed and the unit floated from Murmansk to the site some 100 km away. Bernstein (1974) describes the installation and reports that it is performing well after three years of operation.

601

The Severn Estuary, UK

The potential of the Severn Estuary for tidal power production has long been recognised because of the large tidal ranges (>12 m). Feasibility studies were carried out as long ago as 1926 when Gibson (see Allen, 1947) constructed a scale hydraulic model to investigate the effects of a tidal barrage at the Seven Stones (Fig. 18.10). Interest was renewed with the oil crisis of 1973 and Shaw (1974) reaffirmed the possibility for the provision of a significant amount of power. A large number of studies have been instigated in recent years, and these have been summarised in the report of the Severn Barrage Committee (1981). A number of complex basin schemes were considered but the preferred solution would now be a single barrage. This barrage would run from Lavernock Point to Brean Down as shown in Figure 18.10. This barrage would be 19 km long and contain 160 turbines, each rated at 25 MW. The cost estimate for the scheme in 1982 was £5,600 million, which was considered economic with other energy costs rising (see Anon., 1982). The annual power production would be 13TWh which was estimated as potentially saving ~8 million tons of coal per year. An additional inner basin was proposed as a future development to increase the output by about 55 per cent. Based on these and related barrage studies a similar scheme to that of the Severn has been con-

Figure 18.10: Location of proposed Severn Barrage showing Seven Stones and Outer Barrage options

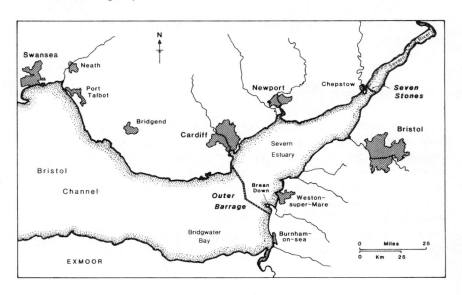

Source: After Shaw (1974).

602

templated for operation at Strangford Lough, Northern Ireland (Northern Ireland Economic Council, 1981).

Annapolis Bay, Canada

The Bay of Fundy experiences the largest tidal ranges in the world and its potential for tidal power has long been recognised. One advantage of this area is that a number of possible basins can be formed in the embayments around the Bay (Fig. 18.11). This is in contrast to most other possible locations, where only a single scheme is feasible. A reassessment of the Bay's potential by Clark (1977) has led to a pilot study located at Annapolis Bay. This study involves the construction of a 20 MW demonstration unit at a cost of $46 million (Spec. Comm. on Alternative Energy, 1981). An existing tidal barrage is being utilised with a power house being constructed on Hog's Island. The single Straflo turbine is 7.6 m in diameter and is capable of generating power when differential heads exceed 1.4 m. This is the first time that this turbine has been used in sea water and it is

Figure 18.11: Bay of Fundy, Canada: possible tidal power installations

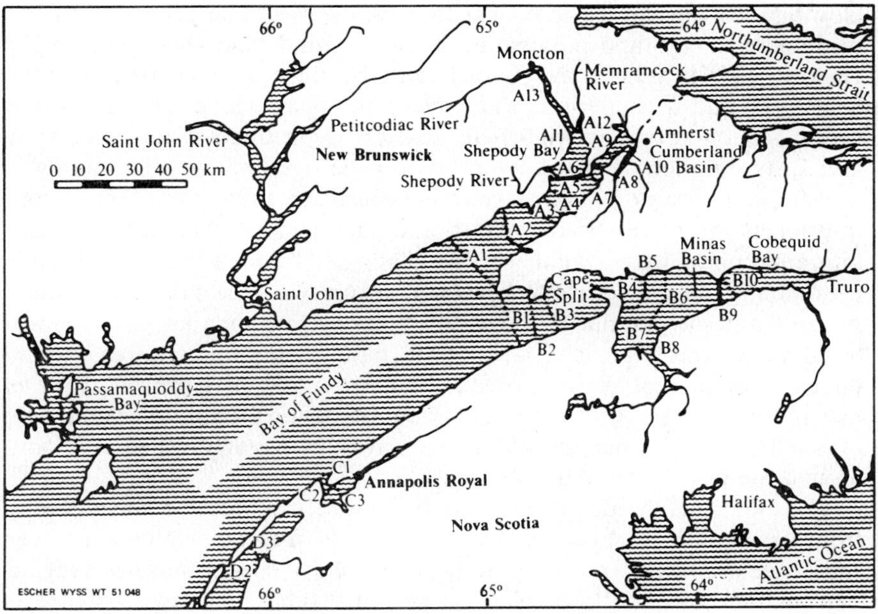

Source: Douma *et al.* (1982).

seen as an important demonstration test for full-scale development else-where in the Bay (Douma *et al.*, 1982). In the longer term it is proposed that the electricity generated at Fundy would be exported to the United States which will make the project financially viable.

Jiangxia, China

Interest in tidal power in China has a long tradition and many potential sites are being investigated. A prototype tidal generator of only 40 kW was completed in 1959. This prototype was a single-basin scheme using ebb tide generation only. A larger scheme operating in both flood and ebb tide for generation was completed at Jiangxia in 1981. This new scheme is rated at 500 kW and has a 2.5 m diameter bulb turbine. A 670 m-long earth and rock dam impounds a reservoir which is utilised for mariculture purposes. The construction cost for this scheme was $1 million in 1981.

WAVES

Solar energy incident on the earth causes differential heating of the atmos-phere resulting in circulation patterns which we know as winds. Wind blowing over the surface of the ocean causes waves to be generated by the transfer of energy and momentum from the wind field (Kinsman, 1965; Harvey, 1976). Waves are characterised by their wavelength (L), wave height (H) and wave period (T) as shown in Figure 18.12. The size of the waves generated by any wind field depends upon three factors: the wind speed, its duration, and the fetch or distance over which wind energy is transferred to waves. For any given wind speed there is a limit to the energy transfer as the waves become steep and start to form whitecaps. At this saturation point, excess wind energy is dissipated by wave breaking and this condition is called a *fully developed sea*. Waves leave the storm or generat-ing area and moving out of the wind's influence are then known as *swell*. Swell waves travel great distances with little energy loss. Barber (1963) quotes examples of waves measured off the coast of Cornwall, UK as having travelled 11,000 km from the South Atlantic Ocean. Measurements of swell off California showed these waves to have travelled half way around the world from Australia.

Coastlines on the downwind side of large oceans receive wave energy from the whole of that ocean. Waves effectively integrate wind energy over the whole ocean and transfer it to the shore. Wave energy thus has a higher density than wind or direct solar energy and also has greater persistence as it does not only depend on local conditions. The highest wind belts occur between 40° and 60° of latitude and thus the western seaboard of Europe,

Figure 18.12: Sinusoidal wave profile

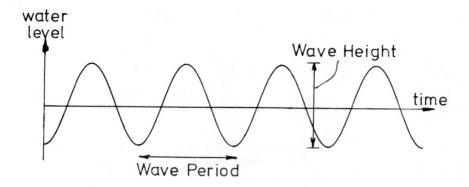

which faces a large ocean, should experience a large energy resource. Countries which lie in the trade wind belts (0°-30° latitude) may not experience as high a wave energy resource but its firmness may be better due to the wind persistence in these areas. (On a separate but related theme, discussion of the impact of wave energy dissipation on coastal dynamics—coastal processes may be found in Orford, this volume, Chapter 13 (see also Derbyshire *et al.*, 1979; Hails and Carr, 1975).)

Wave energy calculation

The power transferred within a wave train may be calculated using simple linear wave theory (Kinsman, 1965). Power transmitted by a sinusoidal wave as shown in Figure 18.12 is given by

$$P = 0.98 \; H^2 T \; kW/m \; \text{width of coastline or wave energy device.}$$

Waves in the real ocean are irregular with continuous variations of wave height and wave length as shown in Figure 18.13. We can represent these conditions as a summation of individual sine waves through the concept of the wave spectrum (Kinsman, 1965). The total power in the irregular sea is given by

$$P = 0.49 H^2_{CHR}. \; Te$$

where H_{CHR} is a characteristic wave height and Te is the energy period (see

605

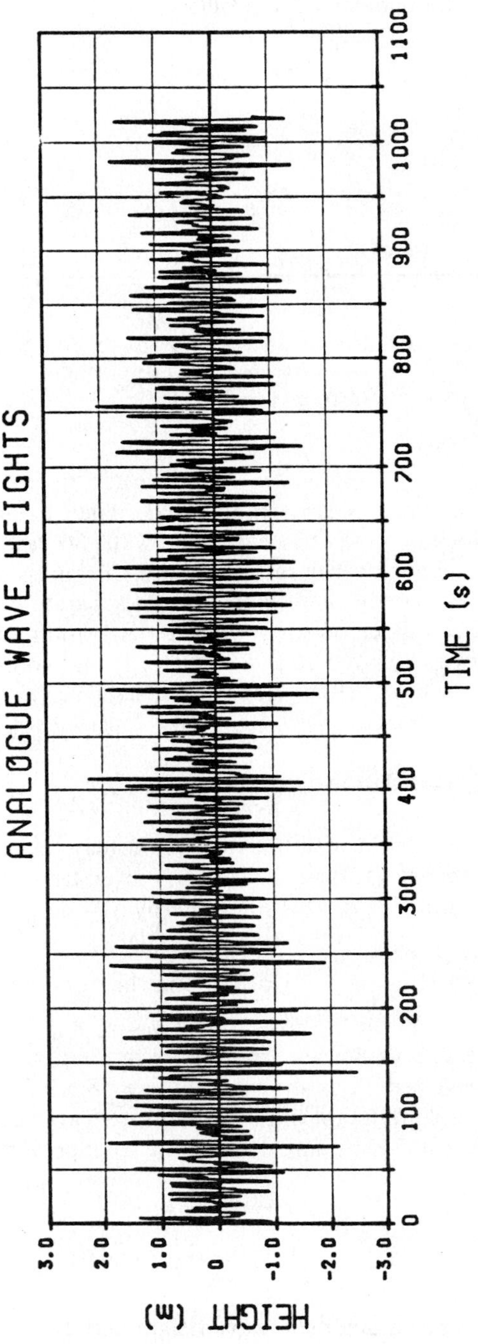

Figure 18.13: Irregular wave train recorded at sea

Mollison, 1980 for further definition). As indicated, wave motion consists of vertical surface water displacements which can be observed propagating across the oceans and also movements of the water particles below the surface. The transmission of wave energy is, however, confined to a layer ~0.25 L in thickness below the surface. Thus, for common wave conditions this will be in the upper 40 m of the ocean.

It has been estimated that the total wave energy resource in the oceans at any one time is 2 TW (Wick and Castel, 1978; Scobie and Leishman, 1975). This is equal to about one-tenth of the projected world energy demand for the year 2050 (Isaacs, 1979). The total resource capable of utilisation will be lower than this because of geographical locations with low power densities and low persistence in many areas. Further, measurements of wave climate are essential if a meaningful assessment of resource is to be made at a specific location. Wave activity can vary dramatically over relatively short distances of shoreline because of seabed refraction and energy dissipation.

WAVE ENERGY CONVERTORS

In contrast to other types of renewable energy resource utilisation there are a large number of concepts for wave energy conversion. No large-scale devices are as yet commercially available and very few have been tested at sea, even at prototype scale. The nature of the wave energy resource and the concepts for wave energy devices mean that units need to be of medium to large scale (multi-kilowatt to multi-megawatt size). However, small-scale devices are viable in terms of navigational considerations or for sea-water pumping in high equivalent cost systems.

There have been several hundred patents awarded for wave energy convertors (WEC) although few have been constructed (Scobie and Leishman, 1975; McCormick, 1981). Damy (1982) shows that of 184 patents awarded in France between 1969 and 1978, over 100 were drawn up in the two years following the 1973 oil crisis. Many proposed devices show a basic lack of understanding of wave motion by their inventors whilst others stand little chance of survival in storm wave conditions. Serious development programmes have been in progress for less than ten years in the countries now to be discussed. Nevertheless, it is technically feasible to construct efficient wave energy convertors which will perform well at sea.

Device classification

The essential element of all wave energy convertors is an 'interface unit' which responds to variations in wave pressure. Damping of the motion of

607

this interface is used to abstract energy from the waves. Optimal hydro-dynamic design of the interface element is the key to high absorption efficiency. The apparent large number of concepts for wave energy con-vertors can be classified within a few basic generic types. Devices may be classified according to their interface element, their power take-off system or their mode of deployment. Interface elements fall into two categories: active devices and passive devices. There are three modes of deployment: terminator, attenuator and point absorber.

Active devices

These are devices which move in response to wave action and do useful mechanical work against a stable reference frame. Mechanisms of this type form the bulk of concepts proposed for wave energy convertors. One of the problems here is to provide the reference frame, especially in deep-water locations. The simplest option is to connect the structure to the seabed which usually results in an expensive device. Submerged devices are subject to lower peak forces and seabed connection can then be used effectively as in the Bristol Cylinder (see later discussion and Fig. 18.14). Inertial reaction is an alternative method and this can be provided by a large floating structure as in the Cockerell Raft (Grove-Palmer, 1982). The most effective solution is to use phase differences in the wave action to balance forces. Long spine type devices are designed to do this by averaging torques over their length as a result of short crested seas. Early versions of Salter's Duck used this principle.

Passive devices

The whole structure in these devices remains stationary to form the refer-ence frame and relative movement of the water surface is used to do work. These devices can either be floating or fixed to the seabed. Floating devices are moored in position and remain relatively fixed, either by virtue of their own inertia or of the mooring system stiffness. Oscillating Water Column (OWC) systems and Overtopping Channel systems are examples of this type. The 'KAIMEI' and 'Lanchester Clam' (see later discussion) are examples of floating devices of this type. The NEL.OWC is an example of a seabed fixed system.

Mode of deployment

Devices may be further sub-classified by their mode of deployment. *Terminator* systems are those which are deployed perpendicular to the pre-dominant wave direction. These devices intercept all incident wave energy and must be assymetric in form to achieve high absorption efficiency. High wave forces are thus experienced and seabed-fixed devices must be designed to resist these. Floating devices can include mooring compliance to reduce these extreme forces and submerged devices experience lower forces due to the attenuation of wave motion with depth. *Attenuator* systems lie parallel to the predominant wave direction and absorb wave energy progressively along their length. Wave forces on this type of device are consequently lower, as was demonstrated in the 'KAIMEI' experiment in Japan. Point absorbers are individual devices which are small in relation to the wavelength of the waves. They will be deployed in large arrays and interactive effects allow them to absorb energy from a sea area many times the device size. This type of system offers low overall costs in terms of device volume per given equivalent area of energy absorption.

Systems

In the same way that there are large numbers of device types there is a corresponding variety of power take-off systems. The optimum power take-off has not been identified as no large-scale demonstration units have been constructed and little operational experience is available. The power take-off in any realistic wave energy device will have two main stages: the primary conversion of wave energy into useful mechanical energy by damping the interface element and the secondary conversion of this mechanical energy into some useful form of power. A speed multiplying device must be incorporated into any system as the low frequency of wave motion is not compatible with most applications. Also, because of second-by-second variations in wave energy input, some means of smoothing the output power is necessary. It is not generally possible to smooth power output over long periods of time as energy storage methods are expensive.

The common application of wave energy devices is in the generation of electricity, although non-electrical applications are discussed below. Some power take-off systems have endeavoured to convert device motion into electrical energy in a single stage. McCormick (1981) and Mogridge (1980) describe systems involving piezoelectric materials, a protonic conversion fuel cell and electromagnetic methods. Calculations of the performance of these devices, together with limited trials, however, show low conversion efficiency. A number of self-rectifying air turbines to cope with oscillating flow have been proposed for air chamber devices. The most successful system which has been constructed to date is the Wells turbine which used symmetrical aerofoil blades. A full description of its principle of operation is given by Long (1978) and a 40 kW tandem unit has been constructed by the Fuji Corporation in Japan. Other turbines are described by McCormick (1981) and the next phase of the International Energy Agency's (IEA) sea trial on 'KAIMEI' will test a number of systems.

609

Potential applications

Large-scale power stations

In the event of wave power stations supplying electricity in the future to mainland national grid distribution, systems will have to consist of several hundred individual devices. Most large grid systems are capable of accepting up to 15 per cent of power from fluctuating sources. Control of individual devices is necessary to optimise output and devices must be capable of shutting down if the grid loses load.

Isolated communities and offshore islands

Small-scale wave energy devices between 40 kW and 400 kW can be cost effective if the design can achieve a reduction in structure costs. Shore based OWC chambers can do this and the demonstration unit at Sanze, Japan is an example. Alternatively, there is often an existing diesel power generating system. Here medium-scale (100 kW-2 MW) wave energy convertors can be integrated with the system and used in a fuel saving mode with diesel generators providing security of supply. This application is relevant to island communities and in developing countries where the cost of electricity is high and so wave generated electricity is very competitive. In certain circumstances the existence of the WEC in a small grid system may improve its overall stability (Sulley, 1984).

Sea-water pumping and desalination using wave energy

There are many applications which involve sea-water pumping and one of the few successful wave powered systems was used for this purpose for forty years in Monaco at the Institute of Oceanography. Intensive aquaculture using artificially induced upwelling has been carried out (Eurocean, 1979), where nutrient-rich water pumped from the deep ocean is used in lagoons for fish or shellfish culture. Studies using waste water from OTEC systems are also in progress. The application of wave power to desalination problems uses a similar technology to that of water pumping. Pleass (1978) has developed a system which uses reverse osmosis in the desalination process. Smoothing of the wave energy here is achieved through storage of the product water.

Hydrogen production

Applications here of wave power to produce hydrogen for island communities where end use is for domestic heating or cooking can be viable. Hydrogen produced by electrolysis can be utilised in most modern appliances without reforming and could be competitive with bottled gas (LPG). An integrated wave-diesel system could incorporate an electrolyser to provide storage for wave energy (Harris and Highgate, 1978).

WAVE ENERGY CONVERTORS: CASE HISTORIES

Research and development into wave energy convertors in recent years has been carried out mainly in the UK, Norway and Japan and some examples of proposed systems are now given. However, only two prototype systems of reasonable size have been constructed to date, although these have provided some power to a public utility system.

France

A number of successful wave energy projects were operated in France during the early part of this century. In 1920 Bouchaux-Praceique (Palme, 1920) constructed an oscillating water column type device by tunnelling through the cliff near his home. Using a novel type of self-rectifying turbine he was able to generate up to 1 kW for his house. The Institut Océano-graphique, Monaco installed a wave powered pump in 1921 (Romanovsky, 1950) to replace an earlier system (Savonius, 1931). A float moving under wave action was used to actuate a pump which delivered seawater continu-ously under a head of 20 m. Tests were carried out by Richard (1913) which showed the device capable of pumping to a height of 64 m with a flow of 2,400 litres/hour. More recently in 1979 Gauthier and Martinais drew up a report which set out the essential points of a programme to study wave energy utilisation (Damy, 1982). This programme was adopted by the Government and the Centre National pour l'Exploitation des Océans (CNEXO) was allocated a budget to proceed. It was envisaged that wave energy would not make a major contribution to power production in main-land France but could be significant for small isolated communities and remote islands. A competition was held in 1981 by Agence Nationale pour la Valorisation de la Recherche (ANVAR) and CNEXO, when over 200 proposals were received. Three main devices were selected and the subse-quent development programme has received 3.5 million francs since 1979.

United Kingdom

The United Kingdom has been very active in promoting wave energy development in recent years. The estimated recoverable resource around the coastline is in excess of 21 GW which is equivalent to about one-third of existing installed capacity. A large number of concepts for wave power devices were developed and the review of Scobie and Leishman (1975) was the first attempt at an assessment of their feasibility. In 1976 the Energy Technology Support Unit (ETSU) at Harwell was given overall manage-ment of the British Wave Energy Programme by the Department of Energy

(Grove-Palmer, 1982). The Wave Energy Steering Committee (WESC) and a number of Technical Advisory Groups (TAG) were set up and about £15 million has been spent on development up to 1983. Nine major devices were investigated, together with a number of generic studies, including the resource assessment at one particular site. Rendel, Palmer and Tritton (RPT) were retained as consultants to ETSU with an overall brief to assist in the design process and to make an assessment of each device. A 2 GW power station at South Uist was taken as the reference design for economic assessment. Results of this assessment were presented to the Department of Energy in 1983 and the Government committee ACORD recommended that further large-scale expenditure was unjustified. It was therefore decided to reduce the expenditure to about £0.5 million per annum for fundamental studies and generic research. Seven devices were taken through to final reference design and costing, and some of these are now described.

The Queen's University Belfast Oscillating Water Column (OWC) device

This is a point absorber type device which developed from a simple buoy system (Whittaker et al., 1978). A novel self-rectifying air turbine developed by Wells (1977) forms the basis for the power take-off. Originally, the device was a floating system but has now evolved into a seabed fixed structure. A series of originally six J-shaped chambers around the periphery of a caisson have now been replaced by four. This makes the device more directionally biased with three of the chambers being relatively short whilst the fourth is long. The bandwidth of this new device is consequently expected to be large because of multi-resonant effects. Figure 18.14a shows the omnidirectional device and Figure 18.14b the directional device, which is 64 m in diameter with a power rating of 3.3 MW.

The Bristol Cylinder

This device is the result of theoretical work by Evans (1976) and has a high hydrodynamic efficiency. The submerged cylinder oscillating, out of phase, in two orthogonal directions can theoretically absorb all of the incident wave energy (Fig. 18.15). Pumps fitted in the mooring rodes deliver water under pressure to a centralised generator facility. Each device is 100 m long and has a rating of 5.2 MW. Problems in designing a high pressure, closed circuit pump system have made this device relatively uneconomic. The device team is more optimistic about the solution to this problem and suggest that productivity could be increased.

Sea Energy Associates' (SEA) 'Clam' device

The 'Clam' device, developed at Lanchester Polytechnic, consists of air-filled flexible bags attached to a floating spine (Fig. 18.16). These bags 'breathe' air into the hollow spine in response to wave action (Bellamy,

Figure 18.14: The Queen's University of Belfast Oscillating Water Column (OWC) Device. (a) The omnidirectional device with multiple chambers. (b) The directional device with four chambers

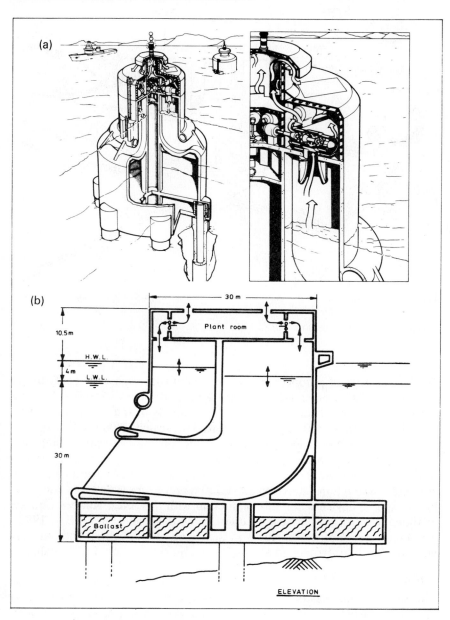

Source: (a) Grove-Palmer (1982), (b) Whittaker.

Figure 18.15: The Bristol cylinder

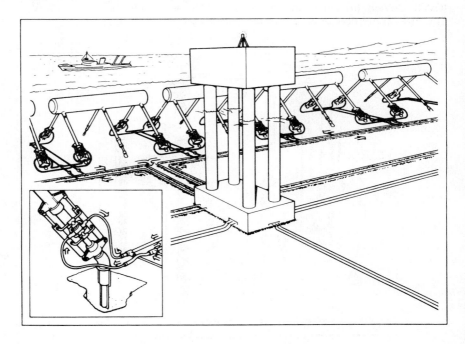

Source: Grove-Palmer (1982).

1982). Air is thus pumped in a closed circuit through self-rectifying air turbines. Wells turbines are proposed and model tests have been carried out on the design of these turbines. All elements of the device have been investigated and a 1:10 scale model tested in 'scaled' Atlantic Ocean conditions in Loch Ness, Scotland. Each device in the full scale design of a 2 GW station is 290 m long and rated at 10 MW. Assessment of this device suggests that it could be the most cost-effective within the programme, although question marks still remain on the viability of the flexible bags. The device team, in conjunction with the Avon Rubber Co., are confident that the air bags can be constructed to specification (SEA, 1982). The final version of this device is in the shape of a doughnut with air bags around the circumference to allow energy absorption omnidirectionally.

The Edinburgh 'Duck'

The Salter 'Duck' was one of the earliest systems to suggest that electricity production from sea waves would be an economic proposition (Salter, 1974). A series of cam-shaped floats which roll in response to waves are

Figure 18.16: The SEA 'Clam' wave energy converter

Source: Sea Energy Associates — Lanchester publicity brochure.

mounted on a spine (Fig. 18.17). The shape of the float is such that no waves are generated as a result of the device movement, which results in high efficiency. In the earlier designs relative motion between the 'duck' and spine cylinder was used to pump hydraulic fluid in the power take-off system. Recently, Salter (1982) has proposed a novel power take-off system involving a pair of gyroscopes and extensive model testing has been carried out. It is envisaged that this system would be 'intelligent' and alter its characteristics to optimise output for changing wave conditions.

A 1:10 scale model consisting of 20 'ducks' fixed along a 50 m spine beam has been tested in Loch Ness (Bellamy, 1978). The 2 GW station will consist of spine units 90 m long, each carrying 2 'ducks' and connected by Hooke's universal joints. Each 'duck' will be rated at 2.2 MW and will have a spine diameter of 14 m. A full assessment by the consultants has not been possible because they suggest that the success of the system relies heavily on future technology. Economic viability requires a high degree of component reliability. Further, two key elements, the Hooke joints and spine support bearings, need to be demonstrated as feasible. The power take-off system is, however, identified as solving a number of problems for other wave energy devices and merits further development.

Figure 18.17: The Salter 'Duck'. (a) Detail of power take-off mechanism. (b) Artist's impression of the full-scale device

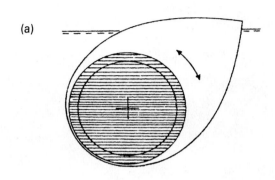

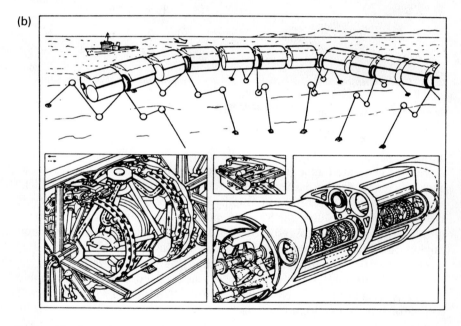

Source: After Grove-Palmer (1982).

The National Engineering Laboratory (NEL) Oscillating Water Column (OWC)

The NEL has been developing OWC type devices since 1976 (Meir, 1978). A number of configurations have been tested at model scale and mathematical descriptions of devices developed. Earlier designs were conceived as a floating terminator and the final configuration (Moody and

Elliot, 1982) was 263 m long with a power rating of 9 MW. Preliminary studies of a floating attenuator configuration (Moody, 1979) showed that a long narrow structure offered best efficiency. Tank testing of various configurations, including the final design, showed poor performance. The attenuator was subsequently dropped from the programme (NEL, 1981).

A study on a seabed fixed OWC system was initiated in 1979 to improve the performance of the attenuator and the final device selected is shown in Figure 18.18. Each unit of the breakwater is 64 m long, consisting of four oscillating columns with a total rating of 3.1 MW. Power take-off is achieved with standard Francis or axial flow air turbines. These turbines are not able to cope with the reversing air flow from the column and so a series of flow rectifying valves are necessary. The NEL seabed fixed OWC has been developed into the 'Breakwater Module'. This unit consists of four chambers, each 15 m square, with a total rating of 5 MW. A study for a prototype on the island of Lewis, Scotland has been carried out by the NEL Breakwater Working Group, which consists of eleven industrial partners (Elliot, 1984). The Island of Lewis has an installed capacity of 21 MW and would be suitable for integration with this system. A site-specific detailed design study was completed in 1985. The results of this study were not sufficiently optimistic to allow further funding to be allocated.

Norway

Wave focusing

Although wave energy is one of the highest grade renewable energies its density is still relatively low. A concept to concentrate wave energy and reduce the number of machines necessary for power conversion has been proposed by Mehlum (1979). Focusing of wave energy is achieved using a 'Water Wave Lens'. Lens elements consist of horizontal plates moored below the surface and research, based in an open-air wave basin near Oslo, has concentrated on the problem of their optimal shape (Mehlum, 1982). The power take-off for this system consists of a converging concrete channel which causes the wave heights to increase. As the waves propagate along the channel water is forced to spill over the walls into a reservoir. The stored water is returned to the sea through a conventional low head turbine to generate electricity. Impedance matching between the converging channel and focused wave energy in model tests shows high efficiency. The cost of a large-scale lens system is high and suitable sites limited, so Norwave A.S. has been investigating natural focusing by seabed features. Optimum location of the wave energy conversion channel can be determined using computer analysis of wave propagation (Norwave, 1983).

Source: National Engineering Laboratories.

Kvaerner OWC

Kvaerner Engineering has developed an OWC which has a harbour constructed at its mouth (Ambli *et al.*, 1982). This harbour effectively increases output by widening the bandwidth of the device. Interaction between the OWC and natural modes of harbour oscillation increases power output by up to 200 per cent (Count and Evans, 1984). Kvaerner has designed a full size prototype which is 9.4 m wide and 10.6 m long, to be sited in 7 m water depth. Each unit is fitted with a Wells turbine generator with peak output of 300 kW. The consultants' report (Royal Ministry for Petroleum and Energy, 1982) states that this is the most promising of all Norwegian devices.

Medium-scale prototypes of both preferred Norwegian devices have been constructed at Toftestallen, on an island near Bergen. Natural features have been utilised to form a reservoir and a concrete converging channel constructed. Figure 18.19 shows the layout of this system which has a 400 kW generator in the dam. Close to this reservoir a bench has been blasted in the cliff face and a concrete chamber 10 m square constructed to form an OWC, which is fitted with a 300 kW Wells turbine generator. Both systems were commissioned in November 1985 and supply electricity to the local grid.

Figure 18.19: Overtopping channel system and storage reservoir at Toftestallen, Norway

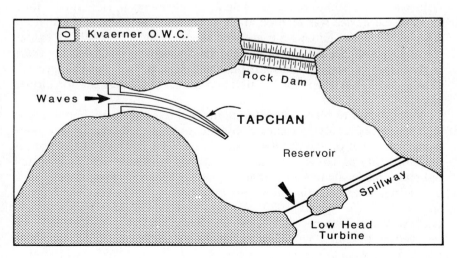

Japan

Japan is an island nation almost wholly dependent on imported fuel. Thus research programmes are under way into all aspects of energy diversification and the development of new, renewable energy sources. Over 20 billion yen was spent by the Science and Technology Agency (STA) in 1983, although wave energy projects only receive a small percentage of this. The wave energy resource in Japan varies from about 25 kW m^{-1} on the Pacific Coast to 11 kW m^{-1} on the Japan Sea coast. Research has been in progress for over thirty years into wave energy development (Masuda, 1971). Successful generating units for navigation buoys and lighthouse power have been manufactured and over 1,000 have been sold (Ryokuseisha, 1983). A number of Research and Development projects are in progress and are now described.

Japan Marine Science and Technology Centre (JAMSTEC)

Research into wave energy began at JAMSTEC in 1974 with experiments in the wave tank (Ishii et al., 1982) and Masuda (1971, 1972) had been involved with a number of field experiments which resulted in a small commercial wave generator. At JAMSTEC he was instrumental in the planning and construction of the large-scale generating system 'KAIMEI' (Masuda et al., 1978) and research concentrated on OWC conversion systems built into a floating structure. Various configurations were tested but it was found that the attenuator was most efficient. In 1977 a 2 MW ship-shaped attenuator was constructed at Ishikawajima-Harima Heavy Industry factory at Aloi. This device is 80 m long and 12 m wide, having a displacement of 500 tons and housing 22 air chambers open at the bottom. In 1978 the device was moored at Yura, on the Japan Sea, and tested for ten months. These initial sea trials were encouraging and 'KAIMEI' subsequently became the subject of an IEA co-operative project. During these trials electricity was fed into the mainland grid, a 'first' for this project. During 1985 'KAIMEI' was refitted and towed to the test site for further trials.

All of the air turbines fitted during the first sea trials were impulse type with a series of rectifying valves. Subsequently a large number of tests have been carried out on the development of the Wells Turbine. Model tests at the University of Tokyo (Suzuki et al., 1984) in steady flow, investigated optimum blade shape/configuration and resulted in formulation of a design method. Oscillating flow tests have been carried out at the University of Osaka (Setoguchi, 1983) and Fuji Electric Co. have built a 40 kW tandem Wells Turbine unit. This 40 kW generator will eventually be installed in 'KAIMEI' having in the meantime undergone testing by JAMSTEC at Sanze near the Yura test site. A shore-based OWC steel chamber has been constructed by MITSUI (Shipbuilding) Co. and output has reached 40 kW consistently over the winter months. Experience is now being gained with operating characteristics and performance of turbine materials in the open marine environment.

Active control of wave energy devices to increase output was first proposed by Budal (1979) in Norway. Subsequent investigations at JAM-STEC (Miyazaki *et al.*, 1983) have shown that phase control of the OWC will increase output at low frequency by up to ten times. Tests are in progress at the Sanze site and these will continue on the 'KAIMEI' sea trial. Overall the Japanese target for wave produced electricity is 50¥ kWh⁻¹. It is known that the structure of 'KAIMEI' is not optimum for Japan Sea conditions. Nevertheless, model tests and theoretical calculations will allow results of the field trial to be implemented, confident that final cost estimates will be below the target price.

CONCLUSION

Dynamic variations in sea-surface level with periods shorter than half a day represent a large energy resource. The technology for the utilisation of the resource has progressed to a stage where there are only a few obstacles to full development. One of the disadvantages, however, is that the distribution of significant energy quantities is not globally uniform and that conversion will only be feasible in a limited number of countries. The economics of energy production from sea-surface variations will improve as conventional fuel supplies become exhausted. A true economic evaluation of ocean energy is difficult with so few working power plants for comparison, but figures have been quoted above where such an assessment is possible. Initially, these systems will only be competitive in high-cost regions of remote communities, but their use will be extended as conventional (fossil) fuel costs rise, as necessitated by the inevitable exhaustion of fossil fuel resources. One severe disadvantage of tidal power development, however, is that it involves a large capital investment at one time. Wave energy development by contrast can be developed progressively with a consequent easing of the investment burden. It is expected that nuclear fusion will hold the key to the world's future energy requirements, but technological barriers to working fusion power generators still remain. In the transition period from fossil fuels to fusion power there could well be an 'energy gap' which could be filled by ocean energy resources.

REFERENCES

Airy, G.B. (1861) 'Tides and Waves', *Encyclopaedia Metropolitana*, 5, 241-396.
Allen, J. (1947), *Scale Models in Hydraulic Engineering*, Longman, London.

Ambli, N., Bønke, K., Malmo, O and Reitan, A. (1982) 'The Kvaerner multi-resonant OWC', in H. Berge (ed.), *Wave Energy Utilisation*, Proc. Second Int. Symp. on Wave Energy Utilisation, Trondheim, Tapir Press, Trondheim, pp. 275-96.

Anon. (1982) 'Higher nuclear power costs favour tidal scheme', *Water Power and Dam Construction*, March issue, 5.

Banal, M. and Bichon, A. (1981) 'Tidal energy in France. The Rance tidal power station: some results after 15 years of operation', in *Proc. Second Int. Conf. on Wave and Tidal Energy*, Cambridge, UK, British Hydromechanics Res. Assoc., Cranfield, pp. B1-1-B1-11.

Barber, N.F. (1963) *Water Waves*, Wykeham Publications, London.

Bellamy, N.W. (1978) 'The Loch Ness trials of the DUCK', *Proc. Heathrow Wave Energy Conference*, HMSO, London, pp. 120-30.

—— (1982) 'Development of the SEA "Clam" wave energy convertor', in H. Berge (ed.), *Wave Energy Utilisation*, Proc. Second Int. Symp. on Wave Energy Utilisation, Trondheim, Tapir Press, Trondheim, pp. 175-90.

Berge, H. (1982) *Wave Energy Utilisation*, Proc. Second Int. Symp. on Wave Energy Utilisation, Trondheim, Tapir Press, Trondheim.

Bernstein, L.B. (1974) 'Kislogubskaya: a small station generating great expectations', *Water Power*, 26, 172-7.

BHRA (1978) *Proceedings of First International Conference on Wave and Tidal Energy*, Cambridge, UK, British Hydromechanics Res. Ass., Cranfield.

BHRA (1982) *Proceedings of Second International Conference on Wave and Tidal Energy*, Cambridge, UK, British Hydromechanics Res. Ass., Cranfield.

Bruun, P. (1980) *Port Engineering*, Gulf Pub. Co., Houston.

Budal, K. and Falnes, J. (1979) 'Interacting point absorbers with controlled motion', in B. Count (ed.), *Power From the Sea*, Academic Press, London and New York, pp. 381-401.

—— and Falnes, J. (1983) 'Status 1983 of the Norwegian Buoy Project: Open report', Norwegian Tech. Hogescole (unpublished report).

Charlier, R.H. (1978) 'Tidal power plants: sites, history and geographical distribution', in *Proc. First Int. Conf. on Wave and Tidal Energy*, British Hydromechanics Res. Ass., Cranfield, pp. A1-1–B1-1.

Clare, R. and Oakley, A.J. (1981) 'Towing and Positioning of caissons in a tidal barrage', in *Proc. Second Int. Conf. on Wave and Tidal Energy*, Cambridge, UK, British Hydromechanics Res. Ass., Cranfield, pp. 177-91.

Clark, R.H. (1977) *Re-assessment of Fundy Tidal Power: A Report of the Bay of Fundy Tidal Power Review Board and Management Committee*, Ministry of Supply and Services, Ottawa.

Count, B. (ed.) (1980) *Power from Sea Waves*, Academic Press, London and New York.

—— and Evans, D.V. (1984) 'The harbour concept in wave energy', in *Energy Options*, Proc. Fourth Int. Conf. on Energy Options, IEE Publication 233, Peter Peregrinus, London, pp. 211-15.

—— , Fry, R. and Haskell, J. (1983) 'An experimental investigation of the harbour concept in wave energy', CEGB Report TRPD/M/1298/N82, Marchwood Engineering Laboratories, Southampton.

Damy, G. (1982) 'Production d'énergie à partir de la houle. Bilan provisoire après 2 ans de déroulement du programme', *Rapport TD1/CCTRME/82-31/GD-PV* CNEXO, Brest (unpublished report).

—— and Gauthier, M. (1981) 'Production d'énergie à partir de la houle', CNEXO, Brest (unpublished report).

Darwin, G.H. (1898) *The Tides and Kindred Phenomena of the Solar System*, Greenman, San Francisco.

Defant, A. (1961) *Physical Oceanography*, 2 vols, Pergamon Press, London.

Derbyshire, E., Gregory, K.J. and Hails, J.R. (1979) *Geomorphological Processes*, Butterworth, London.

Douma, A., Stewart, G.D. and Meier, W. (1982) 'Straflo turbine at Annapolis Royal: first tidal power plant in the Bay of Fundy', *Escher Wyss News*, *1/1981-1/1982*, 3-10.

Elliot, G. (1984) 'The NEL Breakwater — Britain's proposed first wave power station', in *Energy Options*, Proc. Fourth Int. Conf. on Energy Options, IEE Publication 233, Peter Peregrinus, London, pp. 202-6.

Eurocean, (1979) *Report No. 1*, Eurocean, Monaco.

Evans, D.V. (1976) 'Wave power absorption by oscillating bodies', *J. Fluid Mechanics*, *77*, 1-25.

Gibrat, R. (1966) *L'Energie des marées*, Presses Universitaires de France, Paris.

Gibson, R.A. and Wilson, E.M. (1978) 'Studies in retiming tidal energy', in *Proc. First Int. Conf. on Wave and Energy*, Cambridge, UK, British Hydromechanics Res. Ass., Cranfield, pp. H1-H21.

Grove-Palmer, C.O.J. (1982) 'Wave energy in the UK: a review of the programme June, 1975-March, 1982', in Berge, H. (ed.), *Wave Energy Utilisation*, Proc. Second Int. Symp. on Wave Energy Utilisation, Trondheim, Tapir Press, Trondheim, pp. 23-54.

Hails, J.R. (1977) 'Applied geomorphology in coastal zone planning and management', in J.R. Hails (ed.), *Applied Geomorphology: A Perspective of the Contribution of Geomorphology to Interdisciplinary Studies and Environmental Management*, Elsevier, Amsterdam, pp. 317-68.

—— and Carr, A. (eds) (1975) *Nearshore Sediment Dynamics and Sedimentation: An Interdisciplinary Review*, Wiley, London.

Harris, R.I. and Highgate, D.J. (1978) 'Energy supplies for remote communities' in *Proc. Wind, Wave and Water Conference, Dublin 1978*, Paper 2, Session 2, Irish Management Inst., Dublin.

Harvey, J.G. (1976) *Atmosphere and Ocean: Our Fluid Environment*, Artemis Press, Horsham.

Heathrow Conference (1978) *Proceedings of Wave Energy Conference*, HMSO, London.

Hendershott, M.C. (1977) 'Numerical models of ocean tides', in D. Goldberg, I.N. McCave, J.J. O'Brien and J.H. Steele (eds), *The Sea*, *6*, Wiley-Interscience, New York, pp. 47-95.

—— (1982) 'The effect of solid earth deformations on global area tides', *Geophys. J. Roy. Astron. Soc.*, *29*, 389-403.

IEE (1984) *Energy Options*, Proceedings of the Fourth International Conference on Energy Options, IEE Publication 233, Peter Peregrinus, London.

Isaacs, J.D. (1979) 'Ideas and some developments of wave power conversion', in *Proc. First Int. Conf. on Wave Energy Utilisation, Gothenburg*, Chalmers University, Gothenburg, pp. 204-22.

Ishii, S., Miyazaki, T., Masuda, Y. and Kai, G. (1982) 'Reports and future plans for the Kaimei project', in H. Berge (ed.), *Wave Energy Utilisation*, Proc. Second Int. Symp. on Wave Energy Utilisation, Trondheim, Tapir Press, Trondheim, pp. 305-21.

Iversen, L.C. and Lillebekken, P.M. (1983) 'Buoy of Type N2 in the sea', Report Inst. for Experimental Physics, NTH, Trondheim, Norway (unpublished report).

Kinsman, B. (1965) *Wind Waves*, Prentice-Hall, Englewood Cliffs, NJ.

Laplace, P.S. (1799) *Traité de Méchanique Céleste*, Crapelet, Paris.

Leliavsky, S. (1982) *Hydro-electric Engineering for Civil Engineers*, Chapman & Hall, London.

Ley, W. (1954) *Engineers Dreams*, Viking Press, New York.

Long, A.E. (1978) 'The Belfast Device', in *Heathrow Wave Energy Conference*, HMSO, London, pp. 61-3.

McCormick, M. (1978) 'Wave energy conversion in a random sea', in *Proc. Thirteenth Intersociety Energy Conversion Conference, San Diego, Calif.*, Marine Tech. Soc., pp. 63-71.

—— (1981) *Ocean Wave Energy Conversion*, Wiley, Chichester and New York.

McMillan. D.H. (1966) *The Tides*, C.R. Books, London.

Masuda, Y. (1971) 'Wave activated generator for robot weather buoy and other use', *Colloq. Int. sur l'Exploitation des Océans, Bordeaux, France, V-72-05*, 1-17.

—— (1972) 'Study of wave activated generators and future view as an island power source', in *Second Int. Ocean Development Conference*, Tokyo, Japan, Soc. of Engineers, Tokyo, pp. 38-42.

Mehlum, E. (1982) 'Recent developments in the focusing of wave energy', in Berge, H. (ed.), *Wave Energy Utilisation*, Proc. Second Int. Symp. on Wave Energy Utilisation, Trondheim, Tapir Press, Trondheim, pp. 419-40.

Mehlum, E. and Stamnes, J. (1979) 'Power production based on focussing of swells', *First Int. Conf. on Wave Energy Utilisation, Gothenburg*, Chalmers University, Gothenburg, pp. 29-35.

Meir, R. (1978) 'The development of the oscillating water column', in *Proc. of the Heathrow Wave Energy Conference*, HMSO, London, pp. 35-44.

Miyazaki, T., Yokomizo, H., Hotta, H. and Washio, Y. (1983) 'A study of phase control of air flow in order to increase output power of OWC wave generators', Japan Marine Science Technology Centre, Natsushima, Japan (unpublished report).

Mogridge, G.R. (1980) *A Review of Wave Power Technology*, Report LTR-HY-74, NRC, Ottawa.

Mollison, D. (1980) 'Prediction of device performance', in Count, B. (ed.), *Power from Sea Waves*, Academic Press, London and New York, pp. 135-74.

Moody, G. (1979) 'NEL oscillating water column: recent developments', in *First Int. Conf. on Wave Energy Utilisation, Gothenburg*, Chalmers University, Gothenburg, pp. 283-97.

—— and Elliot, G. (1982) 'The development of the NEL Breakwater WEC', in H. Berge (ed.), *Wave Energy Utilisation*, Proc. Second Int. Symp. on Wave Energy Utilisation, Trondheim, Tapir Press, Trondheim, pp. 421-51.

NEL (1981), 'Progress report on the floating attenuator design', National Engineering Laboratories, East Kilbride, (unpublished report).

Neumann, C. and Pierson, W.J. (1966) *Principles of Physical Oceanography*, Prentice-Hall, Englewood Cliffs, NJ.

Northern Ireland Economic Council (1981) 'Strangford Lough tidal energy', *Northern Ireland Economic Council Rept. 24*, 1-56.

Norwave (1983) 'Wave Track — a system for wave refraction calculation', Norwave Company Report, Oslo (unpublished report).

Palme, A. (1920) 'A wave motion turbine', *Power 52, 18*, 700-1.

Pethick, J. (1984) *An Introduction to Coastal Geomorphology*, Edward Arnold, London.

Pleass, C.M. (1978) 'Use of a wave power system for desalination', in *Proc. First Int. Conf. on Wave and Tidal Energy*, Cambridge, UK, Paper D1, British

Hydromechanics Res. Ass., Cranfield.

Richard, J. (1933) 'Sur l'utilisation des mouvements de la mer par l'Ondo-pompe', *Bull. Inst. d'Océanographique Monaco, 625*, 2-14.

Romanovksy, P. (1950) 'L'Energie des mers: est-elle utilisable?', *Science et Vie*, May issue, 279-83.

Royal Ministry for Petroleum and Energy (1982) *Nye Fornybare Energikilder i Norge Stortinsmelding*, White Paper to Norwegian Parliament, Govt. Publ. Office, Oslo.

Salter, S.H. (1974) 'Wave Power', *Nature, 249*, 720-4.

—— (1982) 'Use of gyros as power take-off', in H. Berge (ed.), *Wave Energy Utilisation*, Proc. Second Int. Symp. on Wave Energy Utilisation, Trondheim, Tapir Press, Trondheim, pp. 99-116.

Savonius, S.J. (1931) 'The S-Rotor and its applications', *Mechanical Engineering, 53*, 333-8.

Scobie, G. and Leishman, J.M. (1975) 'The development of wave power', Economic Assessment Unit, NEL, East Kilbride, (unpublished report).

Scott, D.B. and Greenburg, D.A. (1983) 'Relative sea-level rise and tidal development in the Fundy tidal system', *Can. J. Earth Sci., 20*, 1554-64.

SEA (1982) 'The SEA Lanchester Wave Energy Programme — Bag report', Report LJD October, Lanchester Polytechnic, Coventry (unpublished report).

Settoguchi, M. (1983) 'Tests on a Wells turbine in oscillating flow', *J. Soc. Naval Architecture of Japan, 152*, 84-92.

Severn, B. and Campbell, R.O. (1978) 'Prefabricated caissons for tidal power development', in *Proc. First Int. Conf. on Wave and Tidal Energy*, Cambridge, UK, British Hydromechanics Res. Assoc., Cranfield, pp. G1/1-G1/12.

Severn Barrage Committee (1981) *Tidal Power from the Severn Estuary, 1 & 2*, Energy Paper 46, HMSO, London.

Shaw, T.L. (1974) 'Tidal energy from the Severn estuary', *Nature, 249*, 730-3.

Simeons, C. (1980) *Hydro-power*, Pergamon Press, Oxford.

Special Committee on Alternative Energy (1981) *Energy Alternatives*, Report to the Parliament of Canada, Can. Govt. Publ., Ottawa.

Sulley, J.L., Moffat, A.M. and Barlow, J.M. (1984) 'Technical aspects of integrating wind and wave generation sources to small island systems', *Energy Options*, Proc. Fourth Int. Conf. on Energy Options, IEE Publication 233, Peter Peregrinus, London, pp. 207-11.

Suzuki, M. Arakawa, C. and Tagori, T. (1984) 'Fundamental studies on the Wells turbine for wave power generation: first report on the effect of solidity and self-starting', Report No. 83-0070 Dept. of Mechanical Engineering, Tokyo University (unpublished report).

Wailes, R. (1941) 'Tide mills in England and Wales', *Int. Inst. Engineers J. and Record of Trans., 51*, 91-114.

Wave Energy Utilisation (1979) *Proceedings of First Conference on Wave Energy*, Gothenburg, Chalmers University, Gothenburg.

Wick, G.L. and Castel, D. (1978) 'The Isaacs wave energy pump — field tests off the coast of Hawaii', *Ocean Engineering, 5*, 235-42.

Part Six

Conclusions

19

Sea-surface Changes: Where Do We Go From Here?

R.J.N. Devoy

THE VALUE OF STUDYING SEA LEVELS

Geophysical applications

'The response of the earth to transient loads of ice or water provides the most explicit information available about the non-elastic properties of the mantle.' This statement from Crittenden (1967) encapsulates one of the most important applications of sea-level studies: an understanding of the structure and dynamics of the earth's crust. More specifically, 'delevelled shorelines are eloquent testimony to the outer structure of the earth and its behaviour under stress' (Pirazzoli and Grant, 1986). As such the analysis of sea-level data presents an effective means of investigating the rheological properties of the earth and recent lithospheric deformation.

Evidence of sea-level change is ubiquitous, to be found in the form of marine washing limits, shoreline notches and platforms, fossil beach structures, in coral reefs — micro atolls and in estuaries — submerged sedimentary sequences from the Thames to the Mekong. These and present shorelines represent the position of gravitational equilibrium between the lithosphere and hydrosphere. Detail of the factors involved and their interaction in determining this position have been given in Chapters 1, 3, 8 and 10 (see Fig. 1.3). Based on knowledge of the pattern of changing ice cover and water distribution on the earth since lateglacial time, global isostatic models have been developed for prediction of the varying sea-level–crustal responses to these changes in earth surface load (Walcott, 1972; Clark *et al.*, 1978). Using observed sea-level histories as a control, these and related geophysical models of earth behaviour can be re-defined to allow adjustment of the calculated crustal–mantle factors involved until cause and effect are matched (see Chapter 3; Cathles, 1975; Peltier and Andrews, 1976; Clark, 1980). As a result both sea-level and crustal movements at any point on the earth's surface can be calculated for changes in size (growth/shrinkage) of complex ice masses. Prediction matches observation reasonably well and six broad zones of differential movement have been defined (see Fig. 10.4). These models and the allied analysis of sea-level data have shown for the first time why the direction and magnitude of Late Quaternary crustal movements, and consequently the associated broad patterns of regional sea-level change, appear to vary so much over the earth. Explanation of disparate movements in seemingly comparable areas is now partially provided, for example the emergence of the south Pacific islands versus the submergence of Micronesia (Pirazzoli and Grant, 1986). Sea-level studies can, therefore, present a wealth of information on the internal and external influences upon the lithosphere. The delevelling of former marine and lake shorelines (Crittenden, 1963), resulting from glacial unloading and water loading, has given important information on the processes of lithospheric deformation on plate margins and interiors (Faure et al., 1980; Martin et al., 1979-80; Chappell et al., 1982; Quinlan and Beaumont, 1983). Further, differences between predicted and observed shoreline-crustal deformation, interpreted as the result of regional variations in earth crustal/rheological factors, such as crustal thickness and temperature and mantle viscosity, have resulted in new insights into the properties of mantle viscosity, configuration and behaviour (Mörner, 1980).

Apart from the effects of glacier–water loading shorelines also register the results of other earth processes. Major regional disparities in shoreline data (time–altitude patterns) may represent significant temporal and areal changes of the geoid, a subject worthy of further study (Mörner, 1983; Tooley, 1985). Alternatively, sea-level evidence from deep oceanic environments, such as the Pacific with its myriad of islands each recording the local pattern of sea-level change, allows definition of processes of thermo- and volcano-isostatic deformation. Increases in density of oceanic crust following its production results in isostatic subsidence of the ocean floor, and thus gradual submergence of oceanic islands, in the Pacific as elsewhere, as they move away from ridge spreading centres. Subsidence is increased in tropical waters by the growth of mantling coral reefs which progressively thicken to maintain their sea-level position. Variations in this pattern, from changes in crustal density and local isostatic movements, through interaction with crustal 'hot spots' (Detrick and Crough, 1978) or extrusion of lavas and consequent isostatic loading comparable to the effects of ice build-up (Walcott, 1970), may produce a complex cycle of island emergence and submergence (Coudray and Montaggioni, 1982; Scott and Rotando, 1983). The mapping and quantification of crustal deformation rates from the resultant shoreline evidence again reflects on the geophysical properties of the earth. Cores from coral reefs in the Marshall Islands and Mururoa Atoll give subsidence rates of 0.2 mm a^{-1} over 60 Ma (Menard and Ladd, 1963) and 0.12 mm a^{-1} over 7 Ma (Labeyrie et al., 1969) respectively, data which has subsequently helped refine global isostatic models. In the Society Islands observation of the 3,000 BP shoreline has been used here to define the pattern of crustal flexure (Pirazzoli, 1983), showing a slight but progressive decrease in emergence (~0.5 m) along a transect from Maupiti-Bora Bora-Tahiti.

By comparison, crustal movements in areas of lithospheric plate *convergence* may also be recorded in sea-level data. In southwestern Japan phases of land uplift since the Late Quaternary are evidenced by series of raised marine terraces (Yoshikawa *et al.*, 1981). Dating and levelling of these shorelines has facilitated establishment of the recurrence intervals for uplift, ranging from 90-264a to 1,000-2,000a (Yonekura, 1975; Matsuda *et al.*, 1978). From the Mediterranean application of archaeological and other data in dating shorelines has allowed definition here of the complex pattern of delevelling (Flemming, 1978). Between southern Greece and Crete an independently active crustal block ~150 km long has been identified from sea-level change studies (Thommeret *et al.*, 1981; Pirazzoli *et al.*, 1982). Shorelines in such areas of crustal convergence may be used, therefore, in a variety of ways; to determine the age, spatial limits and pattern of crustal uplift and also to aid prediction of the probable areas of future large magnitude earthquakes-associated scale of crustal uplift. In southwest Japan, for example, shoreline and associated evidence suggest that the next great earthquake will occur in the Tokai district (Ando, 1975; Yonekura, 1975).

Socio-economic interests

Moving away from the sphere of geophysics, the application of sea-level studies to other aspects of environmental change has also gathered momentum. As discussed in Chapter 16, the publication of Vail *et al.*'s (1977) approach to seismic stratigraphy has stimulated the whole field of the geological-commercial application of sea levels. In terms of climate prediction a possible man-induced increase in atmospheric CO_2 (National Academy of Sciences, 1983) has brought a new awareness in science of the linkage between global temperature change—ice melt and sea-level rise (Barth and Titus, 1984). The possible economic and social repercussions of this on coastal communities and physical infrastructure, if not on all mankind, has gripped the attention of many scientists (Hoffman *et al.*, 1983). Consequently organisations such as the Environmental Protection Agency (EPA) in the United States, IGCP-200 and the IGU are encouraging research and pursuing publicity programmes to increase public awareness and a planned response to the potential hazards (see Chapter 15). The EPA has undertaken a range of publications with this aim in view (Titus *et al.*, 1985; Hull and Titus, 1986), holding a conference in Washington in 1986 on the 'Health and Environmental Effects of Ozone

Modification and Climatic Change'. Similarly, in Delft, the Netherlands, an international group of scientists met in 1986 to plan a long-term project to study the 'Impact of Sea-level Rise on Society' (ISOS), aimed at assessing the nature of the problems (Fig. 19.1) and the proposal of plans for 'decision makers' on the protection of the coastline. In America and Europe coastal engineers, charged with the physical defence of the coastline, are becoming seriously concerned with the possibility of an acceleration in sea-level rise and are actively researching the phenomenon (see Chapter 14; Sorensen and Weggel, 1986), although sea-level change has long been a parameter in coastal defence structure design (Horner, 1972).

Interdisciplinary research

A measure of the consequent concern now engendered in sea-level research can be seen from the number of organisations engaged in evaluating recent/continuing sea-level movements. The International Association of Geodesists (IAG) are developing links with sea-level work undertaken through IGCP-200 (Shennan, 1986). The Inter-governmental Oceanographic Commission (IOC) in conjunction with the World Meteorological Organisation (WMO), is concerned with developing a co-ordinated system of global sea-level measurements, through such groups as the Integrated Global Ocean Services System (IGOSS) (Wyrtki and Nakahara, 1984). The activities of the Permanent Service for Mean Sea Level (PSMSL, IOS, Bidston, UK) engaged in the analysis of worldwide tide gauge data, also falls under the auspices of the IOC, as does the work of the Committee on Climate Change and the Ocean (CCCO) (Shennan, 1986). In terms of space geodesy, NOAA (National Oceanic and Atmospheric Administration), NASA (National Aeronautics and Space Administration) and others in America are using Very Long Baseline Interferiometry (VLBI) and Global Positioning System (GPS) techniques to measure deformations of earth surface shape accurate to ± 1 cm, or finer (Fig. 19.2). Under projects such as POLARIS (Polar-motion Analysis by Radio Interferiometric Surveying) and IRIS (International Radio Interferiometric Surveying), geodynamic networks are being integrated with tide gauge systems as a check on localised crustal deformation, and a direct monitoring of sea-surface movements (Carter *et al.*, 1985).

REFLECTIONS

A complete summary of chapter contents is both unnecessary and impossible! Through its six sections this book has attempted to present an interdisciplinary view of the geological perspectives of sea surface studies, ranging from an examination of the causes, methodology, spatial–temporal

Figure 19.1: Scheme of factors influencing the decision makers'
(human/socio-economic) response to climate/sea-level changes

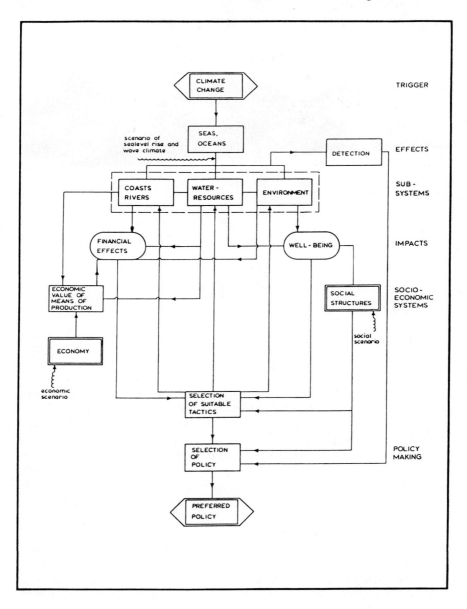

Source: After Wind, Vrevgdenhril and Goemans — Delft Hydraulics Laboratory (1986).

Figure 19.2: Depiction of linked usage of fixed VLBI, mobile VLBI and GPS systems for monitoring the stability of tide gauge stations and earth surface movements

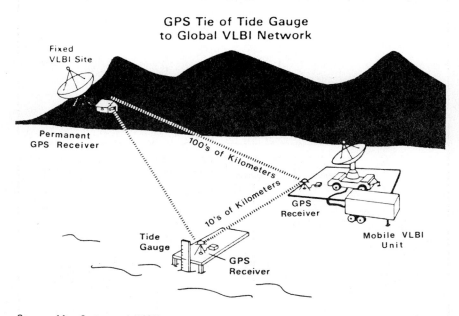

Source: After Carter *et al.* (1985).

patterns through to the application of sea-level changes. The subject and its linkage with related disciplines, such as geophysics, geodesy and ocean-ography, is vast. As such the approaches presented here cannot form a complete review of the subject, being conditioned by the individual per-spectives and experience of contributors. Nevertheless, wherever possible referencing to further unexplored literature has been made, providing readers with a way forward. A particular gap, however, has been the wealth of material produced by researchers in the USSR and its neighbours, which has barely been touched upon. A useful introduction to this may be gained from *Fluctuations of the Sea and Ocean Level for the Past 15,000 Years* (Kaplin *et al.*, 1982), although other original material (in translation) is often difficult to obtain. In Western literature, books, journal articles and results of symposia are readily available and reference to these is best found through the journals *Quaternaria* and *Quaternary Research* and newsletters such as *Sea-Level* (1979-82), *Nivmer* (1978-82), *Litoralia* (INQUA) and IGCP-200 (1986).

Finally, what will the future hold? The current watchwords are 'inter-disciplinary co-operation'. The operation of the hydrosphere is intimately linked to those of the lithosphere and atmosphere, and its examination must thus involve work upon these other environments. Similarly, in tackling the collection and analysis of sea-level information isolated data must now be seen as of limited value in simulating, for example, regional scale sea-level movements, climate linkages and geoid behaviour. This is not to say that the co-ordinated collection of basic information at the local level is redundant, not at all. Major areas of the world still lack data. However, good global coverage of the oceans would need tens of thousand of reliable measurements. A solution to data collection is suggested by Pirazzoli and Grant (1986) in the selection of large-scale transects, taken for example perpendicular to ice sheets, continental margins and active plate boundaries, with a view to testing developed models. In other spheres, what are the likely approaches to be? Satellite geodesy and sea surface monitoring must be an important element. The coupling of these results with those from geophysical and climate-oceanographic models, palaeotidal work and numerical treatment of sea-level data will continue to develop. In the face of this sophistication, however, it is perhaps salutary to reflect on the experience of past theories and paradigms in the earth sciences, such as that of the 'Deluge' (Diluvial theory) explanation of Quaternary events. As Velikovsky (1956) describes, the sea has figured prominently, if somewhat unacceptably now, in some writers' interpretation of geological phenomena. Hypothesis and explanation which may appear as reasonable now may be replaced in future by equally convincing paradigms. The behaviour of the sea surface past and present is complex; it is as well then that we keep an open mind in its study.

REFERENCES

Ando, M. (1975) 'Source mechanism and tectonic significance of historical earthquakes along the Nankai Trough, Japan', *Tectonophys.*, *27*, 119-40.

Barth, M.C. and Titus, J.G. (1984) *Greenhouse Effect and Sea-level Rise*, Van Nostrand Reinhold, New York.

Carter, W.E., Robertson, D.S., Pyle, T.E. and Diamante, J. (1985) 'The application of geodetic radio interferiometric surveying to the monitoring of sea level', *IAMAP/IAPSO Symposium on Variations of Sea Level*, 7-8 Aug. 1985, Honolulu, Hawaii.

Cathles III, L.M. (1975) *The Viscosity of the Earth's Mantle*, Princeton University Press, Princeton, NJ.

Chappell, J., Rhodes, E.G., Thom, B.G. and Wallensky, E. (1982) 'Hydro-isostasy and sea-level isobase of 5,500 BP in North Queensland, Australia', *Mar. Geol.*, *49*, 81-90.

Clark, J.A. (1980) 'A numerical model of worldwide sea-level changes on a visco-elastic earth', in N.-A. Mörner, (ed.), *Earth Rheology, Isostasy and Eustasy*, Wiley, Chichester and New York, pp. 525-34.

——— , Farrell, W.E. and Peltier, W.R. (1978) 'Global changes in postglacial sea level: a numerical calculation', *Quat. Res.*, *9*, 265-87.

Coudray, J. and Montaggioni, L. (1982) 'Coraux et récifs coralliens de la province Indo-Pacifique: répartition géographique et altitudinale en relation avec la tectonique globale', *Bull. Soc. Géol. France, 24*, 981-93.

Crittenden, M.D., Jr (1963) 'New data on the isostatic deformation of Lake Bonneville', *US Geol. Surv. Prof. Paper, 454-E*, 1-31.

—— (1967) 'Viscosity and finite strength of the mantle as determined by water and ice loads', *Geophys. J. Roy. Astron. Soc., 14*, 261-79.

Delft Hydraulics Laboratory (1986) *Impact of Sea-level Rise on Society (ISOS)*, Project Planning Document, Delft, The Netherlands.

Detrick, R.S. and Crough, S.T. (1978) 'Island subsidence, hot spots and lithospheric thinning', *J. Geophys. Res., 83*, 1236-44.

Faure, H., Fontes, J.C., Hébrard, L., Monteillet, J. and Pirazolli, P.A. (1980) 'Geoidal change and shore level tilt along Holocene estuaries: Senegal river area, West Africa', *Science, 210*, 421-3.

Flemming, N.C. (1978) 'Holocene eustatic changes and coastal tectonics in the northeast Mediterranean: implications for models of crustal consumption', *Phil. Trans. Roy. Soc. Lond., A., 289*, 405-58.

Hoffman, J., Keyes, D. and Titus, J. (1983) *Projecting Future Sea-Level Rise: Methodology, Estimates to the Year 2100, and Research Needs*, 2nd rev. edn, Government Printing Office, Washington, DC.

Horner, R.W. (1972) 'Current proposals for the Thames barrier and the organization of the investigations', *Phil. Trans. Roy. Soc. Lond., A., 272*, 179-85.

Hull, C.H.J. and Titus, J.G. (eds) (1986) *Greenhouse Effect, Sea-Level Rise and Salinity in the Delaware Estuary*, US Environmental Protection Agency, Washington, DC.

Kaplin, P.A., Klige, P.K. and Chepalyga, A.L. (1982) *Fluctuations of the Sea and Ocean Level for the Past 15,000 Years*, Nauka, Moscow.

Labeyrie, J., Lalou, C. and Delibrias, G. (1969), 'Etude des transgressions marines sur l'atoll de Mururoa par la datation des différents niveaux de corail', *Cahiers du Pacifique, 13*, 59-68.

Litoralia, newsletter for the INQUA Commission on Quaternary Shorelines, ed. D.R. Grant, c/o 5 Birchview Court, Nepean, K2G 3M7, Canada.

Martin, L., Suguio, K., Flexor, J.M., Bittencourt, A. and Vilas-Boas, G. (1979-80) 'Le Quaternaire marin Brésilien (littoral pauliste, sud fluminense et bahianais)', *Cah. ORSTOM, sér. Geol., 11*, 95-124.

Matsuda, T., Ota, Y., Ando, M., and Yonekura, N. (1978) 'Fault mechanism and recurrence time of major earthquakes in Southern Kanto district, Japan, as deduced from coastal terrace data', *Bull. Geol. Soc. Am., 89*, 1610-18.

Menard, H.W. and Ladd, H.P. (1963) 'Oceanic islands, sea mounts, guyots and atolls', in A.E. Maxwell (ed.), *The Sea*, Interscience, New York, pp. 365-87.

Mörner, N.-A. (1980) *Earth Rheology, Isostasy and Eustasy*, Wiley, Chichester and New York.

—— (1983) 'Differential Holocene sea-level changes over the globe: evidence from glacial eustasy, geoidal eustasy and crustal movements', in *Int. Symp. on Coastal Evolution in the Holocene*, Tokyo, pp. 93-6.

National Academy of Sciences (1983) *'Changing Climate'. Report of the Carbon Dioxide Assessment Committee*, National Academy Press, Washington, D.C.

Nivmer (1978-82), a publication for IGCP Project 61, Nos. 1-8, Montrouge, France.

Peltier, W.R. and Andrews, J.T. (1976) 'Glacial isostatic adjustment, I: The forward problem', *Geophys. J. Roy. Astron. Soc., 46*, 669-705.

Pirazzoli, P.A. (1983) 'Mise en évidence d'un flexure active de la lithosphère dans l'archipel de la Société (Polynésie française), d'après la position des rivages de la

fin de l'Holocène', *Comptes Rendus Acad. Sci. Paris, II, 296*, 695-8.
—— and Grant, D.R. (1986) 'Lithospheric deformation deduced from ancient shorelines', *Litoralia, 12*, 8-11.
——, Thommeret, Y., Thommeret, J., Laborel, J. and Montaggioni, L.F. (1982) 'Crustal block movements from Holocene shorelines: Crete and Antikythira (Greece)', *Tectonophys., 86*, 27-43.
Quinlan, G. and Beaumont, C. (1983) 'The deglaciation of Atlantic Canada as reconstructed from the postglacial relative sea-level record', *Can. J. Earth Sci., 19*, 2232-46.
Scott, G.A.J. and Rotondo, G.M. (1983) 'A model to explain the differences between Pacific plate island-atoll types', *Coral Reefs, 1*, 139-50.
Sea-Level (1979-82), Information Bulletin of IGCP Project 61, Nos. 1-8, Dept. of Geography, University of Durham.
Shennan, I. (ed.) (1986) *IGCP Project 200: Newsletter and Annual Report*, Dept. of Geography, University of Durham.
Sorensen, R.M. and Weggel, J.R. (1986) 'Impact of accelerated sea-level rise on shore stabilization works', abstract in R.W. Carter and R.J.N. Devoy (org.), *The Sedimentary and Hydrodynamic Consequences of Sea-level Change, Conference, Cork, Ireland*.
Thommeret, Y., Thommeret, J., Laborel, J., Montaggioni, L.F. and Pirazzoli, P.A. (1981) 'Late Holocene shoreline changes and seismo-tectonic displacements in western Crete (Greece)', *Zeit. Geomorph. Suppl.-Bd., 40*, 127-49.
Titus, J.G., Leatherman, S.P., Everts, C.H. and Kriebel, D.L. (1985) *Potential Impacts of Sea-Level Rise on the Beach at Ocean City*. US Environmental Protection Agency, Washington, DC.
Tooley, M.J. (1985) 'Sea levels', *Prog. Phys. Geogr., 9*, 113-20.
Vail, P.R., Mitchum, R.M., Jr, Todd, R.G., Widmier, J.M., Thompson III, S., Sangree, J.B., Bubb, J.N. and Hatlelid, W.G. (1977) 'Seismic stratigraphy and global changes of sea-level', in C.E. Payton (ed.), *Seismic Stratigraphy: Applications to Hydrocarbon Exploration*, Am. Ass. Petrol. Geol., Tulsa, Okla, pp. 49-212.
Velikovsky, I. (1956) *Earth in Upheaval*, Gollancz, London.
Wallcott, R.I. (1970) 'Flexural rigidity, thickness and viscosity of the lithosphere', *J. Geophys. Res., 75*, 3941-54.
—— (1972) 'Past sea levels, eustasy and deformation of the earth', *Quat. Res., 2*, 1-14.
Wyrtki, K. and Nakahara, S. (1984) *Monthly Maps of Sea-Level Anomalies in the Pacific 1975-1981*, Inst. Geophys., University of Hawaii, Honolulu.
Yonekura, N. (1975) 'Quaternary tectonic movements in the outer arc of southwest Japan, with special reference to seismic crustal deformation', *Bull. Dept. Geogr. Uni. Tokyo, 7*, 19-71.
Yoshikawa, T., Kaizuka, S. and Ota, Y. (1981) *The Landforms of Japan*, University of Tokyo Press, Tokyo.

Appendix I

GEOLOGICAL TIMESCALE

	ERA	PERIODS		AGE (Ma)
			Epochs	
PHANEROZOIC	CENOZOIC	Quaternary	Holocone	
			Pleistocene	
				2.5
		Tertiary	Pliocene	
			Miocene	
			Oligocene	
			Eocene	50
			Paleocene	
	MESO-ZOIC	Cretaceous		65
		Jurassic		140
		Triassic		200
				230
	PALAEOZOIC	Permian		280
		Pennsylvanian = L. Carboniferous		
		Mississippian = E. Carboniferous		
		Devonian		345
		Silurian		390
		Ordovician		430
		Cambrian		500
CRYPTOZOIC	PROT-ERO-ZOIC	Metazoans		600
	ARCHEOZOIC	Fungi		2300
		Blue-green Algae		2700
		First Traces of Life (Bacteria)		3300
		Origin of the Earth		

Appendix II

Estimated age (ka BP)	PLEISTOCENE	NORTH AMERICA	WESTERN ALASKA (Marine Events)	NORTHERN EUROPE	BRITAIN	MEDITERRANEAN (Marine Stages)	BLACK SEA	CHINA
10	HOLOCENE	HOLOCENE	Krusensternian	Holocene	Holocene	'Nizza' Marine Inundation	Modern Black Sea	Holocene
35	Late	WISCONSIN	Woronzofian	WEICHSEL	DEVENSIAN	'Marine Removal'	New Euxine	Malan Loess
75–110	Late	Sangamon	Pelukian	Eem	Ipswichian	Tyrrhenian	Karangat	
125 ?		ILLINOIAN		SAALE	WOLSTONIAN	'Marine Removal'		
	Middle	Yarmouth	Kotzebuan	Holstein	Hoxnian	?	Uzunlar	
	Middle	KANSAN		? ELSTER	ANGLIAN	Roman Regression	Old Euxine	Choukoutien
400	Middle	Aftonian	Einahnuhtan	? Cromer Complex	Cromerian Beestonian Pastonian	Crotonian	Chauda	
700		NEBRAS-KAN	Anvilian	? MENAPIAN	—	Sicilian	Guria	Nihowan
900	Early		Beringian	? Waal	—			
	Early			EBURONIAN	BAVENTIAN	Emilian	Kuyalnik	Villa-franchian Deposits
1.7 Ma	Early			Tiglian	Antian Thurnian Ludhamian			
2–2.5 Ma (Plio-/Pleistocene Boundary)				PRE-TIGLIAN	?WALTONIAN	Calabrian		

Correlation chart of selected Quaternary stratigraphic sequences referred to in the text (Sources — Nilsson, T. (1983), *The Pleistocene*. Dordrecht, Reidel; Flint, R.F. (1971), *Glacial and Quaternary Geology*. New York, Wiley; Bowen, D.Q. (1978), *Quaternary Geology*. Oxford, Pergamon). In North America, Northern Europe and Britain names shown in capitals are cold or glacial stages, and those in lower case, temperate stages. Differences in the timing and spatial recording of major environmental events reflected in stratigraphic changes make the construction of such charts difficult. The terminology and correlations shown are open to dispute and revision. The chart should be used as a tentative guide only.

Index